AGROCHEMICALS DESK REFERENCE
Environmental Data

John H. Montgomery

LEWIS PUBLISHERS
Boca Raton   Ann Arbor   London   Tokyo

Library of Congress Cataloging-in-Publication Data

Montgomery, John H. (John Harold), 1955-
   Agrochemicals desk reference: environmental data/John H. Montgomery.
      p. cm.
   Includes bibliographical references (p.   ) and indexes.
   1. Agricultural chemicals--Environmental aspects--Handbooks,
manuals, etc. 2. Water, Underground--Pollution--Handbooks, manuals,
etc. 3. Agricultural chemicals--Handbooks, manuals, etc.
I. Title.
TD427.A35M66 1993
628.1'684--dc20 92-36124
ISBN 0-87371-738-4

LEWIS PUBLISHERS
121 South Main Street, Chelsea, Michigan 48118

PRINTED IN THE UNITED STATES OF AMERICA 1 2 3 4 5 6 7 8 9 0
Printed on acid-free paper

Preface

The use of pesticides, herbicides, fungicides, and other agricultural chemicals has increased significantly over the last several decades. These "agrochemicals" have varying degrees of stability in the environment. Despite their degradation in the environment by a variety of mechanisms, the parent compounds and their degradates persist long enough to adversely impact soils and/or groundwater. As a consequence, the contamination of groundwater by agrochemicals has gained considerable and more serious attention in recent years. The sources of agrochemicals adversely impacting the groundwater environment include, but are not limited to, hazardous waste sites, municipal landfills, forested land areas, agricultural operations (e.g., farmlands, pesticide-treated forests), domestic practices (e.g., fertilizers, herbicides, etc.), and non-point source areas.

The protection of groundwater from agrochemicals and other pollutants requires that regulators and public and private interests cooperate. To this end, professionals from environmental consultants to the local fire officers need reliable, accurate and readily accessible data to accomplish these tasks. Unfortunately, the necessary data are scattered throughout many reference books, papers and journals. This book is designed to include, in one reference, all the information on widely used agrochemicals needed by those involved in the protection and remediation of the groundwater environment. Its format is easy to use by a wide spectrum of professionals. The data fields have been selected to fulfill the minimum technical requirements of the user based on the extensive experience of the author, his colleagues, and others.

This book should be useful to government agencies, environmental scientists, emergency response teams and cleanup contractors (for its physicochemical properties, health and exposure data, and uses category); environmental personnel from consulting and industrial firms (for its health and exposure data and physico-chemical data); chemical engineers, soil scientists, and industrial hygienists (for its synonym index, uses category, symptoms of exposure and physicochemical data); real estate developers, insurance underwriters and environmental attorneys (for its uses category and synonym index).

The book is based on more than 740 references. Most of the citations reviewed from the documented literature and included in this book pertained to the fate and transport of agrochemicals in the subsurface environment. This book is broad enough and comprehensive enough to serve its purpose as a desk reference, but small enough to be taken out into the field. The publisher and authors would appreciate hearing from readers regarding corrections, suggestions, or comments for revisions in future editions.

The author is grateful to the staff of Lewis Publishers, in particular, Mr. Skip DeWall, Ms. Elise Hoffman, and Ms. Vivian Collier for their invaluable contributions and suggestions during the preparation of this book. The author also extends thanks to the many anonymous reviewers for their comments and suggestions on draft proofs.

Introduction

The compounds profiled include those herbicides, insecticides, and fumigants most commonly found in the groundwater environment. The number of compounds included in this book was narrowed down based on information available in the documented literature. Compounds profiled in this book include pesticides having the potential for contaminating groundwater in New Jersey (Murphy and Fenske, 1987), compounds for which physicochemical data are available from the U.S. Department of Agriculture (1990), and those compounds which have been identified as organic Priority Pollutants promulgated by the U.S. Environmental Protection Agency (U.S. EPA) under the Clean Water Act of 1977 (40 CFR 136, 1977).

The compound headings are those commonly used by the U.S. EPA and many agricultural organizations. Positional and/or structural prefixes set in italic type are not an integral part of the chemical name and are disregarded in alphabetizing. These include *asym-, sym-, n-, sec-, cis-, trans-*, α-, β-, γ-, o-, *m-, p-, N-, S-*, etc.

Synonyms: These are listed alphabetically following the convention used for the compound headings. Compounds in boldface type are the Chemical Abstracts Service (CAS) names listed in the eighth or ninth Collective Index. If no synonym appears in boldface type, then the compound heading is the CAS assigned name. Synonyms include chemical names, common or generic names, trade names, registered trademarks, government codes and acronyms. All synonyms found in the literature are listed.

Although synonyms were retrieved from several references, most of them were retrieved from the Registry of Toxic Effects of Chemical Substances (RTECS, 1985).

Structure: This is given for every compound regardless of its complexity. The structural formula is a graphic representation of atoms or group(s) of atoms relative to each other. Clearly, the limitation of structural formulas is that they depict these relationships in two dimensions.

DESIGNATIONS

Chemical Abstracts Service (CAS) Registry Number: This is a unique identifier assigned by the American Chemical Society to chemicals recorded in the CAS Registry System. This number is used to access various chemical databases such as the Hazardous Substances Data Bank (HSDB), CAS Online, Chemical Substances Information Network and many others. This entry is also useful to conclusively identify a substance regardless of the assigned name.

Department of Transportation (DOT) Designation: This is a four-digit number assigned by the U.S. Department of Transportation (DOT) for hazardous materials and is identical to the United Nations identification number (which is preceded by the letters UN). This number is required on shipping papers, on placards or orange panels on tanks, and on a label or package containing the material. These numbers are widely used for personnel responding to emergency situations, e.g., overturned tractor trailers, in which the identification of the transported material is quickly and easily determined. Additional information may be obtained through the U.S. Department of Transportation, Research and Special Programs Administration, Materials Transportation Bureau, Washington, DC 20590.

Molecular Formula (mf): This is arranged by carbon, hydrogen and remaining elements in alphabetical order in accordance with the system developed by Hill (1900). Molecular formulas are useful in identifying isomers (i.e., compounds with identical molecular formulas) and are required if one wishes to calculate the formula weight of a substance.

Formula Weight (fw): This is calculated to the nearest hundredth using the empirical formula and the 1981 Table of Standard Atomic Weights as reported in Weast (1986). Formula weights are required for many calculations, such as converting weight/volume units, e.g., mg/L or g/L, to molar units (mol/L); with density for calculating molar volumes; and for estimating Henry's law constants.

Registry of Toxic Effects of Chemical Substances (RTECS) Number: Many compounds are assigned a unique accession number consisting of two letters followed by seven numerals. This number is needed to quickly and easily locate additional toxicity and health-based data which are cross-referenced in the RTECS (1985). Contact the National Institute for Occupational Safety and Health (NIOSH), U.S. Department of Health and Human Services, 4676 Columbia Parkway, Cincinnati, OH 45226 for additional information.

PHYSICAL AND CHEMICAL PROPERTIES

Appearance and Odor: The appearance, including the physical state (solid, liquid, or gas) of a chemical at room temperature (20-25 °C) is provided. If the compound can be detected by the olfactory sense, the odor is noted. Unless noted otherwise, the information provided in this category is for the pure substance and was obtained from many sources (CHRIS Hazardous Chemical Data, 1984; Hazardous Substances Data Bank, 1989; Hawley, 1981; Sax, 1984; Sax and Lewis, 1987; Toxic and Hazardous Industrial Chemicals Safety Manual for Handling and Disposal with Toxicity and Hazard Data, 1986; Sittig, 1985; Verschueren, 1983; Windholz et al., 1983).

Boiling Point (bp): This is defined as the temperature at which the vapor pressure of a liquid equals the atmospheric pressure. Unless otherwise noted, all boiling points are reported at one atmosphere pressure (760 mmHg). Although not used in environmental assessments, boiling points for aromatic compounds have been found to be linearly correlated with aqueous solubility (Almgren et al., 1979). Boiling points are also useful in assessing entry of toxic substances into the body. Body contact with high-boiling liquids is the most common means of entry into the body whereas the inhalation route is the most common for low-boiling liquids (Shafer, 1987).

Dissociation Constant (K_a): In an aqueous solution, an acid (HA) will dissociate into the carboxylate anion (A^-) and hydrogen ion (H^+) and may be represented by the general equation:

$$HA_{(aq)} \rightleftharpoons H^+ + A^- \qquad [1]$$

At equilibrium, the ratio of the products (ions) to the reactant (non-ionized electrolyte) is related by the equation:

$$K_a = ([H^+] [A^-]/[HA]) \qquad [2]$$

where K_a is the dissociation constant. This expression shows that K_a increases if there is increased ionization and vice versa. A strong acid (weak base) such as hydrochloric acid ionizes readily and has a large K_a, whereas a weak acid (or stronger base) such as benzoic acid ionizes to a lesser extent and has a lower K_a. The dissociation constants for weak acids are sometimes expressed as K_b, the dissociation constant for the base, and both are related to the dissociation constant for water by the expression:

$$K_w = K_a + K_b \qquad [3]$$

where K_w is the dissociation constant for water (10^{-14} at 25 °C), K_a is the acid dissociation constant, K_b is the base dissociation constant.

The dissociation constant is usually expressed as $pK_a = -\log_{10}K_a$. Equation [3] becomes:

$$pK_w = pK_a + pK_b \qquad [4]$$

When the pH of the solution and the pK_a are equal, 50% of the acid will have dissociated into ions. The percent dissociation of an acid or base can be calculated if the pH of the solution and the pK_a of the compound are known (Guswa et al., 1984):

For organic acids: $$\alpha_a = [100/(1 + 10^{(pH-pKa)})]$$ [5]

For organic bases: $$\alpha_b = [100/(1 + 10^{(pKw-pKb-pH)})]$$ [6]

where α_a is the percent of the organic acid that is nondissociated, α_b is the percent of the organic base that is nondissociated, pK_a is the $-\log_{10}$ dissociation constant for an acid, pK_w is the $-\log_{10}$ dissociation constant for water (14.00 at 25 °C), pK_b is the $-\log_{10}$ dissociation constant for base ($pK_b = pK_w-pK_a$), and pH is the $-\log_{10}$ hydrogen ion activity (concentration) of the solution.

Since ions tend to remain in solution, the degree of dissociation will affect processes such as volatilization, photolysis, adsorption and bioconcentration (Howard, 1989).

Henry's Law Constant (K_H): Sometimes referred to as the air–water partition coefficient, the Henry's law constant is defined as the ratio of the partial pressure of a compound in air to the concentration of the compound in water at a given temperature under equilibrium conditions. If the vapor pressure and solubility of a compound are known, this parameter can be calculated at 1 atm (760 mmHg) as follows:

$$K_H = Pfw/760S$$ [7]

where K_H is Henry's law constant (atm·m^3/mol), P is the vapor pressure (mmHg), S is the solubility in water (mg/L), and fw is the formula weight (g/mol).
    Henry's law constant can also be expressed in dimensionless form and may be calculated using one of the following equations:

$$K_H' = K_H/RK   \quad or \quad   K_H' = S_a/S$$ [8]

where K_H' is Henry's law constant (dimensionless), R is the ideal gas constant (8.20575 x 10^{-5} atm·m^3/mol·K), K is the temperature of water (degrees Kelvin), S_a is the solute concentration in air (mol/L), S is the aqueous solute concentration (mol/L).
    It should be noted that estimating Henry's law constant assumes that the gas obeys the ideal gas law and the aqueous solution behaves as an ideally dilute solution. The solubility and vapor pressure data inputted into the equations are valid only for the pure compound and must be in the same standard state at the same temperature.
    The major drawback in estimating Henry's law constant is that both the solubility and the vapor pressure of the compound are needed in equation [7]. If one or both these parameters are unknown, an empirical equation based on quantitative structure-activity relationships (QSAR) may be used to estimate

Henry's law constants (Nirmalakhandan and Speece, 1988). In this QSAR model, only the structure of the compound is needed. From this, connectivity indexes (based on molecular topology), polarizability (based on atomic contributions) and the propensity of the compound to form hydrogen bonds can easily be determined. These parameters, when regressed against known Henry's law constants for 180 organic compounds, yielded an empirical equation that explained more than 98% of the variance in the data set having an average standard error of only 0.262 logarithm units. Henry's law constant may also be estimated using the bond or group contribution method developed by Hine and Mookerjee (1975). The constants for the bond and group contributions were determined using experimentally determined Henry's law constants for 292 compounds. The authors found that those estimated values significantly deviating from observed values (particularly for compounds containing halogen, nitrogen, oxygen, and sulfur substituents) could be explained by "distant polar interactions", i.e., interactions between polar bonds or structural groups.

A more recent study for estimating Henry's law constants using the bond contribution method was provided by Meylan and Howard (1991). In this study, the authors updated and revised the method developed by Hine and Mookerjee (1975) due to new experimental data that have become available since 1975. Bond contribution values were determined for 59 chemical bonds based on known Henry's law constants for 345 organic compounds. A good statistical fit [correlation coefficient $(r^2) = 0.94$] was obtained when the bond contribution values were regressed against known Henry's law constants for all compounds. For selected chemicals classes, r^2 increased slightly to 0.97.

Henry's law constants provided an indication of the relative volatility of a substance. According to Lyman et al. (1982), if $K_H < 10^{-7}$ atm·m^3/mol, the substance has a low volatility. If K_H is $> 10^{-7}$ but $< 10^{-5}$ atm·m^3/mol, the substance will volatilize slowly. Volatilization becomes an important transfer mechanism in the range $10^{-5} < H < 10^{-3}$ atm·m^3/mol. Values of $K_H > 10^{-3}$ atm·m^3/mol indicate volatilization will proceed rapidly.

The rate of volatilization will also increase with an increase in temperature. ten Hulscher et al. (1992) studied the temperature dependence of Henry's law constants for three chlorobenzenes, three chlorinated biphenyls, and six polynuclear aromatic hydrocarbons. They observed that over the temperature range of 10 to 55 °C, Henry's law constant was doubled for every 10 °C increase in temperature. This temperature relationship should be considered when assessing the role of chemical volatilization from large surface water bodies whose temperatures are generally higher than those typically observed in groundwater.

Hydrolysis Half-Life (H-t$_{\frac{1}{2}}$): The hydrolysis half-life of a chemical is the time that it takes to reach one-half or 50% of its original concentration. The rate of chemical hydrolysis is highly dependent upon the compound's solubility, temperature, and pH. Since other environmental factors such as photolysis, adsorption, volatility (i.e., Henry's law constants), and adsorption can affect the rate of hydrolysis, these factors are virtually eliminated by performing hydrolysis experiments under carefully controlled laboratory conditions. The hydrolysis half-lives reported in the literature were calculated using experimentally determined hydrolysis rate constants.

Ionization Potential (IP): The ionization potential of a compound is defined as the energy required to remove a given electron from the molecule's atomic orbit (outermost shell) and is expressed in electron volts (eV). One electron volt is equivalent to 23,053 cal/mol.

Knowing the ionization potential of a contaminant is required in determining the appropriate photoionization lamp for detecting that contaminant or family of contaminants. Photoionization instruments are equipped with a radiation source (ultraviolet lamp), pump, ionization chamber, an amplifier and a recorder (either digital or meter). Generally, compounds with ionization potentials smaller than the radiation source (UV lamp rating) being used will readily ionize and will be detected by the instrument. Conversely, compounds with ionization potentials higher than the lamp rating will not ionize and will not be detected by the instrument.

Log K_{oc}: The soil/sediment partition or sorption coefficient is defined as the ratio of adsorbed chemical per unit weight of organic carbon to the aqueous solute concentration. This value provides an indication of the tendency of a chemical to partition between particles containing organic carbon and water. Compounds that bind strongly to organic carbon have characteristically low solubilities, whereas compounds with low tendencies to adsorb onto organic particles have high solubilities.

Non-ionizable chemicals that sorb onto organic materials in an aquifer (i.e., organic carbon) are retarded in their movement in groundwater. The sorbing solute travels at linear velocity that is lower than the groundwater flow velocity by a factor of R_d, the retardation factor. If the K_{oc} of a compound is known, the retardation factor may be calculated using the following equation from Freeze and Cherry (1974) for unconsolidated sediments:

$$R_d = V_w/V_c = [1 + (BK_d/n_e)] \tag{9}$$

where R_d is the retardation factor (unitless), V_w is the average linear velocity of groundwater (e.g., ft/day), V_c is the average linear velocity of contaminant (e.g., ft/day), B is the average soil bulk density (g/cm^3), n_e is the effective porosity (unitless), K_d is the distribution (sorption) coefficient (cm^3/g).

By definition, K_d is defined as the ratio of the concentration of the solute on the solid to the concentration of the solute in solution. This can be represented by the Freundlich equation:

$$K_d = VM_S/MM_L = C_S/C_L^n \qquad [10]$$

where V is the volume of the solution (cm^3), M_S is the mass of the sorbed solute (g), M is the mass of the porous medium (g), M_L is mass of the solute in solution (g), C_S is the concentration of the sorbed solute (g/cm^3), C_L is the concentration of the solute in the solution (g/cm^3), and n is a constant.

Values of n are normally between 0.7 and 1.1 although values of 1.6 have been reported (Lyman et al., 1982). If n is unknown, it is assumed to be unity and a plot of C_S versus C_L will be linear. The distribution coefficient is related to K_{oc} by the equation:

$$K_{oc} = K_d/f_{oc} \qquad [11]$$

where f_{oc} is the fraction of naturally occurring organic carbon in soil.

Sometimes K_d is expressed on an organic-matter basis and is defined as:

$$K_{om} = K_d/f_{om} \qquad [12]$$

where f_{om} is the fraction of naturally occurring organic matter in soil. The relationship between K_{oc} and K_{om} is defined as:

$$K_{om} = 0.58K_{oc} \qquad [13]$$

where the constant 0.58 is assumed to represent the fraction of carbon present in the soil or sediment organic matter (Allison, 1965).

For fractured rock aquifers in which the porosity of the solid mass between fractures is insignificant, Freeze and Cherry (1974) report the retardation equation as:

$$R_d = V_w/V_c = [1 + (2K_A/b)] \qquad [14]$$

where K_A is the distribution coefficient (cm) and b is the aperture of fracture (cm).

To calculate the retardation factors for ionizable compounds such as acids and bases, the fraction of un-ionized acid (α_a) or base (α_b) needs to be determined (see **Dissociation Constant**). According to Guswa et al. (1984), if it is assumed only the un-ionized portion of the acid is adsorbed onto the soil, the retardation factor for the acid becomes:

$$R_a = [1 + (\alpha_a BK_d/n_e)] \qquad [15]$$

However, for a base they assume that the ionized portion is exchanged with a monovalent ion and the un-ionized portion of the base is adsorbed hydrophobically. Therefore, the retardation factor for the base is:

$$R_b = \{1 + [(\alpha_b BK_d)/n_e] + [CECB(1-\alpha_b)]/(100\Sigma z^+ n_e)\} \qquad [16]$$

where CEC is the cation exchange capacity of the soil (cm^3/g) and Σz^+ is the sum of all positively charged particles in the soil (milliequivalents/cm^3). Guswa et al. (1984) report that the term Σz^+ is approximately 0.001 for most agricultural soils.

Correlations between K_{oc} and bioconcentration factors in fish and beef have shown a log-log linear relationship (Kenaga, 1980) as well as solubility of organic compounds in water (Abdul et al., 1987; Means et al., 1980). Moreover, the log K_{oc} has been shown to be related to molecular connectivity indices (Govers et al., 1984; Gerstl and Helling, 1987; Koch, 1983; Meylan et al., 1992; Sabljić and Protić, 1982; Sabljić, 1984, 1987) and high performance liquid chromatography (HPLC) capacity factors (Haky and Young, 1984; Hodson and Williams, 1988; Szabo et al., 1990, 1990a).

In instances where experimentally determined K_{oc} values are not available, they can be estimated using recommended regression equations as cited in Lyman et al. (1982) or Meylan et al. (1992). All the K_{oc} estimations are based on regression equations in which the solubility or the K_{ow} of the substance is known.

Log K_{ow}: The K_{ow} of a substance is the n-octanol/water partition coefficient and is defined as the ratio of the solute concentration in the water-saturated n-octanol phase to the solute concentration in the n-octanol-saturated water phase. Values of K_{ow} are therefore unitless.

The partition coefficient has been recognized as a key parameter in predicting the environmental fate of organic compounds. The log K_{ow} has been shown to be linearly correlated with log bioconcentration factors (BCF) in aquatic organisms (Davies and Dobbs, 1984; de Wolf et al., 1992; Isnard and Lambert, 1988), in fish (Davies and Dobbs, 1984; Kenaga, 1980; Isnard and Lambert, 1988; Neely et al., 1974; Ogata et al., 1984; Oliver and Niimi, 1985), log soil/sediment partition coefficients (K_{oc}) (Chiou et al., 1979; Kenaga and Goring, 1980), log of the solubility of organic compounds in water (Banerjee et al., 1980; Chiou et al., 1977, 1982; Hansch et al., 1968; Isnard and Lambert, 1988; Miller et al., 1984, 1985; Tewari et al., 1982; Yalkowsky and Valvani, 1979, 1980), molecular surface area (Camilleri et al., 1988; Funasaki et al., 1985; Miller et al., 1984; Woodburn et al., 1992; Yalkowsky and Valvani, 1979, 1980), molar refraction (Yoshida et al., 1983), molecular connectivity indices (Govers et al., 1984; Woodburn et al., 1992), reversed-phase liquid chromatography (RPLC) retention factors (Khaledi and Breyer, 1989; Woodburn et al., 1992), RPLC capacity factors (Braumann, 1986; Minick et al., 1989), HPLC capacity factors (Brooke et al., 1986; Carlson et al., 1975; DeKock and Lord, 1987; Eadsforth, 1986; Hammers et al., 1982; Harnisch et al., 1983; Kraak et al., 1986; Miyake et al., 1982, 1987, 1988; Szabo et al., 1990), HPLC retention times (Burkhard and Kuehl, 1986; Mirrlees et al., 1976; Sarna et al., 1984; Veith and Morris, 1978; Webster et al., 1985), reversed-phase thin layer chromatography retention parameters (Bruggeman et al., 1982), gas chromatography retention indices (Valkó et al., 1984), distribution coefficients (Campbell et al., 1983), solvatochromic parameters (Sadek et al., 1985), biological responses (Schultz et al., 1989), log of the n-hexane/water and L-a-phosphatidycholine dimyristol partitioning coefficients (Gobas et al., 1988), molecular descriptors and physicochemical properties (Bodor et al., 1989; Warne et al., 1990), linear solvation energy relationships (LSER) (Kamlet et al., 1988), and substituent constants which are based on empirically derived atomic or group constants and structural factors (Hansch and Anderson, 1967; Hansch et al., 1972). Variables needed for employing the LSER method have recently been presented by Hickey and Passino-Reader (1991).

For ionizable compounds (e.g., acids, amines, and phenols), K_{ow} values are a function of pH. Unfortunately, many investigators have neglected to report the pH of the solution at which the K_{ow} was determined. If a K_{ow} value is used for an ionizable compound for which the pH is known, both values should be noted.

Melting Point (mp): The melting point of a substance is defined as the temperature at which a solid substance undergoes a phase change to a liquid. The reverse process, the temperature at which a liquid freezes to a solid, is called the freezing point. For a given substance, the melting point is identical to the freezing point.

Unless noted otherwise, all melting points are reported at the standard pressure of 1 atmosphere (760 mmHg). Although the melting point of a substance is not directly used in predicting its behavior in the environment, it is useful in determining the phase in which the substance would be found under typical conditions.

Photolysis Half-Life (P-$t_{1/2}$): The photolysis half-life of a chemical is the time required for the parent chemical to reach one-half or 50% of its original concentration. Chemicals will undergo photolysis if they can absorb sunlight. Photolysis can occur in air, soil, water, and plants. The rate of photolysis is dependent upon the pH, temperature, presence of sensitizers, sorption to soil, depth of the compound in soil and water. Lyman et al. (1982) present an excellent overview of the photolysis process.

Solubility in Organics (S_o): The presence of small quantities of solvents can enhance a compound's solubility in water (Nyssen et al., 1987). Consequently, its fate and transport in soils, sediments, and groundwater will be changed due to the presence of these cosolvents. For example, soils contaminated with compounds having low water solubilities tend to remain bound to the soil by adsorbing onto organic carbon and/or by interfacial tension with water. A solvent introduced to an unsaturated soil environment (e.g., a surface spill, leaking aboveground tank, etc.) may come in contact with existing soil contaminants. As the solvent interacts with the existing contamination, it may mobilize it, thereby facilitating its migration. Consequently, the organic solvent can facilitate the leaching of contaminants from the soil to the water table. Therefore, the presence of cosolutes must be considered when predicting the fate and transport of contaminants in the unsaturated zone, the water table, and surface water bodies.

Solubility in Water (S_w): The water solubility of a compound is defined as the saturated concentration of the compound in water at a given temperature and pressure. This parameter is perhaps the most important factor in estimating a chemical's fate and transport in the aquatic environment. Compounds with high water solubilities tend to desorb from soils and sediments (i.e., they have low K_{oc} values), are less likely to volatilize from water, and are susceptible to biodegradation. Conversely, compounds with low solubilities tend to adsorb onto soils and sediments (have high K_{oc}), volatilize more readily from water, and bioconcentrate in aquatic organisms. The more soluble compounds commonly enter the water table more readily than their less soluble counterparts.

The water solubility of a compound varies with temperature, pH (particularly, ionizable compounds such as acids and bases), and other dissolved constituents, e.g., inorganic salts (electrolytes) and organic chemicals including naturally occurring organic carbon, such as humic and fulvic acids. At a given temperature, the variability/discrepancy of water-solubility measurements documented by investigators may be attributed to one or more of the following: (1) purity of the compound, (2) analytical method employed, (3) particle size (for solid solubility determinations only), (4) adsorption onto the container and/or suspended solids, (5) time allowed for equilibrium conditions to be reached, (6) losses due to volatilization, and (7) chemical transformations (e.g., hydrolysis).

The water solubility of chemical substances has been related to bioconcentration factors (BCF), soil/sediment partition coefficients (K_{oc}) (Abdul et al., 1987; Means et al., 1980), n-octanol/water partition coefficients (K_{ow}) (Chiou et al., 1977, 1982; Hansch et al., 1968; Miller et al., 1984, 1985; Yalkowsky and Valvani, 1979, 1980), HPLC capacity factors (Hafkenscheid and Tomlinson, 1981; Whitehouse and Cooke, 1982), molecular descriptors and physicochemical properties (Warne et al., 1990), soil organic matter K_{om} (Chiou et al., 1983), total molecular surface area (Amidon et al., 1975; Hermann, 1972; Lande and Banerjee, 1981; Lande et al., 1985), the compound's molecular structure (Nirmalakhandan and Speece, 1988a, 1989), boiling points (Almgren et al., 1979), and for homologous series of hydrocarbons or classes of organic compounds-carbon number (Bell, 1973; Krzyzanowska and Szeliga, 1978; Mitra et al., 1977; Robb, 1966) and molar volumes (Lande and Banerjee, 1981). With the exception of the molecular structure-solubility relationship, regression equations generated from the other relationships have demonstrated a log-log linear relationship for these properties.

The reported regression equations are useful in estimating the solubility of a compound in water if experimental values are not available. In addition, the solubility of a compound may be estimated if an experimentally determined Henry's law constant is available (Kamlet et al., 1987).

Unless otherwise noted, all reported solubilities were determined using distilled and/or deionized water. For some compounds, solubilities were determined using groundwater, natural seawater or artificial seawater.

Specific Density (ρ): The specific density, also known as relative density, is defined as:

$$\rho = d_s/d_w \tag{17}$$

where d_s is the density of a substance (g/mL or g/cm^3) and d_w is the density of distilled water (g/mL or g/cm^3). Values of specific density are unitless and are reported in the form ρ at T_s/T_w where ρ is the specific density of the substance, T_s is the temperature of substance at the time of measurement (°C) and T_w is the water temperature (°C).

For example, the value 1.1750 at 20/4 °C indicates a specific density of 1.1750 for the substance at 20 °C with respect to water at 4 °C. At 4 °C, the density of water is exactly 1.0000 g/mL (g/cm^3). Therefore, the specific density of a substance is equivalent to the density of the substance relative to the density of water at 4 °C.

The density of a hydrophobic substance enables it to sink or float in water. Density values are especially important for liquids migrating through the unsaturated zone and encountering the water table as "free product." Generally, liquids that are less dense than water "float" on the water table. Conversely, organic liquids that are more dense than water commonly "sink" through the water table, e.g., dense nonaqueous phase liquids (DNAPLs) such as chloroform, dichloroethane, and tetrachloroethylene.

Hydrophilic substances, on the other hand, behave differently. Acetone, which is less dense than water, does not float on water because it is freely miscible with it in all proportions. Therefore, the solubility of a substance must be considered in assessing its behavior in the subsurface.

Transformation Products: Chemicals released in the environment are susceptible to several degradation pathways. These include chemical (i.e., hydrolysis, oxidation, reduction, dealkylation, dealkoxylation, decarboxylation, methylation, isomerization, and conjugation), photolysis or photooxidation and biodegradation. Compounds transformed by one or more of these processes may result in the formation of more toxic or less toxic substances. In addition, the transformed product(s) will behave differently than the parent compound due to changes in their physicochemical properties. Many researchers focus their attention on transformation rates rather than the transformation products. Consequently, only limited data exist on the transitional and resultant end products. Where available, compounds that are transformed into identified products are listed.

In addition to chemical transformations occurring under normal environmental conditions, abiotic degradation products are also included. Types of abiotic transformation processes or treatment technologies fall into two categories -- physical and chemical. Types of physical processes used in removing or eliminating hazardous wastes include sedimentation, centrifugation, flocculation, oil/water separation, dissolved air flotation, heavy media separation, evaporation, air stripping, steam stripping, distillation, soil flushing, chelation, liquid-liquid extraction, supercritical extraction, filtration, carbon adsorption, reverse osmosis, ion exchange and electrodialysis. This information can be useful in evaluating abiotic degradation as a possible remedial measure. Chemical processes include neutralization, precipitation, hydrolysis (acid or base catalyzed), photolysis, oxidation-reduction, oxidation by hydrogen peroxide, alkaline chlorination, electrolytic oxidation, catalytic dehydrochlorination, and alkali metal dechlorination.

Most of the abiotic chemical transformation products reported in this book are limited to only three processes: hydrolysis, photooxidation and chemical oxidation-reduction. These processes are the most widely studied and reported in the literature. Detailed information describing the above technologies, their availability/limitation and company sources is available (U.S. Environmental Protection Agency, 1987).

Vapor Density (vap d): The vapor density of a substance is defined as the ratio of the mass of vapor per unit volume. An equation for estimating vapor density is readily derived from a varied form of the ideal gas law:

$$PV = MRK/fw \qquad [18]$$

where P is the vapor pressure (atm), V is the volume (L), M is the mass (g), R is the ideal gas constant (8.20575×10^{-2} atm·L/mol·K), and K is the temperature (degrees Kelvin). Recognizing that the density of a substance is defined as:

$$d = M/V \qquad [19]$$

Substituting this equation into the equation [18], rearranging, and simplifying results in an expression to determine the vapor density (g/L):

$$p = Pfw/RK \qquad [20]$$

At standard temperature (293.15K) and pressure (1 atm), equation [20] simplifies to:

$$p = fw/24.47 \qquad [21]$$

The specific vapor density of a substance relative to air is determined using:

$$p_v = fw/24.47p_{air} \qquad [22]$$

where p_v is the specific vapor density of a substance (unitless) and p_{air} is the vapor density of air (g/L).

The specific vapor density, p_v, is simply the ratio of the vapor density of the substance to that of air under the same pressure and temperature. According to Weast (1986), the vapor density of dry air at 20 °C and 760 mmHg is 1.204 g/L. At 25 °C, the vapor density of air decreases slightly to 1.184 g/L. Calculated specific vapor densities are reported relative to air (set equal to 1) only for compounds which are liquids at room temperature (i.e., 25 °C). These are reported in addition to the calculated vapor densities.

Vapor Pressure (vp): The vapor pressure of a substance is defined as the pressure exerted by the vapor (gas) of a substance when it is under equilibrium conditions. It provides a semi-quantitative rate at which it will volatilize from soil and/or water. The vapor pressure of a substance is a required input parameter for calculating the air-water partition coefficient (see **Henry's Law Constant**), which in turn is used in estimating the volatilization rate of compounds from groundwater to the unsaturated zone and from surface water bodies to the atmosphere.

FIRE HAZARDS

Flash Point (fl p): The flash point is defined as the minimum temperature at which a substance releases ignitable flammable vapors in the presence of an ignition source. e.g., spark or flame. Flash points may be determined by two methods --- Tag closed cup (ASTM method D56) or Cleveland open cup (ASTM method D93). Unless otherwise noted, all flash point values represent closed cup method determinations. Flash point values determined by the open cup method are slightly higher than those determined by the closed cup method; however, the open cup method is more representative of actual conditions.

According to Sax (1984), a material with a flash point of 100 °F or less is considered dangerous, whereas a material having a flash point greater than 200 °F is considered to have a low flammability. Substances with flash points within this temperature range are considered to have moderate flammabilities.

Lower Explosive Limit (LEL): The minimum concentration (vol % in air) of a flammable gas or vapor required for ignition or explosion to occur in the presence of an ignition source (see also **Flash Point**).

Upper Explosive Limit (UEL): The maximum concentration (vol % in air) of a flammable gas or vapor required for ignition or explosion to occur in the presence of an ignition source (see also **Flash Point**).

HEALTH HAZARD DATA

Immediately Dangerous to Life or Health (IDLH): According to the National Institute of Occupational Safety and Health (1987), the IDLH level ". . . for the purpose of respirator selection represents a maximum concentration from which, in the event of respirator failure, one could escape within 30 minutes without experiencing any escape-impairing or irreversible health effects." Concentrations are reported in parts per million (ppm) or milligrams per cubic meter (mg/m^3).

Exposure Limits: The permissible exposure limits (PEL) in air, set by the Occupational Health and Safety Administration (OSHA), can be found in the Code of Federal Regulations (General Industry Standards for Toxic and Hazardous Substances, 1977). Unless noted otherwise, the PEL are 8-hr time-weighted average (TWA) concentrations. If NIOSH (1987) and/or the American Conference of Governmental Industrial Hygienists (ACGIH) has published recommended exposure limits, these are also included. The ACGIH's recommended exposure limits, commonly known as threshold limit values (TLV), are subdivided into three exposure classes (Threshold Limit Values and Biological Exposure Indices for 1987-1988). The TLVs, which are updated annually, are defined as follows:

Threshold Limit Value-Time Weighted Average (TLV-TWA) - the time-weighted average (TWA) concentration for a normal 8-hr workday and a 40-hr workweek, to which nearly all workers may be repeatedly exposed, day after day, without adverse effect.

Threshold Limit Value-Short Term Exposure Limit (TLV-STEL) - the concentration to which workers can be exposed continuously for a short period of time without suffering from 1) irritation, 2) chronic or irreversible tissue damage, or 3) narcosis of sufficient degree to increase the likelihood of accidental injury, impair self-rescue or materially reduce work efficiency, and provided that the daily TLV-TWA is not exceeded. It is not a separate independent exposure limit; rather, it supplements the TWA limit where there are recognized acute toxic effects from a substance whose toxic effects are primarily of a chronic nature. STELs are recommended only where toxic effects have been reported from high short-term exposures in either humans or animals.

A STEL is defined as a 15-minute time-weighted average exposure which should not be exceeded at any time during a workday even if the 8-hr TWA is within the TLV. Exposures at the STEL should not be longer than 15 minutes and should not be repeated more than four times per day. There should be at least 60 minutes between successive exposures at the STEL. An averaging period other than 15 minutes may be recommended when this is warranted by observed biological effects.

Threshold Limit Value-Ceiling (TLV-C) - the concentration that should not be exceeded during any part of the working exposure.

For additional information from OSHA, write to Technical Data Center, U.S. Department of Labor, Washington, DC 20210. The ACGIH's address is 6500 Glenway Ave., Building D-7, Cincinnati, OH 45211-7881.

Formulations Types: Types of formulations, e.g., emulsifiable concentrates, water or oil solubles, water dispersible liquids and granules, are provided. These were obtained from several sources (Ashton and Monaco, 1991; Hartley and Kidd, 1987; Worthing and Walker, 1991).

Toxicology: Information on toxicity to aquatic life was obtained primarily from the Royal Society of Chemistry (Hartley and Kidd, 1987) and the Chemical Hazard Response Information System (CHRIS) Manual (1984). Information on toxicity to rats and/or mice were obtained from Ashton and Monaco (1991), Hartley and Kidd (1987), and RTECS (1985). The absence of toxicity data does not imply that toxic effects do not exist.

Uses: Descriptions of specific agricultural uses are based on one or more of the following sources - HSDB (1989), CHRIS Manual (1984), Sittig (1985), and Verschueren (1983). This information is useful in attempting to identify potential sources of the industrial and environmental contamination.

REFERENCES

Abdul, S.A., T.L. Gibson, and D.N. Rai. "Statistical Correlations for Predicting the Partition Coefficient for Nonpolar Organic Contaminants between Aquifer Organic Carbon and Water," *Haz. Waste Haz. Mater.*, 4(3):211-222 (1987).

Allison, L.E. "Organic Carbon" in *Methods of Soil Analysis, Part 2.*, Black, C., Evans, D., White, J., Ensminger, L., and F. Clark, eds. (Madison, WI: American Society of Agronomy, 1965), pp. 1367-1378.

Almgren, M., F. Grieser, J.R. Powell, and J.K. Thomas. "A Correlation between the Solubility of Aromatic Hydrocarbons in Water and Micellar Solutions, with Their Normal Boiling Points," *J. Chem. Eng. Data*, 24(4):285-287 (1979).

Amidon, G.L., S.H. Yalkowsky, S.T. Anik, and S.C. Valvani. "Solubility of

Nonelectrolytes in Polar Solvents. V. Estimation of the Solubility of Aliphatic Monofunctional Compounds in Water Using a Molecular Surface Area Approach," *J. Phys. Chem.*, 79(21):2239-2246 (1975).

Ashton, F.M. and T.J. Monaco. *Weed Science* (New York: John Wiley & Sons, Inc., 1991), 466 p.

Banerjee, S., S.H. Yalkowsky, and S.C. Valvani. "Water Solubility and Octanol/Water Partition Coefficients of Organics. Limitations of the Solubility-Partition Coefficient Correlation," *Environ. Sci. Technol.*, 14(10):1227-1229 (1980).

Bell, G.H. "Solubilities of Normal Aliphatic Acids, Alcohols and Alkanes in Water," *Chem. Phys. Lipids*, 10:1-10 (1973).

Bodor, N., Z. Gabanyi, and C.-K. Wong. "A New Method for the Estimation of Partition Coefficient," *J. Am. Chem. Soc.*, 111(11):3783-3786 (1989).

Braumann, T. "Determination of Hydrophobic Parameters By Reversed-Phase Liquid Chromatography: Theory, Experimental Techniques, and Application in Studies on Quantitative Structure-Activity Relationships," *J. Chromatogr.*, 373:191-225 (1986).

Brooke, D.N., A.J. Dobbs, and N. Williams. "Octanol:Water Partition Coefficients (P): Measurement, Estimation, and Interpretation, Particularly for Chemicals with P >10 [5]," *Ecotoxicol. Environ. Saf.*, 11(3):251-260 (1986).

Bruggeman, W.A., J. Van Der Steen, and O. Hutzinger. "Reversed-Phase Thin-Layer Chromatography of Polynuclear Aromatic Hydrocarbons and Chlorinated Biphenyls. Relationship with Hydrophobicity as Measured by Aqueous Solubility and Octanol-Water Partition Coefficient," *J. Chromatogr.*, 238:335-346 (1982).

Burkhard, L.P. and D.W. Kuehl. "*n*-Octanol/Water Partition Coefficients by Reverse Phase Liquid Chromatography/Mass Spectrometry for Eight Tetrachlorinated Planar Molecules," *Chemosphere*, 15(2):163-167 (1986).

Camilleri, P., S.A. Watts, and J.A. Boraston. "A Surface Area Approach to Determination of Partition Coefficients," *J. Chem. Soc., Perkin Trans. 2*, (September 1988), pp 1699-1707.

Campbell, J.R., R.G. Luthy, and M.J.T. Carrondo. "Measurement and Prediction of Distribution Coefficients for Wastewater Aromatic Solutes," *Environ. Sci. Technol.*, 17(10):582-590 (1983).

Carlson, R.M., R.E. Carlson, and H.L. Kopperman. "Determination of Partition Coefficients by Liquid Chromatography," *J. Chromatogr.*, 107:219-223 (1975).

Chiou, C.T., V.H. Freed, D.W. Schmedding, and R.L. Kohnert. "Partition Coefficients and Bioaccumulation of Selected Organic Chemicals," *Environ. Sci. Technol.*, 11(5):475-478 (1977).

Chiou, C.T., L.J. Peters, and V.H. Freed. "A Physical Concept of Soil-Water Equilibria for Nonionic Organic Compounds," *Science (Washington, DC)*, 206(4420):831-832 (1979).

Chiou, C.T., P.E. Porter, and D.W. Schmedding. "Partition Equilibria of Nonionic

Organic Compounds between Organic Matter and Water," *Environ. Sci. Technol.*, 17(4):227-231 (1983).

Chiou, C.T., D.W. Schmedding, and M. Manes. "Partitioning of Organic Compounds in Octanol-Water Systems," *Environ. Sci. Technol.*, 16(1):4-10 (1982).

"CHRIS Hazardous Chemical Data, Vol. 2," U.S. Department of Transportation, U.S. Coast Guard, U.S. Government Printing Office (November, 1984).

Davies, R.P. and A.J. Dobbs. "The Prediction of Bioconcentration in Fish," *Water Res.*, 18(10):1253-1262 (1984).

DeKock, A.C. and D.A. Lord. "A Simple Procedure for Determining Octanol-Water Partition Coefficients using Reverse Phase High Performance Liquid Chromatography (RPHPLC)," *Chemosphere*, 16(1):133-142 (1987).

de Wolf, W., J.H.M. de Bruijn, W. Sienen, and J.L.M. Hermens. "Influence of Biotransformation on the Relationship between Bioconcentration Factors and Octanol-Water Partition Coefficients," *Environ. Sci. Technol.*, 26(6):1197-1201 (1992).

Eadsforth, C.V. "Application of Reverse-Phase H.P.L.C. for the Determination of Partition Coefficients," *Pestic. Sci.*, 17(3):311-325 (1986).

Freeze, R.A. and J.A. Cherry. *Groundwater* (Englewood Cliffs, NJ: Prentice-Hall, Inc., 1974), 604 p.

Funasaki, N., S. Hada, and S. Neya. "Partition Coefficients of Aliphatic Ethers - Molecular Surface Area Approach," *J. Phys. Chem.*, 89(14):3046-3049 (1985).

"General Industry Standards for Toxic and Hazardous Substances," U.S. Code of Federal Regulations, 29 CFR 1910.1000, Subpart Z (January 1977).

Gerstl, Z. and C.S. Helling. "Evaluation of Molecular Connectivity as a Predictive Method for the Adsorption of Pesticides in Soils," *J. Environ. Sci. Health*, B22(1):55-69 (1987).

Gobas, F.A.P.C., J.M. Lahittete, G. Garofalo, W.Y. Shiu, and D. Mackay. "A Novel Method for Measuring Membrane-Water Partition Coefficients of Hydrophobic Organic Chemicals: Comparison with 1-Octanol-Water Partitioning," *J. Pharm. Sci.*, 77(3):265-272 (1988).

Govers, H., C. Ruepert, and H. Aiking. "Quantitative Structure-Activity Relationships for Polycyclic Aromatic Hydrocarbons: Correlation between Molecular Connectivity, Physico-Chemical Properties, Bioconcentration and Toxicity in *Daphnia Pulex*," *Chemosphere*, 13(2):227-236 (1984).

"Guidelines Establishing Test Procedures for the Analysis of Pollutants," U.S. Code of Federal Regulations, 40 CFR 136, 44(233):69464-69575.

Guswa, J.H., W.J. Lyman, A.S. Donigan, Jr., T.Y.R. Lo, and E.W. Shanahan. *Groundwater Contamination and Emergency Response Guide* (Park Ridge, NJ: Noyes Publications, 1984), 490 p.

Hafkenscheid, T.L. and E. Tomlinson. "Estimation of Aqueous Solubilities of Organic Non-Electrolytes Using Liquid Chromatographic Retention Data," *J. Chromatogr.*, 218:409-425 (1981).

Haky, J.E. and A.M. Young. "Evaluation of a Simple HPLC Correlation Method for the Estimation of the Octanol-Water Partition Coefficients of Organic Compounds," *J. Liq. Chromatogr.*, 7(4):675-689 (1984).

Hammers, W.E., G.J. Meurs, and C.L. De Ligny. "Correlations between Chromatographic Capacity Ratio Data on Lichrosorb RP-18 and Partition Coefficients in the Octanol-Water System," *J. Chromatogr.*, 247:1-13 (1982).

Hansch, C. and S.M. Anderson. "The Effect of Intramolecular Hydrophobic Bonding on Partition Coefficients," *J. Org. Chem.*, 32:2853-2586 (1967).

Hansch, C., A. Leo, and D. Nikaitani. "On the Additive-Constitutive Character of Partition Coefficients," *J. Org. Chem.*, 37(20):3090-3092 (1972).

Hansch, C., J.E. Quinlan, and G.L. Lawrence. "The Linear Free-Energy Relationship between Partition Coefficients and Aqueous Solubility of Organic Liquids," *J. Org. Chem.*, 33(1):347-350 (1968).

Harnisch, M., H.J. Mockel, and G. Schulze. "Relationship between Log P_{ow} Shake-Flask Values and Capacity Factors Derived from Reversed-Phase High Performance Liquid Chromatography for *n*-Alkylbenzene and Some OECD Reference Substances," *J. Chromatogr.*, 282:315-332 (1983).

Hartley, D. and H. Kidd, Eds. *The Agrochemicals Handbook*, 2nd ed. (England: Royal Society of Chemistry, 1987).

Hawley, G.G. *The Condensed Chemical Dictionary* (New York: Van Nostrand Reinhold Co., 1981), 1135 p.

Hazardous Substances Data Bank. National Library of Medicine, Toxicology Information Program (1989).

Hermann, R.B. "Theory of Hydrophobic Bonding. II. The Correlation of Hydrocarbon Solubility in Water with Solvent Cavity Surface Area," *J. Phys. Chem.*, 76(19):2754-2759 (1972).

Hickey, J.P. and D.R. Passino-Reader. "Linear Solvation Energy Relationships: "Rules of Thumb" for Estimation of Variable Values," *Environ. Sci. Technol.*, 25(10):1753-1760 (1991).

Hill, E.A. "On a System of Indexing Chemical Literature; Adopted by the Classification Division of the U.S. Patent Office," *J. Am. Chem. Soc.*, 22(8):478-494 (1900).

Hine, J. and P.K. Mookerjee. "The Intrinsic Hydrophobic Character of Organic Compounds. Correlations in Terms of Structural Contributions," *J. Org. Chem.*, 40(3):292-298 (1975).

Hodson, J. and N.A. Williams. "The Estimation of the Adsorption Coefficient (K_{oc}) for Soils by High Performance Liquid Chromatography," *Chemosphere*, 19(1):67-77 (1988).

Howard, P.H. *Handbook of Environmental Fate and Exposure Data for Organic Chemicals - Volume I. Large Production and Priority Pollutants* (Chelsea, MI: Lewis Publishers, Inc., 1989), 574 p.

Isnard, S. and S. Lambert. "Estimating Bioconcentration Factors from Octanol-Water Partition Coefficient and Aqueous Solubility," *Chemosphere*, 17(1):21-34 (1988).

Kamlet, M.J., R.M. Doherty, M.H. Abraham, P.W. Carr, R.F. Doherty, and R.W. Taft. "Linear Solvation Energy Relationships. 41. Important Differences between Aqueous Solubility Relationships for Aliphatic and Aromatic Solutes," *J. Phys. Chem.*, 91(7):1996-2004 (1987).

Kamlet, M.J., R.M. Doherty, P.W. Carr, D. Mackay, M.H. Abraham, and R.W. Taft. "Linear Solvation Energy Relationships. 44. Parameter Estimation Rules That Allow Accurate Prediction of Octanol/Water Partition Coefficients and Other Solubility and Toxicity Properties of Polychlorinated Biphenyls and Polycyclic Aromatic Hydrocarbons," *Environ. Sci. Technol.*, 22(5):503-509 (1988).

Kenaga, E.E. "Correlation of Bioconcentration Factors of Chemicals in Aquatic and Terrestrial Organisms with Their Physical and Chemical Properties," *Environ. Sci. Technol.*, 14(5):553-556 (1980).

Kenaga, E.E. and C.A.I. Goring. "Relationship between Water Solubility, Soil Sorption, Octanol-Water Partitioning and Concentration of Chemicals in Biota," in *Aquatic Toxicology, ASTM STP 707*, Eaton, J.G., P.R. Parrish, and A.C. Hendricks, Eds. (Philadelphia, PA: American Society for Testing and Materials, 1980), pp. 78-115.

Khaledi, M.G. and E.D. Breyer. "Quantitation of Hydrophobicity with Micellar Liquid Chromatography," *Anal. Chem.*, 61(9):1040-1047 (1989).

Koch, R. "Molecular Connectivity Index for Assessing Ecotoxicological Behaviour of Organic Compounds," *Toxicol. Environ. Chem.*, 6(2):87-96 (1983).

Kraak, J.C., H.H. Van Rooij, and J.L.G. Thus. "Reversed-Phase Ion-Pair Systems for the Prediction of n-Octanol-Water Partition Coefficients of Basic Compounds by High-Performance Liquid Chromatography," *J. Chromatogr.*, 352:455-463 (1986).

Krzyzanowska, T. and J. Szeliga. "A Method for Determining the Solubility of Individual Hydrocarbons," *Nafta*, 28:414-417 (1978).

Lande, S.S. and S. Banerjee. "Predicting Aqueous Solubility of Organic Nonelectrolytes from Molar Volume," *Chemosphere*, 10(7):751-759 (1981).

Lande, S.S., D.F. Hagen, and A.E. Seaver. "Computation of Total Molecular Surface Area from Gas Phase Ion Mobility Data and its Correlation with Aqueous Solubilities of Hydrocarbons," *Environ. Toxicol. Chem.*, 4(3):325-334 (1985).

Lyman, W.J., W.F. Reehl, and D.H. Rosenblatt. *Handbook of Chemical Property Estimation Methods: Environmental Behavior of Organic Compounds* (New York: McGraw-Hill, Inc., 1982).

Means, J.C., S.G. Wood, J.J. Hassett, and W.L. Banwart. "Sorption of Polynuclear Aromatic Hydrocarbons by Sediments and Soils," *Environ. Sci. Technol.*, 14(2):1524-1528 (1980).

Meylan, W. and P.H. Howard. "Bond Contribution Method for Estimating Henry's Law Constants," *Environ. Toxicol. Chem.*, 10(10):1283-1293 (1991).

Meylan, W., P.H. Howard, and R.S. Boethling. "Molecular Topology/Fragment

Contribution Method for Predicting Soil Sorption Coefficients," *Environ. Sci. Technol.*, 26(8):1560-1567 (1992).

Miller, M.M., S. Ghodbane, S.P. Wasik, Y.B. Tewari, and D.E. Martire. "Aqueous Solubilities, Octanol/Water Partition Coefficients, and Entropies of Melting of Chlorinated Benzenes and Biphenyls," *J. Chem. Eng. Data*, 29(2):184-190 (1984).

Miller, M.M., S.P. Wasik, G.-L. Huang, W.-Y. Shiu, and D. Mackay. "Relationships between Octanol-Water Partition Coefficient and Aqueous Solubility," *Environ. Sci. Technol.*, 19(6):522-529 (1985).

Minick, D.J., D.A. Brent, and J. Frenz. "Modeling Octanol-Water Partition Coefficients by Reversed-Phase Liquid Chromatography," *J. Chromatogr.*, 461:177-191 (1989).

Mirrlees, M.S., S.J. Moulton, C.T. Murphy, and P.J. Taylor. "Direct Measurement of Octanol-Water Partition Coefficient by High-Pressure Liquid Chromatography," *J. Med. Chem.*, 19(5):615-619 (1976).

Mitra, A., R.K. Saksena, and C.R. Mitra. "A Prediction Plot for Unknown Water Solubilities of Some Hydrocarbons and Their Mixtures," *Chem. Petro-Chem. J.*, 8:16-17 (1977).

Miyake, K., F. Kitaura, N. Mizuno, and H. Terada. "Phosphatidylcholine-Coated Silica as a Useful Stationary Phase for High-Performance Liquid Chromatographic Determination of Partition Coefficients between Octanol and Water," *J. Chromatogr.*, 389(1):47-56 (1987).

Miyake, K., N. Mizuno, and H. Terada. "Effect of Hydrogen Bonding on the High-Performance Liquid Chromatographic Behaviour of Organic Compounds. Relationship between Capacity Factors and Partition Coefficients," *J. Chromatogr.*, 439:227-235 (1988).

Miyake, K. and H. Terada. "Determination of Partition Coefficients of Very Hydrophobic Compounds by High-Performance Liquid Chromatography on Glyceryl-Coated Controlled-Pore Glass," *J. Chromatogr.*, 240(1):9-20 (1982).

Murphy, E. and R. Fenske. "Pesticide Use in New Jersey: Implications for Groundwater Quality," Office of Science and Research, New Jersey Department of Environmental Protection, 1987), 214 p.

Neely, W.B., D.R. Branson, and G.E. Blau. "Partition Coefficient to Measure Bioconcentration Potential of Organic Chemicals in Fish," *Environ. Sci. Technol.*, 8(13):1113-1115 (1974).

"NIOSH Pocket Guide to Chemical Hazards," U.S. Department of Health and Human Services, U.S. Government Printing Office (1987), 241 p.

Nirmalakhandan, N.N. and R.E. Speece. "QSAR Model for Predicting Henry's Constant," *Environ. Sci. Technol.*, 22(11):1349-1357 (1988).

Nirmalakhandan, N.N. and R.E. Speece. "Prediction of Aqueous Solubility of Organic Compounds Based on Molecular Structure," *Environ. Sci. Technol.*, 22(3):328-338 (1988a).

Nirmalakhandan, N.N. and R.E. Speece. "Prediction of Aqueous Solubility of Organic Compounds Based on Molecular Structure. 2. Application to PNAs,

PCBs, PCDDs, etc.," *Environ. Sci. Technol.*, 23(6):708-713 (1989).

Nyssen, G.A., E.T. Miller, T.F. Glass, C.R. Quinn II, J. Underwood, J., and D.J. Wilson. "Solubilities of Hydrophobic Compounds in Aqueous-Organic Solvent Mixtures," *Environ. Monit. Assess.*, 9(1):1-11 (1987).

Ogata, M., K. Fujisawa, Y. Ogino, and E. Mano. "Partition Coefficients as a Measure of Bioconcentration Potential of Crude Oil in Fish and Sunfish," *Bull. Environ. Contam. Toxicol.*, 33(5):561-567 (1984).

Oliver, B.G. and A.J. Niimi. "Bioconcentration Factors of Some Halogenated Organics for Rainbow Trout: Limitations in Their Use for Prediction of Environmental Residues," *Environ. Sci. Technol.*, 19(9):842-849 (1985).

"Registry of Toxic Effects of Chemical Substances," U.S. Department of Health and Human Services, National Institute for Occupational Safety and Health (1985), 2050 p.

Robb, I.D. "Determination of the Aqueous Solubility of Fatty Acids and Alcohols," *Aust. J. Chem.*, 18:2281-2285 (1966).

Sabljić, A. "On the Prediction of Soil Sorption Coefficients of Organic Pollutants from Molecular Structure: Application of Molecular Topology Model," *Environ. Sci. Technol.*, 21(4):358-366 (1987).

Sabljić, A. "Predictions of the Nature and Strength of Soil Sorption of Organic Pollutants by Molecular Topology," *J. Agric. Food Chem.*, 32(2):243-246 (1984).

Sabljić, A. and M. Protić. "Relationship between Molecular Connectivity Indices and Soil Sorption Coefficients of Polycyclic Aromatic Hydrocarbons," *Bull. Environ. Contam. Toxicol.*, 28(2):162-165 (1982).

Sadek, P.C., P.W. Carr, R.M. Doherty, M.J. Kamlet, R.W. Taft, and M.H. Abraham. "Study of Retention Processes in Reversed-Phase High-Performance Liquid Chromatography by the Use of the Solvatochromic Comparison Method," *Anal. Chem.*, 57(14):2971-2978 (1985).

Sarna, L.P., P.E. Hodge, and G.R.B. Webster. "Octanol-Water Partition Coefficients of Chlorinated Dioxins and Dibenzofurans by Reversed-Phase HPLC Using Several C_{18} Columns," *Chemosphere*, 13(9):975-983 (1984).

Sax, N.I. *Dangerous Properties of Industrial Materials* (New York: Van Nostrand Reinhold Co., 1984), 3124 p.

Sax, N.I. and R.J. Lewis, Sr. *Hazardous Chemicals Desk Reference* (New York: Van Nostrand Reinhold Co., 1987), 1084 p.

Schultz, T.W., S.K. Wesley, and L.L. Baker. "Structure-Activity Relationships for Di and Tri Alkyl and/or Halogen Substituted Phenols," *Bull. Environ. Contam. Toxicol.*, 43(2):192-198 (1989).

Shafer, D. *Hazardous Materials Training Handbook* (Madison, CT: Bureau of Law and Business, Inc., 1987), 206 p.

Sittig, M. *Handbook of Toxic and Hazardous Chemicals and Carcinogens* (Park Ridge, NJ: Noyes Publications, 1985), 950 p.

Szabo, G., S.L. Prosser, and R.A. Bulman. "Adsorption Coefficient (K_{oc}) and HPLC Retention Factors of Aromatic Hydrocarbons," *Chemosphere*,

21(4/5):495-505 (1990).

Szabo, G., S.L. Prosser, and R.A. Bulman. "Determination of the Adsorption Coefficient (K_{oc}) of Some Aromatics for Soil by RP-HPLC on Two Immobilized Humic Acid Phases," Chemosphere, 21(6):777-788 (1990a).

Szabo, G., S.L. Prosser, and R.A. Bulman. "Prediction of the Adsorption Coefficient (K_{oc}) for Soil by a Chemically Immobilized Humic Acid Column using RP-HPLC," Chemosphere, 21(6):729-739 (1990).

ten Hulscher, Th.E.M., L.E. van der Velde, and W.A. Bruggeman. "Temperature Dependence of Henry's Law Constants for Selected Chlorobenzenes, Polychlorinated Biphenyls and Polycyclic Aromatic Hydrocarbons," Environ. Toxicol. Chem., 11(11):1595-1603 (1992).

Tewari, Y.B., M.M. Miller, S.P. Wasik, and D.E. Martire. "Aqueous Solubility of Octanol/Water Partition Coefficient of Organic Compounds at 25.0 °C," J. Chem. Eng. Data, 27(4):451-454 (1982).

Threshold Limit Values and Biological Exposure Indices for 1987-1988 (Cincinnati, OH: American Conference of Governmental Industrial Hygienists, 1987), 114 p.

Toxic and Hazardous Industrial Chemicals Safety Manual for Handling and Disposal with Toxicity and Hazard Data (Tokyo, Japan: International Technical Information Institute, 1986), 700 p.

U.S. EPA. "A Compendium of Technologies Used in the Treatment of Hazardous Wastes," Office of Research and Development, U.S. EPA Report-625/8-87-014 (1987), 49 p.

U.S. Department of Agriculture. Agricultural Research Service Pesticide Properties Database. Systems Research Laboratory, Beltsville, MD (1990).

Valkó, K., O. Papp, and F. Darvas. "Selection of Gas Chromatographic Stationary Phase Pairs for Characterization of the 1-Octanol-Water Partition Coefficient," J. Chromatogr., 301:355-364 (1984).

Veith, G.D. and R.T. Morris. "A Rapid Method for Estimating Log P for Organic Chemicals," U.S. EPA Report-600/3-78-049 (1978), 15 p.

Verschueren, K. Handbook of Environmental Data on Organic Chemicals (New York: Van Nostrand Reinhold Co., 1983), 1310 p.

Warne, M. St.J., D.W. Connell, D.W. Hawker, and G. Schüürmann. "Prediction of Aqueous Solubility and the Octanol-Water Partition Coefficient for Lipophilic Organic Compounds Using Molecular Descriptors and Physicochemical Properties," Chemosphere, 21(7):877-888 (1990).

Weast, R.C., Ed. CRC Handbook of Chemistry and Physics, 67th ed. (Boca Raton, FL: CRC Press, Inc., 1986), 2406 p.

Webster, G.R.B., K.J. Friesen, L.P. Sarna, and D.C.G. Muir. "Environmental Fate Modeling of Chlorodioxins: Determination of Physical Constants," Chemosphere, 14(6/7):609-622 (1985).

Whitehouse, B.G. and R.C. Cooke. "Estimating the Aqueous Solubility of Aromatic Hydrocarbons by High Performance Liquid Chromatography," Chemosphere,

11(8):689-699 (1982).

Windholz, M., S. Budavari, R.F. Blumetti, and E.S. Otterbein, Eds., *The Merck Index*, 10th ed. (Rahway, NJ: Merck and Co., 1983), 1463 p.

Woodburn, K.B., J.J. Delfino, and Rao, P.S.C. "Retention of Hydrophobic Solutes on Reversed-Phase Liquid Chromatography Supports: Correlation with Solute Topology and Hydrophobicity Indices," *Chemosphere*, 24(8):1037-1046 (1992).

Worthing, C.R. and S.B. Walker, Eds. *The Pesticide Manual - A World Compendium*, 9th ed. (Great Britain: British Crop Protection Council, 1991), 1141 p.

Yalkowsky, S.H. and S.C. Valvani. "Solubilities and Partitioning 2. Relationships between Aqueous Solubilities, Partition Coefficients, and Molecular Surface Areas of Rigid Aromatic Hydrocarbons," *J. Chem. Eng. Data*, 24(2):127-129 (1979).

Yalkowsky, S.H. and S.C. Valvani. "Solubility and Partitioning I: Solubility of Nonelectrolytes in Water," *J. Pharm. Sci.*, 69(8):912-922 (1980).

Yoshida, K., S. Tadayoshi, and F. Yamauchi. "Relationship between Molar Refraction and *n*-Octanol/Water Partition Coefficient," *Ecotoxicol. Environ. Saf.*, 7(6):558-565 (1983).

Bibliography

The books listed below were used in the preparation of the Agrochemicals Desk Reference. Most of the physicochemical properties included for each compound can be readily obtained from these and many other sources. Individual citations were not provided; however, information on transformation products, soil sorption and aquatic toxicity data for the majority of the compounds is widely scattered in many journal articles throughout the literature. For this reason, individual citations have been provided.

Ashton, F.M. and T.J. Monaco. *Weed Science* (New York: John Wiley & Sons, Inc., 1991), 466 p.

"CHRIS Hazardous Chemical Data, Vol. 2," U.S. Department of Transportation, U.S. Coast Guard, U.S. Government Printing Office (November, 1984).

Cremlyn, R.J. *Agrochemicals - Preparation and Mode of Action* (New York: John Wiley & Sons, Inc., 1991), 396 p.

Dean, J.A. *Handbook of Organic Chemistry* (New York: McGraw-Hill, Inc., 1987), 957 p.

Hartley, D. and H. Kidd, Eds. *The Agrochemicals Handbook*, 2nd ed. (England: Royal Society of Chemistry, 1987).

Hawley, G.G. *The Condensed Chemical Dictionary* (New York: Van Nostrand Reinhold Co., 1981), 1135 p.

Keith, L.H. and D.B. Walters. *National Toxicology Program's Chemical Solubility Compendium*, (Chelsea, MI: Lewis Publishers, Inc., 1992), 437 p.

Keith, L.H. and D.B. Walters. *National Toxicology Program's Chemical Data Compendium - Volume II, Chemical and Physical Properties*, (Chelsea, MI: Lewis Publishers, Inc., 1992), 1642 p.

Montgomery, J.H. *Groundwater Chemicals Desk Reference - Volume 2* (Chelsea, MI: Lewis Publishers, Inc., 1991), 944 p.

Montgomery, J.H. and L.M. Welkom. *Groundwater Chemicals Desk Reference* (Chelsea, MI: Lewis Publishers, Inc., 1990), 640 p.

Murphy, E. and R. Penske. *Pesticides in New Jersey: Implications in Groundwater Quality* (Trenton, NJ: New Jersey Department of Environmental Protection, Office of Science and Research, 1987), 214 p.

"NIOSH Pocket Guide to Chemical Hazards," U.S. Department of Health and Human Services, U.S. Government Printing Office (1987), 241 p.

Que Hee, S.S. and R.G. Sutherland. *The Phenoxyalkanoic Herbicides. Vol. 1. Chemistry, Analysis, and Environmental Pollution* (Boca Raton, FL: CRC Press, Inc, 1981), 321 p.

"Registry of Toxic Effects of Chemical Substances," U.S. Department of Health and Human Services, National Institute for Occupational Safety and Health (1985), 2050 p.

Sax, N.I. and R.J. Lewis, Sr. *Hazardous Chemicals Desk Reference* (New York: Van Nostrand Reinhold Co., 1987), 1084 p.

Sittig, M. *Handbook of Toxic and Hazardous Chemicals and Carcinogens* (Park

Ridge, NJ: Noyes Publications, 1985), 950 p.

Threshold Limit Values and Biological Exposure Indices for 1987-1988 (Cincinnati, OH: American Conference of Governmental Industrial Hygienists, 1987), 114 p.

Verschueren, K. *Handbook of Environmental Data on Organic Chemicals* (New York: Van Nostrand Reinhold Co., 1983), 1310 p.

Windholz, M., S. Budavari, R.F. Blumetti, and E.S. Otterbein, Eds., *The Merck Index*, 10th ed. (Rahway, NJ: Merck & Co., 1983), 1463 p.

Worthing, C.R., and S.B. Walker, Eds. *The Pesticide Manual - A World Compendium*, 9th ed. (Great Britain: British Crop Protection Council, 1991), 1141 p.

Abbreviations and Symbols

Å	angstrom
α	alpha
α_a	percent of acid that is nondissociated
α_b	percent of base that is nondissociated
$\approx$	approximately equal to
ACGIH	American Conference of Governmental Industrial Hygienists
ASTM	American Society for Testing and Materials
asym	asymmetric
atm	atmosphere
b	aperture of fracture
B	average soil bulk density (g/cm^3)
β	beta
BCF	bioconcentration factor
bp	boiling point
C	ceiling
°C	degrees Centigrade (Celsius)
cal	calorie
CAS	Chemical Abstracts Service
CEC	cation exchange capacity (meq/L unless noted otherwise)
CERCLA	Comprehensive Environmental Response, Compensation and Liability Act
CHRIS	Chemical Hazard Response Information System
cm	centimeter
C_L	concentration of solute in solution
C_S	concentration of sorbed solute
d	day(s)
DOT	Department of Transportation (U.S.)
d_s	density of a substance
d_w	density of water
δ	delta
EC_{50}	concentration necessary for 50% of the aquatic species tested showing abnormal behavior
et al.	and others
eV	electron volts
°F	degrees Fahrenheit
fl p	flash point
f_{oc}	fraction of organic carbon
fw	formula weight
γ	gamma
g	gram
gal	gallon
GC/MS	gas chromatography/mass spectrometry
>	greater than

$\geq$	greater than or equal to
hr	hour(s)
HPLC	high performance liquid chromatography
HSDB	Hazardous Substances Data Bank
H-t$_{1/2}$	hydrolysis half-life
IDLH	immediately dangerous to life or health
IP	ionization potential
K	kelvin (°C + 273.15)
K_a	acid dissociation constant
K_A	distribution coefficient (cm)
K_b	base dissociation constant
K_d	distribution coefficient (cm^3/g)
kg	kilogram
K_H	Henry's law constant (atm·m^3/mol·K)
$K_{H'}$	Henry's law constant (dimensionless)
K_{oc}	soil/sediment partition coefficient (organic carbon basis)
K_{om}	soil/sediment partition coefficient (organic matter basis)
K_{ow}	n-octanol/water partition coefficient
kPa	kilopascal
K_w	dissociation constant for water (10^{-14} at 25 °C)
<	less than
$\leq$	less than or equal to
L	liter
lb	pound
LC$_{50}$	lethal concentration necessary to kill 50% of the aquatic species tested
LC$_{100}$	lethal concentration necessary to kill 100% of the aquatic species tested
LD$_{50}$	lethal dose necessary to kill 50% of the mammals tested
lel	lower explosive limit
m	meter
m-	meta (as in m-dichlorobenzene)
M	molarity (moles/liter)
M	mass
meq	milliequivalents
mg	milligram
min	minute(s)
mL	milliliter
M_L	mass of sorbed solute
mmHg	millimeters of mercury
mmol	millimole
mo	month(s)
mol	mole

mp	melting point
M_S	mass of solute in solution
mV	millivolt
N	normality (equivalents/liter)
n-, N-	normal (as in n-propyl, N-nitroso)
n_e	effective porosity
ng	nanogram
NIOSH	National Institute for Occupational Safety and Health
nm	nanometer
o-	ortho (as in o-dichlorobenzene)
OSHA	Occupational Safety and Health Administration
ρ	specific density (unitless)
p-	para (as in p-dichlorobenzene)
P	pressure
Pa	pascal
p_{air}	vapor density of air
PEL	permissible exposure limit
pH	$-\log_{10}$ hydrogen ion activity (concentration)
pK_a	$-\log_{10}$ dissociation constant of an acid
pK_b	$-\log_{10}$ dissociation constant of a base
pK_w	$-\log_{10}$ dissociation constant of water
ppb	parts per billion (μg/L)
ppm	parts per million (mg/L)
P-$t_{1/2}$	photolysis half-life
p_v	specific vapor density
QSAR	quantitative structure-activity relationships
R	ideal gas constant (8.20575 x 10^{-5} atm·m^3/mol)
R_a	retardation factor for an acid
R_b	retardation factor for a base
RCRA	Resource Conservation and Recovery Act
R_d	retardation factor
RTECS	Registry of Toxic Effects of Chemical Substances
S	solubility
S_a	solute concentration in air (mol/L)
SARA	Superfund Amendments and Reauthorization Act
sec-	secondary (as in sec-butyl)
S_o	solubility in organics
S_w	solubility in water
sp.	species
spp.	species (plural)
STEL	short-term exposure limit
sym	symmetric
t-	tertiary (as in t-butyl; but $tert$-butyl)

$t_{1/2}$	half-life
TLV	threshold limit value
TOC	total organic carbon (mg/L)
T_s	temperature of a substance
T_w	temperature of water
TWA	time-weighted average
μ	micro (10^{-6})
μg	microgram
uel	upper explosive limit
unsym	unsymmetric
U.S. EPA	U.S. Environmental Protection Agency
UV	ultraviolet
V, vol	volume
vap d	vapor density
V_c	average linear velocity of contaminant (e.g., ft/day)
vp	vapor pressure
V_w	average linear velocity of groundwater (e.g., ft/day)
W	watt
λ	wavelength
wk	week
wt	weight
yr	year(s)
z^+	positively charged species (milliequivalents/cm^3)

Contents

xliv   CONTENTS

AGROCHEMICALS DESK REFERENCE
Environmental Data

ACEPHATE

Synonyms: Acetylphosphoramidothioic acid *O,S*-dimethyl ester; *N*-Acetylphosphoramidothioic acid *O,S*-dimethyl ester; Chevron RE 12420; *O,S*-Dimethylacetylphosphoroamidothioate; ENT 27822; *N*-(Methoxy(methylthio)phosphinoyl)acetamide; Orthene; Orthene-755; Ortho 12420; Ortran; Ortril; RE 12420; 75 SP.

Structure:

$$CH_3CNH-P\begin{smallmatrix}O\\||\end{smallmatrix}\begin{smallmatrix}O\\||\end{smallmatrix}\begin{smallmatrix}OCH_3\\SCH_3\end{smallmatrix}$$

Designations: CAS: 30560-19-1; mf: $C_4H_{10}NO_3PS$; fw: 183.16; RTECS: TB4760000.

Properties: Colorless to white solid. Mp: 64-68 °C (impure), 93 °C; ρ: 1.35 at 20/4 °C; H-t$_{1/2}$ at 40 °C: 60 hr (pH 9), 710 hr (pH 3); K_H: 5.2 x 10^{-13} atm·m^3/mol at 20-24 °C (approximate - calculated from water solubility and vapor pressure); log K_{oc}: 0.48; log K_{ow}: -1.87 (calculated); S_o (g/L): acetone (>100), ethanol (<50); S_w: 790 g/L at 20 °C; vp: 1.7 x 10^{-6} mmHg at 24 °C.

Transformation Products

Soil: Methamidophos and carbon dioxide were identified as the major soil metabolites (Hartley and Kidd, 1987).

Plant: Acephate is quickly absorbed, translocated, and transformed in pine seedlings (Werner, 1974) and cotton plants (Bull, 1979). The chemical was metabolized via cleavage of the amide bond to form methamidophos (*O,S*-dimethyl phosphoramidothioate) and an unknown, but insecticidally active compound, which were identified in the roots, stems, and leaves (Werner, 1974). Methamidophos was also found in cotton leaves following a single application of acephate. Four additional degradation products were formed - two of which were tentatively identified as *O,S*-dimethyl phosphorothioate and *S*-methyl acetylphosphoramidothioate. The amount of methamidophos and the four products represented about 9 and 5% of the applied amount, respectively (Bull, 1979).

Chemical/Physical: Emits toxic fumes of phosphorus, nitrogen, and sulfur oxides when heated to decomposition (Sax and Lewis, 1987).

Symptoms of Exposure: Slight irritation of eyes and skin.

Formulation Types: Wettable powder; water-soluble powder; encapsulated granules.

Toxicity: LC_{50} (96 hr) for rainbow trout >1 g/L, channel catfish 2.23 g/L, largemouth black bass 1.725 g/L, bluegill sunfish 2.05 g/L, and goldfish 9.55 g/L (Hartley and Kidd, 1987); acute oral LD_{50} for male and female rats is 945 and 866 mg/kg, respectively (Hartley and Kidd, 1987), 700 mg/kg (RTECS, 1985).

Uses: Contact and systemic insecticide for control of sucking and chewing insects in cotton, ornamentals, forestry, tobacco, fruits, vegetables, and other crops.

ACROLEIN

Synonyms: Acraldehyde; Acrylaldehyde; Acrylic aldehyde; Allyl aldehyde; Aqualin; Aqualine; Biocide; Crolean; Ethylene aldehyde; Magnacide; NSC 8819; Propenal; **2-Propenal**; Prop-2-en-1-al; 2-Propen-1-one; RCRA waste number P003; Slimicide; UN 1092.

Structure:

$$CH_2=CHCHO$$

Designations: CAS: 107-02-8; DOT: 1092 (inhibited); mf: C_3H_4O; fw: 56.06; RTECS: AS1050000.

Properties: Colorless to yellow, watery liquid with a very sharp, pungent, irritating odor. Mp: -86.9 °C; bp: 52.7 °C; ρ: 0.847 at 15.6 °C, 0.8410 at 20/4 °C; fl p: -18 °C (open cup), -25 °C (closed cup); lel: 2.8%; uel: 31%; H-t½: 3.5 d (pH 5), 1.5 d (pH 7), 4 hr (pH 10); K_H: 4.4 x 10^{-6} atm·m³/mol at 25 °C; IP: 10.10 eV; log K_{oc}: -0.31; log K_{ow}: -0.10; S_o: completely miscible with lower alcohols, acetone, benzene, ethyl ether, hydrocarbons; S_w: 200 g/L at 25 °C; vap d: 2.29 g/L at 25 °C, 1.94 (air = 1); vp: 220 mmHg at 20 °C.

Transformation Products

Biological: Microbes in site water degraded acrolein to β-hydroxypropionaldehyde (Kobayashi and Rittman, 1982).

Photolytic: Photolysis products include carbon monoxide, ethylene, free radicals, and a polymer (Calvert and Pitts, 1966). Anticipated products from the reaction of acrylonitrile with ozone or hydroxyl radicals in the atmosphere are glyoxal, formaldehyde, formic acid, and carbon dioxide (Cupitt, 1980). The major product reported from the photooxidation of acrolein with nitrogen oxides is formaldehyde with a trace of glyoxal (Altshuller, 1983).

Chemical/Physical: Wet oxidation of acrolein at 320 °C yielded formic and acetic acids (Randall and Knopp, 1980). May polymerize in the presence of light and explosively in the presence of concentrated acids (Worthing and Hance, 1991) forming disacryl, a white plastic solid (Humburg et al., 1989; Windholz et al., 1983). In distilled water, acrolein was hydrolyzed to β-hydroxypropionaldehyde (Burczyk et al., 1968; Reinert and Rodgers, 1987).

3

4   ACROLEIN

Exposure Limits: NIOSH REL: IDLH 5 ppm; OSHA PEL: TWA 0.1 ppm, STEL 0.3 ppm; ACGIH TLV: TWA 0.1 ppm, STEL 0.3 ppm.

Symptoms of Exposure: Severe irritation of eyes, skin, mucous membranes; abnormal pulmonary function; delayed pulmonary edema, chronic respiratory disease.

Formulation Types: Liquid (includes an inhibitor such as hydroquinone to prevent polymerization).

Toxicity: EC_{50} (96 hr) for oysters 55 μg/L (salt water); EC_{50} (24 hr) for salmon 80 μg/L (fresh water); LC_{50} (24 hr): for bluegill sunfish 0.079 mg/L, mosquito fish 0.39 mg/L, rainbow trout 0.15 mg/L, and shiners 0.04 mg/L; LC_{50} (48 hr): for oysters 0.56 mg/L and shrimps 0.10 mg/L (Worthing and Hance, 1991); acute oral LD_{50} for rats 46 mg/kg (Ashton and Monaco, 1991), 25,100 μg/kg (RTECS, 1985).

Uses: Contact herbicide and algicide; injected in water for the control of submerged and floating weeds in irrigation ditches and canals.

ACRYLONITRILE

Synonyms: Acritet; Acrylon; Acrylonitrile monomer; An; Carbacryl; Cyanoethylene; ENT 54; Fumigrain; Miller's fumigrain; Nitrile; Propenenitrile; **2-Propenenitrile**; RCRA waste number U009; TL 314; UN 1093; VCN; Ventox; Vinyl cyanide.

Structure:

$$CH_2=CHCN$$

Designations: CAS: 107-13-1; DOT: 1093; mf: C_3H_3N; fw: 53.06; RTECS: AT5250000.

Properties: Clear, colorless, watery liquid with a sweet irritating odor resembling peach pits. Slowly turns yellow on exposure to visible light. Mp: -83 °C; bp: 77.5-79 °C; ρ: 0.8060 at 20/4 °C; fl p: -1 °C; lel: 3.05%; uel: 17.0 ± 0.5%; $H-t_{1/2}$: 1,220 yr at 25 °C and pH 7; K_H: 1.10 x 10^{-4} atm·m^3/mol at 25 °C; IP: 10.91 eV; log K_{oc}: -1.13 (calculated); log K_{ow}: -0.92 to 1.20; S_o: soluble in ethanol, ethyl ether, acetone, benzene, carbon tetrachloride, toluene; miscible with alcohol and chloroform; S_w: 80 g/L at 25 °C; vap d: 2.17 g/L at 25 °C, 1.83 (air = 1); vp: 83 mmHg at 20 °C.

Transformation Products

Biological: Degradation by the microorganism *Nocardia rhodochrous* yielded ammonium ion and propionic acid, the latter being oxidized to carbon dioxide and water (DiGeronimo and Antoine, 1976).

Chemical/Physical: In an aqueous solution at 50 °C, UV light converted acrylonitrile to carbon dioxide. After 24 hr, the concentration of acrylonitrile was reduced 24.2% (Knoevenagel and Himmelreich, 1976).

Ozonolysis of acrylonitrile in the liquid phase yielded formaldehyde and the tentatively identified compounds glyoxal, an epoxide of acrylonitrile, and acetamide (Munshi et al., 1989). In the gas phase, cyanoethylene oxide was reported as an ozonolysis product (Munshi et al., 1989a). Anticipated products from the reaction of acrylonitrile with ozone or hydroxyl radicals in air included formaldehyde, formic acid, HC(O)CN, and cyanide ions (Cupitt, 1980). Wet oxidation of acrylonitrile at 320 °C yielded formic and acetic acids (Randall and Knopp, 1980). Incineration or heating to decomposition releases toxic nitrogen oxides (Sittig, 1985) and cyanides (Lewis, 1990). Polymerizes readily in the absence of oxygen or on exposure to visible light (Windholz et al., 1983).

Exposure Limits: NIOSH REL: TWA 1 ppm, 15-min C 10 ppm; OSHA PEL: TWA 2 ppm, 15-min C 10 ppm; ACGIH TLV: TWA 2 ppm.

Symptoms of Exposure: Asphyxia, eye irritation, headache, sneezing, nausea, vomiting, weakness, light-headedness, skin vesiculation, scaling dermatitis.

Formulation Types: Liquid.

Toxicity: LC_{100} (24 hr) all fish 100 mg/L (fresh water); acute oral LD_{50} for rats 78 mg/kg (Verschueren, 1983).

Uses: Grain fumigant.

ALACHLOR

Synonyms: Alanex; Alochlor; Bronco; Bullet; Cannon; **2-Chloro-*N*-(2,6-diethyl-phenyl)-*N*-(methoxymethyl)acetamide;** 2-Chloro-2′,6′-diethyl-*N*-(methoxy-methyl)acetanilide; CP 50144; Lariat; Lasso; Lasso II; Lasso EC; Lazo; Metachlor; Methachlor; Pillarzo.

Structure:

$$\text{CH}_2\text{CH}_3$$
$$\text{CH}_2\text{OCH}_3$$
$$-\text{N}$$
$$\text{COCH}_2\text{Cl}$$
$$\text{CH}_2\text{CH}_3$$

Designations: CAS: 15972-60-8; mf: $C_{14}H_{20}ClNO_2$; fw: 269.77; RTECS: AE1225000.

Properties: Odorless, cream-colored solid or crystals. Mp: 39.5-41.5 °C; bp: 100 °C at 0.02 mmHg (decomposes at 105 °C); ρ: 1.133 at 25/15.6 °C; K_H: 6.12 x 10^{-8} atm·m^3/mol at 25 °C (approximate - calculated from water solubility and vapor pressure); log K_{oc}: 1.63-2.28; log K_{ow}: 2.64, 2.90; S_o: soluble in acetone, benzene, chloroform, ethanol, ethyl ether, ethyl acetate; S_w: 242 mg/L at 25 °C; vp: 3.10 x 10^{-5} mmHg at 25 °C.

Soil properties and adsorption data

Soil	K_d (mL/g)	f_{oc} (%)	K_{oc} (mL/g)	pH
Drummer silty clay	3.70	1.97	188	--
Dupo silt loam	0.88	0.70	126	--
Lintonia loam	0.35	0.41	86	--
Sand	0.30	0.41	73	6.5
Silt	0.90	0.70	129	8.1
Spinks sand loam	1.30	1.39	94	--

Source: U.S. Department of Agriculture, 1990.

Transformation Products
Soil: Degradation products identified in an upland soil after 80 d of incubation include 7-ethyl-1-hydroxyacetyl-2,3-dihydroindole, 8-ethyl-2-hydroxy-1-(methylmethoxy)-1,2,3,4-tetrahydroquinone, 2′,6′-diethyl-2-hydroxy-*N*-(methoxymethyl)acetanilide, and 9-ethyl-1,5-dihydro-1-(methoxymethyl)-5-methyl-1,4-benzoxazepin-2-(3*H*)-one (Chou, 1977). In an upland soil, the microorganism

Rhizoctonia solani degraded alachlor to unidentified water soluble products (Lee, 1986). Degradation of alachlor by soil fungi gave 2-chloro-2',6'-diethylacetanilide, 2,6-diethylaniline, 1-chloroacetyl-2,3-dihydro-7-ethylindole, 2',6'-diethyl-*N*-methoxymethylaniline, and chloride ions (Tiedje and Hagedorn, 1975). Novick et al. (1986) reported that alachlor and its metabolites may not be mineralized in soils pretreated with the herbicide but may persist for long periods of time. They concluded that leaching of these metabolites to groundwater will be degraded but very slowly. At concentrations 0.073 and 10 µg/mL, <8% of ^{14}C ring-labeled alachlor was mineralized in 30 d (Novick et al., 1986). 2-Chloro-2',6'-diethyl-acetanilide was the major metabolite that formed following the incubation of alachlor in a Sawyer fine sandy loam at 0% relative humidity (Hargrove and Meikle, 1971). Persistence in soil is approximately 6-10 wk (Hartley and Kidd, 1987) but is somewhat shorter in sandy soils low in organic matter (Ashton and Monaco, 1991). The half-lives in soil containing 6 and 15% moisture were 23 and 5.7 d, respectively (Walker and Brown, 1985).

Plant: In plants, alachlor is absorbed, translocated, and transformed into glutathione (GSH) conjugates (Breaux et al., 1987; O'Connell, 1988). Four hours after treating corn with ^{14}C-labeled alachlor, 46% was converted to GSH conjugates and 42% was unreacted alachlor (O'Connell, 1988).

Surface Water: 2,6-Dichloroaniline, 2-chloro-2',6'-diethylacetanilide, and 2-hydroxy-2',6'-diethylacetanilide were reported as possible degradation products of alachlor that were identified in the Mississippi River and its tributaries (Pereira and Rostad, 1990).

Chemical/Physical: Hydrolyzes in strongly acidic or alkaline solutions (Windholz et al., 1983) to give methanol, chloroacetic acid, formaldehyde, and 2,6-diethyl-aniline (Sanborn et al., 1977; Sittig, 1985). Alachlor decomposed in 5M hydrochloric acid at 46 °C. After 72 hr, 2-chloro-2',6'-diethylacetanilide was the major decomposition product identified (Hargrove and Merkle, 1971). Emits toxic fumes of nitrogen oxides and chlorine when heated to decomposition (Sax and Lewis, 1987).

Alachlor, applied as a thin film on borosilicate glass, underwent photodegradation by sunlight via four major photodegradative pathways. These included dechlorination, *N*-dealkylation, *N*-deacylation, and cyclized *N*-dealkylated products (Cessna and Muir, 1991). Major photoproducts reported were 2',6'-diethylacetanilide, 1-chloro-2',6'-diethylacetanilide, 2,6-diethylaniline, chloroacetic acid, 2',6'-diethyl-*N*-methoxymethylaniline, and 1-chloroacetyl-2,3-dihydro-7-ethylindole (Chesters et al., 1989). Photolysis of alachlor in aqueous solutions was also studied by Somich et al. (1988) and reported by Hapeman-Somich (1991) and Somich et al. (1988). The photolysis study was performed using a 220-mL reactor equipped with a medium pressure lamp (λ ≤240 nm). Two major photoproducts identified were hydroxyalachlor and an unreported lactam. Other products identified were norchloralachlor, 2',6'-diethylacetanilide, and 2-hydroxy-2',6'-diethyl-*N*-methylacetanilide. Degradation was rapid (H-t½ = 1.6 min) and appeared to follow first-order kinetics. The decrease in pH during the reaction, from 6.6 to 2.5, indicated the formation of an acid, possibly hydrochloric (Somich et al., 1988).

Formulation Types: Emulsifiable concentrate (4 lb/gal); microscopic capsules (4 lb/gal); granules (15%).

Toxicity: LC_{50} (96 hr) for technical grade - rainbow trout 1.8 mg/L, bluegill sunfish 2.8 mg/L (Hartley and Kidd, 1987); for Lasso EC formulation the LC_{50} (48 hr) for *Daphnia magna* is 35 mg/L and the LC_{50} (96 hr) for rainbow trout and bluegill sunfish are 4.2 and 6.4 mg/L, respectively (Humburg et al., 1989); acute oral LD_{50} of technical alachlor for rats 930 mg/kg (Ashton and Monaco, 1991), Lasso 2,416 mg/kg, Lasso EC 1,000 mg/kg, Lasso II >5,010 mg/kg, Bronco 3,152 mg/kg (Humburg et al., 1989), 1,200 mg/kg (RTECS, 1985).

Uses: Preemergence, early postemergence or soil-incorporated herbicide used to control most annual grasses and many annual broadleaf weeds in beans, corn, cotton, milo, peanuts, peas, soybeans, sunflower, and certain woody ornamentals.

ALDICARB

Synonyms: Aldecarb; Ambush; Carbanolate; ENT 27093; **2-Methyl-2-(methyl-thio)propanal** *O*-((methylamino)carbonyl)oxime; 2-Methyl-2-(methylthio)-propionaldehyde *O*-(methylcarbamoyl)oxime; NCI-C08640; OMS 771; RCRA waste number P070; Temic; Temik; Temik G10; Temik 10 G; UC 21149; Union Carbide 21149; Union Carbide UC-21149.

Structure:

$$CH_3SC(CH_3)_2CH=NOCONHCH_3$$

Designations: CAS: 116-06-3; DOT: 2757; mf: $C_7H_{14}N_2O_2S$; fw: 190.25; RTECS: UE2275000.

Properties: Colorless crystals with a faint sulfurous odor. Mp: 99-100 °C; bp: decomposes; ρ: 1.195 at 25/4 °C; H-t$_{1/2}$ (pH-buffered distilled water at 20 °C): 131 d (pH 3.95), 559 d (pH 6.02), 324 d (pH 7.96), 55 d (pH 8.85), 6 d (pH 9.85); K_H: 1.45 x 10^{-9} atm·m^3/mol at 20-25 °C (approximate - calculated from water solubility and vapor pressure); log K_{oc}: 0.85-1.67; log K_{ow}: 0.70-1.13; S_o (wt %): acetone (35), benzene (15), chlorobenzene (15), chloroform (35), ethyl ether (20), isopropane (20), methylene chloride (30), toluene (10), xylene (5); S_w: 6.0 g/L at 25 °C; vp: 3.47 x 10^{-5} mmHg at 25 °C.

Soil properties and adsorption data

Soil	K_d (mL/g)	f_{oc} (%)	K_{oc} (mL/g)	pH	CEC (meq/100 g)
Arredondo sand	0.20	0.80	25	6.80	--
Astatula sand	0.08	0.17	48	6.29	--
Astatula sand	0.03	0.12	25	5.89	--
Batcombe silt loam	0.87	2.05	42	6.10	--
Cecil sandy loam	0.18	0.90	20	5.60	--
Clarion soil	0.78	2.64	29	5.00	21.02
Eustis fine sand	0.17	0.69	25	5.40	--
Fine sand (0-15 cm)	0.39	1.04	38	6.15	--
Fine sand (15-30 cm)	0.11	0.29	38	5.65	--
Fine sand (90-105 cm)	0.03	0.52	6	5.80	--
Fine sand (105-120 cm)	0.03	0.52	6	5.95	--
Harps soil	1.13	3.80	29	7.30	37.83

Soil	K_d (mL/g)	f_{oc} (%)	K_{oc} (mL/g)	pH	CEC (meq/100 g)
Peat soil	4.16	18.36	22	6.98	77.34
Rothamsted Farm	0.60	1.51	40	5.10	--
Sarpy fine sandy loam	0.19	0.51	37	7.30	5.71
Thurman loamy fine sand	0.22	1.07	21	6.83	6.10
Webster silty clay loam	0.76	3.97	20	7.30	--
Woburn sandy loam	0.06	0.78	8	7.00	--
Woburn sandy loam	0.30	3.43	9	6.34	--

Source: Bilkert and Rao, 1985; Briggs, 1981; Bromilow et al., 1980; Felsot and Dahm, 1979; Lord et al., 1980; U.S. Department of Agriculture, 1990.

Transformation Products
Biological: Jones (1976) reported several fungi degraded aldicarb to water-soluble constituents, namely, methyl(methylsulfonyl)propionamide and methyl(methyl-sulfonyl)propanol. The fungi tested, in order of effectiveness of degrading aldicarb, were *Gliocladium catenulatum*, *Penicillium multicolor* = *Cunninghamella elegans*, *Rhizoctonia* sp., and *Trichoderma harzianum* (Jones, 1976).

Soil: In soils, aldicarb quickly degrades via oxidation to 2-methyl-2-(methyl-sulfonyl)propionaldehyde-*O*-(methylcarbamoyl)oxime (aldicarb sulfoxide), 2-methyl 2-(methylsulfonyl)propionaldehyde-*O*-(methylcarbamoyl)oxime (aldicarb sulfone) (Andrawes et al., 1971; Bull et al., 1968; Coppedge et al., 1967; Day, 1991; Lemley et al., 1988; Macalady et al., 1986; Miles, 1991; Ou et al., 1985; Smelt et al., 1978; Zhong et al., 1986), aldicarb sulfoxide oxime, aldicarb sulfone oxime, and aldicarb sulfoxide nitrile, TLC polar products, and two unidentified compounds (Ou et al., 1985). Aldicarb sulfoxide, aldicarb sulfone, and water soluble noncarbamate compounds were found in field soils 2 yr after aldicarb application (Andrawes et al., 1971). In both soils and water, chemical and biological mediated reactions can transform aldicarb to the corresponding sulfoxide and sulfone via oxidation (Alexander, 1981). Reduction of aldicarb in natural waters and sediments yields 2-methyl-2-methyl thiopropionaldehyde and 2-methyl-2-methyl thio-propionitrile (Wolfe, 1992). The rate of microbial degradation of aldicarb to its metabolites, aldicarb sulfoxide and aldicarb sulfone in soils was essentially the same in soils showing enhanced carbofuran degradation. In addition, the persistence of these compounds were not dramatically altered under these conditions (Racke and Coats, 1988). Metabolites identified in fallow, sandy loam soils include the oxidation products aldicarb sulfone and aldicarb sulfoxide (Bromilow and Leistra, 1980). The primary degradative pathway of aldicarb in surface soils is oxidation by microorganisms (Zhong et al., 1986). Aldicarb is rapidly converted to the sulfoxide in the presence of oxidizing agents (Hartley and Kidd, 1987) and microorganisms

(Zhong et al., 1986). Further oxidation to the sulfone by microorganisms occurs at a much slower rate (Zhong et al., 1986). Rajagopal et al. (1989) used numerous compounds to develop a proposed pathway of degradation of aldicarb in soil. These compounds included aldicarb oxime, N-hydroxymethyl aldicarb, N-hydroxymethyl aldicarb sulfoxide, N-demethyl aldicarb sulfoxide, N-demethyl aldicarb sulfone, aldicarb sulfoxide, aldicarb sulfone, N-hydroxymethyl aldicarb sulfone, aldicarb oxime sulfone, aldicarb sulfone aldehyde, aldicarb sulfone alcohol, aldicarb nitrile sulfone, aldicarb sulfone amide, aldicarb sulfone acid, aldicarb oxime sulfoxide, aldicarb sulfoxide aldehyde, aldicarb sulfoxide alcohol, aldicarb nitrile sulfoxide, aldicarb sulfoxide amide, aldicarb sulfoxide acid, elemental sulfur, carbon dioxide, and water. Mineralization was more rapid in aerobic surface soils than in either aerobic or anaerobic subsurface soils (Ou et al., 1985). The reported half-life in soil was 70 d (Jury et al., 1987). Bromilow et al. (1980) reported the half-life of aldicarb in soil to be 9.9 d at 15 °C and pH 6.34-7.0. In aerobic soils, aldicarb degrades rapidly (H-$t_{1/2}$ = 7 d) releasing carbon dioxide. Mineralization half-lives for the incubation of aldicarb in aerobic and anaerobic soils were 20-361 and 223-1,130 d, respectively. At an application rate of 20 ppm, the half-lives of aldicarb in clay, silty clay loam and fine sandy loam were 9, 7, and 12 d, respectively (Coppedge et al., 1967). Other soil metabolites may include acids, amides, and alcohols (Hartley and Kidd, 1987).

Groundwater: In Floridan groundwater, aldicarb was converted to aldicarb sulfoxide under aerobic conditions. Conversely, under anaerobic conditions, oxidative metabolites (aldicarb sulfoxide and aldicarb sulfone) reverted back to the parents compound (aldicarb) (Miles and Delfino, 1985). In sterile anaerobic groundwater at pH 8.2, aldicarb slowly hydrolyzed to the aldicarb oxime. In a microorganism-enriched groundwater at pH 6.8, aldicarb rapidly degraded to aldicarb nitrile (Trehy et al., 1984).

Plant: In plants, aldicarb is rapidly metabolized to the corresponding sulfoxide, sulfone (Andrawes et al., 1973; Cremlyn, 1991), and water soluble noncarbamate compounds (Andrawes et al., 1973). In cotton, however, Coppedge et al. (1967) found that aldicarb was rapidly metabolized to aldicarb sulfoxide but further oxidation to aldicarb sulfone was much slower. At moderate temperatures, aldicarb is completely oxidized to aldicarb sulfoxide in cotton plants within 4-9 d. This metabolite is hydrolyzed to the corresponding oxime which was reported as the principal metabolite in the cotton plant (Metcalf et al., 1966). In a later study, Bartley et al. (1970) identified up to 10 metabolites in cotton plants at harvest time. Aldicarb degraded primarily via a reductive pathway forming conjugates of 2-methyl-2-(methylsulfinyl)propanol and lesser quantities of conjugated 2-methyl-2-(methylsulfonyl)propanol, 2-methyl-2-(methylsulfonyl)propionaldehyde oxime and 2-methyl-2-(methylsulfinyl)propionaldehyde oxime. Oxidation also occurred and this led to the formation of nonconjugated 2-methyl-2-(methylsulfinyl)-

propionamide, 2-methyl-2-(methylsulfinyl)propionic acid, and 2-methyl-2-(methylsulfonyl)propionic acid (Bartley et al., 1970).

Chemical/Physical: Hansen and Spiegel (1983) studied the hydrolysis rate of aldicarb, aldicarb sulfoxide, and aldicarb sulfone in aqueous buffer solutions. At a given temperature, the rate of hydrolysis increases rapidly above pH 7.5. The reported hydrolysis half-lives of aldicarb are as follows: 4,580 d at pH 5.5 and 5 °C, 3,240 d at pH 5.5 and 15 °C, 1,950 d at pH 7.5 and 5 °C, 1,900 d at pH 7.5 and 15 °C, 1,380 d at pH 8.5 and 5 °C, and 170 d at pH 8.5 and 15 °C (Hansen and Spiegel, 1983). The reported hydrolysis half-lives of aldicarb in water at pH 4.5 and 22 °C and pH 7.0 and 25 °C were 175 and 245 d, respectively (Chapman and Cole, 1982). Aldicarb degrades rapidly in the chlorination of drinking water forming aldicarb sulfoxide which subsequently degrades to aldicarb sulfone, (chloromethyl)sulfonyl species, and N-chloroaldicarb sulfoxide (Miles, 1991a).

Bank and Tyrrell (1985) studied the Cu^{2+}-promoted decomposition of aldicarb in aqueous solution over the pH region 2.91-5.51. 2-Methyl-2-(methylthio)propionitrile and 2-methyl-2-(methylthio)propanal formed at yields of 82 and 18%, respectively. Emits toxic fumes of nitrogen and sulfur oxides when heated to decomposition (Sax and Lewis, 1987).

Formulation Types: Granules.

Toxicity: LC_{50} (96 hr) for rainbow trout 0.88 mg/L and bluegill sunfish 1.5 mg/L (Hartley and Kidd, 1987); LC_{50} (72 hr) for bluegill sunfish 100 μg/L (Day, 1991); acute oral LD_{50} for rats 930 μg/kg (Hartley and Kidd, 1987), 650 μg/kg (RTECS, 1985).

Uses: Systemic insecticide, acaricide, nematocide.

ALDRIN

Synonyms: Aldrec; Aldrex; Aldrex 30; Aldrite; Aldrosol; Altox; Compound 118; Drinox; ENT 15949; Hexachlorohexahydro-*endo,exo*-dimethanonaphthalene; **1,2,3,4,10,10-Hexachloro-1,4,4a,5,8,8a-hexahydro-1,4:5,8-dimethanonaphthalene;** 1,2,3,4,10,10-Hexachloro-1,4,4a,5,8,8a-hexahydro-1,4-*endo,exo*-5,8-dimethanonaphthalene; 1,2,3,4,10,10-Hexachloro-1,4,4a,5,8,8a-hexahydro-*exo*-1,4-*endo*-5,8-dimethanonaphthalene; 1,4,4a,5,8,8a-Hexahydro-1,4-*endo,exo*-5,8-dimethanonaphthalene; HHDN; NA 2761; NA 2762; NCI-C00044; Octalene; RCRA waste number P004; Seedrin; Seedrin liquid.

Structure:

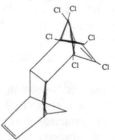

Designations: CAS: 309-00-2; DOT: 2761; mf: $C_{12}H_8Cl_6$; fw: 364.92; RTECS: IO2100000.

Properties: White, odorless crystals when pure; technical grades are tan to dark brown with a mild chemical odor. Mp: 104 °C (pure), 49-60 °C (technical); bp: 145 °C at 2 mmHg; ρ: 1.70 at 20/4 °C; fl p: nonflammable; H-t$_{1/2}$: 760 d at 25 °C and pH 7; K_H: 4.96 x 10^{-4} atm·m^3/mol at 25 °C; log K_{oc}: 2.61, 4.69; log K_{ow}: 5.17-7.4; P-t$_{1/2}$: 113.49 hr (absorbance λ = 227.0 nm, concentration on glass plates = 6.7 μg/cm^2); S_o (g/L): acetone (5-10), benzene (>600), dimethyl sulfoxide (1-5), 95% ethanol (<1), xylene (>600); S_w: 17-180 μg/L at 25 °C; vp: 2.31 x 10^{-5} mmHg at 20 °C.

Soil properties and adsorption data

Soil	K_d (mL/g)	f_{oc} (%)	K_{oc} (mL/g)	pH
Batcombe silt loam	996	2.05	48,585	6.1
Rothamsted Farm	730	1.51	48,344	5.1

Source: Briggs, 1981; Lord et al., 1980.

Transformation Products
Biological: Dieldrin is the major metabolite formed from the microbial degradation of aldrin via epoxidation (Kearney and Kaufman, 1976; Korte et al., 1962). Dieldrin may further degrade to photodieldrin (Kearney and Kaufman,

1976). Aldrin was found to be very persistent in an agricultural soil. Fifteen years after application of aldrin (20 lb/acre), 5.8% of the applied amount was recovered as dieldrin and 0.2% was recovered as photodieldrin (Lichtenstein et al., 1971). A pure culture of the marine alga namely *Dunaliella* sp. degraded aldrin to dieldrin and the diol at yields of 23.2 and 5.2%, respectively (Patil et al., 1972).

Soil: Patil and Matsumura (1970) reported 13 of 20 soil microorganisms were able to degrade aldrin to dieldrin under laboratory conditions.

Plant: Photoaldrin and photodieldrin formed when aldrin was codeposited on bean leaves and exposed to sunlight (Ivie and Casida, 1971). Dieldrin and 1,2,3,4,7,8-hexachloro-1,4,4a,6,7,7a-hexahydro-1,4-*endo*-methyleneindene-5,7-dicarboxylic acid were identified in aldrin-treated soil on which potatoes were grown (Klein et al., 1973).

Surface Water: When raw water obtained from the Little Miami River in Ohio containing aldrin (10 μg/L) was placed in a sealed glass jar and exposed to sunlight or artificial fluorescent light for 8 wk, about 80% was converted to dieldrin (Eichelberger and Lichtenberg, 1971).

Photolytic: Photolysis of 0.33 ppb aldrin in San Francisco Bay water by sunlight produced photodieldrin ($t_{1/2}$ = 1.1 d) (Singmaster, 1975). When an aqueous solution containing aldrin was photooxidized by UV light at 90-95 °C, 25, 50, and 75% degraded to carbon dioxide after 14.1, 28.2, and 109.7 hr, respectively (Knoevenagel and Himmelreich, 1976). Aldrin in a hydrogen peroxide solution (5 μM) was irradiated by UV light (λ = 290 nm). After 12 hr, the aldrin concentration was reduced 79.5%. Dieldrin, photoaldrin and an unidentified compound were reported as metabolites (Draper and Crosby, 1984). After a short-term (<1 hr) exposure to sunlight, aldrin on silica gel chromatoplates was converted to photoaldrin. Photodecomposition was accelerated by several photosensitizing agents (Ivie and Casida, 1971a). Photodegradation of aldrin by sunlight for 1 mo yielded the following products: dieldrin, photodieldrin, photoaldrin, and a polymeric substance (Rosen and Sutherland, 1967). Photolysis of solid aldrin using a high pressure mercury lamp with a pyrex filter (λ >300 nm) yielded a polymeric substance with small amounts of photoaldrin, dieldrin, hydrochloric acid, and carbon dioxide (Gäb et al., 1974). Sunlight and UV light can convert aldrin to photoaldrin (Georgacakis and Khan, 1971). When aldrin vapor (5 mg) in a reaction vessel was irradiated by a sunlamp for 45 hr, 14-34% degraded to dieldrin (50-60 μg) and photodieldrin (20-30 μg). However, when the aldrin vapor concentration was reduced to 1 μg and irradiation time extended to 14 d, 60% degraded to dieldrin (0.63 μg), photodieldrin (0.02 μg), and photoaldrin (0.02 μg) (Crosby and Moilanen, 1974). Photooxidation of aldrin in water is accelerated when hydrogen peroxide is present. Products identified include dieldrin, photoaldrin, and possibly a hydroperoxide (Draper and Crosby, 1984). When an aqueous solution of aldrin (0.07 μM) in natural water samples collected from California and Hawaii were irradiated (λ <220 nm) for 36 hr, 25% was photooxidized to dieldrin (Ross and Crosby, 1985). In an aqueous solution containing peracetic acid, aldrin was transformed to dieldrin in the dark (Ross and Crosby, 1975).

Chemical/Physical: Aldrin is oxidized in the presence of oxygen forming dieldrin (Saravanja-Bozanic et al., 1977). Photoaldrin was formed when a benzene solution containing aldrin and benzophenone as a sensitizer was exposed to UV light (λ = 268-356 nm) (Rosen and Carey, 1968).

When heated to decomposition, toxic chlorides are released (Lewis, 1990). Slowly releases hydrogen chloride during storage (Hartley and Kidd, 1987).

Exposure Limits: NIOSH REL: lowest detectable limit; OSHA PEL: TWA 0.25 mg/m^3; ACGIH TLV: TWA 0.25 mg/m^3.

Symptoms of Exposure: Headache, dizziness; nausea, vomiting, malaise; myoclonic jerks of limbs; clonic, tonic convulsions; coma; hematuria, azotemia.

Formulation Types: Wettable powder; emulsifiable concentrate; dustable powder; granules.

Toxicity: LC_{50} (96 hr) for American eel 5 ppb, mummichog 4-8 ppb, striped killifish 17 ppb, Atlantic silverside 13 ppb, striped mullet 100 ppb, bluehead 12 ppb, northern puffer 36 ppb, fathead minnow 28 μg/L, bluegill sunfish 13 μg/L, rainbow trout 17.7 μg/L, coho salmon 45.9 μg/L, chinook 7.5 μg/L, striped bass 10 μg/L, pumpkinseed 20 μg/L, and white perch 42 μg/L; LC_{50} (48 hr) for mosquito fish 36 ppb; LC_{50} (24 hr) for bluegill sunfish 260 ppb (Verschueren, 1983); acute oral LD_{50} for rats is 38-67 mg/kg (Hartley and Kidd, 1987).

Uses: Formerly as insecticide and fumigant; manufacture and use has been discontinued in the U.S.

ALLIDOCHLOR

Synonyms: Alidochlor; CDAA; CDAAT; 2-Chloro-*N,N*-diallylacetamide; 2-Chloro-*N,N*-di-2-propenylacetamide; α-Chloro-*N,N*-diallylacetamide; CP 6343; Diallylchloroacetamide; *N,N*-Diallylchloroacetamide; *N,N*-Diallyl-2-chloroacetamide; *N,N*-Diallyl-α-chloroacetamide; NCI-C04035; Radox; Randox; Randox T.

Structure:

$$CH_2 = CHCH_2 \Big\rangle N\overset{\displaystyle O}{\overset{\|}{C}}CH_2Cl$$
$$CH_2 = CHCH_2$$

Designations: CAS: 93-71-0; mf: $C_8H_{12}ClNO$; fw: 173.65; RTECS: AB5250000.

Properties: Amber liquid. Mp: <25 °C; bp: 92 °C at 2 mmHg; log K_{oc}: 1.83 (calculated); log K_{ow}: 0.97 (calculated); S_o: soluble in ethanol, *n*-hexane, xylene; S_w: 1.97 g/L at 25 °C; vap d: 7.10 g/L at 25 °C, 6.02 (air = 1).

Transformation Products

Plant: Allidochlor is translocated in plants to chloroacetic acid and diallylamine. The diallylamine is further transformed to carbon dioxide. The acid undergoes further degradation to glycollic acid which breaks down to glyoxalic acid. Glyoxalic acid undergoes further degradation to give formic acid, glycine, and carbon dioxide (Cremlyn, 1991).

Chemical/Physical: Emits very toxic fumes of phosphorus oxides and chlorine when heated to decomposition (Sax and Lewis, 1987).

Symptoms of Exposure: Strong irritant.

Toxicity: Acute oral LD_{50} for rats 700 mg/kg (RTECS, 1985).

Uses: Selective preemergence herbicide used to control annual grass weeds and some broadleaf weeds in maize, millet, soybeans, sorghum, sugarcane, vegetables, and ornamentals.

AMETRYN

Synonyms: A 1093; Amtrex; Ametryne; Crisatine; 2-Ethylamino-4-isopropyl-amino-6-methylmercapto-s-triazine; 2-Ethylamino-4-isopropylamino-6-methyl-thio-s-triazine; 2-Ethylamino-4-isopropylamino-6-methylthio-1,3,5-triazine; N-Ethyl-N'-isopropyl-6-methylthio-1,3,5-triazine-2,4-diyldiamine; **N-Ethyl-N'-(1-methylethyl)-6-(methylthio)-1,3,5-triazine-2,4-diamine;** Evik; Evik 80W; G 34,162; Gesapax; 2-Methylmercapto-4-ethylamino-s-triazine; 2-Methylthio-4-ethylamino-6-isopropylamino-s-triazine.

Structure:

Designations: CAS: 834-12-8; mf: $C_9H_{17}N_5S$; fw: 227.35; RTECS: XY9100000.

Properties: Colorless to white crystalline powder. Mp: 84-86 °C; ρ: 1.19 at 20/4 °C; fl p: nonflammable; pK_a: 4.1; H-$t_{1/2}$: 32 d (pH 1), >200 d (pH 13); K_H: 1.36 x 10^{-9} atm·m^3/mol at 20 °C (approximate - calculated from water solubility and vapor pressure); log K_{oc}: 2.23-2.44; log K_{ow}: 2.63-3.07; S_o (g/L at 20 °C): soluble in acetone (500), methylene chloride (600), n-hexane (14), methanol (450), toluene (400); S_w: 185 mg/L at 20 °C; vp: 8.4 x 10^{-7} mmHg at 20 °C.

Soil properties and adsorption data

Soil	K_d (mL/g)	f_{oc} (%)	K_{oc} (mL/g)	pH	CEC (meq/100 g)
Aguadilla loamy sand	3.08	1.44	214	7.4	10.0
Aguirre clay loam	2.76	0.75	368	9.0	14.3
Alonso clay	7.11	1.84	386	5.1	13.8
Altura loam	2.55	2.13	120	8.0	27.6
Bayamón sandy clay loam	3.20	0.98	326	4.7	5.0
Catalina clay	2.78	1.09	255	4.7	11.8
Cataño sand	2.08	1.21	172	7.9	6.9
Cayaguá sandy loam	3.97	1.15	345	5.2	7.3
Cialitos clay loam	13.32	2.82	472	5.4	18.6
Coloso clay loam	9.16	2.13	430	5.7	23.0
Coto clay	2.51	1.84	136	7.7	14.0
Fe clay loam	3.72	1.96	190	7.5	27.6
Fortuna silty clay loam	18.23	1.90	959	5.4	23.3

Soil	K_d (mL/g)	f_{oc} (%)	K_{oc} (mL/g)	pH	CEC (meq/100 g)
Fraternidad clay	3.34	1.21	276	6.3	36.0
Fraternidad clay	8.14	2.42	336	5.9	58.0
Guanicá clay	4.04	2.77	146	8.1	52.1
Humata silty clay loam	4.70	0.98	480	4.5	10.1
Josefa silty loam	7.16	1.90	376	6.0	16.8
Juncos silty clay	10.90	1.55	703	6.2	13.4
Mabí clay	6.49	2.26	287	7.0	55.2
Mabí clay loam	5.58	2.82	198	5.7	31.0
Mercedita silty clay	2.38	1.38	173	8.1	19.9
Moca clay	13.61	2.19	621	5.8	31.0
Múcara loam	1.05	1.90	550	5.8	19.6
Nipe clay loam	9.52	3.06	311	5.7	11.9
Pandura sandy loam	3.63	1.15	316	5.7	7.7
Río Pedras silty clay	3.20	2.02	158	4.9	11.5
San Anton loam	4.33	1.55	279	6.7	26.1
Sand	0.88	0.35	253	5.6	--
Sandy loam	4.84	1.74	278	6.1	--
Silty loam	3.78	1.68	225	6.9	--
Silty loam	4.97	2.90	171	7.0	--
Silty loam	2.81	1.22	230	7.0	--
Toa loam	9.72	1.16	837	5.3	13.0
Toa sandy loam	2.08	0.34	612	6.0	8.0
Talante sandy loam	6.19	0.80	774	5.1	4.0
Vega Alta sandy loam	5.14	2.02	254	5.0	5.6
Via loam	8.49	1.32	643	5.1	39.9

Source: Liu et al., 1970; U.S. Department of Agriculture, 1990.

Transformation Products
Biological: Cook and Hütter (1982) reported that bacterial cultures were capable of degrading ametryne forming the corresponding hydroxy derivative (hydroxy-ametryne).

Soil: Although no products were reported, the half-life in soil is 70–120 d (Worthing and Hance, 1991).

Plant: Ametryn is metabolized by tolerant plants into nontoxic hydroxy and dealkylated derivatives (Humburg et al., 1989).

Photolytic: The dye-sensitized photodecomposition of ametryn was studied in aqueous, aerated solutions (Rejto et al., 1983). When an aqueous ametryn solution was irradiated in sunlight for several hours, 2-(methylthio)-4-(isopropylamino)-6-amino-s-triazine and 2-(methylthio)-4-(isopropylamino)-6-acetamido-s-triazine formed in yields of 55 and 2.6%, respectively (Rejto et al., 1983). Further irradiation of the solution led to the formation of 2-(methylthio)-4,6-diamino-s-triazine which eventually decomposed to unidentified products (Rejto et al., 1983). The UV (λ = 253.7 nm) photolysis of ametryn in water, methanol, ethanol, n-butanol, and benzene yielded the 2-H analog 4-ethylamino-6-isopropylamino-s-triazine. Photodegradation was not observed at wavelengths >300 nm (Pape and Zabik, 1970).

Chemical/Physical: Hydrolyzes to the 6-hydroxy analog, especially in the presence of strong acids and alkalies (Hartley and Kidd, 1987).

Symptoms of Exposure: Eye and skin irritant.

Formulation Types: Wettable concentrate (80%); suspension concentrate.

Toxicity: LC_{50} (96 hr) for rainbow trout 8.8 mg/L, bluegill sunfish 4.1 mg/L, goldfish 14.1 mg/L, carp <1.0 mg/L (Hartley and Kidd, 1987), oyster >1.0 ppm (Humburg et al., 1989); acute oral LD_{50} for rats 1,110 mg/kg (Hartley and Kidd, 1987), 508 mg/kg (RTECS, 1985).

Uses: Herbicide used to control broadleaf and grass weeds in corn, sugarcane, certain citrus subtropical fruits (bananas, pineapple), and in noncropland. Preharvest and postharvest desiccant used in potatoes to control both crop and weeds.

AMINOCARB

Synonyms: A 363; Bay 44646; Bayer 5080; Bayer 44646; 4-Dimethylamine-*m*-cresyl methylcarbamate; 4-Dimethylamino-3-cresyl methylcarbamate; **4-(Dimethylamino)-3-methylphenol methylcarbamate**; 4-(Dimethylamino)-*m*-tolyl methylcarbamate; ENT 25784; Matacil; Mitacil.

Structure:

$$(CH_3)_2N-\underset{CH_3}{\overset{}{\bigcirc}}-O-\overset{O}{\overset{\|}{C}}NHCH_3$$

Designations: CAS: 2032-59-9; mf: $C_{11}H_{16}N_2O_2$; fw: 208.26; RTECS: FC0175000.

Properties: Colorless to white crystals. Mp: 93-94 °C; log K_{oc}: 1.92 (calculated); log K_{ow}: 1.73; S_o: soluble in most polar organic solvents; S_w: 872, 915, and 1,360 mg/L at 10, 20, and 30 °C, respectively.

Transformation Products

Chemical/Physical: Aminocarb is hydrolyzed in purified water to 4-(dimethylamino)-3-methylphenol which is then converted to 2-methyl-1,4-benzoquinone. This compound was then oxidized to form the following compounds: 6-(dimethylamino)-2-methyl-1,4-benzoquinone, 6-(methylamino)-2-methyl-1,4-benzoquinone, 5-(dimethylamino)-2-methyl-1,4-benzoquinone, and 5-(methylamino)-2-methyl-1,4-benzoquinone (Leger and Mallet, 1988). Emits toxic fumes of nitrogen oxides when heated to decomposition (Sax and Lewis, 1987).

Plant/Surface Water: Several transformation products reported by Day (1991) include 4-amino-*m*-tolyl-*N*-methylcarbamate (AA), 4-amino-3-methylphenol (AC), 4-formamido-*m*-tolyl-*N*-methylcarbamate (FA), *N*-(4-hydroxy-2-methylphenyl)-*N*-methylformamide (FC), 4-methylformamido-*m*-tolyl-*N*-methylcarbamate (MFA), 4-methylamino-*m*-tolyl-*N*-methylcarbamate (MAA), 3-methyl-4-(methylamino)phenyl-*N*-methylcarbamate (MAC), phenol, methylamine, and carbon dioxide. MAA was not detected in natural water but was detected in fish tissues following exposure to aminocarb-treated water in the laboratory. The metabolites FA, AC, and MAC were detected in Canadian forests treated with aminocarb but the metabolites AA, MAA, and FC were not detected (Day, 1991).

Formulation Types: Wettable powder.

Toxicity: Acute oral LD_{50} for rats 30-50 mg/kg (Hartley and Kidd, 1987), for male and female rats, 40 and 38 mg/kg, respectively (Windholz et al., 1983).

21

Uses: Nonsystemic, broad-spectrum insecticide used to control the spruce budworm in forests; molluscicide.

AMITROLE

Synonyms: Amazol; Amerol; Aminotriazole; 2-Aminotriazole; 3-Aminotriazole; 3-Amino-s-triazole; 2-Amino-1,3,4-triazole; 3-Amino-1,2,4-triazole; **3-Amino-1H-1,2,4-triazole**; Amino triazole weed killer 90; Aminotriazole-spritzpulver; Amitol; Amitril; Amitril T.L.; Amitrol; Amitrol 90; Amitrol T; Amizol; Amizol D; Amizol F; AT; AT-90; ATA; AT liquid; Azaplant; Azolan; Azole; Campaprim A 1544; Cytrol; Cytrol Amitrole-T; Cytrole; Diurol; Diurol 5030; Domatol; Domatol 88; Elmasil; Emisol; Emisol 50; Emisol F; ENT 25455; Fenamine; Fenavar; Herbidal total; Herbizole; Kleer-lot; Orga-414; Radoxone TL; Ramizol; RCRA waste number U011; Simazol; Triazolamine; 1H-1,2,4-Triazol-3-amine; UN 2588; USAF XR-22; Vorox; Vorox AA; Vorox SS; Weedar ADS; Weedar AT; Weedazin; Weedazin arginit; Weedazol; Weedazol GP2; Weedazol super; Weedazol T; Weedazol TL; Weedex granulat; Weedoclor; X-all liquid.

Structure:

H
N–N
N
NH$_2$

Designations: CAS: 61-82-5; mf: $C_2H_4N_4$; fw: 84.08; RTECS: XZ3850000.

Properties: Colorless to white, odorless crystalline solid. Mp: 157–159 °C; ρ: 1.138 at 20/4 °C; fl p: nonflammable; K_H: 1.63 x 10^{-15} atm·m^3/mol at 20 °C (approximate - calculated from water solubility and vapor pressure); log K_{oc}: 1.73–2.31; log K_{ow}: -0.15; S_o: soluble in chloroform, ethanol, methanol; S_w: 280 g/L at 20 °C (pH = 7.0); vp: 4.13 x 10^{-9} mmHg at 20 °C.

Soil properties and adsorption data

Soil	K_d (mL/g)	f_{oc} (%)	K_{oc} (mL/g)	pH
Sand	0.68	0.46	147	5.6
Sandy loam	3.52	1.74	202	6.1
Silty loam	3.79	3.42	111	6.8
Silty loam	1.57	2.90	54	7.0

Source: U.S. Department of Agriculture, 1990.

Transformation Products
Soil: When radiolabeled amitrole-5-^{14}C was incubated in a Hagerstown silty clay

23

loam, 50 and 70% of the applied amount evolved as $^{14}CO_2$ after 3 and 20 d, respectively. In autoclaved soil, however, no $^{14}CO_2$ was detected (Kaufman et al., 1968). The average persistence in soils is 2-4 wk (Hartley and Kidd, 1987).

Plant: Amitrole is transformed in plants to form the conjugate β-(3-amino-1,2,4-triazol-1-yl)-α-alanine (Humburg et al., 1989) and/or 3-(3-amino-s-triazole-1-yl)-2-aminopropionic acid (Duke et al., 1991). Amitrole is metabolized in Canada thistle (*Cirsium arvense* L.) to three unknown compounds which were more phytotoxic than amitrole (Herrett and Bagley, 1964).

Chemical/Physical: Reacts with acids and bases forming soluble salts (Hartley and Kidd, 1987). Emits toxic fumes of nitrogen oxides when heated to decomposition (Sax and Lewis, 1987); however, incineration with polyethylene results in more than 99% decomposition (Sittig, 1985). An aqueous solution of amitrole has been shown to decompose in the following free radical systems: Fenton's reagent, UV irradiation, and riboflavin-sensitized photodecomposition (Plimmer et al., 1967). Amitrole-5-^{14}C reacted with Fenton's reagent to give radiolabeled carbon dioxide, unlabeled urea, and unlabeled cyanamide. Significant degradation of amitrole was observed when an aqueous solution was irradiated by a sunlamp ($\lambda = 280$-310 nm). In addition to ring compounds, it was postulated that other products may have formed from the polymerization of amitrole free radicals (Plimmer et al., 1967).

Exposure Limits: OSHA PEL: TWA 0.2 mg/m^3; ACGIH TLV: TWA 0.2 mg/m^3.

Formulation Types: Water soluble concentrate.

Toxicity: LC$_{50}$ (48 hr) for bluegill sunfish 100 ppm and coho salmon 325 mg/L (Verschueren, 1983); acute oral LD$_{50}$ for rats 1,100-2,500 mg/kg (Verschueren, 1983).

Uses: Nonselective, foliage-applied herbicide used in uncropped land and orchards to control certain grasses and to kill perennial weeds. It is also effective on poison ivy, poison oak, and aquatic weeds.

ANILAZINE

Synonyms: Anilazin; B 622; Bortrysan; (*o*-Chloroanilo)dichlorotriazine; 2-Chloro-*N*-(4,6-dichloro-1,3,5-triazin-2-yl)aniline; 2,4-Dichloro-6-*o*-chloroanilo-*s*-triazine; 2,4-Dichloro-6-(2-chloroanilo)-1,3,5-triazine; **4,6-Dichloro-*N*-(2-chlorophenyl)-1,3,5-triazin-2-amine**; Direz; Dyrene; Dyrene 50W; ENT 26058; Kemate; NCI-C08684; Triasyn; Triazin; Triazine; Zinochlor.

Structure:

Designations: CAS: 101-05-3; mf: $C_9H_5Cl_3N_4$; fw: 275.54; RTECS: XY7175000.

Properties: White to tan crystals. Mp: 159-160 °C; ρ: 1.8 at 20/4 °C; H-t$_{1/2}$: 730 hr (pH 4), 790 hr (pH 7), 22 hr (pH 9); K_H: 1.12 x 10^{-6} atm·m^3/mol at 20 °C (approximate - calculated from water solubility and vapor pressure); log K_{oc}: 3.48; log K_{ow}: 3.01; S_o (g/L) at 20 °C: *n*-hexane (1.7), methylene chloride (90), 2-propanol (8), at 30 °C: acetone (100), chlorobenzene (60), toluene (50), xylene (40); S_w: 8 mg/L at 20 °C; vp: 2.48 x 10^{-5} mmHg at 20 °C.

Transformation Products
Soil: Though no products were identified, the reported half-life in soil is approximately 12 hr (Hartley and Kidd, 1987).

Plant: In plants, one or both of the chlorine atoms on the triazine ring may be replaced by thio or amino groups (Hartley and Kidd, 1987).

Chemical/Physical: Releases chlorine gas when hydrolyzed (Worthing and Hance, 1991). Anilazine is subject to hydrolysis but the products were not reported (Windholz et al., 1983).

Exposure Limits: The German Research Society's recommended Maximum Allowable Concentration is 0.2 mg/m^3.

Formulation Types: Suspension concentrate; wettable powder.

Toxicity: LC$_{50}$ (96 hr) for bluegill sunfish, goldfish, and carp <1.0 mg/L; LC$_{50}$ (48 hr) for rainbow trout 150 µg/L (Hartley and Kidd, 1987); acute oral LD$_{50}$ for rats >5,000 mg/kg (Hartley and Kidd, 1987), 2,700 mg/kg (RTECS, 1985).

Uses: Nonsystemic, foliar fungicide used in potatoes, tomatoes, wheat, barley, and ornamentals.

ANTU

Synonyms: Anturat; Bantu; Chemical 109; Krysid; **1-Naphthalenylthiourea;** 1-(1-Naphthyl)-2-thiourea; α-Naphthylthiourea; *N*-1-Naphthylthiourea; α-Naphthyl-thiocarbamide; Rattrack.

Structure:

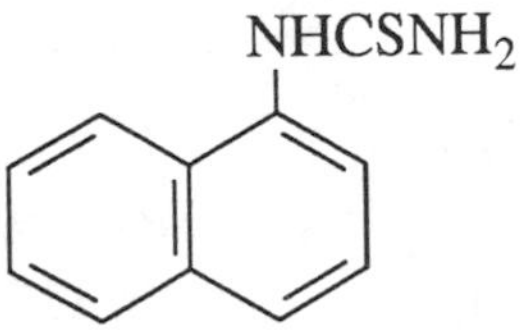

NHCSNH$_2$

Designations: CAS: 86-88-4; DOT: 1651; mf: $C_{11}H_{10}N_2S$; fw: 202.27; RTECS: YT9275000.

Properties: Colorless to gray, odorless solid. Bitter taste. Mp: 198 °C; bp: decomposes; ρ: 1.895 (calculated); H-t$_{1/2}$: 361 d at 25 °C and pH 7; S$_o$ (g/L at 25 °C): acetone (24.3), triethylene glycol (86); S$_w$: 600 mg/L at 25 °C; vp: ≈ 0 mmHg at 20 °C.

Transformation Products
Chemical/Physical: Emits very toxic fumes of nitrogen and sulfur oxides when heated to decomposition (Lewis, 1990).

Exposure Limits: NIOSH REL: IDLH 100 mg/m^3; OSHA PEL: TWA 0.3 mg/m^3.

Symptoms of Exposure: Vomiting; dyspnea; cyanosis; coarse pulmonary rales after ingestion of large doses.

Formulation Types: Tracking powder.

Toxicity: Acute oral LD$_{50}$ for Norwegian rats 6-8 mg/kg (Hartley and Kidd, 1987).

Use: Rodenticide.

ASPON

Synonyms: A 42; ASP 51; Bis-*O,O*-di-*n*-propylphosphorothionic anhydride; E 8573; ENT 16894; NPD; Propylthiopyrophosphate; Stauffer ASP-51; Tetra-*n*-propyl dithionopyrophosphate; **Tetrapropyl dithiopyrophosphate**; Tetra-*n*-propyl dithiopyrophosphate.

Structure:

$$CH_3CH_2CH_2O \underset{CH_3CH_2CH_2O}{\overset{\overset{S}{\|}}{\diagdown}} P - O - P \underset{OCH_2CH_2CH_3}{\overset{\overset{S}{\|}}{\diagup}} OCH_2CH_2CH_3$$

Designations: CAS: 3244-90-4; mf: $C_{12}H_{28}O_5P_2S_3$; fw: 378.46; RTECS: XN4550000.

Properties: Liquid. Mp: <25 °C; H-t$_{1/2}$: 32 d at 40 °C and pH 7; log K_{oc}: 3.80-3.88; log K_{ow}: 2.15 (calculated); S_o: 160 mg/L at room temperature; vap d: 15.47 g/L at 25 °C, 13.11 (air = 1).

Toxicity: Toxic to fish. Acute oral LD_{50} for rats 450 mg/kg (RTECS, 1985).

Uses: Insecticide used to control chinch bugs in lawns.

ASULAM

Synonyms: Asilan; Asulfox F; Asulox; Asulox 40; Jonnix; MB 9057; Methyl *N*-(4-aminobenzenesulfonyl)carbamate; **Methyl((4-aminophenyl)sulfonyl)carbamate**; Methyl sulfanilylcarbamate.

Structure:

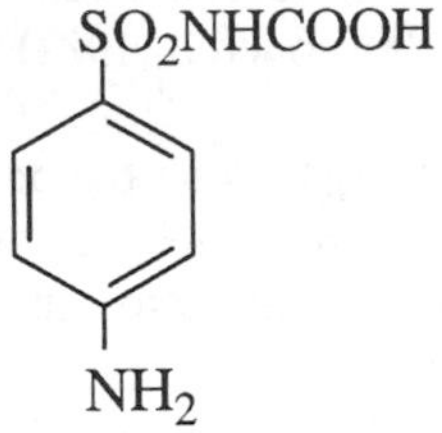

SO$_2$NHCOOH

NH$_2$

Designations: CAS: 3337-71-1; mf: C$_8$H$_{10}$N$_2$O$_4$S; fw: 230.24; RTECS: FD1190000.

Properties: Odorless, colorless crystals or white crystalline powder. Mp: 142-144 °C (decomposes); fl p: nonflammable; pK$_a$: 4.82 (imido group); K$_H$: 1.1 x 10^{-12} atm·m^3/mol at 25 °C (approximate - calculated from water solubility and vapor pressure); log K$_{oc}$: 1.80-2.16; log K$_{ow}$: 0.763; S$_o$ (g/L): acetone (300), *N,N*-dimethylformamide (>800), ethanol (180), methanol (290), 2-butanone (280); S$_w$: 5 g/L at 20-25 °C; vp: 1 x 10^{-8} mmHg at 25 °C.

Soil properties and adsorption data

Soil	K$_d$ (mL/g)	f$_{oc}$ (%)	K$_{oc}$ (mL/g)	pH
Sand	0.80	0.58	138	6.8
Sand	1.00	3.19	31	6.9
Sandy loam	1.35	2.15	63	7.0
Sandy loam	2.72	1.86	146	7.2

Source: U.S. Department of Agriculture, 1990.

Transformation Products
Soil: It is not persistent in soils since its half-life is approximately 6-14 d (Hartley and Kidd, 1987).

Photolytic: The reported photolytic half-lives of asulam in water at pH 3 and 9 were 2.5 and 9 d, respectively (Humburg et al., 1989).

Chemical/Physical: Forms water-soluble salts (Hartley and Kidd, 1987). When heated to 75 °C, asulam decomposed to sulfanilic acid, carbamic acid, and sulfanilamide. At 90 °C, 4-nitro- and 4-nitrosobenzene sulfonic acids were released (Rajagopal et al., 1984).

29

Formulation Types: Aqueous solution of the sodium salt (3.34 lb/gal).

Toxicity: LC_{50} (96 hr) for goldfish, rainbow trout, channel catfish >5 g/L, harlequin fish >1.7 g/L, and bluegill sunfish >3 g/L (Worthing and Hance, 1991); acute oral LD_{50} for rats is >8,000 mg/kg (Ashton and Monaco, 1991), 2,000 mg/kg (RTECS, 1985), >5,000 mg/kg (potassium salt) (Verschueren, 1983), and mice 5,000 mg/kg (potassium salt) (Verschueren, 1983).

Uses: Systemic, pre- and postemergence herbicide used to control several perennial grasses and certain broadleaf weeds such as brackenfern, crabgrass, itchgrass, paragrass, tansy ragwort, and wild mustard, in alfalfa, uncropped land, certain ornamentals, and turf.

ATRAZINE

Synonyms: A 361; Aatrex; Aatrex 4L; Aatrex 4LC; Aatrex nine-o; Aatrex 80W; Aktikon; Aktikon PK; Aktinit A; Aktinit PK; Argezin; Atazinax; Atranex; Atrasine; Atratol A; Atrazin; Atred; Atrex; Candex; Cekuzina-T; 2-Chloro-4-ethylamineisopropylamine-s-triazine; 1-Chloro-3-ethylamino-5-isopropyl-amino-s-triazine; 1-Chloro-3-ethylamino-5-isopropylamino-2,4,6-triazine; 2-Chloro-4-ethylamino-6-isopropylamino-s-triazine; 2-Chloro-4-ethylamino-6-isopropylamino-1,3,5-triazine; **6-Chloro-N^2-ethyl-N^4-isopropyl-1,3,5-triazine-2,4-diamine**; 2-Chloro-4-(2-propylamino)-6-ethylamino-s-triazine; Crisatrina; Crisazine; Cyazin; Farmco atrazine; Fenamin; Fenamine; Fenatrol; G 30,027; Geigy 30027; Gesaprim; Gesoprim; Griffex; Hungazin; Hungazin PK; Inakor; Oleogesaprim; Primatol; Primatol A; Primaze; Radazin; Radizine; Shell atrazine herbicide; Strazine; Triazine A; Triazine A 1294; Vectal; Vectal SC; Weedex A; Wonuk; Zeazin; Zeazine.

Structure:

$$\text{NHCH(CH}_3)_2$$
$$\text{NHCH}_2\text{CH}_3$$

Designations: CAS: 1912-24-9; DOT: 1609; mf: $C_8H_{14}ClN_5$; fw: 215.68; RTECS: XY5600000.

Properties: Colorless powder or white crystalline solid. Mp: 171-174 °C, 175-177 °C; bp: decomposes; ρ: 1.187 at 20/4 °C; fl p: nonflammable; pK_a: 1.62 at 20 °C, 1.70 at 21 °C; H-t$_{1/2}$: 1,771 yr at 25 °C and pH 7; K_H: 3.04 x 10^{-9} atm·m^3/mol at 20 °C (approximate - calculated from water solubility and vapor pressure); log K_{oc}: 1.95-2.71; log K_{ow}: 2.33-2.80; P-t$_{1/2}$: 58.67 hr (absorbance λ = 240.0 nm, concentration on glass plates = 6.7 μg/cm^2); S_o (g/kg) at 20 °C: ethyl acetate (18), n-octanol (10), n-pentane (0.36), at 27 °C: chloroform (52), dimethyl sulfoxide (183), ethyl acetate (28), ethyl ether (12), methanol (18); S_w: 28 mg/L at 20 °C, 33 mg/L at 27 °C; vp: 3.0 x 10^{-7} mmHg at 20 °C, 1.4 x 10^{-6} mmHg at 30 °C.

Soil properties and adsorption data

Soil	K_d (mL/g)	f_{oc} (%)	K_{oc} (mL/g)	pH	Salinity	TOC (mg/L)	CEC[a]
Bates silty loam	0.80	0.80	100	6.5	--	--	9.3
Baxter silty clay loam	2.30	1.21	190	6.0	--	--	11.2

continued

Soil	K_d (mL/g)	f_{oc} (%)	K_{oc} (mL/g)	pH	Salinity	TOC (mg/L)	CEC[a]
Begbroke silty loam	1.00	1.11	90	7.1	--	--	--
Boyce loam	0.78	1.27	61	8.0	--	--	20.0
Cecil loamy sand	0.89	0.90	99	5.6	--	--	6.8
Chehalis sandy loam	0.83	0.98	84	5.2	--	--	16.9
Chillum sandy loam	4.00	2.54	157	4.6	--	--	7.6
Choptank River, MD	--	--	4,860	--	9.92	108.5	--
Choptank River, MD	--	--	13,600	--	1.24	98.6	--
Choptank River, MD	--	--	8,540	--	1.50	65.5	--
Choptank River, MD	--	--	6,990	--	14.20	104.6	--
Choptank River, MD	--	--	4,840	--	17.00	59.3	--
Choptank River, MD	--	--	7,930	--	5.71	74.2	--
Clarksville silty clay loam	1.70	0.80	212	5.7	--	--	5.7
Collombey	0.86	1.28	67	7.8	--	--	--
Cumberland silty loam	1.40	0.69	203	6.4	--	--	6.5
Dark sandy loam	12.30	12.00	102	6.3	--	--	18.0
Deschutes sandy loam	0.68	0.51	132	5.9	--	--	12.9
Drummer clay loam	6.30	3.63	174	8.0	--	--	40.0
Drummer clay loam	12.60	3.63	347	3.9	--	--	40.0
Drummer clay loam	6.50	3.63	179	6.0	--	--	40.0
Drummer clay loam	8.20	3.63	226	5.3	--	--	40.0
Drummer clay loam	10.70	3.63	295	4.7	--	--	40.0
Eldon silty loam	2.50	1.73	144	5.9	--	--	12.9
Eustis fine sand	0.62	0.56	111	5.6	--	--	5.2
Evouettes	1.98	2.09	95	6.1	--	--	--
Gerald silty loam	3.20	1.55	206	4.7	--	--	11.0
Gila silty clay	0.65	0.63	103	8.0	--	--	29.1
Glendale sandy clay loam	0.62	0.50	124	7.4	--	--	--
Grundy silty clay loam	4.80	2.07	232	5.6	--	--	13.5
Hagerstown silty clay loam	3.70	2.48	149	5.5	--	--	12.5
Hickory Hill silt	7.07	3.27	216	--	--	--	--
Kaipoioi LBF	9.30	16.70	56	5.4	--	--	--
Kapaa HFL	2.90	5.70	51	4.4	--	--	--
Keith fine sandy loam	1.49	1.67	89	6.3	--	--	--
Knox silty loam	3.60	1.67	216	5.4	--	--	18.8
Kula RP	6.00	12.70	47	5.9	--	--	--
Lakeland silty loam	1.00	1.90	53	6.2	--	--	2.9

Soil	K_d (mL/g)	f_{oc} (%)	K_{oc} (mL/g)	pH	Salinity	TOC (mg/L)	CEC[a]
Lebanon silty loam	2.20	1.04	212	4.9	--	--	7.7
Lindley loam	2.60	0.86	302	4.7	--	--	6.9
Lintonia loamy sand	0.60	0.34	176	5.3	--	--	3.2
Loam	2.03	1.45	140	6.6	--	--	--
Marian silty loam	2.20	0.80	275	4.6	--	--	9.9
Marshall silty clay loam	4.50	2.42	186	5.4	--	--	21.3
Menfro silty loam	1.70	1.38	123	5.3	--	--	9.1
Metolius sandy loam	0.62	0.80	77	7.1	--	--	18.5
Molokai LHL	2.30	2.30	100	6.3	--	--	--
Monona silty clay loam	1.92	1.67	115	5.8	--	--	--
Newtonia silty loam	1.80	0.92	196	5.2	--	--	8.8
Oswego silty clay loam	2.70	1.67	162	6.4	--	--	21.0
Powder silty loam	1.20	1.67	72	7.7	--	--	31.9
Putnam silty loam	1.90	1.09	174	5.3	--	--	12.3
Quincy sandy loam	0.26	0.28	93	8.0	--	--	11.0
Rhinebeck silty clay loam	1.98	3.13	63	6.7	--		
Salix loam	2.30	1.21	190	6.3	--	--	17.9
Sand	0.42	0.46	91	5.6	--	--	--
Sandy loam	0.99	1.74	57	6.1	--	--	--
Sandy loam	1.00	1.93	52	7.1	--	--	11.0
Sarpy loam	2.20	0.75	293	7.1	--	--	14.3
Sharkey clay	3.10	1.44	215	5.0	--	--	28.2
Sharpsburg silty clay loam	3.89	2.19	178	5.2	--	--	--
Shelby loam	3.20	2.07	154	4.3	--	--	20.1
Silty loam	1.46	1.22	120	7.0	--	--	--
Summit silty clay	5.60	2.82	198	4.8	--	--	35.1
Union silty loam	4.10	1.04	394	5.4	--	--	6.8
Valentine loamy fine sand	0.77	0.80	96	5.9	--	--	--
Valois silty loam	1.98	1.64	121	5.9	--	--	--
Vetroz	2.88	3.25	89	6.7	--	--	--
Wabash clay	3.70	1.27	291	5.7	--	--	40.3
Waverley silty loam	3.00	1.15	261	6.4	--	--	12.8
Webster silty clay loam	6.03	3.87	156	7.3	--	--	54.7
Wehadkee silty loam	1.80	1.09	165	5.6	--	--	10.2
Woodburn silty loam	1.45	1.32	110	5.2	--	--	12.8

Source: Brown and Flagg, 1981; Burkhard and Guth, 1981; Colbert et al., 1975;

Dao and Lavy, 1978; Gamerdinger et al., 1991; Green and Obien, 1969; Grover and Hance, 1969; Hance, 1967; Harris, 1966; Leistra, 1970; McGlamery and Slife, 1966; Obien and Green, 1969; Rao and Davidson, 1979; Talbert and Fletchall, 1965; U.S. Department of Agriculture, 1990. a) meq/100 g.

Transformation Products
Biological: Twelve fungi in a basal salts medium and supplemented with sucrose degraded atrazine via monodealkylation. Degradation by *Aspergillus fumigatus* and *Penicillium janthinellum* yielded 2-chloro-4-amino-6-isopropylamino-*s*-triazine and *Rhizopus stolonifer* yielded 2-chloro-4-ethylamino-6-amino-*s*-triazine (Paris and Lewis, 1973).

Deethylatrazine was the major metabolite found in an alluvial aquifer in south central Ohio. Ring-labeled atrazine was recalcitrant to degradation by soil and aquifer microorganisms under aerobic conditions. Aerobic incubations of sediments mineralized <0.1-1.5% [*ethyl-2-*^{14}C]atrazine to carbon dioxide. The rate of mineralization was higher in shallow sediments (5.2-5.6 m) than in deep sediments (17.7-18.1 m). This suggests that microorganisms in the shallow sediments were better adapted at degrading atrazine than microorganisms in the deeper sediments (McMahon et al., 1992).

Soil: Atrazine undergoes hydrolysis in soil forming hydroxyatrazine (Armstrong et al., 1967; Cremlyn, 1991; Esser et al., 1975; Goswami and Green, 1971; Harris, 1967; Helling, 1971; Khan, 1978; Nair and Schnoor, 1992; Obien and Green, 1969; Somasundaram et al., 1991; Skipper et al., 1967). The rate of hydrolysis, which followed first-order kinetics, was faster in low pH soils (Best and Weber, 1974; Harris et al., 1969; Hiltbold and Buchanan, 1977; Lowder and Weber, 1982; Obien and Green, 1969). Atrazine hydrolysis is also catalyzed with additions of small amounts of sterilized soil (Armstrong et al., 1967) or by the presence of fulvic and humic acids (Armstrong et al., 1967; Khan, 1978; Junk et al., 1980; Li and Felbeck, 1972; Lowder and Weber, 1982). In addition, the rate of hydrolysis in soils was more a function of the organic carbon content (f_{oc}) rather than amount of clay present (Anderson et al., 1980a; Armstrong et al., 1967). Li and Felbeck (1972) reported that the half-lives of atrazine at 25 °C and pH 4 with and without fulvic acid (2%) were 1.73 and 244 d, respectively. The hydrolysis half-lives in a 5 mg/L fulvic acid solution and 25 °C at pHs of 2.9, 4.5, 6.0, and 7.0 were 34.8, 174, 398, and 742 d, respectively. The only product identified was 2-ethylamino-4-hydroxy-6-isopropylamino-1,3,5-triazine (Khan, 1978). The primary degradative pathway appears to be chemical (i.e., hydrolysis) rather than microbial (Armstrong et al., 1967; Best et al., 1974; Gaynor et al., 1981; Geller, 1980; Gormly and Spalding, 1979; Lowder and Weber, 1982; Skipper et al., 1967). In soil-water suspensions under both aerobic and anaerobic conditions, ^{14}C ring-labeled atrazine degraded very slowly (H-t$_{1/2}$ >90 d). In aerobic soils, only 0.59% evolved as ^{14}CO$_2$

(Goswami and Green, 1971). Geller (1980) reported that atrazine degradation in soil and water was probably due to chemical rather than biological processes. Muir and Baker (1978) reported, however, that dechlorination was a major metabolic process in soils and that 2-chloro-4-ethylamino-6-amino-s-triazine and 2-chloro-4-amino-6-isopropylamino-s-triazine were the major metabolites formed. Though atrazine is very persistent in soils, mineralization of ^{14}C ring-labeled atrazine to $^{14}CO_2$ was detected by Skipper and Volk (1972). Nine years after applying uniformly ^{14}C ring-labeled atrazine to a mineral soil in Germany, 50% of the ^{14}C residues were in the bound (nonextractable) form. Residues detected included atrazine (0.11 ppm), deethylatrazine (trace), deisopropylatrazine (trace), hydroxyatrazine (0.10 ppm), deethylhydroxyatrazine (0.13 ppm), and deisopropylhydroxyatrazine (0.07 ppm). Except for deethylatrazine, these metabolites were found in higher concentrations in humic soils (Capriel et al., 1985). The half-lives in soil was reported to be about 6-10 wk (Hartley and Kidd, 1987). In biologically active soils, the chlorine atom at the C-6 position is replaced by a hydroxyl group and hydroxyatrazine is formed (Beynon et al., 1972). Pure cultures of soil fungi were capable of degrading atrazine to two or more metabolites. These included 2-chloro-4-amino-6-isopropylamino-s-triazine and 2-chloro-4-ethyl-amino-6-amino-s-triazine plus the suspected intermediates 2-chloro-4-hydroxy-6-ethylamino-s-triazine, 2-hydroxy-4-amino-6-ethylamino-s-triazine, and 2,4-dihydroxy-6-ethylamino-s-triazine (Kaufman and Kearney, 1970). In a soil-core microcosm study, Winkelmann and Klaine (1991) observed that the concentration of atrazine decreased exponentially over a 6-mo period. Metabolites identified in soil included DEA, deisopropylatrazine, DAA, and hydroxyatrazine. The half-life in soil is 71 d (Jury et al., 1987). Under laboratory conditions, the half-lives of atrazine in a Hatzenbühl soil (pH 4.8) and Neuhofen soil (pH 6.5) at 22 °C were 53 and 113 d, respectively (Burkhard and Guth, 1981). Atrazine degradation products identified in soil were deethylatrazine, deisopropylatrazine, deethyldeisopropyl-atrazine, and hydroxyatrazine (Patumi et al., 1981). Microbial attack of atrazine gave deethylated atrazine and deisopropyl atrazine as major and minor metabolites, respectively (Sirons et al., 1973).

Plant: In tolerant plants, atrazine is readily transformed to hydroxyatrazine which may further degrade via dealkylation of the side chains and subsequent hydrolysis of the amino groups with some evolution of carbon dioxide (Humburg et al., 1989). In corn juice, atrazine was converted to hydroxyatrazine (Montgomery and Freed, 1964). In both roots and shoots of young bean plants, atrazine underwent monodealkylation forming 2-chloro-4-amino-6-isopropylamino-s-triazine. This metabolite was found to be less phytotoxic than atrazine (Shimabukuro, 1967). Roots of the marsh grass *Spartina alterniflora* were placed in an aqueous atrazine solution (5 x 10^{-5} to 5 x 10^{-8} M) for 2 d. The atrazine was readily absorbed by the roots and translocated to the shoots. Metabolites identified by TLC were 2-chloro-4-amino-6-ethylamino-s-triazine,   2-chloro-4-amino-6-isopropylamino-s-triazine, and 2-chloro-4,6-diamino-s-triazine (Pillai et al., 1977). Atrazine was found to be persistent in a peach orchard soil several years after nine consecutive annual application of the herbicide. After 2 and 3.5 yr following the last application of atrazine, metabolites identified in the soil included deethylatrazine, hydroxyatrazine, deethylhydroxyatrazine, and deisopropylhydroxyatrazine. These metabolites were also found in oat plants grown in the same soil, but they were translocated into the plant where they underwent detoxification via conjugation (Khan and Marriage, 1977). Lamoureux et al. (1970) studied the metabolism and detoxification of atrazine in treated sorghum leaf sections. They found a third major pathway for the metabolism of atrazine which led to the formation of S-(4-ethylamino-6-isopropylamino-2-s-triazino)glutathione (III) and γ-L-glutamyl-S-(4-ethylamino-6-isopropylamino-2-s-triazino)-L-cysteine (IV). It was suggested that compound III was formed by the enzymatic catalyzation of glutathione with atrazine. Compound IV formed from the reaction of compound III with a carboxypeptidase or by the enzymatically catalyzed condensation of γ-L-glutamyl-L-cysteine with atrazine.

Surface Water: Desethyl- and desisopropylatrazine were degradation products of atrazine that were identified in the Mississippi River and its tributaries (Pereira and Rostad, 1990). Under laboratory conditions, atrazine in distilled water and river water was completely degraded after 21.3 and 7.3 hr, respectively (Mansour et al., 1989).

Photolytic: In aqueous solutions, atrazine is converted exclusively to hydroxyatrazine by UV light (λ = 253.7 nm) (Khan and Schnitzer, 1978; Pape and Zabik, 1970) and natural sunlight (Pape and Zabik, 1972). Irradiation of atrazine in methanol, ethanol, and n-butanol afforded Atratone (2-methoxy-4-ethylamino-6-isopropylamino-s-triazine), 2-ethoxy-s-triazine analog, and 2-n-butoxy-s-triazine analog, respectively (Pape and Zabik, 1970). Atrazine did not photodegrade when irradiated in methanol, ethanol, or in water at wavelengths >300 nm (Pape and Zabik, 1970). Hydroxyatrazine, two de-N-alkyl and the de-N,N'-dialkyl analogs of atrazine were produced in the presence of fulvic acid (0.01%). The P-t$_{1/2}$ of atrazine in aqueous solution containing fulvic acid remained essentially unchanged but increased at lower pHs (Khan and Schnitzer, 1978). Similar results were achieved in the work by Minero et al. (1992) who reported that the photolysis (λ >340 nm) of aqueous solution of atrazine containing dissolved humic acid (10 ppm organic carbon) was three times higher in the presence of humic acid. Dehalogenation was the significant degradative pathway but some dealkylation and deamination also occurred (Minero et al., 1992). An aqueous solution (15 °C) of atrazine (10 mg/L) containing acetone (1% by volume) as a photosensitizer was exposed to UV light (λ ≥290 nm). The reported photolysis half-lives with and without the sensitizer were 25 and 4.9 hr, respectively (Burkhard and Guth, 1976). Photoproducts formed were hydroxytriazines, two de-

N-alkyl and the de-N,N'-dialkyl analogs (Burkhard and Guth, 1976). Pelizzetti et al. (1990) studied the aqueous photocatalytic degradation of atrazine (ppb level) using simulated sunlight (λ >340 nm) and titanium dioxide as a photocatalyst. Atrazine rapidly degraded from 2 ppb to <0.1 ppb in a few minutes. The following intermediates were identified via HPLC and/or GC/MS: 2,4-diamino-6-hydroxy-N-ethyl-N'-(1-methylethyl)-1,3,5-triazine, ammeline (2,4-diamino-6-hydroxy-1,3,5-triazine), ammelide (2-diamino-4,6-dihydroxy-1,3,5-triazine), 2,4-diamino-6-chloro-1,3,5-triazine, 2,4-diamino-6-chloro-N-(1-methylethyl)-1,3,5-triazine, 2,4-diamino-6-chloro-N-ethyl-1,3,5-triazine, 2-amino-4-chloro-6-hydroxy-1,3,5-triazine, 2-chloro-4,6-dihydroxy-1,3,5-triazine, cyanuric acid (2,4,6-trihydroxy-1,3,5-triazine), and 2-acetylamino-4-amino-6-chloro-N-(1-methylethyl)-1,3,5-triazine. Complete degradation of atrazine gave cyanuric acid (2,4,6-trihydroxy-1,3,5-triazine), chloride, and nitrate ions. Mineralization of cyanuric acid to carbon dioxide was not observed (Pelizzetti et al., 1990). Nearly identical photodegradation products were reported by Pelizzetti et al. (1992). They studied the photocatalytic degradation of atrazine in water containing a suspension of titanium dioxide (0.5 g/L) as the catalyst. Irradiation was carried out in Pyrex glass cells using an 1500-W Xenon lamp (λ cutoff = 340 nm). The major photoprocesses of degradation were alkyl chain oxidation and subsequent oxidation leading to the formation of hydroxy derivatives. Dehalogenation was also observed but this was considered a minor degradative pathway. Cyanuric acid was the major photoproduct formed. There was only a partial conversion or mineralization to carbon dioxide (Pelizzetti et al., 1992). The dye-sensitized photodecomposition of atrazine was studied in aqueous, aerated solutions. When the solution was irradiated in sunlight for several hours, 2-chloro-4-(isopropylamino)-6-amino-s-triazine and 2-chloro-4-(isopropylamino)-6-acetamido-s-triazine formed in yields of 70 and 7%, respectively (Rejto et al., 1983). Further irradiation of the solution led to the formation 2-chloro-4,6-diamino-s-triazine which eventually degraded to unidentified products (Rejto et al., 1983). Hydroxyatrazine was the major intermediate compound formed when atrazine (100 μg/L) in both oxygenated estuarine water (Jones, 1982; Mansour et al., 1989) and estuarine sediments were exposed to sunlight. The rate of degradation was slightly higher in water ($t_{1/2}$ = 3-12 d) than in sediments ($t_{1/2}$ = 1-4 wk) (Jones, 1982).

Chemical/Physical: The hydrolysis half-lives in aqueous buffered solutions at 25 °C and pHs 1, 2, 3, 4, 11, 12, and 13 were reported to be 3.3, 14, 58, 240, 100, 12.5, and 1.5 d, respectively (Armstrong et al., 1967). Atrazine is stable in slightly acidic or basic media, but is hydrolyzed to hydroxy derivatives by alkalies and strong mineral acids (Windholz et al., 1983). Atrazine reacts with strong mineral acids forming hydroxyatrazine (Montgomery and Freed, 1964).

In the presence of hydroxy or perhydroxy radicals generated from Fenton's reagent, atrazine undergoes oxidative dealkylation in aqueous solutions (Kaufman and Kearney, 1970). Major products identified by mass spectrometry included deisopropylatrazine (2-chloro-4-ethylamino-6-amino-s-triazine), 2-chloro-4-amino-6-isopropylamino-s-triazine (DEA), and the dealkylated dealkylatrazine (DAA, 2-chloro-4,6-diamino-s-triazine) (Kaufman and Kearney, 1970).

The primary ozonation by-products of atrazine (15 µg/L) in natural surface water and synthetic water were deethylatrazine, deisopropylatrazine, a didealkylated atrazine (2-chloro-4,6-diamino-s-triazine), a deisopropylatrazine amide (4-acetamido-4-amino-6-chloro-s-triazine), 2-amino-4-hydroxy-6-isopropylamino-s-triazine, and an unknown compound. The types of compounds formed were pH dependent. At high pH, low alkalinity, or in the presence of hydrogen peroxide, hydroxyl radicals formed from ozone yielded s-triazine hydroxy analogs vis hydrolysis of the Cl-Cl bond. At low pH and low alkalinity, which minimized the production of hydroxy radicals, dealkylated atrazine and an amide were the primary by-products formed (Adams and Randtke, 1992).

Products reported from the combustion of atrazine at 900 °C include carbon monoxide, carbon dioxide, hydrochloric acid, and ammonia (Kennedy et al., 1972, 1972a). At 250 °C, however, atrazine decomposes to yellows flakes which were tentatively identified as primary or secondary amines (Stojanovic et al., 1972). Emits toxic fumes of chlorides and nitrogen oxides when atrazine is heated to decomposition (Lewis, 1990).

Exposure Limits: OSHA PEL: TWA 5 mg/m^3; ACGIH TLV: TWA 5 mg/m^3.

Symptoms of Exposure: Dermatitis; severely irritates skin, eyes, nose and throat.

Formulation Types: Liquid concentrate (4 lb/gal); wettable powder (80%); dispersible granules (85.5%).

Toxicity: LC_{50} (96 hr) for rainbow trout 8.8 mg/L, bluegill sunfish 16 mg/L, carp 76 mg/L, perch 16 mg/L, catfish 7.6 mg/L, and guppies 4.3 mg/L (Hartley and Kidd, 1987); LC_{50} (48 hr) for rainbow trout 12.6 ppm (Sanborn et al., 1977); acute oral LD_{50} of the 80% formulation for rats is 5,100 mg/kg (Ashton and Monaco, 1991), 750 mg/kg (RTECS, 1985).

Uses: Preemergence and postemergence herbicide for control of some annual grasses and broadleaf weeds in corn, fallow land, rangeland, sorghum, noncropland, certain tropical plantations, evergreen nurseries, fruit crops, and lawns.

AZINPHOS-METHYL

Synonyms: Bay 9027; Bay 17147; Bayer 9027; Bayer 17147; Benzotriazine-dithiophosphoric acid dimethoxy ester; Carfene; Cotneon; Cotnion methyl; Crysthion 2L; Crysthyon; DBD; *S*-(3,4-Dihydro-4-oxobenzo(α)(1,2,3)triazin-3-ylmethyl)-*O,O*-dimethyl phosphorodithioate; *S*-(3,4-Dihydro-4-oxo-1,2,3-benzotriazin-3-ylmethyl)-*O,O*-dimethyl phosphorodithioate; *O,O*-Dimethyl-*S*-(benzaziminomethyl) dithiophosphate; *O,O*-Dimethyl-*S*-(1,2,3-benzotriazinyl-4-keto)methyl phosphorodithioate; *O,O*-Dimethyl-*S*-(3,4-dihydro-4-keto-1,2,3-benzotriazinyl-3-methyl) dithiophosphate; Dimethyldithiophosphoric acid *N*-methylbenzazimide ester; *O,O*-Dimethyl-*S*-(4-oxo-3*H*-1,2,3-benzotriazine-3-methyl) phosphorodithioate; *O,O*-Dimethyl-*S*-(4-oxobenzotriazino-3-methyl) phosphorodithioate; *O,O*-Dimethyl-*S*-(4-oxo-1,2,3-benzotriazino-3-methyl) thio-thionophosphate; *O,O*-Dimethyl *S*-((4-oxo-1,2,3-benzotriazin-3-yl)methyl) phosphorodithioate; *O,O*-Dimethyl *S*-((4-oxo-1,2,3-benzotriazin-3(4*H*)-yl)methyl) phosphorodithioate; ENT 23233; Gothnion; Gusathion; Gusathion 20; Gusathion 25; Gusathion K; Gusathion M; Gusathion methyl; Guthion; 3-(Mercaptomethyl)-1,2,3-benzotriazin-4(3*H*)-one-*O,O*-dimethyl phosphorodithioate-*S*-ester; Methylazinphos; *N*-Methylbenzazimide, dimethyldithiophosphoric acid ester; Methyl guthion; Metiltriazotion; NA 2783; NCI-C00066; **Phosphorodithioic acid *O,O*-dimethyl ((4-oxo-1,2,3-benzotriazin-3(4*H*)-yl)methyl) ester;** Phosphorodithioic acid *O,O*-dimethyl ester, *S*-ester with 3-mercaptomethyl-1,2,3-benzotriazin-4(3*H*)-one; R 1582.

Structure:

Designations: CAS: 86-50-0; DOT: 2783; mf: $C_{10}H_{12}N_3O_3PS_2$; fw: 317.33; RTECS: TE1925000.

Properties: Colorless crystals or brown waxy solid having a very low odor theshold in water (0.2 μg/kg). Mp: 72.4 °C; bp: decomposes >200 °C; ρ: 1.44 at 20/4 °C; fl p: nonflammable; H-t$_{1/2}$ (pH 8.6): 36.4, 27.9, and 7.2 d at 6, 25, and 40 °C, respectively; K_H: 7.50 x 10^{-9} atm·m^3/mol at 20 °C; log K_{oc}: 2.47-3.53; log K_{ow}: 2.69, 2.75; S_o: soluble in benzene, carbon tetrachloride, chloroform, ethylbenzene, methylene chloride, 2-propanol, toluene, xylene, and many other organic solvents; S_w: 9.5, 20.9, 29, and 43.6 mg/L at 10, 20, 25, and 30 °C, respectively; vp: 1.6 x 10^{-6} mmHg at 20 °C.

Soil properties and adsorption data

Soil	K_d (mL/g)	f_{oc} (%)	K_{oc} (mL/g)	pH
Clayey loam	9.9	0.29	3,414	6.0
Cohansey sand	96.7	2.55	3,789	--
Loamy sand	7.6	1.62	469	6.6
Sandy loam	3.3	1.10	300	6.4
Silty loam	16.8	2.90	579	7.9
Silty loam	11.0	1.04	1,058	5.5
Silty loam	28.5	2.67	1,067	5.4

Source: Reduker et al., 1988; U.S. Department of Agriculture, 1990.

Transformation Products
Soil: The principal degradation products in soil and by selected soil microorganisms are benzazimide, thiomethylbenzazimide, bis(benzazimidyl-methyl)disulfide, and anthranilic acid. Benzazimide is further transformed only by *Pseudomonas* sp. DSM 5030 at a sufficient rate to 5-hydroxybenzazimide (Engelhardt and Wallnöfer, 1983). When radiolabeled azinphos-methyl was incubated in soil, 50 and 93% of the applied amount degraded to carbon dioxide after 44 and 197 d, respectively (Engelhardt et al., 1984). The presence of benzamide, salicylic acid and $^{14}CO_2$ from [carbonyl-^{14}C]- and [ring-U-^{14}C]azin-phosmethyl indicated that the 1,2,3-benzotriazinone ring is cleaved. The key intermediates in the degradation of azinphos-methyl in soil were reported to be 3-(mercaptomethyl)-1,2,3-benzotriazin-4(3*H*)-one, 3-((methylthio)methyl)-1,2,3-benzotriazin-4(3*H*)-one, 3-((methylsulfonyl)methyl)-1,2,3-benzotriazin-4(3*H*)-one, and 3-((methylsulfinyl)methyl)-1,2,3-benzotriazin-4(3*H*)-one (Engelhardt et al., 1984). Other metabolites identified included *O,O*-dimethyl *S*-((4-oxo-1,2,3-benzotriazin-4(3*H*)-yl)methyl) phosphorodithioate, bis(3-(thiomethyl)-1,2,3-benzotriazin-4(3*H*)-one), 3,3'-thiobis(methylene)-1,2,3-benzotriazin-4(3*H*)-one, 3-methyl-1,2,3-benzotriazin-4(3*H*)-one, 3,3'-oxybis(methylene)-bis(1,2,3-benzo-triazin-4(3*H*)-one), 3-(hydroxymethyl)-1,2,3-benzotriazin-4(3*H*)-one, 1,2,3-benzotriazin-4(3*H*)-one, *S*-methyl *S*-((oxo-3*H*-1,2,3-benzotriazin-3-yl)methyl) dithiophosphate, *O*-methyl *S*-((oxo-3*H*-1,2,3-benzotriazin-3-yl)methyl) dithio-phosphate, (4-oxo-3*H*-1,2,3-benzotriazin-3-yl)methanesulfonic acid, and anthra-nilic acid (Engelhardt et al., 1984). In a dry soil (2-3% moisture content), the $t_{1/2}$ were 484, 88, and 32 d at 6, 25, and 40 °C, respectively. In a soil containing 50% moisture, the half-lives were much shorter: 64, 13, and 5 d at 6, 25, and 40 °C, respectively (Yaron et al., 1974).

Chemical/Physical: At temperatures >200 °C, azinphos-methyl decomposes (Windholz et al., 1983) emitting toxic fumes of phosphorus, nitrogen, and sulfur oxides (Sax and Lewis, 1987; Lewis, 1990). Decomposed by alkaline solutions to give anthranilic acid and other products (Sittig, 1985).

Azinphosphos-methyl on glass plates was decomposed by UV light (λ = 2537 Å) to anthranilic acid, benzazimide, methyl benzazimide sulfide, N-methyl benzazimide, and an unidentified water-soluble compound (Liang and Lichtenstein, 1972).

Exposure Limits: OSHA PEL: TWA 0.2 mg/m^3; ACGIH TLV: TWA 0.2 mg/m^3.

Symptoms of Exposure: Miosis, eye ache, chest wheezing, rhinorrhea of the forehead, cyanosis, salivation, laryngeal spasm, nausea, vomiting, anorexia, diarrhea, sweating, paralysis, convulsions, low blood pressure.

Formulation Types: Emulsifiable concentrate; wettable powder; dustable powder.

Toxicity: LC$_{50}$ (96 hr) for rainbow trout 20 μg/L, bluegill sunfish 4.6 μg/L, guppies 100 μg/L (Hartley and Kidd, 1987), goldfish >1,000 μg/L (Worthing and Hance, 1991), fathead minnow 93 μg/L, minnow 240 μg/L, goldfish 4.3 mg/L, bluegill sunfish 5.2 μg/L (Katz, 1961), largemouth bass 52 μg/L, rainbow trout 14 μg/L, brown trout 4 μg/L, coho salmon 17 μg/L, perch 13 μg/L, channel catfish 3.29 mg/L, and black bullhead 3.5 mg/L (Macek and McAllister, 1970); acute oral LD$_{50}$ for rats 4-20 mg/kg (Hartley and Kidd, 1987), 7 mg/kg (RTECS, 1985).

Uses: Nonsystemic insecticide and acaricide for control of insects pests in blueberry, grape, maize, vegetable, cotton, and citrus crops.

BENDIOCARB

Synonyms: Bencarbate; **2,2-Dimethyl-1,3-benzodioxol-4-ol methylcarbamate;** 2,2-Dimethyl-1,3-benzodioxol-4-ol *N*-methylcarbamate; 2,2-Dimethylbenzo-1,3-dioxol-4-ol methylcarbamate; Dycarb; Ficam; Ficam D; Ficam ULV; Ficam W; Garvox; 2,3-Isopropylidenedioxyphenyl methylcarbamate; Multamat; Multimet; NC 6897; Niomil; Rotate; Tattoo; Turcam.

Structure:

Designations: CAS: 22781-23-3; mf: $C_{11}H_{13}NO_4$; fw: 223.23; RTECS: FC1140000.

Properties: Colorless to white crystals. Mp: 129-130 °C; ρ: 1.25 at 20/4 °C; pK_a: 8.8; H-t$_{1/2}$: 4 d at 25 °C and pH 7; K_H: 3.6 x 10^{-6} atm·m^3/mol at 20 °C (approximate - calculated from water solubility and vapor pressure); log K_{oc}: 2.76; log K_{ow}: 1.70; S_o (g/kg at 25 °C): acetone (200), benzene (40), chloroform (200), 1,4-dioxane (200), *n*-hexane (0.35), methylene chloride (200); S_w: 40 mg/L at 20 °C; vp: 4.95 x 10^{-6} mmHg at 20 °C.

Transformation Products
Soil: Though no products were identified, the reported half-life in soil is several days to a few weeks (Hartley and Kidd, 1987).

Chemical/Physical: Hydrolyzes to 2,3-isopropylidenedioxyphenol, methylamine, and carbon dioxide (Worthing and Hance, 1991).

Formulation Types: Wettable powder; dustable powder; granules; suspension concentrate; seed treatment; aerosol.

Toxicity: LC$_{50}$ (96 hr) for rainbow trout 1.55 mg/L (Hartley and Kidd, 1987); acute oral LD$_{50}$ for rats 40-156 mg/kg (Hartley and Kidd, 1987).

Uses: Contact insecticide used to control beetles, wireworms, flies, wasps, and mosquitoes in beets and maize.

BENOMYL

Synonyms: Arilate; BBC; Benlat; Benlate; Benlate 50; Benlate 50W; Benomyl 50W; BNM; **(1-((Butylamino)carbonyl)-1H-benzimidazol-2-yl)carbamic acid methyl ester;** 1-(Butylcarbamoyl)-2-benzimidazolecarbamic acid methyl ester; D 1991; Du Pont 1991; F 1991; Fungicide 1991; Fundasol; Fundazol; MBC; Methyl (1-((butyl-amino)carbonyl)-1H-benzimidazol-2-yl)carbamate; Methyl 1-(butylcarbamoyl)-2-benzimidazolylcarbamate; Tersan 1991.

Structure:

$$\text{CONHCH}_2\text{CH}_2\text{CH}_2\text{CH}_3$$

$$\text{NHCOOCH}_3$$

Designations: CAS: 17804-35-2; mf: $C_{14}H_{18}N_4O_3$; fw: 290.62; RTECS: DD6475000.

Properties: Colorless to white crystalline solid with a faint, acrid odor. Mp: decomposes before melting; bp: decomposes; log K_{oc}: 3.28 (calculated); log K_{ow}: 1.40-3.11; S_o (g/kg at 25 °C): acetone (18), chloroform (94), N,N-dimethyl-formamide (53), ethanol (4), n-heptane (10), xylene (10), and many other organic solvents except oils; S_w: 2.8 mg/L at 25 °C (stable only at pH 7); vp: negligible at room temperature.

Transformation Products

Biological: Mixed cultures can grow on benomyl as the sole carbon source. It was proposed that benomyl degraded to butylamine and methyl 2-benzimidazole-carbamate (MBC), the latter undergoing further degradation to 2-aminobenz-imidazole then to carbon dioxide and other products (Fuchs and de Vries, 1978).

Chemical/Physical: In aqueous solutions, especially in the presence of acids, benomyl hydrolyzes to to the strongly fungicidal methyl-2-benzimidazole-carbamate (carbendazim) (Clemons and Sisler, 1969; Cremlyn, 1991; Worthing and Hance, 1991; Zbozinek, 1984) and butyl isocyanate (Worthing and Hance, 1991; Zbozinek, 1984). The latter is unstable in water and decomposes to butylamine and carbon dioxide (Zbozinek, 1984). In highly acidic and alkaline aqueous solutions (pH <1 and pH >11), benomyl is completely converted to 3-butyl-2,4-dioxo[1,2-a]-s-triazinobenzimidazole (STB) with smaller quantities of methyl 2-benzimi-dazolecarbamate (MBC). In addition 1-(2-benzimidazolyl)-3-n-butylurea was identified but only under highly alkaline conditions (Singh and Chiba, 1985). Emits toxic fumes of nitrogen oxides when heated to decomposition (Lewis, 1990; Sax and Lewis, 1987).

Soil: In soil and water, benomyl is transformed to methyl-2-benzimidazole and 2-aminobenzimidazole (Rajagopal et al., 1984; Ramakrishna et al., 1979; Rhodes and Long, 1974). Benomyl is easily hydrolyzed in soil to methyl-2-benzimidazole carbamate (Li and Nelson, 1985). On bare soil and turf, benomyl degraded to methyl 2-benzimidazolecarbamate and 2-aminobenzimidazole. Most of the residues were immobile in soils containing organic matter and were limited to the top four inches of soil (Baude et al., 1974). In a loamy garden soil, benomyl was degraded by two fungi and four strains of bacteria to nonfungistatic compounds (Helweg, 1972). The reported half-life in soil is 6-12 mo (Hartley and Kidd, 1987).

Plant: On apple foliage treated with a Benlate formulation, benomyl was transformed to MBC. Benomyl dissipated quickly and the reported half-life on foliage was 3-7 d (Chiba and Veres, 1981).

Exposure Limits: OSHA PEL: TWA 10 mg/m^3 (total dust), 5 mg/m^3 (respirable fraction); ACGIH TLV: TWA 10 mg/m^3.

Symptoms of Exposure: Eye and skin irritant.

Formulation Types: Wettable powder.

Toxicity: LC_{50} (96 hr) for rainbow trout 0.17 mg/L and goldfish 4.2 mg/L (Hartley and Kidd, 1987); acute oral LD_{50} for rats >10,000 mg/kg (Hartley and Kidd, 1987), 10 mg/kg (RTECS, 1985).

Uses: Post harvest systemic fungicide used to control fungi and mildew on cotton, roses, soft fruits, tomatoes, cucumbers, and other vegetables.

BENTAZONE

Synonyms: BAS 351-H; Basagran; Bendioxide; Bentazon; 3-Isopropyl-1*H*-2,1,3-benzothiadiazin-4(3*H*)-one-2,2-dioxide; **3-(1-Methylethyl)-1*H*-2,1,3-benzothiadiazin-4(3*H*)-one-2,2-dioxide.**

Structure:

Designations: CAS: 25057-89-0; mf: $C_{10}H_{12}N_2O_3S$; fw: 240.28; RTECS: DK9900000.

Properties: Odorless, colorless crystals to white crystalline powder. Mp: 137-139 °C; bp: decomposes at 200 °C; ρ: 1.47; fl p: nonflammable (>100 °C); pK_a: 3.4; K_H: <4.7 x 10^{-11} atm·m³/mol at 20 °C (approximate - calculated from water solubility and vapor pressure); log K_{oc}: no data found; however, it does not leach below the plow layer; log K_{ow}: -0.46; S_o (g/kg at 20 °C): acetone (1,507), benzene (33), chloroform (180), cyclohexane (0.2), ethanol (861), ethyl acetate (650), ethyl ether (616), olive oil (27); S_w: 500 mg/L at 20 °C; vp: <7.5 x 10^{-8} mmHg at 20 °C.

Transformation Products
Plant: Undergoes hydroxylation of the aromatic ring and subsequent conjugation in plants (Hartley and Kidd, 1987; Otto et al., 1978) forming 6- and 8-hydroxybentazone compounds (Otto et al., 1978).

Soil: Under aerobic conditions, bentazone was reported to degrade to 6- and 8-hydroxybentazone compounds. In addition, anthranilic acid and isopropylamide were reported as soil hydrolysis products (Otto et al., 1978). Persistence in soil is less than 6 wk (Hartley and Kidd, 1987).

Photolytic: Humburg et al. (1989) reported that 30% degradation of bentazone occurred when exposed to UV light (λ = 200-400 nm); however, no photodegradation product(s) were reported. The natural sunlight and simulated sunlight irradiation of bentazone was studied in aqueous solution (500 ppm), on soil, and as thin films on Pyrex plates (Nilles and Zabik, 1975). The natural sunlight irradiation of a saturated aqueous solution of bentazone for 115 hr yielded six major photoproducts including *o*-amino-*N*-isopropylbenzamide, *o*-nitro-*N*-isopropylbenzamide, and *o*-nitroso-*N*-isopropylbenzamide. These three compounds were also observed following the simulated irradiation of bentazone as thin films of glass. The simulated sunlight irradiation of bentazone on soil produced six major photoproducts including *o*-nitro-*N*-isopropylbenzamide and *o*-nitroso-*N*-isopropylbenzamide (Nilles and Zabik, 1975).

Symptoms of Exposure: Ingestion causes apathy, ataxia, anorexia, prostration, tremors, occasional vomiting and diarrhea.

Formulation Types: Soluble concentrate; emulsifiable concentrate (4 lb/gal).

Toxicity: LC_{50} (96 hr) for rainbow trout 190 mg/L and bluegill sunfish 616 mg/L (Hartley and Kidd, 1987); acute oral LD_{50} for rats is 1,100 mg/kg (RTECS, 1985).

Uses: Selective, contact, postemergence herbicide used to control a variety of annual and perennial broadleaf weeds in most grass and legume crops.

α-BHC

Synonyms: Benzenehexachloride-α-isomer; α-Benzenehexachloride; ENT 9232; α-HCH; α-Hexachloran; α-Hexachlorane; α-Hexachlorcyclohexane; α-Hexachlorocyclohexane; 1,2,3,4,5,6-Hexachloro-α-cyclohexane; 1α,2α,3β,4α,5β,6β-Hexachlorocyclohexane; α-1,2,3,4,5,6-Hexachlorocyclohexane; α-Lindane; TBH.

Structure:

Designations: CAS: 319-84-6; DOT: 2761; mf: $C_6H_6Cl_6$; fw: 290.83; RTECS: GV3500000.

Properties: Brownish-to-white crystalline solid with a phosgene-like odor (technical grade). Mp: 159.1 °C; bp: 288 °C; ρ: ≈ 1.87; fl p: nonflammable; K_H: 5.3 x 10^{-6} atm·m^3/mol at 20 °C (approximate - calculated from water solubility and vapor pressure); log K_{oc}: 3.279; log K_{ow}: 3.46-3.89; P-t$_{1/2}$: 91.09 hr (absorbance λ ≤300 nm, concentration on glass plates = 6.7 μg/cm^2); S_o: soluble in ethanol, benzene, chloroform, cod liver oil (22.1, 23.6, 31.2 g/L at 4, 12, and 20 °C, respectively), n-octanol (6.4, 9.4, and 14.9 g/L at 4, 12, and 20 °C, respectively); S_w: 1.63 mg/L at 20 °C; vp: 2.5 x 10^{-5} mmHg at 20 °C.

Transformation Products

Soil: Under aerobic conditions, indigenous microbes in contaminated soil produced pentachlorocyclohexane. However, under methanogenic conditions, α-BHC was converted to chlorobenzene, 3,5-dichlorophenol, and the tentatively identified compound 2,4,5-trichlorophenol (Bachmann et al., 1988). *Clostridium sphenoides* biodegraded α-BHC to δ-3,4,5,6-tetrachloro-1-cyclohexane (Heritage and MacRae, 1977a).

Photolytic: When an aqueous solution containing α-BHC was photooxidized by UV light at 90-95 °C, 25, 50, and 75% degraded to carbon dioxide after 4.2, 24.2, and 40.0 hr, respectively (Knoevenagel and Himmelreich, 1976).

Chemical/Physical: Emits very toxic chloride fumes when heated to decomposition (Lewis, 1990).

Formulation Types: Emulsifiable concentrate; wettable powder; fumigant; suspension concentrate; granules; dustable powder.

Toxicity: LC_{50} (96 hr) for guppies >1.4 mg/L (Verschueren, 1983); acute oral LD_{50} for rats 177 mg/kg (RTECS, 1985).

Use: Not produced commercially in the U.S. and its sale is prohibited by the U.S. EPA.

β-BHC

Synonyms: *trans*-α-Benzenehexachloride; β-Benzenehexachloride; Benzene-*cis*-hexachloride; ENT 9233; β-HCH; β-Hexachlorobenzene; 1α,2β,3α,4β,5α,6β-**Hexachlorocyclohexane**; β-Hexachlorocyclohexane; 1,2,3,4,5,6-Hexachloro-β-cyclohexane; 1,2,3,4,5,6-Hexachloro-*trans*-cyclohexane; β-1,2,3,4,5,6-Hexachlorocyclohexane; β-Isomer; β-Lindane; TBH.

Structure:

Designations: CAS: 319-85-7; DOT: 2761; mf: $C_6H_6Cl_6$; fw: 290.83; RTECS: GV4375000.

Properties: Solid. Mp: 311.7 °C; bp: sublimes; ρ: 1.89 at 19/4 °C; fl p: nonflammable; K_H: 2.3 x 10^{-7} atm·m^3/mol at 20-25 °C (approximate - calculated from water solubility and vapor pressure); log K_{oc}: 3.322-3.553; log K_{ow}: 3.80-4.50; P-t$_{1/2}$: 151.80 hr (absorbance λ ≤300 nm, concentration on glass plates = 6.7 μg/cm^2); S_o: soluble in ethanol, benzene, and chloroform; S_w: 240 ppb at 25 °C; vp: 2.8 x 10^{-7} mmHg at 20 °C.

Transformation Products
Soil: No biodegradation of β-BHC was observed under denitrifying and sulfate-reducing conditions in a contaminated soil collected from the Netherlands (Bachmann et al., 1988).

Chemical/Physical: Emits very toxic fumes of chlorides, hydrochloric acid, and phosgene when heated to decomposition (Lewis, 1990).

Formulation Types: Emulsifiable concentrate; wettable powder; fumigant; suspension concentrate; granules; dustable powder.

Toxicity: Acute oral LD$_{50}$ for rats 6,000 mg/kg (RTECS, 1985).

Use: Insecticide.

δ-BHC

Synonyms: δ-Benzenehexachloride; ENT 9234; δ-HCH; δ-Hexachlorocyclo-hexane; δ-1,2,3,4,5,6-Hexachlorocyclohexane; δ-(aeeeee)-1,2,3,4,5,6-Hexachlorocyclohexane; 1α,2α,3α,4β,5β,6β-Hexachlorocyclohexane; 1,2,3,4,5,6-Hexachloro-δ-cyclohexane; δ-Lindane; TBH.

Structure:

Designations: CAS: 319-86-8; DOT: 2761; mf: $C_6H_6Cl_6$; fw: 290.83; RTECS: GV4550000.

Properties: Solid with a musty-like odor. Mp: 140.8 °C; bp: 60 °C at 0.36 mmHg; ρ: ≈ 1.87; fl p: nonflammable; $H-t_{1/2}$: 191 d (pH 7), 11 hr (pH 9); K_H: 2.5 x 10^{-7} atm·m^3/mol at 20-25 °C (approximate - calculated from water solubility and vapor pressure); log K_{oc}: 3.279; log K_{ow}: 2.80, 4.14 2.80; $P-t_{1/2}$: 153.84 hr (absorbance λ ≤300 nm, concentration on glass plates = 6.7 μg/cm^2); S_o: soluble in acetone, ethanol, benzene, and chloroform; S_w: 21.3 ppm at 25 °C; vp: 1.7 x 10^{-5} mmHg at 20 °C.

Transformation Products

Biological: Dehydrochlorination of δ-BHC by a *Pseudomonas* sp. under aerobic conditions was reported by Sahu et al. (1992). They also reported that when deionized water containing δ-BHC was inoculated with *Pseudomonas* sp., the concentration of δ-BHC decreased to undetectable levels after eight d with concomitant formation of chloride ions and δ-pentachlorocyclohexane.

Chemical/Physical: δ-BHC dehydrochlorinates in the presence of alkalies (Worthing and Hance, 1991). Emits very toxic chloride fumes when heated to decomposition (Lewis, 1990).

Formulation Types: Emulsifiable concentrate; wettable powder; fumigant; suspension concentrate; granules; dustable powder.

Toxicity: Acute oral LD_{50} for rats 1,000 mg/kg (RTECS, 1985).

Use: Insecticide.

BIFENOX

Synonyms: MC-4379; Methyl 5-(2,4-dichlorophenoxy)-2-nitrobenzoate; Modown.

Structure:

Designations: CAS: 42576-02-3; mf: $C_{14}H_9Cl_2NO_5$; fw: 342.14; RTECS: DG7890000.

Properties: Yellow-tan crystalline solid. Mp: 84-86 °C; bp: not applicable; ρ: 1.155; fl p: nonflammable; K_H: 1.09 x 10^{-7} atm·m^3/mol at 25 °C; log K_{oc}: 2.24-4.39; log K_{ow}: 4.48; S_o (g/kg at 25 °C): acetone (400), ethanol (<50), chlorobenzene (400), xylene (300), kerosene and other aliphatic solvents (<10); S_w: 350 μg/L at 25 °C; vp: 2.40 x 10^{-6} mmHg at 30 °C.

Soil properties and adsorption data

Soil	K_d (mL/g)	f_{oc} (%)	K_{oc} (mL/g)	pH
Loam	36.0	2.26	1,593	6.7
Loamy sand	79.0	0.81	9,729	6.9
Sand	2.6	0.17	1,529	7.5
Sandy clay loam	156.0	0.64	24,451	7.3
Silty loam	408.0	1.16	35,172	7.4

Source: U.S. Department of Agriculture, 1990.

Transformation Products

Soil: Bifenox degrades in soil forming 5-(2',4'-dichlorophenoxy)-2-nitrobenzoic acid and methyl 5-(2,4-dichlorophenoxy)anthranilate (Hartley and Kidd, 1987). The average half-life in soils is 7-14 d (Hartley and Kidd, 1987; Humburg et al., 1989).

Plant: Rapidly undergoes ring hydroxylation and subsequent conjugation in rice plants (Ashton and Monaco, 1991).

Photolytic: The UV photolysis (λ = 300 nm) of bifenox in various solvents were studied by Ruzo et al. (1980). In water, 2,4-dichloro-3'-(carboxymethyl)-4'-hydroxydiphenyl ether and 2,4-dichloro-3'-(carboxymethyl)-4'-aminodiphenyl ether were identified. In cyclohexane, 2,4-dichloro-4'-nitrodiphenyl ether and methyl formate were the major products. In methanol, a dichloromethoxy phenol was identified. Photodegradation occurred via reductive dechlorination, decarboxymethylation, nitro group reduction, and cleavage of the ether linkage (Ruzo et al., 1980).

Formulation Types: Emulsifiable concentrate; wettable powder; granules; suspension concentrate; flowable powder (4 lb/gal).

Toxicity: LC_{50} (96 hr) for rainbow trout 0.87 mg/L and bluegill sunfish 0.64 mg/L (Hartley and Kidd, 1987); acute oral LD_{50} for rats is 6,400 mg/kg (RTECS, 1985).

Uses: Selective preemergence or postemergence herbicide used to effectively control a wide variety of broadleaf weeds such as bindweed, jimsonweed, kochia, mustards, pigweeds, sesbania, smartweed, velvetleaf, in tolerant crops (corn, grain sorghum, maize, rice, and soybeans).

BIPHENYL

Synonyms: Bibenzene; 1,1'-Biphenyl; Diphenyl; Lemonene; Phenylbenzene.

Structure:

Designations: CAS: 92-52-4; mf: $C_{12}H_{10}$; fw: 154.21; RTECS: NU8050000.

Properties: Colorless to white scales or crystals with a pleasant but peculiar odor. Mp: 71 °C; bp: 255.9 °C; ρ: 0.8660 at 20/4 °C; fl p: 112.8 °C; lel: 0.6% at 100 °C; uel: 5.8% at 155 °C; K_H: 1.93-4.15 x 10^{-4} atm·m^3/mol at 25 °C; IP: 8.27 eV; log K_{oc}: 3.04-3.32; log K_{ow}: 3.16-4.09; S_o: soluble in alcohol, benzene, ether, and triolein (199.3 and 415 g/kg at 23 and 37 °C, respectively); S_w: 5.94-7.48 ppm at 25 °C; vp: 5.84 x 10^{-4} mmHg at 20.70 °C.

Soil properties and adsorption data

Soil	K_d (mL/g)	f_{oc} (%)	K_{oc} (mL/g)	pH	CEC (meq/100 g)
Clay loam	15.6	1.42	1,100	5.91	12.4
Clay loam	124.0	10.40	1,200	4.89	35.0
Light clay	31.8	1.51	2,100	5.18	13.2
Light clay	56.8	3.23	1,800	5.26	28.3
Sandy loam	90.0	7.91	1,100	5.51	26.3

Source: Kishi et al., 1990.

Transformation Products
Biological: Reported biodegradation products include 2,3-dihydro-2,3-dihydroxybiphenyl, 2,3-dihydroxybiphenyl, 2-hydroxy-6-oxo-6-phenylhexa-2,4-dienoate, 2-hydroxy-3-phenyl-6-oxohexa-2,4-dienoate, 2-oxopenta-4-enoate, phenylpyruvic acid (Verschueren, 1983), 2-hydroxybiphenyl, 4-hydroxybiphenyl, and 4,4'-hydroxybiphenyl (Smith and Rosazza, 1974). Under aerobic conditions, *Beijerinckia* sp. degraded biphenyl to *cis*-2,3-dihydro-2,3-dihydroxybiphenyl. In addition, *Oscillatoria* sp. and *Pseudomonas putida* degraded biphenyl to 4-hydroxybiphenyl and benzoic acid, respectively (Kobayashi and Rittman, 1982). In activated sludge, 15.2% mineralized to carbon dioxide after 5 d (Freitag et al., 1985).

53

Photolytic: A carbon dioxide yield of 9.5% was achieved when biphenyl adsorbed on silica gel was irradiated with light (λ >290 nm) for 17 hr. Irradiation of biphenyl (λ >300 nm) in the presence of nitrogen monoxide resulted in the formation of 2- and 4-nitrobiphenyl (Fukui et al., 1980).

Chemical/Physical: The aqueous chlorination of biphenyl at 40 °C over a pH range of 6.2 to 9.0 yielded *o*-chlorobiphenyl and *m*-chlorobiphenyl (Snider and Alley, 1979). In an acidic aqueous solution (pH = 4.5) containing bromide ions and a chlorinating agent (sodium hypochlorite), 4-bromobiphenyl formed as the major product. Minor products identified include 2-bromobiphenyl, 2,4-, and 4,4'-dibromobiphenyl (Lin et al., 1984).

Exposure Limits: NIOSH REL: IDLH 300 mg/m^3; OSHA PEL: TWA 0.2 ppm; ACGIH TLV: TWA 0.2 ppm.

Symptoms of Exposure: Irritation of throat, eyes; headache; nausea; fatigue; numbness in limbs, liver damage (Humburg et al., 1989).

Formulation Types: Solution.

Toxicity: Acute oral LD$_{50}$ for rats 3,280 mg/kg (RTECS, 1985).

Uses: Fungistat for oranges; plant disease control.

BIS(2-CHLOROETHYL)ETHER

Synonyms: Bis(β-chloroethyl)ether; Chlorex; 1-Chloro-2-(β-chloroethoxy)-ethane; Chloroethyl ether; 2-Chloroethyl ether; (β-Chloroethyl)ether; DCEE; Dichlorodiethyl ether; 2,2'-Dichlorodiethyl ether; β,β'-Dichlorodiethyl ether; Dichloroether; Dichloroethyl ether; α,α'-Dichloroethyl ether; Di(β-chloroethyl)-ether; Di(2-chloroethyl)ether; *sym*-Dichloroethyl ether; 2,2'-Dichloroethyl ether; Dichloroethyl oxide; ENT 4504; **1,1'-Oxybis(2-chloroethane)**; RCRA waste number U025; UN 1916.

Structure:

$$ClCH_2CH_2-O-CH_2CH_2Cl$$

Designations: CAS: 111-44-4; DOT: 1916; mf: $C_4H_8Cl_2O$; fw: 143.01; RTECS: KN0875000.

Properties: Colorless, clear liquid with a strong fruity odor. Mp: -52.2 to -24.5 °C; bp: 178.5 °C; ρ: 1.2199 at 20/4 °C; fl p: 55 °C; H-t$_{1/2}$: 2.5 yr at 25 °C and pH 7; K_H: 1.3 x 10^{-5} atm·m^3/mol; IP: 9.85 eV; log K_{oc}: 1.15; log K_{ow}: 1.12, 1.58; S_o: soluble in acetone, ethanol, benzene, and ether; S_w: 10.7 g/L at 20 °C; vap d: 5.84 g/L at 25 °C, 4.94 (air = 1); vp: 0.71 mmHg at 20 °C.

Transformation Products
Chemical/Physical: Emits chlorinated acids when incinerated (Sittig, 1985). Emits very toxic chloride fumes when heated to decomposition (Lewis, 1990).

Symptoms of Exposure: Inhalation may cause irritation.

Formulation Types: Fumigant.

Toxicity: Acute oral LD$_{50}$ for rats 75-105 mg/kg (Verschueren, 1983).

Uses: Soil fumigant; acaricide.

BIS(2-CHLOROISOPROPYL)ETHER

Synonyms: BCIE; BCMEE; Bis(β-chloroisopropyl)ether; Bis(2-chloro-1-methyl-ethyl)ether; 1-Chloro-2-(β-chloroisopropoxy)propane; 2-Chloroisopropyl ether; β-Chloroisopropyl ether; (2-Chloro-1-methylethyl)ether; Dichlorodiisopropyl ether; Dichloroisopropyl ether; 2,2'-Dichloroisopropyl ether; NCI-C50044; **2,2'-Oxy-bis(1-chloropropane)**; RCRA waste number U027; UN 2490.

Structure:

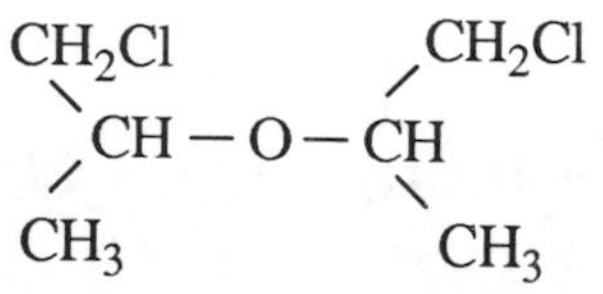

$$
\begin{array}{ccc}
CH_2Cl & & CH_2Cl \\
\diagdown & & \diagup \\
& CH-O-CH & \\
\diagup & & \diagdown \\
CH_3 & & CH_3
\end{array}
$$

Designations: CAS: 108-60-1; mf: $C_6H_{12}Cl_2O$; fw: 171.07; RTECS: KN1750000.

Properties: Colorless to brown oily liquid. Mp: -20 °C; bp: 187 °C; ρ: 1.103 at 20/4 °C; fl p: 85 °C; K_H: 1.1 x 10^{-4} atm·m^3/mol; log K_{oc}: 1.79; log K_{ow}: 2.58; S_o: soluble in acetone, ethanol, benzene, and ether; S_w: 1,700 mg/L at 20 °C; vap d: 6.99 g/L at 25 °C, 5.91 (air = 1); vp: 0.85 mmHg at 20 °C.

Formulation Types: Emulsifiable concentrate; granules.

Toxicity: LC_{50} (48 hr) for carp >40 mg/L (Hartley and Kidd, 1987); acute oral LD_{50} for rats 503.6 mg/kg (Hartley and Kidd, 1987), 240 mg/kg (RTECS, 1985).

Uses: Apparently used as a nematocide in Japan but is not registered in the U.S. for use as a pesticide.

BROMACIL

Synonyms: Borea; Bromax; Bromax 4G; Bromax 4L; Bromazil; 5-Bromo-3-*sec*-butyl-6-methyluracil; **5-Bromo-6-methyl-3-(1-methylpropyl)-2,4(1*H3H*)-pyrimidinedione**; 5-Bromo-6-methyl-3-(1-methylpropyl)uracil; Cynogan; Du Pont herbicide 976; Eerex granular weed killer; Eerex water soluble concentrate weed killer; Herbicide 976; Hyvar; Hyvarex; Hyvar X; Hyvar X Bromacil; Hyvar XL; Hyvar X weed killer; Hyvar X-WS; Krovar I; Krovar II; Nalkil; Uragan; Uragon; Urox B; Urox B water soluble concentrate weed killer; Uron HX; Urox HX granular weed killer; Weed Blast.

Structure:

Designations: CAS: 314-40-9; mf: $C_9H_{13}BrN_2O_2$; fw: 261.12; RTECS: YQ9100000.

Properties: Colorless to white, odorless, crystalline solid. Mp: 157.5-160.0 °C; ρ: 1.55 at 25/4 °C; fl p: active ingredient and dry formulations are nonflammable; pK_a: <7; K_H: 1.05 x 10^{-10} atm·m^3/mol at 25 °C (approximate - calculated from water solubility and vapor pressure); log $K_{oc} \approx$ 1.51; log K_{ow}: 1.84-2.04; S_o (g/L at 25 °C): acetone (167), acetonitrile (71), ethanol (134), xylene (32); S_w: 815 mg/L at 25 °C; vp: 2.48 x 10^{-7} mmHg at 25 °C.

Soil properties and adsorption data

Soil	K_d (mL/g)	f_{oc} (%)	K_{oc} (mL/g)	pH
Adkins loamy sand	0.30	0.40	76	7.3
Baldwin Lake	2.89	1.00	289	8.0-9.0
Basinger fine sand	0.57	0.61	93	5.8
Bet Dagan I	0.10	0.40	25	7.9
Bet Dagan II	0.32	1.01	32	7.8
Big Bear Lake	3.93	7.90	50	6.7-7.2
Boca fine sand	0.76	1.67	46	7.1
Castle Lake	6.35	14.50	44	4.8-5.8
Chobee fine sandy loam	0.76	1.39	55	7.2
Clear Lake	5.26	13.30	40	5.7-6.8

continued

Soil	K_d (mL/g)	f_{oc} (%)	K_{oc} (mL/g)	pH
Delta (Hill Slough)	2.72	4.40	62	6.6-7.2
Gilat	0.16	0.55	29	7.8
Holopaw fine sand	0.33	0.50	66	6.1
Hula-1	19.75	29.88	65	6.3
Hula-2	5.05	7.85	64	6.9
Jenks Lake	1.20	3.00	40	5.3-6.2
Keyport silt loam	1.50	1.20	125	5.4
Mivtachim	0.03	0.06	50	8.5
Mockingbird Canyon	0.56	2.10	27	6.5-7.4
Neve Ya'ar	0.39	1.18	33	7.7
Neve Ya'ar	1.12	1.22	89	7.3
Oxidized Hula-2	1.71	3.08	55	6.9
Oxidized Neve Ya'ar	0.60	0.47	126	7.3
Pineda fine sand	0.57	1.22	93	7.1
Riviera fine sand	0.47	0.94	50	6.2
Sa'ad	0.63	0.56	112	7.6
San Joaquin Marsh	0.68	2.60	26	7.3-8.1
Semiahmoo mucky peat	21.33	27.80	77	5.4
Shefer	0.24	0.72	33	7.2
Wabasso sand	0.57	0.61	93	6.6

Source: Angemar et al., 1984; Corwin and Farmer, 1984; Madhun et al., 1986; Reddy et al., 1992; Rhodes et al., 1970.

Transformation Products

Soil: Metabolites reported in soil included 5-bromo-3-(3-hydroxy-1-methyl-propyl)-6-methyluracil, 5-bromo-3-*sec*-butyl-6-hydroxymethyluracil, 5-bromo-3-(2-hydroxy-1-methylpropyl)-6-methyluracil, and carbon dioxide (Gardiner et al., 1969). In the laboratory, 25.3% of ^{14}C-bromacil degraded in soil to carbon dioxide after 9 wk but mineralization in the field was not observed (Gardiner et al., 1969). To a neutral sandy loam soil maintained at a soil water holding capacity of 60%, 2.88 ppm of $2-^{14}C$-bromacil was applied. After 600 d, 22.1% (0.64 ppm) of the applied amount was converted to $^{14}CO_2$ (Wolf and Martin, 1974). The evolution of $^{14}CO_2$ was significantly reduced when the soil water holding capacity was maintained at 100%, i.e., <0.5% $^{14}CO_2$ after 145 d (Wolf and Martin, 1974). Residual activity in soil is limited to approximately 7 mo (Hartley and Kidd, 1987). The reported half-life in soil is 60 d (Alva and Singh, 1991); however, Gardiner et al. (1969) reported a half-life of 5-6 mo in a silt loam.

Plant: Bromacil is slowly absorbed and translocated in plants. Gardiner et al. (1969)

reported 17% of the herbicide was translocated to the stems and leaves. A 6-hydroxymethyl metabolite was reported as the major metabolite (Gardiner et al., 1969). In orange seedlings, 5-bromo-3-*sec*-butyl-6-hydroxymethyluracil was reported as a major metabolite (Humburg et al., 1989).

Photolytic: When a dilute aqueous solution (1-10 mg/L) of bromacil was exposed to sunlight for 4 mo, the *N*-dealkylated photoproduct, 5-bromo-6-methyluracil, formed in small quantities. This compound is less stable than bromacil and upon further irradiation, the debrominated product, 6-methyluracil, was formed (Moilanen and Crosby, 1974). Acher and Dunkelblum (1979) studied the dye-sensitized photolysis of aerated aqueous solutions of bromacil using sunlight as the irradiation source. After 1 hr, a mixture of diastereoisomers of 3-*sec*-butyl-5-acetyl-5-hydroxyhydantoin formed in an 83% yield. In a subsequent study (1980), another minor intermediate was identified: a 5,5′- photoproduct of 3-*sec*-butyl-6-methyluracil. In this study, the rate of photooxidation increased with pH. The most effective sensitizers were riboflavin (10 ppm) and methylene blue (2-5 ppm) (Acher and Saltzman, 1980).

Chemical/Physical: When bromacil was heated at 900 °C, carbon monoxide, carbon dioxide, chlorine, hydrochloric acid, and ammonia were produced (Kennedy et al., 1972). At 250 °C, bromacil changed from a white crystalline solid to a black solid which was tentatively identified as 3-*sec*-butyl methyluracil (Stojanovic et al., 1972). Emits very toxic fumes of bromides and nitrogen oxides when heated to decomposition (Lewis, 1990).

Exposure Limits: OSHA PEL: TWA 1.0 ppm; ACGIH TLV: TWA 1 ppm.

Symptoms of Exposure: May irritate of eyes, nose, throat, and skin.

Formulation Types: Soluble concentrate (2 and 4 lb/gal); wettable powder (80%); granules (4%).

Toxicity: LC_{50} (48 hr) for rainbow trout 75 mg/L, bluegill sunfish 71 mg/L, and carp 164 mg/L (Hartley and Kidd, 1987); acute oral LD_{50} for rats is 5,200 mg/kg (Ashton and Monaco, 1991), 641 mg/kg (RTECS, 1985).

Uses: Herbicide applied to soil to control a wide variety of annual and perennial grasses, broadleaf weeds, and general vegetation on uncropped land. It is also used for selective weed control in apple, asparagus, cane fruit, hops, and citrus crops.

BROMOXYNIL

Synonyms: Brittox; Brominal; Brominex; Brominil; Broxynil; Buctril; Buctril 20; Buctril 21; Buctril industrial; Butilchlorofos; Chipco Buctril; Chipco crab-kleen; 2,6-Dibromo-4-cyanophenol; 3,5-Dibromo-4-hydroxybenzonitrile; **3,5-Dibromo-4-hydroxyphenylcyanide**; ENT 20852; 4-Hydroxy-3,5-dibromobenzonitrile; MB 10064; M&B 10064; ME4 Brominal; Nu-lawn weeder; Oxytril M; Partner.

Structure:

Designations: CAS: 1689-84-5; mf: $C_7H_3Br_2NO$; fw: 276.93; RTECS: DI3150000.

Properties: Odorless, colorless crystals or a light tan to creamy powder. Technical grades may impart a slight odor. Mp: 190 °C, 194-195 °C; bp: sublimes at 135 °C and 0.15 mmHg; fl p: >82 °C (ester formulation); pK_a: 4.06; K_H: 1.4 x 10^{-6} atm·m^3/mol at 20-25 °C (approximate - calculated from water solubility and vapor pressure); log K_{oc}: 2.48 (calculated); log K_{ow}: <2; S_o (g/L at 20 °C): acetone (170); benzene (10), N,N-dimethylformamide (610), ethanol (70), methanol (90), cyclohexanone (170), tetrahydrofuran (410); S_w: 130 mg/L at 25 °C; vp: 4.8 x 10^{-6} mmHg at 20 °C.

Transformation Products

Soil: In soils, *Klebsiella pneumoniae* metabolized bromoxynil to 3,5-dibromo-4-hydroxybenzoic acid and ammonia (McBride et al., 1986). The half-life in soil is approximately 10 d (Hartley and Kidd, 1987).

Plant: In plants, the cyano group is hydrolyzed to an amido group which is subsequently oxidized to a carboxylic acid. Hydrolyzes to hydroxybenzoic acid (Hartley and Kidd, 1987). In plants, bromoxynil may be hydrolyzed to a benzoic acid (Humburg et al., 1989). Bromoxynil-resistant cotton was recently developed by inserting a *bxn* gene cloned from the soil bacterium *Klebsiella ozaenae*. This gene, which encodes a specific nitrolase, converted bromoxynil to its primary metabolite 3,5-dibromo-4-hydroxybenzoic acid (Stalker et al., 1988).

Biological: Duke et al. (1991) reported that bromoxynil can be converted 3,5-dibromo-4-hydroxybenzoic acid by a microbial nitrolase.

Chemical/Physical: Emits toxic fumes of nitrogen oxides and bromine when heated to decomposition (Sax and Lewis, 1987). Reacts with bases forming water-soluble salts (Worthing and Hance, 1991).

Formulation Types: Suspension concentrate; emulsifiable concentrate.

Toxicity: LC_{50} (48 hr) for harlequin fish 5.0 mg/L (potassium salt) (Hartley and Kidd, 1987); LC_{50} for rainbow trout 100 ppb, goldfish 170 ppb, and catfish 23 ppb (Humburg et al., 1989); acute oral LD_{50} for rats is 190 mg/kg (RTECS, 1985).

Uses: Selective contact foliage-applied herbicide used to control many broadleaf weeds in cereals.

BROMOXYNIL OCTANOATE

Synonyms: Bronate; Buctril; 2,6-Dibromo-4-cyanophenyl octanoate; 3,5-Dibromo-4-octanoyloxybenzonitrile; M&B 10731; RP 16272.

Structure:

Designations: CAS: 1689-99-2; mf: $C_{15}H_{17}Br_2NO_2$; fw: 403.13; RTECS: DI3325000.

Properties: Cream-colored, waxy solid. Mp: 45-46 °C; bp: sublimes at 90 °C and 0.1 mmHg; K_H: 3.16 x 10^{-5} atm·m^3/mol at 25 °C; log K_{oc}: 4.25 (calculated); log K_{ow}: 5.06; S_o (g/L at 25 °C): acetone (100-170), benzene (700), chloroform (800), cyclohexanone (550), methanol (740), methylene chloride (800), toluene (760); S_w: 80 μg/L at 25 °C; vp: 4.80 x 10^{-6} mmHg at 20 °C.

Transformation Products
Chemical/Physical: Emits toxic fumes of nitrogen oxides and bromine when heated to decomposition (Sax and Lewis, 1987). Hydrolyzes, especially at pH >9, to bromoxynil and octanoic acid (Hartley and Kidd, 1987; Worthing and Hance, 1991).

Formulation Types: Emulsifiable concentrate; suspension concentrate.

Toxicity: LC_{50} (48 hr) for rainbow trout 150 μg/L (Hartley and Kidd, 1987), goldfish 460 μg/L, and catfish 63 μg/L (Worthing and Hance, 1991); acute oral LD_{50} for rats 365 mg/kg (Hartley and Kidd, 1987).

Uses: Selective contact foliage-applied herbicide used to control many broadleaf weeds such as bindweed, chickweed, and *Veronica* spp. in cereals.

BUTIFOS

Synonyms: B 1776; Butiphos; Butyl phosphorotrithioate; Chemagro 1,776; Chemagro B1776; Def; Def defoliant; De-green; E-Z-Off; Fos-fall 'A'; Ortho phosphate defoliant; **S,S,S-Tributyl phosphorotrithioate**; *S,S,S*-Tributyl trithiophosphate.

Structure:

$$(CH_3CH_2CH_2CH_2S)_3P$$

Designations: CAS: 78-48-8; mf: $C_{12}H_{27}OPS_3$; fw: 314.52; RTECS: TG5425000.

Properties: Pale yellow liquid with a mercaptan-like odor. Mp: <-25 °C; bp: 150 °C at 0.3 mmHg; ρ: 1.057 at 20/4 °C; K_H: 2.88 x 10^{-4} atm·m^3/mol at 20 °C (approximate - calculated from water solubility and vapor pressure); log K_{oc}: 3.44 (calculated); log K_{ow}: 3.23; S_o: soluble in most organic solvents; S_w: 2.3 mg/L at 20 °C; vap d: 12.85 g/L at 25 °C, 10.89 (air = 1); vp: 1.60 x 10^{-3} mmHg at 20 °C.

Formulation Types: Emulsifiable concentrate.

Toxicity: LC_{50} (96 hr) for rainbow trout <5.0 mg/L and bluegill sunfish 1.0 mg/L (Hartley and Kidd, 1987); acute oral LD_{50} for male and female rats is 233 and 200 mg/kg, respectively (Hartley and Kidd, 1987), 150 mg/kg (RTECS, 1985).

Uses: Defoliant for cotton.

BUTYLATE

Synonyms: **Bis(2-methylpropyl)carbamothioic acid S-ethyl ester;** Butilate; Diisobutylthiocarbamic acid S-ethyl ester; Diisocarb; S-Ethyl bis(2-methylpropyl)-carbamothioate; Ethyl N,N-diisobutylthiocarbamate; S-Ethyl diisobutylthiocarbamate; S-Ethyl N,N-diisobutylthiocarbamate; Ethyl-N,N-diisobutylthiolcarbamate; Genate Plus 6.7EC; R 1910; Stauffer R 1910; Sutan.

Structure:

$$CH_3CH_2S\overset{\overset{\displaystyle O}{\|}}{C}-N\underset{\diagdown}{\overset{\diagup}{}}\begin{matrix} CH_2\overset{\displaystyle CH_3}{\overset{|}{C}}HCH_3 \\ \\ CH_2\underset{\displaystyle CH_3}{\underset{|}{C}}HCH_3 \end{matrix}$$

Designations: CAS: 2008-41-5; mf: $C_{11}H_{23}NOS$; fw: 217.37; RTECS: EZ7525000.

Properties: Clear, colorless to amber, semi-volatile liquid with an aromatic odor. Mp: <25 °C; bp: 138 °C at 21 mmHg; ρ: 0.9402 and 0.9417 at 25/4 and 20/4 °C, respectively; fl p: 110 °C (open cup); K_H: 5.5 x 10^{-6} atm·m^3/mol at 20-25 °C (approximate - calculated from water solubility and vapor pressure); log K_{oc}: 2.73 (calculated); log K_{ow}: 4.15; S_o: miscible with acetone, ethanol, kerosene, methyl isobutyl ketone, xylene, and many other organic solvents at 20 °C; S_w: 45 mg/L at 22-25 °C; vap d: 8.88 g/L at 25 °C, 7.53 (air = 1); vp: 1.3 x 10^{-2} mmHg at 20 °C.

Transformation Products

Soil: Hydrolyzes in soil to ethyl mercaptan, carbon dioxide, and diisobutylamine (Hartley and Kidd, 1987). Somasundaram and Coats (1991) reported butylate in soils is oxidized to the corresponding sulfoxide. The reported half-life in soil is approximately 1.5-10 wk (Worthing and Hance, 1991). The reported half-life of butylate in a loam soil at 21-27 °C was 3 wk (Humburg et al., 1989). Residual activity in soil is limited to approximately 4 mo (Hartley and Kidd, 1987).

Plant: In plants, butylate is metabolized to carbon dioxide, diisobutylamine, fatty acids, conjugates of amines, and other compounds (Hartley and Kidd, 1987; Humburg et al., 1989).

Formulation Types: Granules (10%) with an inert safener; emulsifiable concentrate (6.7 lb/gal).

Toxicity: LC_{50} (96 hr) for rainbow trout 4.2 mg/L and bluegill sunfish 6.9 mg/L (Hartley and Kidd, 1987); acute oral LD_{50} for male and female rats is 4,659 and 5,431 mg/kg, respectively (Hartley and Kidd, 1987), 4,000 mg/kg (RTECS, 1985).

Uses: Selective, systemic, herbicide for the preemergent control annual grasses and broadleaf weeds in corn.

CARBARYL

Synonyms: Arylam; Carbamine; Carbatox; Carbatox 60; Carbatox 75; Carpolin; Carylderm; Cekubaryl; Crag sevin; Denapon; Devicarb; Dicarbam; ENT 23969; Experimental insecticide 7744; Gamonil; Germain's; Hexavin; Karbaspray; Karbatox; Karbosep; Methylcarbamate-1-naphthalenol; Methylcarbamate-1-naphthol; Methyl carbamic acid 1-naphthyl ester; *N*-Methyl-1-naphthylcarbamate; *N*-Methyl-α-naphthylcarbamate; *N*-Methyl-α-naphthylurethan; NA 2757; NAC; **1-Naphthalenol methylcarbamate**; 1-Naphthol-*N*-methylcarbamate; 1-Naphthyl methylcarbamate; 1-Naphthyl-*N*-methylcarbamate; α-Naphthyl-*N*-methylcarbamate; OMS 29; Panam; Ravyon; Rylam; Seffein; Septene; Sevimol; Sevin; Sok; Tercyl; Toxan; Tricarnam; UC 7744; Union Carbide 7744.

Structure:

Designations: CAS: 63-25-2; DOT: 2757; mf: $C_{12}H_{11}NO_2$; fw: 201.22; RTECS: FC5950000.

Properties: Colorless solid or white crystals. Mp: 142.2 °C; ρ: 1.232 at 20/20 °C; fl p: 195 °C; H-t$_{1/2}$ (27 °C): 1,500 d (pH 5), 15 d (pH 7), 0.15 d (pH 9); K_H: 1.27 x 10^{-5} atm·m^3/mol at 20 °C (approximate - calculated from water solubility and vapor pressure); log K_{oc}: 2.02-2.59; log K_{ow}: 2.31-2.81; P-t$_{1/2}$: 51.66 hr (absorbance λ = 257.5 nm, concentration on glass plates = 6.7 μg/cm^2) and approximately 45 hr (buffered solution irradiated at λ >280 nm); S_o: moderately soluble in acetone, cyclohexanone, *N,N*-dimethylformamide, and isophorone; S_w: 72.4, 104, and 130 mg/L at 10, 20, and 30 °C, respectively; vp: 6.578 x 10^{-6} atm at 25 °C.

Soil properties and adsorption data

Soil	K_d (mL/g)	f_{oc} (%)	K_{oc} (mL/g)	pH
Batcombe silt loam	2.13	2.05	104	6.1
Catlin	5.56	2.01	280	6.2
Commerce	1.81	0.68	265	6.7
Loamy sand	2.93	0.58	505	5.3
Rothamsted Farm	1.56	1.51	103	5.1
Sand	2.45	0.23	1,056	7.7
Sand	3.89	2.38	163	5.4

Soil	K_d (mL/g)	f_{oc} (%)	K_{oc} (mL/g)	pH
Sandy clay loam	0.44	1.71	26	8.1
Silty loam	3.29	2.09	157	6.3
Silty loam	4.69	3.07	153	5.0
Tracy	3.96	1.12	353	6.2

Source: Briggs, 1981; Lord et al., 1980; McCall et al., 1981; U.S. Department of Agriculture, 1990.

Transformation Products

Biological: Fourteen soil fungi metabolized methyl-^{14}C-labeled carbaryl via hydroxylation to 1-naphthyl-*N*-hydroxymethylcarbamate, 4-hydroxy-1-naphthylmethylcarbamate, and 5-hydroxy-1-naphthylmethylcarbamate (Bollag and Liu, 1972). Carbaryl was degraded by a culture of *Aspergillus terreus* to 1-naphthylcarbamate ($t_{1/2}$ = 8 d) (Liu and Bollag, 1971a).

Soil: The rate of hydrolysis of carbaryl in flooded soil increased when the soil was pretreated with the hydrolysis product, 1-naphthol (Rajagopal et al., 1986). Carbaryl is hydrolyzed in both flooded and nonflooded soils but the rate is slightly higher under flooded conditions (Rajagopal et al., 1983). When ^{14}C-carbonyl-labeled carbaryl (200 ppm) was added to five different soils and incubated at 25 °C for 32 d, evolution of ^{14}C-carbon dioxide varied from 2.2-37.4% (Kazano et al., 1972). Metabolites identified in soil included 1-naphthol (hydrolysis product) (Ramanand et al., 1988a; Sud et al., 1972), hydroquinone, catechol, pyruvate (Sud et al., 1972), coumarin, carbon dioxide (Kazano et al., 1972), 1-naphthylcarbamate, 1-naphthyl *N*-hydroxymethylcarbamate, 5-hydroxy-1-naphthylmethylcarbamate, 4-hydroxy-1-naphthylmethylcarbamate, and 1-naphthylhydroxymethylcarbamate (Liu and Bollag, 1971, 1971a). 1-Naphthol, a metabolite of carbaryl in soil, was recalcitrant to further degradation by a bacterium tentatively identified as an *Arthrobacter* sp. under anaerobic conditions (Ramanand et al., 1988a). Carbaryl or its metabolite 1-naphthol at normal and ten times the field application rate had no effect on the growth of *Rhizobium* sp. or *Azotobacter chroococcum* (Kale et al., 1989).

Rajagopal et al. (1984) proposed degradation pathway of carbaryl in soil and in microbial cultures included the following compounds: 5,6-dihydrodihydroxy carbaryl, 2-hydroxy carbaryl, 4-hydroxy carbaryl, 5-hydroxy carbaryl, 1-naphthol, *N*-hydroxymethyl carbaryl, 1-naphthyl carbamate, 1,2-dihydroxynaphthalene, 1,4-dihydroxynaphthalene, *o*-coumaric acid, *o*-hydroxybenzalpyruvate, 1,4-naphthoquinone, 2-hydroxy-1,4-naphthoquinone, coumarin, γ-hydroxy-γ-*o*-hydroxyphenyl-α-oxobutyrate, 4-hydroxy-1-tetralone, 3,4-dihydroxy-1-tetralone, pyruvic acid, salicylaldehyde, salicylic acid, phenol, hydroquinone, catechol, carbon dioxide, and water. When carbaryl was incubated at room temperature in a mineral salts medium by soil-enrichment cultures for 30 d, 26.8 and 31.5% of the applied insecticide remained in flooded and nonflooded soils, respectively (Rajagopal et al., 1984a). A *Bacillus* sp. and the enrichment cultures both degraded carbaryl to 1-naphthol. Mineralization to carbon dioxide was negligible (Rajagopal et al., 1984a).

Plant: In plants, the *N*-methyl group may be subject to oxidation or hydroxylation (Kuhr, 1968).

Surface Water: In a laboratory aquaria containing estuarine water, 43% of dissolved carbaryl was converted to 1-naphthol in 17 d at 20 °C (pH = 7.5-8.1). The half-life of carbaryl in estuarine water without mud at 8 °C was 38 d. When mud was present, both carbaryl and 1-naphthol decreased to less than 10% in the estuarine water after 10 d. Based on a total recovery of only 40%, it was postulated that the remainder was evolved as methane (Karinen et al., 1967). The rate of hydrolysis of carbaryl increased with an increase in temperature (Karinen et al., 1967) and in increases of pH above 7.0 (Rajagopal et al., 1984). The presence of a micelle [hexadecyltrimethylammonium bromide (HDATB), 3 x 10^{-3} M] in natural waters greatly enhanced the hydrolysis rate. The hydrolysis half-lives in natural water samples with and without HDATB were 0.12-0.67 and 9.7-138.6 hr, respectively (González et al., 1992). In the dark, carbaryl was incubated in 21 °C water obtained from the Holland Marsh drainage canal. Degradation was complete after 4 wk (Sharom et al., 1980).

In pond water, carbaryl rapidly degraded to 1-naphthol. The latter was further degraded, presumably by *Flavobacterium* sp., into hydroxycinnamic acid, salicylic acid, and an unidentified compound (Hazardous Substances Data Bank, 1989). Four days after carbaryl (30 mg/L and 300 μg/L) was added to Fall Creek water, >60% was mineralized to carbon dioxide. At pH 3, however, <10% was converted to carbon dioxide (Boethling and Alexander, 1979). Under these conditions, hydrolysis of carbaryl to 1-naphthol was rapid. The authors could not determine how much carbon dioxide was attributed to biodegradation of carbaryl and how much was due to the biodegradation of 1-naphthol (Boethling and Alexander, 1979).

Photolytic: Based on data for phenol, a structurally related compound, an aqueous solution containing the 1-naphthoxide ion (3 x 10^{-4} M) in room light would be expected to photooxidize to give 2-hydroxy-1,4-naphthoquinone (Tomkiewicz et al., 1971). 1-Naphthol, methyl isocyanate, and other unidentified cholinesterase inhibitors were reported as products formed from the direct photolysis of carbaryl by sunlight (Wolfe et al., 1976). In an aqueous solution at 25 °C, the photolysis half-life of carbaryl by natural sunlight or UV light ($\lambda = 313$ nm) is 6.6 d (Wolfe et al., 1978a).

Chemical/Physical: Ozonation of carbaryl in water yielded 1-naphthol, naphthoquinone, phthalic anhydride, *N*-formylcarbamate of 1-naphthol (Martin et al., 1983), naphthoquinones, and acidic compounds (Shevchenko et al., 1982). Hydrolysis and photolysis of carbaryl forms 1-naphthol (Rajagopal et al., 1984, 1986; Wauchope and Haque, 1973; Miles et al., 1988; MacRae, 1989; Ramanand et al., 1988a; Somasundaram et al., 1991) and 2-hydroxy-1,4-naphthoquinone (Wauchope and Haque, 1973), respectively. In aqueous solutions, carbaryl hydrolyzes to 1-naphthol (Boethling and Alexander, 1979; Vontor et al., 1972), methylamine, and carbon dioxide (Vontor et al., 1972) especially under alkaline conditions (Wolfe et al., 1978).

Miles et al. (1988) studied the rate of hydrolysis of carbaryl in phosphate-buffered water (0.01 M) at 26 °C with and without a chlorinating agent (10 mg/L hypochlorite solution). The hydrolysis half-lives at pH 7 and 8 with and without chlorine were 3.5 and 10.3 d and 0.05 and 1.2 d, respectively (Miles et al., 1988). The reported hydrolysis half-lives of carbaryl in water at pH 7, 8, 9, and 10 were 10.5 d, 1.3 d, 2.5 hr, and 15.0 min, respectively (Aly and El-Dib, 1971). 1-Naphthol, the major product of carbaryl hydrolysis, dissociates in water to the 1-naphthoxide ion (Wauchope and Haque, 1973). The hydrolysis half-lives of carbaryl in a sterile 1% ethanol/water solution at 25 °C and pHs of 4.5, 6.0, 7.0, and 8.0, were 2,100, 406, 14, and 1.9 d, respectively (Chapman and Cole, 1982).

Releases toxic nitrogen oxides when heated to decomposition (Sax and Lewis, 1987; Lewis, 1990). Products reported from the combustion of carbaryl at 900 °C include carbon monoxide, carbon dioxide, ammonia, and oxygen (Kennedy et al., 1972a).

Exposure Limits: NIOSH REL: 10 hr TWA 5 mg/m^3, IDLH 600 mg/m^3; OSHA PEL: TWA 5 mg/m^3; ACGIH TLV: TWA 5 mg/m^3.

Symptoms of Exposure: Miosis, blurred vision, tearing, nasal discharge, salivation, sweating, abdominal cramps, nausea, vomiting, diarrhea, tremor, cyanosis, convulsions, severe eye and skin irritation.

Formulation Types: Emulsifiable concentrate; wettable powder; granules; suspension concentrate; dustable powder; oil-miscible flowable concentrate.

Toxicity: LC_{50} (96 hr) for rainbow trout 1.3 mg/L, sheepshead minnow 2.2 mg/L (Worthing and Hance, 1991), bluegill sunfish 6.76 mg/L, goldfish 13.2 mg/L (Hartley and Kidd, 1987), catfish 15.8 mg/L, bullhead 20.0 mg/L, carp 5.3 mg/L, bass 6.4, brown trout 2.0 mg/L, coho salmon 0.76 mg/L (Verschueren, 1983); LC_{50}

(24 hr) for bluegill sunfish 3.4 ppm and rainbow trout 3.5 ppm (Verschueren, 1983); acute oral LD_{50} for male and female rats is 850 and 500 mg/kg, respectively (Hartley and Kidd, 1987).

Uses: Contact insecticide used to control most insects on fruits, vegetables, and ornamentals.

CARBOFURAN

Synonyms: Bay 70143; **2,3-Dihydro-2,2-dimethyl-7-benzofuranol methylcarbamate;** 2,2-Dimethyl-7-coumaranyl-*N*-methylcarbamate; 2,2-Dimethyl-2,3-dihydro-7-benzofuranyl-*N*-methylcarbamate; ENT 27164; Furadan; Methyl carbamic acid 2,3-dihydro-2,2-dimethyl-7-benzofuranyl ester; NIA 10242; Niagara 10242.

Structure:

Designations: CAS: 1563-66-2; DOT: 2757; mf: $C_{12}H_{15}NO$; fw: 221.26; RTECS: FB9450000.

Properties: White, odorless, crystalline solid. Mp: 153-154 °C; bp: decomposes >150 °C; ρ: 1.18 at 20/20 °C; fl p: nonflammable; H-$t_{1/2}$ (25 °C): 170, 690, 690, 8.2 and 1.0 wk at pHs of 4.5, 5.0, 6.0, 7.0, and 8.0, respectively; K_H: 3.88 x 10^{-8} atm·m^3/mol at 30-33 °C (approximate - calculated from water solubility and vapor pressure); log K_{oc}: 1.98-2.32; log K_{ow}: 1.60-2.32; P-$t_{1/2}$: 70.64 hr (absorbance λ = 258.0 nm, concentration on glass plates = 6.7 μg/cm^2); S_o: acetone (15%), acetonitrile (14%), benzene (4%), cyclohexane (9%), *N,N*-dimethylformamide (27%), dimethylsulfoxide (25%), methylene chloride (>200 g/L), 2-propanol (20-50 g/L); S_w: 291, 320, and 375 mg/L at 10, 20, and 30 °C, respectively; vp: 2 x 10^{-5} mmHg at 33 °C.

Soil properties and adsorption data

Soil	K_d (mL/g)	f_{oc} (%)	K_{oc} (mL/g)	pH	CEC (meq/100 g)
Bryce silty clay loam	2.22	7.50	30	--	55.5
Bryce-Swygert silty clay	1.40	2.70	52	--	34.4
Catlin	1.20	2.01	60	6.2	--
Chaslcus	0.61	1.62	38	8.1	--
Commerce	1.12	0.68	160	6.7	--
Drummer silty clay loam	1.13	3.10	36	--	24.8
Earth/Humus	1.65	4.16	40	6.5	--

continued

71

Soil	K_d (mL/g)	f_{oc} (%)	K_{oc} (mL/g)	pH	CEC (meq/100 g)
Flanagan silty loam	1.39	3.50	40	--	27.7
Gilford-Hoopeston-Ade sandy loam	0.74	1.20	62	--	7.5
Houghton muck	8.74	16.80	52	--	72.4
Humus	3.48	10.97	32	6.8	--
Kari soil	2.38	5.10	47	2.7	32.2
Limagne	0.72	2.08	35	8.0	--
Loam	1.08	3.02	34	7.1	--
Loamy sand	0.56	1.80	31	6.1	--
Plainfield-Bloomfield sand	0.25	0.40	63	--	1.7
Pokkali soil	2.08	4.12	50	3.6	31.2
Sand	0.10	0.99	10	6.9	--
Sandy loam	1.25	4.23	30	5.6	--
Sediment A	0.33	1.50	22	7.7	--
Sediment B	0.30	1.85	16	7.1	--
Silty loam	0.30	2.26	13	4.9	--
Soil #1 (0-30 cm)	0.48	1.46	33	6.0	--
Soil #1 (30-50 cm)	0.49	0.93	53	6.1	--
Soil #1 (50-91 cm)	0.52	1.24	42	6.1	--
Soil #1 (91-135 cm)	0.18	0.70	26	6.4	--
Soil #1 (135-165 cm)	0.09	0.17	53	6.5	--
Soil #1 (>165 cm)	0.08	0.07	117	6.7	--
Soil #2 (0-27 cm)	0.38	1.50	25	5.9	--
Soil #2 (27-53 cm)	0.15	0.16	95	6.3	--
Soil #2 (53-61 cm)	0.07	0.14	48	6.4	--
Soil #2 (61-81 cm)	0.15	0.28	55	6.4	--
Soil #2 (81-157 cm)	0.29	0.50	58	6.0	--
Soil #2 (>157 cm)	0.51	1.30	39	5.9	--
Tracy	1.07	1.12	95	6.2	--
Versailles	0.26	1.10	24	6.4	--

Source: Felsot and Wilson, 1980; Jamet and Piedallu, 1975; McCall et al., 1981; Singh et al., 1990; Sukop and Cogger, 1992; U.S. Department of Agriculture, 1990.

Transformation Products
Biological: Carbofuran or their metabolites (3-hydroxycarbofuran, and 3-keto-carbofuran) at normal and ten times the field application rate had no effect on *Rhizobium* sp. However, in a nitrogen-free culture medium, *Azotobacter chroococcum* growth was inhibited by carbofuran, 3-hydroxycarbofuran and 3-ketocarbofuran (Kale et al., 1989). Under *in vitro* conditions, 15 of 20 soil fungi

degraded carbofuran to one or more of the following compounds: 3-hydroxy-carbofuran, 3-ketocarbofuran, carbofuran phenol, and 3-hydroxyphenol (Arunachalam and Lakshmanan, 1988).

Soil: Carbofuran phenol is formed from the hydrolysis of carbofuran at pH 7.0. Carbofuran phenol was also found to be the major biodegradation product by *Azospirillum lipoferum* and *Streptomyces* spp. isolated from a flooded alluvial soil (Venkateswarlu and Sethunathan, 1984). The hydrolysis of carbofuran to carbo-furan phenol was catalyzed by the addition of rice straw in an anaerobic flooded soil where it accumulated (Venkateswarlu and Sethunathan, 1979). In an alluvial soil, carbaryl and its analog carbosulfan 2,3-dihydro-2,2-dimethyl-7-benzo-furanyl [(di-*n*-butyl)aminosulfenyl methylcarbamate] both degraded faster at 35 °C than at 25 °C with carbosulfan degrading to carbofuran (Sahoo et al., 1990). An enrichment culture isolated from a flooded alluvial soil (Ramanand et al., 1988) and a bacterium tentatively identified as an *Arthrobacter* sp. (Ramanand et al., 1988a) readily mineralized carbofuran to carbon dioxide at 35 °C. Mineralization was slower at lower temperatures (20-28 °C). Under anaerobic conditions, carbo-furan did not degrade (Ramanand et al., 1988a). The reported half-lives in soil are 1-2 mo (Hartley and Kidd, 1987); 11-13 d at pH 6.5, and 60-75 d for a granular formulation (Ahmad et al., 1979).

Rajagopal et al. (1984) proposed degradation pathway of carbofuran in both soils and microbial cultures included the following compounds: 3-hydroxycarbofuran, 3-ketocarbofuran, carbofuran phenol, 3-hydroxycarbofuran phenol, 3-ketocarbo-furan phenol, 6,7-dihydroxycarbofuran phenol, 3,6,7-trihydroxycarbofuran phenol, 3-keto-6,7-dihydroxycarbofuran phenol, and carbon dioxide. In soils, microorganisms degraded carbofuran to carbofuran phenol (Ou et al., 1982), 3-hydroxycarbofuran then to and 3-ketocarbofuran (Kale et al., 1989; Ou et al., 1982).

Hydrolyzes in soil to carbofuran phenol (Rajagopal et al., 1986; Somasundaram et al., 1989, 1991). Hydrolysis of carbofuran occurs in both flooded and nonflooded soils, but the rate is slightly higher under flooded conditions (Venkateswarlu et al., 1977) especially when the soil is pretreated with the hydrolysis product, carbofuran phenol (Rajagopal et al., 1986).

Plant: Carbofuran is rapidly metabolized in plants to nontoxic products (Cremlyn, 1991). Metcalf et al. (1968) reported that carbofuran undergoes hydroxylation and hydrolysis in plants, insects, and mice. Hydroxylation of the benzylic carbon gives 3-hydroxycarbofuran which is subsequently oxidized to 3-ketocarbofuran. In carrots, carbofuran initially degraded to 3-hydroxycarbofuran. This compound reacted with naturally occurring angelic acid in carrots forming a conjugated metabolite identified as 2,3-dihydro-2,2-dimethyl-7-(((methylamino)carbonyl)-

oxy)-3-benzofuranyl (Z)-2-methyl-2-butenoic acid (Sonobe et al., 1981). Metabolites identified in three types of strawberries (Day-Neutral, Tioga, and Tufts) were 2,3-dihydro-2,2-dimethyl-3-hydroxy-7-benzofuranyl-N-methylcarbamate, 2,3-dihydro-2,2-dimethyl-3-oxo-7-benzofuranyl-N-methylcarbamate, 2,3-dihydro-2,2-dimethyl-3-benzofuranol, 2,3-dihydro-2,2-dimethyl-3,7-benzofuranol, and 2,3-dihydro-2,2-dimethyl-3-oxo-7-benzofuranol (Archer et al., 1977). Oat plants were grown in two soils treated with [^{14}C]carbofuran. Most of the residues recovered in oat leaves were in the form of carbofuran and 3-hydroxycarbofuran. Other metabolites identified were 3-ketocarbofuran, a 3-keto-7-phenol, and a 3-hydroxy-7-phenol (Fuhremann and Lichtenstein, 1980).

Photolytic: 2,3-Dihydro-2,2-dimethylbenzofuran-4,7-diol and 2,3-dihydro-3-keto-2,2-dimethylbenzofuran-7-ylcarbamate were formed when carbofuran dissolved in water was irradiated by sunlight for 5 d (Raha and Das, 1990).

Chemical/Physical: Releases toxic nitrogen oxides when heated to decomposition (Sax and Lewis, 1987; Lewis, 1990). The hydrolysis half-lives of carbofuran in a sterile 1% ethanol/water solution at 25 °C and pHs of 4.5, 5.0, 6.0, 7.0, and 8.0 were 170, 690, 690, 8.2, and 1.0 wk, respectively (Chapman and Cole, 1982).

Exposure Limits: OSHA PEL: TWA 0.1 mg/m^3; ACGIH TLV: TWA 0.1 mg/m^3.

Symptoms of Exposure: Early symptoms include burning eyes, dimming of vision, miosis, loss of accommodation, headache, light-headedness, weakness and nausea. Later symptoms include blurred vision, constriction of pupils, abdominal cramps, excessive salivation, perspiration, and vomiting.

Formulation Types: Wettable powder; granules; suspension concentrate; flowable concentrate.

Toxicity: LC$_{50}$ (96 hr) for rainbow trout 280 μg/L, bluegill sunfish 240 μg/L, and channel catfish 210 μg/L; acute oral LD$_{50}$ for rats 8.2-14.1 mg/kg (corn oil) (Hartley and Kidd, 1987), 5,300 μg/kg (RTECS, 1985), 11 mg/kg (Verschueren, 1983).

Uses: Broad-spectrum, systemic insecticide, nematocide, and acaricide applied in soil to control soil insects and nematodes or on foliage to control insects and mites.

CARBON DISULFIDE

Synonyms: Carbon bisulfide; Carbon bisulphide; Carbon disulphide; Carbon sulfide; Carbon sulphide; Dithiocarbonic anhydride; NCI-C04591; RCRA waste number P022; Sulphocarbonic anhydride; UN 1131; Weeviltox.

Structure:

$$S=C=S$$

Designations: CAS: 75-15-0; DOT: 1131; mf: CS_2; fw: 76.13; RTECS: FF6650000.

Properties: Highly refractive, mobile, clear, colorless to pale yellow liquid; sweet, pleasing, ethereal odor when pure. Technical grades have foul odors. Mp: -111.5 °C; bp: 46.2 °C; ρ: 1.27055 at 15/4 °C, 1.2632 at 20/4 °C; fl p: -30 °C; lel: 1.3%; uel: 50%; K_H: 1.33 x 10^{-2} atm·m^3/mol at 20-22 °C (approximate - calculated from water solubility and vapor pressure); IP: 10.06 eV, 10.080 eV; log K_{oc}: 2.38-2.55 (calculated); log K_{ow}: 1.84, 2.16 (calculated); S_o: miscible with benzene, ethanol, carbon tetrachloride, chloroform, ethyl ether, n-propanol, and many other organic solvents; S_w: 2.3 g/L at 22 °C; vap d: 3.11 g/L at 25 °C, 2.63 (air = 1); vp: 297.5 mmHg at 20 °C.

Transformation Products

Chemical/Physical: Carbon disulfide hydrolyzes in alkaline solutions to carbon dioxide and hydrogen disulfide (Peyton et al., 1976). In an aqueous alkaline solution containing hydrogen peroxide, dithiopercarbonate, sulfide, elemental sulfur, and polysulfides may be expected to form (Elliott, 1990). In an aqueous alkaline solution (pH ≥8), carbon disulfide reacted with hydrogen peroxide forming sulfate and carbonate ions. However, when the pH is lowered to 7-7.4, colloidal sulfur is formed (Adewuyi and Carmichael, 1987). Forms a hemihydrate which decomposes at -3 °C (Keith and Walters, 1992).

Burns with a blue flame releasing carbon dioxide and sulfur dioxide (Windholz et al., 1983). Emits very toxic sulfur oxides when heated to decomposition (Lewis, 1990). Burns with a blue flame releasing carbon dioxide and sulfur dioxide (Windholz et al., 1983).

Carbon disulfide oxidizes in the troposphere producing carbonyl sulfide. The atmospheric half-lives of carbon disulfide and carbonyl sulfide were estimated to be approximately 2 yr and 13 d, respectively (Khalil and Rasmussen, 1984).

Exposure Limits: NIOSH REL: 10 hr TWA 1 ppm, 15-min C 10 ppm, IDLH 500 ppm; OSHA PEL: TWA 4 ppm, C 12 ppm; ACGIH TLV: TWA 10 ppm.

Symptoms of Exposure: Dizziness, headache, poor sleep, fatigue, nervousness; anorexia, weight loss; psychosis; polyneuropathy; Parkinson-like; ocular changes; cardiovascular system; gastrointestinal; burns, dermatitis. Contact with skin causes burning pain, erythema, and exfoliation.

Formulation Types: Dilute solution.

Toxicity: Acute oral LD_{50} for rats 3,188 mg/kg (RTECS, 1985).

Uses: Formerly used as an insecticide.

CARBON TETRACHLORIDE

Synonyms: Benzinoform; Carbona; Carbon chloride; Carbon tet; ENT 4705; Fasciolin; Flukoids; Freon 10; Halon 104; Methane tetrachloride; Necatorina; Necatorine; Perchloromethane; R 10; RCRA waste number U211; Tetrachloormetaan; Tetrachlorocarbon; **Tetrachloromethane**; Tetrafinol; Tetraform; Tetrasol; UN 1846; Univerm; Vermoestricid.

Structure:

$$CCl_4$$

Designations: CAS: 56-23-5; DOT: 1846; mf: CCl_4; fw: 153.82; RTECS: FG4900000.

Properties: Clear, colorless, heavy, liquid with a strong sweetish, ether-like odor and burning taste. Mp: -22.96 °C; bp: 76.5 °C; ρ: 1.5940 at 20/4 °C, 1.585 at 25/4 °C; H-t$_{1/2}$: 40.5 yr at 25 °C and pH 7; K_H: 3.02 x 10^{-2} atm·m^3/mol at 25 °C; IP: 11.47 eV; log K_{oc}: 2.35-2.64; log K_{ow}: 2.73, 2.83; S_o: miscible with acetone, ethanol, benzene, chloroform, carbon disulfide, dimethylsulfoxide, ethyl ether, light petroleum, petroleum ether, solvent naphtha, volatile oils, dehydrated alcohol, fat solvents, and most organic solvents; S_w: 800 mg/L at 20 °C; vap d: 6.29 g/L at 25 °C, 5.31 (air = 1); vp: 90 mmHg at 20 °C, 137 mmHg at 30 °C.

Transformation Products

Biological: Carbon tetrachloride was degraded by denitrifying bacteria forming chloroform (Smith and Dragun, 1984). An anaerobic species of *Clostridium* biodegraded carbon tetrachloride by reductive dechlorination yielding trichloromethane (chloroform), dichloromethane, and unidentified products (Gälli and McCarty, 1989).

Chemical/Physical: Under laboratory conditions, carbon tetrachloride was partially hydrolyzed to chloroform and carbon dioxide. Complete hydrolysis yielded carbon dioxide and hydrochloric acid (Smith and Dragun, 1984). Carbon tetrachloride slowly reacts with hydrogen sulfide in aqueous solution yielding carbon dioxide via the intermediate carbon disulfide. However, in the presence of two micaceous minerals (biotite and vermiculite) and amorphous silica, the rate of transformation is increased. At 25 °C and a hydrogen sulfide concentration of 1 mM, the half-lives of carbon tetrachloride were calculated to be 2,600, 160, and 50 d for the silica, vermiculite, and biotite studies, respectively. In all three studies, the major transformation pathway is the formation of carbon disulfide. This compound is then hydrolyzed to afford carbon dioxide (81-86% yield) and HS⁻. Minor intermediates detected include chloroform (5-15% yield), carbon monoxide (1-2% yield), and a nonvolatile compound tentatively identified as formic acid (3-6% yield) (Kriegman-King and Reinhard, 1992).

Anticipated products from the reaction of carbon tetrachloride with ozone or hydroxyl radicals in the atmosphere are phosgene and chloride radicals (Cupitt, 1980). Phosgene is hydrolyzed readily to hydrochloric acid and carbon dioxide (Morrison and Boyd, 1971). Emits toxic fumes of chlorides and phosgene when heated to decomposition (Lewis, 1990).

Exposure Limits: NIOSH REL: 1 hr C 2 ppm; OSHA PEL: TWA 4 ppm, STEL 12 ppm; ACGIH TLV: TWA 5 ppm.

Symptoms of Exposure: Central nervous system depression; nausea, vomiting; liver, kidney damage; skin irritation.

Formulation Types: Fumigant (mixture containing 1,2-dichloroethane or 1,2-dibromoethane and 1,2-dichloroethane).

Toxicity: LC_{50} (14 d) for guppies 67 ppm; acute oral LD_{50} for rats 2,920 mg/kg (Verschueren, 1983), 2,800 mg/kg (RTECS, 1985).

Uses: Agricultural fumigant.

CARBOPHENOTHION

Synonyms: Acarithion; Akarithion; *S*-((4-Chlorophenyl)thio)methyl *O,O*-diethyl phosphorodithioate; *S*-((*p*-Chlorophenyl)thio)methyl *O,O*-diethyl phosphorodithioate; *S*-(4-Chlorophenylthiomethyl)diethyl phosphorothiolothionate; Dagadip; *O,O*-Diethyl *p*-chlorophenylmercaptomethyl dithiophosphate; *O,O*-Diethyl *S*-(4-chlorophenylthio)methyl dithiophosphate; *O,O*-Diethyl *S*-(*p*-chlorophenylthio)-methyl phosphorodithioate; Endyl; ENT 23708; Garrathion; Lethox; Nephocarp; Oleoakarithion; R 1303; **Phosphorodithioic acid *S*-(((4-chlorophenyl)thio)methyl) *O,O*-diethyl ester;** Stauffer R 1303; Trithion; Trithion miticide.

Structure:

$$CH_3CH_2O \diagdown \overset{\overset{\text{S}}{\|}}{\underset{\diagup}{P}} - SCH_2S - \text{〈phenyl〉} - Cl$$
$$CH_3CH_2O$$

Designations: CAS: 786-19-6; mf: $C_{11}H_{16}ClO_2PS_3$; fw: 342.96; RTECS: TD5250000.

Properties: Colorless to light amber liquid with a mercaptan-like odor. Mp: <25 °C; bp: 82 °C at 0.01 mmHg; ρ: 1.271 at 25/4 °C; K_H: 4.5 x 10^{-7} atm·m³/mol at 20 °C; log K_{oc}: 3.92-4.98; log K_{ow}: 4.75 (calculated); S_o: miscible with most organic solvents and vegetable oils; S_w: 610, 630, and 730 μg/L at 10, 20, and 30 °C, respectively; vap d: 14.02 g/L at 25 °C, 11.88 (air = 1); vp: 8.03 x 10^{-6} mmHg at 20 °C.

Soil properties and adsorption data

Soil	K_d (mL/g)	f_{oc} (%)	K_{oc} (mL/g)	pH
Elkhorn sandy loam	82.3	0.09	95,698	6.0
Hugo gravelly sandy loam	62.8	0.12	52,333	5.5
Sweeney sandy clay loam	54.6	0.65	8,394	6.3
Tierra clay loam	98.6	0.33	29,894	6.2

Source: Rao and Davidson, 1982.

Transformation Products
Soil: Though no products were reported, the half-life was reported to be ≥100 d (Verschueren, 1983).

Chemical/Physical: Oxidizes to the sulfoxide, sulfone, thiol, thiosulfone, and thiosulfoxide (Hartley and Kidd, 1987). Emits toxic fumes of chlorine, phosphorus and sulfur oxides when heated to decomposition (Sax and Lewis, 1987).

Symptoms of Exposure: Headache, weakness, and dizziness.

Formulation Types: Wettable powder; dustable powder; granules; emulsifiable concentrate; seed treatment.

Toxicity: Very toxic to fish (Hartley and Kidd, 1987); acute oral LD_{50} for male and female rats is 30 and 10 mg/kg, respectively (Hartley and Kidd, 1987), 6,800 mg/kg (RTECS, 1985).

Uses: Nonsystemic insecticide and acaricide for controlling mites, aphids, and other insects on deciduous fruit trees.

CARBOXIN

Synonyms: 5-Carboxanilido-2,3-dihydro-6-methyl-1,4-oxathiin; Carboxine; D 735; DCMO; 2,3-Dihydro-5-carboxanilido-6-methyl-1,4-oxatiin; 2,3-Dihydro-6-methyl-1,4-oxathiin-5-carboxanilide; 5,6-Dihydro-2-methyl-1,4-oxathiin-3-carboxanilide; **5,6-Dihydro-2-methyl-N-phenyl-1,4-oxathiin-3-carboxamide;** F 735; Flo pro V seed protectant; Vitavax.

Structure:

Designations: CAS: 5234-68-4; mf: $C_{12}H_{13}NO_2S$; fw: 235.31; RTECS: RP4550000.

Properties: Colorless crystals. Mp: 93-95 °C; ρ: 1.36; H-t$\frac{1}{2}$: <3 d when exposed to light; K_H: 3.4 x 10^{-10} atm·m^3/mol at 20-25 °C (approximate - calculated from water solubility and vapor pressure); log K_{oc}: 2.41; log K_{ow}: 2.17; S_o (g/kg at 25 °C): acetone (600), benzene (150), dimethyl sulfoxide (1,500), methanol (210); S_w: 170 mg/L at 25 °C; vp: 1.88 x 10^{-7} mmHg at 20 °C.

Transformation Products

Biological: The sulfoxidation of carboxin to carboxin sulfoxide by the fungus *Ustilago maydis* was reported by Bollag and Liu (1990).

Soil: Carboxin oxidizes in soil forming carboxin sulfoxide. The half-life in soil was reported to be 24 hr (Worthing and Hance, 1991).

Plant: In plants (barley, cotton, and wheat) and water, carboxin oxidizes to the corresponding sulfoxide (Worthing and Hance, 1991).

Formulation Types: Wettable powder; seed treatment; suspension concentrate.

Toxicity: LC$_{50}$ (96 hr) for rainbow trout 2 mg/L and bluegill sunfish 1.2 mg/L (Hartley and Kidd, 1987); acute oral LD$_{50}$ for rats 3,820 mg/kg (Hartley and Kidd, 1987), 430 mg/kg (RTECS, 1985).

Uses: Systemic plant fungicide.

CHLORAMBEN

Synonyms: ACP-M-728; Ambiben; Amiben; Amiben DS; Amibin; **3-Amino-2,5-dichlorobenzoic acid**; NCI-C00055; Ornamental weeder; Vegaben; Vegiben.

Structure:

Designations: CAS: 133-90-4; mf: $C_7H_5Cl_2NO_2$; fw: 206.02; RTECS: DG1925000.

Properties: Odorless, colorless to white crystals or amorphous powder. Mp: 200-201 °C; fl p: nonflammable; log K_{oc}: 2.28; log K_{ow}: 1.11; S_o (g/L): acetone (233 at 29 °C), benzene (0.2 at 24 °C), chloroform (0.9 at 25 °C); N,N-dimethylformamide (1,206 at 20-25 °C), ethanol (173 at 25 °C), ethyl ether (70 at 20-25 °C), methanol (223 at 20-25 °C), 2-propanol (113 at 25 °C); S_w: 700 mg/L at 25 °C; vp: 7 x 10^{-3} mmHg at 100 °C.

Soil properties and adsorption data

Soil	K_d (mL/g)	f_{oc} (%)	K_{oc} (mL/g)	pH	CEC (meq/100 g)
Anselmo sandy loam	0.10	0.98	10.20	7.0	7.0
Crowley sandy loam[a]	4.50	0.92	48.91	6.5	12.6
Ella loamy sand	0.24	0.92	26.09	3.8	--
Fargo clay	0.30	4.56	6.58	7.9	30.1
Gallion fine sandy loam[a]	1.10	0.35	31.43	6.2	2.8
Keith sandy loam	0.30	1.67	17.96	6.2	11.6
Kewaunee clay	0.11	2.19	5.02	6.4	--
Monona sandy loam	0.50	2.42	20.66	5.8	17.5
Poygan silty clay	0.24	5.70	4.21	7.2	--
Sharkey clay[a]	6.50	0.75	86.67	6.7	33.4
Sharpsburg sandy clay loam	1.00	2.54	39.37	5.8	24.5
Taloka sandy loam[a]	1.90	0.40	47.50	6.4	4.6

Source: Schliebe et al., 1965; Talbert et al., 1970; Wildung et al., 1968. a) Adsorption data for the methyl ester.

Transformation Products
Soil: In soils, chloramben is degraded by microorganisms but no products were

82

identified (Humburg et al., 1991). The main degradative pathway of chloramben in soil is decarboxylation and subsequent mineralization to carbon dioxide. The calculated half-lives in Ella loamy sand, Kewaunee clay, and Poygan silty clay were 120-201, 182-286, and 176-314 d, respectively (Wildung et al., 1968). Persistence in soil is 6-8 wk (Hartley and Kidd, 1987).

Plant: Degrades in plants to *N*-glucoside, glucose ester, conjugates, and insoluble residues (Ashton and Monaco, 1991).

Photolytic: Plimmer and Hummer (1969) studied the irradiation of chloramben in water (2-4 mg/L) under a 450-W mercury vapor lamp (λ >2800 Å) for periods of 2-20 hr. Chloride ion was released and a complex mixture of colored products was observed. It was postulated that amino free radicals reacted with each other via polymerization and oxidation processes (Plimmer and Hummer, 1969). The experiment was repeated except the solution contained sodium bisulfite as an inhibitor under a nitrogen atmosphere. Oxidation did not occur and loss of the 2-chloro substituent gave 3-amino-5-chlorobenzoic acid (Plimmer and Hummer, 1969). Chloramben (sodium salt) in aqueous solutions (100 mg/L) was rapidly photodegraded in outdoor sunlight and under 360-W mercury arc lamp (Crosby and Leitis, 1969). In sunlight, the solution became yellow-brown. Subsequent analysis by GLC did not resolve any compounds other than chloramben. However, analysis by TLC indicated at least 12 unidentified products were formed. These products were reportedly formed via replacement of chlorine by a hydroxy group, reductive dechlorination, and abstraction of hydrogen from the amine group (oxidation). No photodegradation products could be identified in the solutions irradiated with the mercury arc lamp (Crosby and Leitis, 1969).

Chemical/Physical: Emits toxic fumes of nitrogen oxides and chlorine when heated to decomposition (Sax and Lewis, 1987). Forms soluble salts with alkalies.

Exposure Limits: An experimental carcinogen.

Formulation Types: Soluble concentrate (1.8 lb/gal); granules (10%); water-soluble powder or granules (75%).

Toxicity: Nontoxic to fish (Hartley and Kidd, 1987); acute oral LD_{50} for rats is 5,620 mg/kg (Hartley and Kidd, 1987), 3,500 mg/kg (RTECS, 1985).

Uses: Preemergence or preplant herbicide used in many vegetable and field crops to control annual broadleaf weeds and grasses. Also for postemergent control common ragweed, redroot pigweed, smartweed, and velvetleaf.

CHLORDANE

Synonyms: A 1068; Aspon-chlordane; Belt; CD-68; Chlordan; γ-Chlordan; Chloridan; Chlorindan; Chlor kil; Chlorodane; Chlortox; Corodane; Cortilan-neu; Dichlorochlordene; Dowklor; ENT 9932; ENT 25552; HCS 3260; Kypchlor; M 140; M 410; NA 2762; NCI-C00099; Niran; Octachlor; 1,2,4,5,6,7,8,8-Octachlor-2,3,3a,4,7,7a-hexahydro-4,7-methanoindane; Octachlorodihydrodicyclopentadiene; 1,2,4,5,6,7,8,8-Octachloro-2,3,3a,4,7,7a-hexahydro-4,7-methanoindene; **1,2,4,5,6,7,8,8-Octachloro-2,3,3a,4,7,7a-hexahydro-4,7-methano-1H-indene**; 1,2,4,5,6,7,8,8-Octachloro-3a,4,7,7a-hexahydro-4,7-methyleneindane; Octachloro-4,7-methanohydroindane; Octachloro-4,7-methanotetrahydroindane; 1,2,4,5,6,7,8,8-Octachloro-4,7-methano-3a,4,7,7a-tetrahydroindane; 1,2,4,5,6,7,8,8-Octachloro-3a,4,7,7a-tetrahydro-4,7-methanoindan; 1,2,4,5,6,7,8,8-Octachloro-3a,4,7,7a-tetrahydro-4,7-methanoindane; Octaklor; Octaterr; Orthoklor; RCRA waste number U036; SD 5532; Shell SD 5532; Synklor; Tat chlor 4; Topichlor 20; Topiclor; Topiclor 20; Toxichlor; Velsicol 1068.

Structure:

cis   *trans*

Designations: CAS: 57-74-9; DOT: 2762; mf: $C_{10}H_6Cl_8$; fw: 409.78; RTECS: PB9800000.

Properties: Colorless to amber to yellowish-brown viscous liquid with an aromatic, slight pungent odor similar to chlorine. Mp: <25 °C; bp: 175 °C at 2 mmHg; ρ: 1.59-1.63 at 20/4 °C; fl p (kerosene solution): 55.6 °C, 107.2 °C (open cup); lel: 0.7% (kerosene solution); uel: 5% (kerosene solution); K_H: 4.8 x 10^{-5} atm·m^3/mol at 25 °C; log K_{oc}: 4.58-5.57; log K_{ow}: 6.00; S_o: miscible with cyclohexanone, deodorized kerosene, petroleum solvents, 2-propanol, trichloroethylene, aliphatic and aromatic solvents; S_w: 56 ppb at 25 °C; vap d: 16.75 g/L at 25 °C, 14.15 (air = 1); vp: 10^{-6} mmHg at 20 °C.

Soil properties and adsorption data

Soil	K_d (mL/g)	f_{oc} (%)	K_{oc} (mL/g)	pH
Sand	28	0.04	70,000	--
Sand	30	0.04	75,000	--
Silt	190	0.36	52,788	--

84

Soil	K_d (mL/g)	f_{oc} (%)	K_{oc} (mL/g)	pH
Silt	220	0.36	61,111	--

Source: Johnson-Logan et al., 1992.

Transformation Products

Soil: The actinomycete, *Nocardiopsis* sp., isolated from soil extensively degraded pure *cis*- and *trans*-chlordane to dichlorochlordene, oxychlordane, heptachlor, heptachlor *endo*-epoxide, chlordene chlorohydrin, and 3-hydroxy-*trans*-chlordane. Oxychlordane is slowly degraded to 1-hydroxy-2-chlorochlordene (Beeman and Matsumura, 1981). The reported half-life in soil is approximately 1 yr (Hartley and Kidd, 1987).

Chemical/Physical: In an alkaline medium or solvent, carrier, diluent or emulsifier having an alkaline reaction, chlorine will be released (Windholz et al., 1983). Technical grade chlordane passed over a 5% platinum catalyst at 200 °C resulted in the formation of tetrahydrodicyclopentadiene (Musoke et al., 1982).

Emits very toxic fumes of chlorides when heated to decomposition (Lewis, 1990).

Exposure Limits: NIOSH REL: IDLH 500 mg/m³; OSHA PEL: TWA 0.5 mg/m³; ACGIH TLV: TWA 0.5 mg/m³, STEL 2 mg/m³.

Symptoms of Exposure: Blurred vision, confusion, ataxia, delirium, cough, abdominal pain, nausea, vomiting, diarrhea; irritability, tremor, convulsions, anuria.

Formulation Types: Emulsifiable concentrate; wettable powder; granules; suspension concentrate; dustable powder; oil.

Toxicity: LC_{50} (96 hr) for rainbow trout 90 µg/L and bluegill sunfish 70 µg/L (Hartley and Kidd, 1987); EC_{50} (96 hr) for goldfish 0.5 mg/L; acute oral LD_{50} for rats 365-590 mg/kg (Hartley and Kidd, 1987), 343 mg/kg (RTECS, 1985).

Uses: Insecticide and fumigant.

cis-CHLORDANE

Synonyms: α-Chlordane; β-Chlordane; α-1,2,4,5,6,7,8,8-Octachloro-3a,4,7,7a-tetrahydro-4,7-methanoindan.

Structure:

Designations: CAS: 5103-74-2; DOT: 2762; mf: $C_{10}H_6Cl_8$; fw: 409.78; RTECS: PC0175000.

Properties: Solid. Mp: 107.0-108.8 °C; bp: 175 °C (isomeric mixture); H-t½: >197,000 yr at 25 °C and pH 7; K_H: 8.75 x 10^{-4} atm·m³/mol at 23 °C; log K_{oc}: 5.40-6.00; log K_{ow}: 5.93 (calculated); S_o: cod liver oil (89.6 g/L at 4 °C), *n*-octanol (56.6, 62.0, and 85.6 g/L at 4, 12, and 20 °C, respectively), miscible with aliphatic and aromatic solvents; S_w: 51 μg/L at 20-25 °C; vp: 8.3 x 10^{-5} mmHg at 23 °C (calculated).

Transformation Products
Photolytic: Irradiation of *cis*-chlordane by a 450-W high-pressure mercury lamp gave photo-*cis*-chlordane (Ivie et al., 1972).

Chemical/Physical: In an alkaline medium or solvent, carrier, diluent or emulsifier having an alkaline reaction, chlorine will be released (Windholz et al., 1983). Based on a reported hydrolysis half-life of >197,000 yr at 25 °C and pH 7 (Ellington et al., 1988), chemical hydrolysis is not expected to be an environmentally relevant fate process (Lyman et al., 1982). Emits very toxic fumes of chlorides when heated to decomposition (Lewis, 1990).

Symptoms of Exposure: Blurred vision; confusion; ataxia; delirium; cough; abdominal pain, nausea, vomiting, diarrhea; irritability; tremor, convulsions, anuria.

Formulation Types: See Chlordane.

Toxicity: See Chlordane.

Use: Insecticide.

trans-CHLORDANE

Synonyms: α-Chlordan; *cis*-Chlordan; α-Chlordane; α(*cis*)-Chlordane; γ-Chlordane; 1,2,4,5,6,7,8,8-Octachloro-3a,4,7,7a-tetrahydro-4,7-methanoindan.

Structure:

Designations: CAS: 5103-71-9; DOT: 2762; mf: $C_{10}H_6Cl_8$; fw: 409.78; RTECS: PB9705000.

Properties: Solid. Mp: 103.0-105.0 °C; bp: 175 °C (isomeric mixture); K_H: 1.34 x 10^{-3} atm·m^3/mol at 23 °C; log K_{oc}: 5.48, 6.00; log K_{ow}: 8.69, 9.65 (calculated); S_o: *n*-octanol (83.7, 112.9, and 142.1 g/L at 4, 12, and 20 °C, respectively), miscible with aliphatic and aromatic solvents; vp: 9.75 x 10^{-6} mmHg at 30 °C (estimated).

Transformation Products

Photolytic: Irradiation of *trans*-chlordane by a 450-W high-pressure mercury lamp gave photo-*trans*-chlordane (Ivie et al., 1972).

Chemical/Physical: In an alkaline medium or solvent, carrier, diluent or emulsifier having an alkaline reaction, chlorine will be released (Windholz et al., 1983). Emits very toxic fumes of chlorides when heated to decomposition (Lewis, 1990).

Symptoms of Exposure: Blurred vision; confusion; ataxia; delirium; cough; abdominal pain, nausea, vomiting, diarrhea; irritability; tremor, convulsions, anuria.

Formulation Types: See Chlordane.

Toxicity: See Chlordane.

Use: Insecticide.

CHLORDIMEFORM

Synonyms: Acaron; Bermat; C 8514; Carzol; CDM; Chlorfenamidine; *N'*-(4-Chloro-2-methylphenyl)-*N,N*-dimethylmethanimidamide; Chlorophenamidin; Chlorophenamidine; *N'*-(4-Chloro-*o*-tolyl)-*N,N*-dimethylformamidine; Chlorphenamidine; CIBA 8514; CIBA C8514; *N,N*-Dimethyl-*N'*-(2-methyl-4-chlorophenyl)formamidine; ENT 27335; ENT 27567; EP 333; Fundal; Fundal 500; Fundex; Galecron; NSC 190935; RS 141; Schering 36268; SN 36268; Spanon; Spanone.

Structure:

Designations: CAS: 6164-98-3; mf: $C_{10}H_{13}ClN_2$; fw: 196.68; RTECS: LQ4375000.

Properties: Colorless to buff-colored crystals. Mp: 32.0 °C; bp: 156-157 °C at 0.4 mmHg; ρ: 1.105 at 25/4 °C; K_H: 3.38 x 10^{-4} atm·m³/mol at 20 °C (approximate - calculated from water solubility and vapor pressure); log K_{oc}: 2.30 (calculated); log K_{ow}: 1.80, 2.89; S_o: very soluble in acetone, benzene, chloroform, ethyl acetate, *n*-hexane, methanol; S_w: 203 and 270 mg/L, at 10 and 20 °C, respectively; vp: 3.5 x 10^{-4} mmHg at 20 °C.

Transformation Products

Plant: Principal soil and/or plants metabolites are *p*-chloro-*o*-toluidine, *N'*-(4-chloro-*o*-tolyl)-*N*-methylformamidine (desmethylchlorphenamidine), and *N*-formyl-*p*-chloro-*o*-toluidine (Verschueren, 1983). *p*-Chloro-*o*-toluidine, *N'*-(4-chloro-*o*-tolyl)-*N*-methylformamidine (desmethylchlor phenamidine), and *N*-formyl-*p*-chloro-*o*-toluidine were identified in rice grains and straws at concentrations of 3-61, 0.2-1, 10-38, and 80-6,900, 10-180, 67-500 ppb, respectively (Iizuka and Masuda, 1979). Witkonton and Ercegovich (1972) studied the transformation of chlordimeform in six different fruit following foliar spray application. They found 4'-chloro-*o*-formotoluidide was the only major metabolite identified in apples, pears, cherries, plums, strawberries, and peaches.

Chemical/Physical: Reacts with acids forming soluble salts (Hartley and Kidd, 1987). Emits toxic fumes of nitrogen oxides and chlorine when heated to decomposition (Sax and Lewis, 1987).

Symptoms of Exposure: Eye and skin irritant.

Formulation Types: Emulsifiable concentrate; water soluble powder.

Toxicity: LC_{50} (96 hr) for rainbow trout 7.1 mg/L, bluegill sunfish 1.0 mg/L, and barbel 4.5 mg/L (Hartley and Kidd, 1987); acute oral LD_{50} for rats 340 mg/kg (Hartley and Kidd, 1987), 160 mg/kg (RTECS, 1985); 238 mg/kg (Windholz et al., 1983).

Uses: Broad-spectrum acaricide used to against adult mites, eggs, and larvae. Also used as an insecticide against cockroaches, cotton bollworm, and budworm and other pests.

CHLOROBENZILATE

Synonyms: Acar; Acaraben; Acaraben 4E; Akar; Akar 50; Akar 338; Benzilan; Benz-o-chlor; Chlorbenzilat; Chlorbenzilate; Chlorobenzylate; Compound 338; 4,4'-Dichlorobenzilate; 4,4'-Dichlorobenzilic acid ethyl ester; ENT 18596; **Ethyl 4-chloro-α-(4-chlorophenyl)-α-hydroxybenzene** acetate; Ethyl-4,4-dichloro-benzilate; Ethyl-*p,p'*-dichlorobenzilate; Ethyl-4,4'-dichlorodiphenyl glycollate; Ethyl-*p,p'*-dichlorodiphenyl glycollate; Ethyl-2-hydroxy-2,2-bis(4-chlorophenyl)acetate; Folbex; Folbex smoke-strips; G 338; G 23,992; Geigy 338; Kop-mite; NCI-C00408; NCI-C60413; RCRA waste number U038.

Structure:

$$Cl - \underset{\underset{COOCH_2CH_3}{|}}{\overset{\overset{OH}{|}}{C}} - Cl$$

Designations: CAS: 510-15-6; mf: $C_{16}H_{14}Cl_2O_3$; fw: 325.21; RTECS: DD2275000.

Properties: Colorless to yellow, viscous liquid or crystals. Mp: 36-37.5 °C; bp: 415 °C; ρ: 1.2816 at 20/4 °C; K_H: 3.85 x 10^{-8} atm·m^3/mol at 20 °C (approximate - calculated from water solubility and vapor pressure); log K_{oc}: 3.24-3.67; log K_{ow}: 4.58; S_o (kg/kg at 20 °C): acetone (1.0), *n*-hexane (0.6), methanol (1.0), methylene chloride (1.0), *n*-octanol (0.70), toluene (1.0); S_w: 10 mg/L at 20 °C; vp: 9.0 x 10^{-7} mmHg at 20 °C.

Soil properties and adsorption data

Soil	K_d (mL/g)	f_{oc} (%)	K_{oc} (mL/g)	pH
Loam	48.4	2.78	1,741	5.9
Sand	16.4	0.35	4,713	7.7
Sandy loam	20.5	0.75	2,719	7.9
Sandy loam	48.6	2.32	2,095	7.8

Source: U.S. Department of Agriculture, 1990.

Transformation Products

Biological: Rhodotorula gracilis, a yeast isolated from an insecticide-treated soil, degraded chlorobenzilate in a basal medium supplemented by sucrose. Metabolites identified by this decarboxylation process were 4,4'-dichlorobenzilic acid, 4,4'-dichlorobenzophenone, and carbon dioxide (Miyazaki et al., 1969, 1970).

Chemical/Physical: Emits toxic fumes of chlorine when heated to decomposition (Sax and Lewis, 1987).

Symptoms of Exposure: Irritation of eyes and skin.

Formulation Types: Emulsifiable concentrate; wettable powder; fumigant.

Toxicity: LC_{50} (96 hr) for rainbow trout 0.6 mg/L and bluegill sunfish 1.8 mg/L (Hartley and Kidd, 1987); acute oral LD_{50} for rats 2,784–3,880 mg/kg (Hartley and Kidd, 1987), 700 mg/kg (RTECS, 1985).

Uses: Nonsystemic pesticide and acaricide.

CHLOROPICRIN

Synonyms: Acquinite; Chlor-o-pic; Dolochlor; G 25; Larvacide 100; Microlysin; NA 1583; NCI-C00533; Nitrochloroform; Nitrotrichloromethane; Pic-clor; Picfume; Picride; Profume A; PS; S 1; **Trichloronitromethane**; Tri-clor; UN 1580.

Structure:

$$Cl_3CNO_2$$

Designations: CAS: 76-06-2; DOT: 1580; mf: CCl_3NO_2; fw: 164.38; RTECS: PB6300000.

Properties: Colorless, slightly oily liquid with a sharp, penetrating, tear gas-like odor. Mp: -64.5 °C and -69.2 °C (corrected); bp: 111.8 °C; ρ: 1.6558 at 20/4 °C, 1.6483 at 25/4 °C; fl p: detonates; K_H: 2.05 x 10^{-3} atm·m^3/mol; log K_{oc}: 0.82 (calculated); log K_{ow}: 1.03, 2.09; P-t$_{1/2}$: 20 d (simulated atmosphere), 3 d in aqueous solution (sunlight irradiation); S_o: soluble in acetic acid and acetone; miscible with benzene, carbon disulfide, carbon tetrachloride, ethyl ether, methanol, and ethanol; S_w: 2,270 mg/L at 25 °C; vp: 16.9, 23.8, and 33 mmHg at 20, 25, and 30 °C, respectively.

Transformation Products

Biological: Four *Pseudomonas* sp., including *Pseudomonas putida* (ATCC culture 29607) isolated from soil, degraded chlorpicrin by sequential reductive dechlorination. The proposed degradative pathway is chloropicrin → nitrodichloromethane → nitrochloromethane → nitromethane + small amounts of carbon dioxide. In addition, a highly water soluble substance tentatively identified as a peptide, was produced by a nonenzymatic mechanism (Castro et al., 1983).

Photolytic: Photodegrades under simulated atmospheric conditions to phosgene and nitrosyl chloride. Photolysis of nitrosyl chloride yields chlorine and nitrous oxide (Moilanen et al., 1978). When aqueous solution of chloropicrin (10^{-3} M) is exposed to artificial UV light (λ <300 nm), protons, carbon dioxide, hydrochloric and nitric acids are formed (Castro and Belser, 1981).

Chemical/Physical: Releases very toxic fumes of nitrogen oxides and chlorine when heated to decomposition (Sax and Lewis, 1987; Lewis, 1990). Reacts with alcoholic sodium sulfite solutions and ammonia to give methanetrisulfonic acid and guanidine, respectively (Sittig, 1985).

92

Exposure Limits: NIOSH REL: IDLH 4 ppm; OSHA PEL: TWA 0.1 ppm; ACGIH TLV: TWA 0.1 ppm, STEL 0.3 ppm.

Symptoms of Exposure: Eye and skin irritation, lacrimation; cough, pulmonary edema; nausea, vomiting.

Formulation Types: Fumigant (mixture containing methyl bromide, methyl isothiocyanate, 1,2-dichloropropane, and 1,3-dichloro-1-propene).

Toxicity: Acute oral LD_{50} for rats 250 mg/kg (RTECS, 1985).

Uses: Disinfecting cereals and grains; fumigant and soil insecticide; fungicide; rat exterminator.

CHLOROTHALONIL

Synonyms: Bravo; Bravo 6F; Bravo-W-75; Chloroalonil; DAC 2787; Daconil; Daconil 2787; Daconil 2787 flowable fungicide; Dacosoil; 1,3-Dicyanotetrachlorobenzene; Exotherm; Exotherm termil; Forturf; NCI-C00102; Nopcocide; Nopcocide N-96; Nopcocide N40D & N96; Sweep; TCIN; *m*-TCPN; Termil; **2,4,5,6-Tetrachloro-1,3-benzenedicarbonitrile**; 2,4,5,6-Tetrachloro-3-cyanobenzonitrile; Tetrachloroisophthalonitrile; *m*-Tetrachlorophthalonitrile; TPN.

Structure:

Designations: CAS: 1897-45-6; mf: $C_8Cl_4N_2$; fw: 265.89; RTECS: NT2600000.

Properties: Colorless to white, odorless crystals. Mp: 250-251 °C; bp: 350 °C; ρ: 1.8 (pure), 1.7 at 25/4 °C; K_H: 1.96 x 10^{-7} atm·m^3/mol at 25 °C; log K_{oc}: 2.76, 3.14; log K_{ow}: 2.64; S_o (g/kg at 25 °C): acetone (20), cyclohexanone (30), *N,N*-dimethylformamide (30), dimethylsulfoxide (20), kerosene (<10), xylene (80); S_w: 600 µg/L at 25 °C; vp: 9.75 x 10^{-6} mmHg at 40 °C.

Soil properties and adsorption data

Soil	K_d (mL/g)	f_{oc} (%)	K_{oc} (mL/g)	pH
Cohansey sand	25	2.55	980	--

Source: Reduker et al., 1988.

Transformation Products

Plant: Degrades in plants to 4-hydroxy-2,5,6-trichloroisophthalonitrile (Hartley and Kidd, 1987), 1,3-dicyano-4-hydroxy-2,5,6-trichlorobenzene, and 1,3-dicarbamoyl-2,4,5,6-tetrachlorobenzene (Rouchaud et al., 1988). No evidence of degradation products were reported in apple foliage 15 d after application (Gilbert, 1976).

Soil: Metabolites identified in soil were 1,3-dicyano-4-hydroxy-2,5,6-trichlorobenzene, 1,3-dicarbamoyl-2,4,5,6-tetrachlorobenzene, and 1-carbamoyl-3-cyano-4-hydroxy-2,5,6-trichlorobenzene (Rouchaud et al., 1988). The half-life was reported as 4.1 d (Gilbert, 1976) and 1.5-3 mo (Hartley and Kidd, 1987).

Chemical/Physical: Emits toxic fumes of nitrogen oxides, cyanides and chlorine when heated to decomposition (Sax and Lewis, 1987). Chlorothalonil is resistant to hydrolysis under acidic conditions. At pH 9, chlorothalonil (0.5 ppm) hydrolyzed to 4-hydroxy-2,5,6-trichloroisophthalonitrile and 3-cyano-2,4,5,6-tetrachloro-benzamide. Degradation followed first-order kinetics at a rate of 1.8% per day (Szalkowski and Stallard, 1977).

Exposure Limits: An experimental carcinogen.

Formulation Types: Wettable powder; suspension concentrate; fogging concentrate.

Toxicity: LC_{50} (estimated) for bluegill sunfish 380 μg/L, rainbow trout 250 μg/L, and channel catfish 430 μg/L (Hartley and Kidd, 1987); acute oral LD_{50} for albino rats >10,000 mg/kg (Verschueren, 1983).

Uses: Fungicide; bactericide; nematocide.

CHLOROXURON

Synonyms: C 1983; N'-(4-(4-Chlorophenoxy)phenyl)-N,N-dimethylurea; 3-(p-(p-Chlorophenoxy)phenyl)-1,1-dimethylurea; Chloroxifenidim; CIBA 1983; Norex; Tenoran 50W.

Structure:

Cl—⟨benzene ring⟩—O—⟨benzene ring⟩—NHCON(CH$_3$)$_2$

Designations: CAS: 1982-47-4; mf: $C_{15}H_{15}ClN_2O_2$; fw: 290.75; RTECS: UG1490000.

Properties: Odorless, colorless powder or white crystals. Mp: 151-152 °C; ρ: 1.34 at 20/4 °C; fl p: nonflammable; log K_{oc}: 3.12-3.27; log K_{ow}: 3.20; S_o (g/kg): acetone (44), methanol (35), methylene chloride (106), toluene (4); S_w: 3.7 mg/L at 20 °C; vp: 1.79 x 10^{-11} mmHg at 20 °C.

Soil properties and adsorption data

Soil	K_d (mL/g)	f_{oc} (%)	K_{oc} (mL/g)	pH	CEC (meq/100 g)
Great House SP	475	12.00	3,958	6.3	18.0
Loye clay loam	40	2.55	1,569	7.5	--
Rosemaunde sandy clay loam	70	1.76	3,977	6.7	14.0
Toll Farm HP	330	11.70	2,820	7.4	41.0
Transcosed silty clay loam	175	3.69	4,742	6.2	--
Uvrier sandy loam	14	0.75	1,857	7.8	--
Vétroz humus soil	110	8.29	1,327	--	--
Weed Res. sandy loam	120	1.93	6,218	7.1	11.0

Source: Geissbühler et al., 1963; Hance, 1965.

Transformation Products

Soil: Hartley and Kidd (1987) reported 4-(4-chlorophenoxy)aniline as a soil metabolite. Chloroxuron was degraded by microorganisms in humus soil and a sandy loam to form N'-(4-chlorophenoxy)phenyl-N-methylurea, N'-(4-chlorophenoxy)phenylurea, and (4-chlorophenoxy)aniline, and two minor unidentified compounds (Geissbühler et al., 1963a). Residual activity in soil is limited to approximately 4 mo (Hartley and Kidd, 1987).

Plant: In plants, chloroxuron is degraded to monomethylated and demethylated derivatives followed by decarboxylation forming 4-(4-chlorophenoxy)aniline (Humburg et al., 1989).

Photolytic: The UV irradiation of an aqueous solution of chloroxuron for 13 hr resulted in 90% decomposition of the herbicide. Products identified (% yield) were mono- (2.2%) and didemethylated (4.2%) products, and carbon dioxide (64%) (Plimmer, 1970).

Chemical/Physical: Hydrolyzes in strong acids and bases forming 4-(4-chloro-phenoxy)aniline (Hartley and Kidd, 1987). Emits toxic fumes of nitrogen oxides, cyanides, and chlorine when heated to decomposition (Sax and Lewis, 1987).

Symptoms of Exposure: Ingestion may cause depression, locomotion, hyperpnea, gasping, coma, death.

Formulation Types: Wettable powder.

Toxicity: LC_{50} (96 hr) for rainbow trout >100 mg/L, bluegill sunfish 28 mg/L, and crucian carp >150 mg/L (Hartley and Kidd, 1987); LC_{50} (48 hr) for bluegill sunfish 25.0 mg/L (Verschueren, 1983); acute oral LD_{50} for male and female rats is 3,700 and 5,400 mg/kg, respectively (Hartley and Kidd, 1987).

Uses: Postemergence herbicide used to control most annual grasses and broadleaf weeds.

CHLORPROPHAM

Synonyms: Beet-Kleen; Bud-nip; Chlor-IFC; Chlor-IPC; Chloro-IFK; Chloro-IPC; **(3-Chlorophenyl)carbamic acid 1-methylethyl ester;** Chloropropham; CICP; CI-IPC; CIPC; Ebanil; ENT 18060; Fasco Wy-hoe; Furloe; Furloe 4EC; Isopropyl *m*-chlorocarbanilate; Isopropyl-3-chlorocarbanilate; Isopropyl-3-chlorophenyl-carbamate; Isopropyl-*N*-(3-chlorophenyl)carbamate; Isopropyl-*N*-(*m*-chlorophen-yl)carbamate; *O*-Isopropyl-*N*-(3-chlorophenyl)carbamate; Liro CIPC; Metoxon; 1-Methylethyl-3-chlorophenylcarbamate; Nexoval; Prevenol; Prevenol 56; Preventol; Preventol 56; Preweed; Sprout nip; Sprout-nip EC; Spud-nic; Spud-nie; Stopgerme-S; Taterpex; Triherbicide CIPC; Unicrop CIPC; Y 3.

Structure:

$$NHCOOCH(CH_3)_2$$

Designations: CAS: 101-21-3; mf: $C_{10}H_{12}ClNO_2$; fw: 213.67; RTECS: FD8050000.

Properties: Colorless to pale brown crystals with a faint characteristic odor. Mp: 40.7-41.4 °C; bp: 247 °C (decomposes); ρ: 1.180 at 30/4 °C; fl p: high; H-t$\frac{1}{2}$: calculated to be >10,000 d assuming pseudo first-order kinetics; K_H: 2.1 x 10^{-8} atm·m^3/mol at 20-25 °C (approximate - calculated from water solubility and vapor pressure); log K_{oc}: 2.77, 2.91; P-t$\frac{1}{2}$: 121 d at pH 5-9 (calculated); S_o: very soluble in ethanol, 2-propanol, and ketones but miscible with acetone and carbon disulfide; S_w: 89 mg/L at 25 °C; vp: 1 x 10^{-5} mmHg at 25 °C (estimated).

Soil properties and adsorption data

Soil	K_d (mL/g)	f_{oc} (%)	K_{oc} (mL/g)	pH
Soil (0-6 inches)	37	4.9	755	--
Soil (6-12 inches)	26	2.7	962	--
Soil (12-18 inches)	12	1.3	923	--
Soil (18-24 inches)	5	0.8	625	--

Source: Roberts and Wilson, 1965.

Transformation Products

Soil: Hydrolyzes in soil forming 3-chloroaniline (Hartley and Kidd, 1987; Rajagopal et al., 1989). In soil, *Pseudomonas striata* Chester, a *Flavobacterium* sp.,

98

an *Agrobacterium* sp., and an *Achromobacter* sp. readily degraded chlorpropham to 3-chloroaniline. Subsequent degradation by enzymatic hydrolysis yielded carbon dioxide and chloride ions (Rajagopal et al., 1989). Hydrolysis products that may form in soil and in microbial cultures include *N*-phenyl-3-chlorocarbamic acid, 3-chloroaniline, 2-amino-4-chlorophenol, monoisopropyl carbonate, 2-propanol, carbon dioxide, and condensation products (Rajagopal et al., 1989). The reported half-lives in soil at 15 and 29 °C are 65 and 30 d, respectively (Hartley and Kidd, 1987).

Plant: Chlorpropham is rapidly metabolized in plants (Ashton and Monaco, 1991). Metabolites identified in soybean plants include isopropyl-*N*-4-hydroxy-3-chlorophenylcarbamate, 1-hydroxy-2-propyl-3'-chlorocarbanilate, and isopropyl-*N*-5-chloro-2-hydroxyphenylcarbamate (Humburg et al., 1989). Isopropyl-*N*-4-hydroxy-3-chlorophenylcarbamate was the only metabolite identified in cucumber plants (Humburg et al., 1989).

Photolytic: Though no products were identified, the photodegradation rate of chlorpropham in aqueous solution was enhanced in the presence of a surfactant (TMN-10) (Tanaka et al., 1981). In a later study, Tanaka et al. (1985) studied the photolysis of chlorpropham (50 mg/L) in aqueous solution using UV light ($\lambda = 300$ nm) or sunlight. After 10 hr of irradiation, 40% degraded yielding <1% of a hydroxylated biphenyl product (2'-hydroxy-3,4'-biphenylcarbamic acid diisopropyl ester) and hydrogen chloride. The biphenyl product probably formed from the coupling of photoexcited chlorpropham with the intermediate compound isopropyl 3-hydroxycarbanilate (Tanaka et al., 1985).

Chemical/Physical: Emits toxic fumes of phosgene when heated to decomposition (Sax and Lewis, 1987). In a 0.50 N sodium hydroxide solution at 20 °C, chlorpropham was hydrolyzed to aniline derivatives. The half-life of this reaction was 3.5 d (El-Dib and Aly, 1976).

Exposure Limits: An experimental carcinogen and neoplastigen.

Formulation Types: Fogging concentrate; emulsifiable concentrate (3 and 4 lb/gal); granules (10 and 20%); dustable powder.

Toxicity: LC_{50} (48 hr) for bass 10 mg/L (Hartley and Kidd, 1987) and bluegill sunfish 8 mg/L (Verschueren, 1983); acute oral LD_{50} of technical chlorpropham for rats is 3,800 mg/kg (Ashton and Monaco, 1991), pure chlorpropham 1,200 mg/kg (RTECS, 1985).

Uses: Preemergent and postemergent herbicide used to regulate plant growth and control of weeds in carrot, onion, garlic, and other crops.

CHLORPYRIFOS

Synonyms: Brodan; Chlorpyrifos-ethyl; Detmol U.A.; *O,O*-Diethyl-*O*-3,5,6-tri-chloro-2-pyridyl phosphorothioate; Dowco-179; Dursban; Dursban F; ENT 27311; Eradex; Lorsban; NA 2783; OMS 971; **Phosphorothionic acid *O,O*-diethyl *O*-(3,5,6-trichloro-2-pyridyl)ester;** Pyrinex.

Structure:

Designations: CAS: 2921-88-2; DOT: 2783; mf: $C_9H_{11}Cl_3NO_3PS$; fw: 350.59; RTECS: TF6300000.

Properties: White to amber granular crystals with a mild mercaptan-like odor. Mp: 41.5-43.5 °C; bp: begins to decompose at 160 °C; ρ: 1.398 at 43.5/4 °C; H-t$_{1/2}$: 22.8 d (pH 8.1), 35.3 d (pH 6.9), 62.7 d (pH 4.7); K_H: 4.16 x 10^{-6} atm·m^3/mol at 25 °C; log K_{oc}: 3.77-4.13; log K_{ow}: 3.31-5.267; P-t$_{1/2}$: 52.45 hr (absorbance λ = 229.5 nm, concentration on glass plates = 6.7 μg/cm^2), 11.0, 12.2, and 7.8 d in buffered, distilled water (9 x 10^{-7} M, 25 °C) at pH 5.0, 6.9, and 8.0, respectively; S_o (kg/kg): acetone (6.5), benzene (7.9), chloroform (6.3), 2,2,4-trimethylpentane (isooctane) (79), methanol (43), ethanol, *n*-propanol, xylene, and many other organic solvents; S_w: 450, 730, and 1,300 μg/L at 10, 20, and 30 °C, respectively; vp: 1.87 x 10^{-5} mmHg at 25 °C.

Soil properties and adsorption data

Soil	K_d (mL/g)	f_{oc} (%)	K_{oc} (mL/g)	pH	CEC (meq/100 g)
Catlin	99.70	2.01	5,000	6.20	--
Clarion soil	161.81	2.64	6,119	5.00	21.02
Commerce	49.50	0.68	7,300	6.70	--
Harps soil	397.19	3.80	10,450	7.30	37.83
Kanuma high clay	13.40	1.35	995	5.70	--
Sarpy fine sandy loam	28.38	0.51	5,560	7.30	5.71
Thurman loamy fine sand	46.88	1.07	4,395	6.83	6.10
Tracy	65.60	1.12	5,900	6.20	--
Tsukuba clay loam	116.20	4.24	2,740	6.50	--

Source: Felsot and Dahm, 1979; Kanazawa, 1989; McCall et al., 1981.

Transformation Products

Soil: Hydrolyzes in soil to 3,5,6-trichloro-2-pyridinol (Somasundaram et al., 1991). The half-lives in a silt loam and clay loam were 12 and 4 wk, respectively (Getzin, 1981). In another study, Getzin (1981a) reported the hydrolysis half-lives in a Sultan silt loam at 5, 15, 25, 35, and 45 °C were >20, >20, 8, 3, and 1 d, respectively. The only breakdown product identified was the hydrolysis product 3,5,6-trichloro-2-pyridinol. Degrades in soil forming oxychlorpyrifos, 3,5,6-trichloro-2-pyridinol (hydrolysis product), carbon dioxide, and water-soluble products (Racke et al., 1988). Leoni et al. (1981) reported that the major degradation product of chlorpyrifos in soil is 3,5,6-trichloro-2-pyridinol. The reported half-life in soil is 60-100 d (Hartley and Kidd, 1987).

Photolytic: Photolysis of chlorpyrifos in water yielded 3,5,6-trichloro-2-pyridinol. Continued photolysis yielded chloride ions, carbon dioxide, ammonia, and possibly polyhydroxychloropyridines (Dilling et al., 1984).

Surface Water: In an estuary, the half-life of chlorpyrifos was 24 d (Schimmel et al., 1983).

Photolytic: The following photolytic half-lives in water at north 40° latitude were reported: 31 d during midsummer at a depth of 10^{-3} cm; 345 d during midwinter at a depth of 10^{-3} cm; 43 d at a depth of one meter; 2.7 yr during midsummer at a depth of one meter in river water (Dilling et al., 1984). The combined photolysis-hydrolysis products identified in buffered, distilled water were *O*-ethyl *O*-(3,5,6-trichloro-2-pyridyl)phosphorothioate, 3,5,6-trichloro-2-pyridinol, and five radio-active unknowns (Meikle et al., 1983).

Chemical/Physical: Hydrolyzes in water forming 3,5,6-trichloro-2-pyridinol, *O*-ethyl *O*-hydrogen-*O*-(3,5,6-trichloro-2-pyridyl)phosphorothioate, and *O,O*-dihydrogen-*O*-(3,5,6-trichloro-2-pyridyl)phosphorothioate (Meikle and Youngson, 1978). The hydrolysis half-life in three different natural waters was about 48 d at 25 °C (Macalady and Wolfe, 1985). Freed et al. (1979) reported hydrolysis half-lives of 120 and 53 d at pH 6.1 and pH 7.4, respectively, at 25 °C. At 25 °C and a pH range of 1-7, the hydrolysis half-life was about 78 d (Macalady and Wolfe, 1983). However, the alkaline hydrolysis rate of chlorpyrifos in the sediment-sorbed phase were found to be considerably slower (Macalady and Wolfe, 1985). Over the pH range of 9-13, 3,5,6-trichloro-2-pyridinol and *O,O*-diethyl phosphorothioic acid formed as major hydrolysis products (Macalady and Wolfe, 1983). Emits very toxic fumes of chlorides and oxides of nitrogen, phosphorus, and sulfur when heated to decomposition (Lewis, 1990).

Exposure Limits: OSHA PEL: TWA 0.2 mg/m^3; ACGIH TLV: TWA 0.2 mg/m^3, STEL 0.6 mg/m^3.

Formulation Types: Emulsifiable concentrate; wettable powder; granules; suspension concentrate; dustable powder; pellets.

Toxicity: LC_{50} (96 hr) for rainbow trout 3 µg/L (Hartley and Kidd, 1987), estuarine mysid 0.035 µg/L, sheapshead minnow 136 µg/L, longnose killifish 4.1 µg/L, Atlantic silverside 1.7 µg/L, striped mullet 5.4 µg/L (Schimmel et al., 1983), and bluegill sunfish 2.6 µg/L (Verschueren, 1983); acute oral LD_{50} for male and female rats is 163 and 135 mg/kg, respectively (Verschueren, 1983), 82 mg/kg (RTECS, 1985).

Uses: Insecticide used to control insects on a wide variety of crops including fruits, vegetables, ornamentals and forestry.

CHLORSULFURON

Synonyms: 2-Chloro-N-(((4-methoxy-6-methyl-1,3,5-triazin-2-yl)amino)-carbonyl)benzenesulfonamide; 1-((o-Chlorophenyl)sulfonyl)-3-(4-methoxy-6-methyl-s-triazin-2-yl) urea; DPX 4189; Finesse; Glean; Glean 20DF; Telar.

Structure:

Designations: CAS: 64902-72-3; mf: $C_{12}H_{12}ClN_5O_4S$; fw: 357.80; RTECS: YS6640000.

Properties: Odorless, colorless to white crystals. Mp: 174-178 °C; bp: decomposes at 192 °C; fl p: nonflammable; pK_a: 3.6 at 25 °C and pH 7; H-t$\frac{1}{2}$: 4-8 wk at 20 °C and pH 5.7-7.0; K_H: 3.55 x 10^{-16} atm·m^3/mol at 25 °C (approximate - calculated from water solubility and vapor pressure); log K_{oc}: 1.02 (Flanagan silt loam); log K_{ow}: -1.0; pK_a: <7; S_o (g/L at 22 °C): acetone (57), n-hexane (0.01), methanol (14), methylene chloride (102), toluene (3); S_w at 25 °C: 60 mg/L at pH 5, 7,000 mg/L at pH 7; vp: 2.33 x 10^{-11} mmHg at 25 °C.

Transformation Products

Soil: Degrades in soil via hydrolysis followed by microbial degradation forming low molecular weight, inactive compounds (t$\frac{1}{2}$ ≈ 4-6 wk) (Hartley and Kidd, 1987; Cremlyn, 1991). Microorganisms capable of degrading chlorsulfuron are *Aspergillis niger*, *Streptomyces griseolus*, and *Penicillium* sp. (Humburg et al., 1989).

Plant: Chlorsulfuron is metabolized by plants to hydroxylated, nonphytotoxic compounds including 2-chloro-N-(((4-methoxy-6-methyl-1,3,5-triazin-2-yl)-amino)carbonyl)benzenesulfonamide (Duke et al., 1991).

Photolytic: The reported photolysis half-lives of chlorsulfuron in distilled water, methanol, and natural creek water at λ >290 nm were 18, 92, and 18 hr, respectively. In all cases, 2-chlorobenzene sulfonamide, 2-methoxy-4-methyl-6-amino-1,3,5-triazine, and trace amounts of the tentatively identified compound nitroso-2-chlorophenyl-sulfone formed as photoproducts (Herrmann et al., 1985).

Symptoms of Exposure: May cause eye, nose, throat, and skin irritation.

Formulation Types: Water dispersible granules; dry flowable formulation (60%).

Toxicity: LC_{50} (96 hr) for rainbow trout, carp, and bluegill sunfish >250 mg/L (Hartley and Kidd, 1987); acute oral LD_{50} for male and female rats is 5,545 and 6,293 mg/kg, respectively (Hartley and Kidd, 1987).

Uses: Herbicide used to control broadleaf weeds and some annual grass weeds.

CHLORTHAL-DIMETHYL

Synonyms: Chlorothal; DAC 893; Dacthal; Dacthalor; DCPA; Dimethyl 2,3,5,6-tetrachloro-1,4-benzenedicarboxylate; Dimethyl tetrachloroterephthalate; Dimethyl 2,3,5,6-tetrachloroterephthalate; Fatal; **2,3,5,6-Tetrachloro-1,4-benzenedicarboxylic acid dimethyl ester;** Tetrachloroterephthalic acid dimethyl ester; 2,3,5,6-Tetrachloroterephthalic acid dimethyl ester.

Structure:

Designations: CAS: 1861-32-1; mf: $C_{10}H_6Cl_4O_4$; fw: 331.99; RTECS: WZ1500000.

Properties: Colorless to beige crystals with a slight aromatic odor. Mp: 156 °C; bp: decomposes at approximately 360-370 °C; ρ: 1.70 at 20/4 °C; fl p: nonflammable; K_H: 2.2 x 10^{-6} atm·m^3/mol at 25 °C (approximate - calculated from water solubility and vapor pressure); log K_{oc}: 3.81 (calculated); log K_{ow}: 4.87 (calculated); S_o (wt%): acetone (10), benzene (25), carbon tetrachloride (7), 1,4-dioxane (12), toluene (17), xylene (14); S_w: ≈ 500 μg/L at 25 °C; vp: 2.5 x 10^{-6} mmHg at 25 °C.

Transformation Products
Soil: Bartha and Pramer (1967) reported that DCPA was degraded soil microorganisms via cleavage of the herbicide molecule into propionic acid and 3,4-dichloroaniline. The acid was mineralized to carbon dioxide and water and two molecules of 3,4-dichloroaniline were condensed to form 3,3'4,4'-tetrachloroazobenzene. Metabolites identified in soil and turfgrass thatch are monomethyl tetrachloroterephthalate and 2,3,5,6-tetrachloroterephthalic acid (Hartley and Kidd, 1987; Krause and Niemczyk, 1990). Residual activity in soil and the half-life in soil were reported to be approximately 3 mo (Hartley and Kidd, 1987; Worthing and Hance, 1991).

Formulation Types: Granules; wettable powder (75%); suspension concentrate.

Toxicity: Nontoxic to fish (Hartley and Kidd, 1987); acute oral LD_{50} for rats >3,000 mg/kg (Hartley and Kidd, 1987).

Uses: Selective, nonsystemic, preemergent herbicide to control most annual grasses and many broadleaf weeds.

CROTOXYPHOS

Synonyms: Ciodrin; Ciovap; Cyodrin; Cypona E.C.; **3-((Dimethoxyphosphinyl)-oxy)-2-butenoic acid 1-phenylethyl ester;** O,O-Dimethyl O-(1-methyl-2-carboxy-α-phenylethyl)vinyl phosphate; Decrotox; Dimethyl-cis-1-methyl-2-(1-phenylethoxycarbonyl)vinyl phosphate; Dimethyl phosphate of α-methylbenzyl 3-hydroxy-cis-crotonate; Duo-kill; Duravos; ENT 24717; 1-Methylbenzyl-3-(di-methoxyphosphinyloxo)isocrotonate; α-Methyl benzyl-3-(dimethoxyphosphinyl-oxy)-cis-crotonate; α-Methylbenzyl 3-hydroxycrotonate dimethyl phosphate; Pantozol 1; cis-2-(1-Phenylethoxy)carbonyl-1-methylvinyl dimethyl phosphate; SD 4294; Shell SD 4294; Volfazol.

Structure:

Designations: CAS: 7700-17-6; mf: $C_{14}H_{19}O_6P$; fw: 314.28; RTECS: GQ5075000.

Properties: Pale straw-colored liquid. Mp: <25 °C; bp: 135 °C at 0.03 mmHg; ρ: 1.19 at 25/4 °C; K_H: 6.2 x 10^{-9} atm·m^3/mol at 20-25 °C (approximate - calculated from water solubility and vapor pressure); log K_{oc}: 2.23; log K_{ow}: 1.28 (calculated); S_o: soluble in acetone, chloroform, ethanol, chlorinated hydrocarbons; S_w: 1 g/L at 25 °C; vp: 1.4 x 10^{-5} mmHg at 20 °C.

Transformation Products
Chemical/Physical: Emits toxic fumes of phosphorus oxides when heated to decomposition.

Toxicity: LC$_{50}$ (96 hr) for bluegill sunfish 250 μg/L, largemouth bass 1.10 mg/L, rainbow trout 55 μg/L, and channel catfish 2.50 mg/L (Verschueren, 1983); acute oral LD$_{50}$ for rats approximately 125 mg/kg (Verschueren, 1983), 74 mg/kg (RTECS, 1985).

Uses: Insecticide.

CYANAZINE

Synonyms: Bladex; Bladex 90DF; Bladex 4L; Bladex 80WP; 2-Chloro-4-(1-cyano-1-methylethylamino)-6-ethylamino-1,3,5-triazine; 2-(4-Chloro-6-ethyl-amino-*s*-triazine-2-ylamino)-2-methylpropionitrile; 2-((4-Chloro-6-(ethyl-amino)-*s*-triazin-2-yl)amino)-2-methylpropionitrile; **2-((4-Chloro-6-(ethyl-amino)-1,3,5-triazin-2-yl)amino)-2-methylpropanenitrile**; DW 3418; Fortrol; Payze; SD 15418; WL 19805.

Structure:

$$CH_3CH_2N \quad N \quad NHC(CH_3)_2$$
$$N \quad N \quad CN$$
$$Cl$$

Designations: CAS: 21725-46-2; mf: $C_9H_{13}ClN_6$; fw: 240.70; RTECS: UG1490000.

Properties: Colorless to white crystals. Mp: 167.5-169 °C; ρ: 0.35 g/mL (fluffed technical material), 0.45 g/mL (packed technical material); fl p: nonflammable; pK_a: 0.63, 1.1; K_H: 2.78 x 10^{-2} atm·m³/mol at 20-25 °C (approximate - calculated from water solubility and vapor pressure); log K_{oc}: 1.58-2.63; log K_{ow}: 1.80, 2.24; S_o (g/L at 25 °C): acetone (195), benzene (15), carbon tetrachloride (<10), chloro-benzene (<100), chloroform (210), ethanol (45), *n*-hexane (15), methylcyclohexa-none (210), methylene chloride (145), xylene (<100); S_w: 171 mg/L at 25 °C; vp: 1.6 x 10^{-9} mmHg at 20 °C.

Soil properties and adsorption data

Soil	K_d (mL/g)	f_{oc} (%)	K_{oc} (mL/g)	pH	CEC (meq/100 g)
Hickory Hill silt	5.98	3.27	183	--	--
Monona clay loam	4.30	1.67	257	6.5	21.2
Rhinebeck silty clay loam	1.20	3.13	38	6.7	--
Valentine loamy fine sand	3.40	0.80	425	6.6	10.1
Valois silty loam	1.20	1.64	73	5.9	--

Source: Brown and Flagg, 1981; Gamerdinger et al., 1991; Majka and Lavy, 1977.

Transformation Products
Soil/Groundwater: In sandy loam soils (f_{oc} = 0.01), the half-life is 12-15 d. However, in silt loam (f_{oc} = 0.028) and clay loam (f_{oc} = 0.03) soils, the reported half-life is 20-25 d (Humburg et al., 1989). The half-life of cyanazine is longer in

alkaline soils (pH >7.5) than in acidic soils (pH <5.5) (Humburg et al., 1989). Cyanazine amide was identified as a metabolite in groundwater in corn fields (Muir and Baker, 1976).

Plant: In plants, cyanazine is degraded by elimination of the ethyl group, hydration of the cyano group, and the removal and replacement of the chlorine atom by a hydroxyl group (Humburg et al., 1989).

Chemical/Physical: In laboratory tests, the nitrile group was hydrolyzed to the corresponding carboxylic acid. The rate of hydrolysis is faster under higher temperatures and low pHs (Grayson, 1980). The chlorine atom may be replaced by a hydroxyl group forming 2((4-hydroxy-6-(ethylamino)-1,3,5-triazin-2-yl)-amino)-2-methylpropanenitrile (Hartley and Kidd, 1987).

Formulation Types: Wettable powder; granules; suspension concentrate.

Toxicity: LC_{50} (48 hr) for sheepshead minnow 18 mg/L and harlequin fish 10 mg/L (Worthing and Hance, 1991); acute oral LD_{50} for rats 182-334 mg/kg (Hartley and Kidd, 1987), 149 mg/kg (RTECS, 1985).

Uses: Herbicide used for control of annual grasses and broadleaf weeds in cereals, cotton, maize, onions, peanuts, peas, potatoes, soybeans, sugar cane, and wheat fallow.

CYCLOATE

Synonyms: *S*-Ethyl cyclohexylethylcarbamothioate; *S*-Ethyl *N*-ethyl *N*-cyclo-hexylthiolcarbamate; Eurex; Hexylthiocarbam; R 2063; Ro-neet; Ronit.

Structure:

$$NCOSCH_2CH_3$$
$$CH_2CH_3$$

Designations: CAS: 1134-23-2; mf: $C_{11}H_{21}NOS$; fw: 215.37; RTECS: GU7200000.

Properties: Colorless, clear liquid with an aromatic odor. Mp: 11.5 °C; bp: 145-146 °C at 10 mmHg; ρ: 1.0156 at 30/4 °C; fl p: 139 °C (open cup); K_H: 2.4 x 10^{-5} atm·m^3/mol at 20 °C (approximate - calculated from water solubility and vapor pressure); log K_{oc}: 2.54; log K_{ow}: 4.11; S_o at 20 °C: miscible with kerosene, methyl isobutyl ketone; xylene; S_w: 75 mg/L at 20 °C; vap d: 8.80 g/L at 25 °C, 7.46 (air = 1); vp: 6.22 x 10^{-3} mmHg at 20 °C.

Transformation Products
Soil: Though no products were identified, the reported half-life in soil is approximately 4-8 wk (Hartley and Kidd, 1987).

Plant: Cycloate is rapidly metabolized in sugarbeets to carbon dioxide, ethylcyclohexylamine, sugars, amino acids, and other natural constituents (Humburg et al., 1989).

Formulation Types: Granules (10%), emulsifiable concentrate (6 lb/gal).

Toxicity: LC$_{50}$ (96 hr) for rainbow trout 4.5 mg/L (Hartley and Kidd, 1987), for mosquito fish 10 ppm (Humburg et al., 1989); acute oral LD$_{50}$ of technical cycloate for male and female rats is 2,000-3,190 and 3,160-4,100 mg/kg, respectively (Hartley and Kidd, 1987), 1,678 mg/kg (RTECS, 1985).

Uses: Herbicide used to control several broadleaf weeds and many annual grasses in sugar beets, table beets, and spinach.

CYFLUTHRIN

Synonyms: Bay FCR 1272; Baythroid; Baythroid H; **Cyano(4-fluoro-3-phenoxy-phenyl)methyl 3-(2,2-dichloroethenyl)-2,2-dimethylcyclopropanecarboxylate.**

Structure:

Designations: CAS: 68359-37-5; mf: $C_{22}H_{18}Cl_2FNO_3$; fw: 434.30; RTECS: GZ1253000.

Properties: Yellow oil or paste. Mp: 60 °C; ρ: 1.27-1.28 (supercooled at 20 °C); K_H: 9.41 x 10^{-3} atm·m³/mol at 20-25 °C (approximate - calculated from water solubility and vapor pressure); log K_{oc}: 5.13 (calculated); log K_{ow}: 5.91, 6.00; S_o: soluble in 2-propanol, methylene chloride, toluene; S_w: 2.0 μg/L at 25 °C; vp: 3.3 x 10^{-5} mmHg at 20 °C.

Formulation Types: Emulsifiable concentrate; wettable powder; granules; water-oil emulsion.

Toxicity: LC_{50} (96 hr) for rainbow trout 0.6 μg/L, golden orfe 3.2 μg/L, bluegill sunfish 1.5 μg/L, and carp 22 μg/L (Hartley and Kidd, 1987); acute oral LD_{50} for male rats 590 mg/kg (Hartley and Kidd, 1987), 251 mg/kg (RTECS, 1985).

Uses: Insecticide.

CYPERMETHRIN

Synonyms: Ammo; Ardap; Avicade; Barricade; CCN52; (+)-α-Cyano-3-phenoxy-benzyl 2,2-dimethyl-3-(2,2-dichlorovinyl)cyclopropane carboxylate; **Cyano(3-phenoxyphenyl)methyl 3-(2,2-dichloroethenyl)-2,2-dimethylcyclopropane-carboxylate**; Cymbush; Cyperkill; FMC 30980; FMC 45497; FMC 45806; Imperator; JF 5705F; Kafil super; NRDC 149; NRDC 160; NRDC 166; PP 383; Ripcord; Siperin; Stockade; WL 43467.

Structure:

Designations: CAS: 52315-07-8; mf: $C_{22}H_{19}Cl_2NO_3$; fw: 416.30; RTECS: GZ1250000.

Properties: Viscous, yellowish-brown mass or liquid. Mp: 60-80 °C (technical grade); K_H: 1.96 x 10^{-7} atm·m^3/mol at 20-25 °C (approximate - calculated from water solubility and vapor pressure); log K_{oc}: 4.00-4.53; log K_{ow}: 6.60; P-t$_{1/2}$: 179.82 hr (*cis*) (absorbance λ = 227.5 nm, concentration on glass plates = 6.7 μg/cm^2); S_o (g/L): acetone (>450), ethanol (337), *n*-hexane (103); S_w: 4.0 μg/L at 25 °C; vp: 1.43 x 10^{-9} mmHg at 20 °C (extrapolated).

Soil properties and adsorption data

Soil	K_d (mL/g)	f_{oc} (%)	K_{oc} (mL/g)	pH
Loam	2	3.02	66	7.1
Loamy sand	20	1.22	1,639	5.4
Sandy loam	70	1.97	3,553	6.5
Sandy loam	260	1.97	13,198	6.5
Silty loam	4	1.16	345	5.6

Source: U.S. Department of Agriculture, 1990.

Transformation Products
Soil: The major soil metabolite was reported to be 3-phenoxybenzoic acid (Hartley and Kidd, 1987).

Formulation Types: Emulsifiable concentrate; wettable powder; granules.

Toxicity: LC_{50} (96 hr) for brown trout 2.0–2.8 µg/L (Hartley and Kidd, 1987); acute oral LD_{50} for rats 200–800 mg/kg (Hartley and Kidd, 1987).

Uses: Insecticide.

CYROMAZINE

Synonyms: *N*-Cyclopropyl-1,3,5-triazine-2,4,6-triamine; 2-Cyclopropylamino-4,6-diamino-*s*-triazine; CGA 72662; Vetrazine.

Structure:

Designations: CAS: 66215-27-8; DOT: 2763; mf: $C_6H_{10}N_6$; fw: 166.19.

Properties: Colorless crystals. Mp: 219-222 °C; ρ: 1.35 at 20/4 °C; K_H: 6.16 x 10^{-14} atm·m^3/mol at 25 °C (approximate - calculated from water solubility and vapor pressure); log K_{oc}: 3.07 (calculated); log K_{ow}: -0.155; pK$_a$: >7; S$_o$: methanol (17 g/L); S$_w$: 11 g/L at 25 °C (pH 7.5); vp: 3.3 x 10^{-9} mmHg at 25 °C.

Transformation Products
Chemical/Physical: Probably reacts with mineral acids (e.g., hydrochloric acid, sulfuric acid) forming water-soluble salts.

Formulation Types: Water soluble granules or powder; wettable powder.

Toxicity: LC$_{50}$ (96 hr) for rainbow trout and carp >100 mg/L, and bluegill sunfish >90 mg/L (Hartley and Kidd, 1987); acute oral LD$_{50}$ for rats 3,387 mg/kg (Hartley and Kidd, 1987).

Uses: Insect growth regulator.

2,4-D

Synonyms: Agrotect; Agroxone; Amidox; Amoxone; Aqua-kleen; BH 2,4-D; Brush-rhap; B-Selektonon; Chipco turf herbicide 'D'; Chloroxone; Crop rider; Crotilin; D 50; 2,4-D acid; Dacamine; Debroussaillant 600; Decamine; Ded-weed; Ded-weed LV-69; Desormone; Dichlorophenoxyacetic acid; **(2,4-Dichlorophen-oxy)acetic acid**; Dicopur; Dicotox; Dinoxol; DMA-4; Dormone; Emulsamine BK; Emulsamine E-3; ENT 8538; Envert 171; Envert DT; Esteron; Esteron 76 BE; Esteron 44 weed killer; Esteron 99; Esteron 99 concentrate; Esteron brush killer; Esterone 4; Estone; Farmco; Fernesta; Fernimine; Fernoxone; Ferxone; Foredex 75; Formula 40; Hedonal; Herbidal; Ipaner; Krotiline; Lawn-keep; Macrondray; Miracle; Monosan; Moxone; NA 2765; Netagrone; Netagrone 600; NSC 423; Pennamine; Pennamine D; Phenox; Pielik; Planotox; Plantgard; RCRA waste number U240; Rhodia; Salvo; Spritz-hormin/2,4-D; Spritz-hormit/2,4-D; Super D weedone; Transamine; Tributon; Trinoxol; U 46; U 46 D; U 5043; U 46DP; Vergemaster; Verton; Verton D; Verton 2D; Vertron 2D; Vidon 638; Visko-rhap; Visko-rhap drift herbicides; Visko-rhap low volatile 4L; Weedar; Weddar-64; Weddatul; Weed-b-gon; Weedez wonder bar; Weedone; Weedone LV4; Weed-rhap; Weed tox; Weedtrol.

Structure:

Designations: CAS: 94-75-7; DOT: 2765; mf: $C_8H_6Cl_2O_3$; fw: 221.04; RTECS: AG6825000.

Properties: White to pale yellow prismatic crystals with a faint phenolic-like odor. Mp: 140-141 °C (free acid), 179-180 °C (ammonium salt), 157-159 °C (methyl-ammonium salt), 85-87 °C (dimethylammonium salt), 145-147 °C (ethanolamine salt), 142-144 °C (triethanolamine salt). Sodium salt decomposes at 215 °C. Bp: 160 °C at 0.4 mmHg, decomposes at 760 mmHg; ρ: 1.416 at 25/4 °C, 1.565 at 30/4 °C; fl p: nonflammable (free acid and salts); pK_a: 2.64-3.31; K_H: 1.95 x 10^{-2} atm·m^3/mol at 20-25 °C (approximate - calculated from water solubility and vapor pressure); log K_{oc}: 1.68-2.73; log K_{ow}: 1.47-4.88; P-t½: 2-4 d (aqueous solution irradiated at λ = 356 nm); S_o (g/kg at 25 °C unless otherwise noted): acetone (850), benzene (10.7 at 28 °C), carbon disulfide (5.0 at 29 °C), carbon tetrachloride (1), o-dichlorobenzene (4), diesel oil (1), diesel oil + kerosene (3.5), 1,4-dioxane (785 at 31 °C), ethyl ether (270), ethyl alcohol (1,300), n-heptane (1.15), methyl isobutyl ketone (312.7), 2-propanol (316 at 31 °C), toluene (0.67),

114

xylene (5.8), insoluble in petroleum oils; S_w: 890 ppm at 25 °C (free acid), 45 g/L at 20 °C (sodium salt); vp: 4.7 x 10^{-3} mmHg at 20 °C.

Soil properties and adsorption data

Soil	K_d (mL/g)	f_{oc} (%)	K_{oc} (mL/g)	pH	CEC (meq/100 g)
A-horizon	2.21	1.41	160	3.23	4.8
AB-horizon	2.38	5.11	50	3.88	13.0
Agricultural soil	2.48	1.64	150	5.40	14.0
Asquith sandy loam	0.14	1.02	14	7.50	--
B-horizon	5.45	2.58	210	3.59	9.6
C-horizon	2.70	1.82	150	4.07	7.0
C-horizon	0.16	0.09	180	4.95	1.6
C-horizon	0.14	0.15	90	4.21	1.3
Catlin	0.33	2.01	48	6.20	--
Clayey till	0.13	0.13	100	7.64	40.5
Commerce	0.33	0.68	48	6.70	--
Ferrod	4.30	3.56	112	3.88	--
Indian Head loam	0.53	2.34	23	7.80	--
Indian Head loam	0.44	2.35	19	7.80	--
Melfort loam	3.38	6.05	56	5.90	--
Melfort loam	0.99	6.05	16	5.90	--
Meltwater sand	0.27	0.05	540	6.14	1.4
Regina clay	0.31	2.40	13	7.70	--
Regina heavy clay	0.19	2.39	8	7.70	--
Sandy till	0.23	0.06	380	4.71	9.1
Tracy	0.85	1.12	76	6.20	--
Udalf	1.10	0.76	145	7.45	--
Weyburn Oxbow loam	0.61	3.72	16	6.50	--
Weyburn Oxbow loam	0.45	3.72	12	6.50	--

Source: Grover, 1977; Grover and Smith, 1974; Løkke, 1984; McCall et al., 1981; Rippen et al., 1982.

Transformation Products
Biological: 2,4-D degraded in anaerobic sewage sludge to 4-chlorophenol (Mikesell and Boyd, 1985).

Soil: In moist soils, 2,4-D degraded to 2,4-dichlorophenol and 2,4-dichloroanisole as intermediates followed by complete mineralization to carbon dioxide (Smith, 1985). 2,4-Dichlorophenol was reported as a hydrolysis metabolite (Somasundaram et al., 1989, 1991; Somasundaram and Coats, 1991). In a soil pretreated with its hydrolysis metabolite, 80% of the applied [^{14}C]2,4-D mineralized to $^{14}CO_2$ within 4 d. In soils not treated with the hydrolysis product (2,4-dichlorophenol), only 6% of the applied [^{14}C]2,4-D degraded to $^{14}CO_2$ after 4 d (Somasundaram et al., 1989). Steenson and Walker (1957) reported that the soil microorganisms *Flavobacterium peregrinum* and *Achromobacter* both degraded 2,4-D yielding 2,4-dichlorophenol and 4-chlorocatechol as metabolites. The microorganisms *Gloeosporium olivarium*, *Gloeosporium kaki*, and *Schisophyllum communs* also degraded 2,4-D in soil forming 2-(2,4-dichlorophenoxy)ethanol as the major metabolite (Nakajima et al., 1973). Microbial degradation of 2,4-D was more rapid under aerobic conditions ($t_{1/2}$ = 1.8-3.1 d) than under anaerobic conditions ($t_{1/2}$ = 69-135 d) (Liu et al., 1981). In a 5 d experiment, ^{14}C-labeled 2,4-D applied to soil-water suspensions under aerobic and anaerobic conditions gave $^{14}CO_2$ yields of 0.5 and 0.7%, respectively (Scheunert et al., 1987). The reported half-life in soil is 15 d (Jury et al., 1987). Residual activity in soil is limited to approximately 6 wk (Hartley and Kidd, 1987).

Plant: Reported metabolic products in bean and soybean plants include 4-*O*-β-glucosides of 4-hydroxy-2,5-dichlorophenoxyacetic acid, 4-hydroxy-2,3-dichlorophenoxyacetic acid, *N*-(2,4-dichlorophenoxyacetyl)-L-aspartic acid, and *N*-(2,4-dichlorophenocyacetyl)-L-glutamic acid. Metabolites identified in cereals and strawberries include 1-*O*-(2,4-dichlorophenoxyacetyl)-β-D-glucose and 2,4-dichlorophenol, respectively (Verschueren, 1983). 2,4-D was metabolized by soybean cultures forming 2,4-dichlorophenoxyacetyl derivatives of alanine, leucine, phenylalanine, tryptophan, valine, aspartic and glutamic acids (Feung et al., 1971, 1972, 1973). When 2,4-D was applied to resistant grasses, 3-(2,4-dichlorophenoxy)propionic acid formed (Hagin et al., 1970). On bean plants, 2,4-D degraded via β-oxidation and ring hydroxylation to form 2,4-dichloro-4-hydroxyphenoxyacetic acid, 2,3-dichloro-4-hydroxyphenoxyacetic acid (Hamilton et al., 1971), and 2-chloro-4-hydroxyphenoxyacetic acid (Fleeker and Steen, 1971). 2,5-Dichloro-4-hydroxyphenoxyacetic acid was the predominant product identified in several weed species as wells smaller quantities of 2-chloro-4-hydroxyphenoxyacetic acid in wild buckwheat, yellow foxtail, and wild oats (Fleeker and Steen, 1971).

Photolytic: Photolysis of 2,4-D in distilled water using mercury arc lamps (λ = 254 nm) or by natural sunlight yielded 2,4-dichlorophenol, 4-chlorocatechol, 2-hydroxy-4-chlorophenoxyacetic acid, 1,2,4-benzenetriol, and brown polymeric humic acids ($t_{1/2}$ = 50 min) (Crosby and Tutass, 1966). Half-lives of 2-4 d were reported when an aqueous solution was irradiated with UV light (λ = 356 nm) (Baur and Bovey, 1974). Bell (1956) reported that the composition of photo-degradation products formed were dependent upon the initial 2,4-D concentration and pH of the solutions. 2,4-D undergoes reductive dechlorination when various polar solvents (methanol, *n*-butanol, isobutyl alcohol, *t*-butyl alcohol, *n*-octanol, ethylene glycol) are irradiated at wavelengths between 254-420 nm. Photoproducts formed included 2,4-dichlorophenol, 2,4-dichloroanisole, 4-chlorophenol, 2-, and 4-chlorophenoxyacetic acid (Que Hee and Sutherland, 1981).

Chemical/Physical: In a helium pressurized reactor containing ammonium nitrate and polyphosphoric acid at temperatures of 121 and 232 °C, 2,4-D was oxidized to carbon dioxide, water, and hydrochloric acid (Leavitt and Abraham, 1990). Carbon dioxide, chloride, aldehydes, oxalic, and glycolic acids, were reported as ozonation products of 2,4-D in water at pH 8 (Struif et al., 1978). Reacts with alkali metals and amines forming water soluble salts (Hartley and Kidd, 1987). When 2,4-D was heated at 900 °C, carbon monoxide, carbon dioxide, chlorine, hydrochloric acid, and oxygen were produced (Kennedy et al., 1972, 1972a). Emits very toxic chloride fumes when heated to decomposition (Lewis, 1990).

Exposure Limits: NIOSH REL: IDLH 500 mg/m^3; OSHA PEL: TWA 10 mg/m^3; ACGIH TLV: TWA 10 mg/m^3.

Symptoms of Exposure: Weak stupor, hyporflexia, muscle twitch, convulsions, dermatitis.

Formulation Types: Emulsifiable concentrate; water soluble powder or granules; soluble concentrate.

Toxicity: LC$_{50}$ (48 hr) for bluegill sunfish 0.9 ppm, rainbow trout 1.1 ppm (Verschueren, 1983); acute oral LD$_{50}$ for rats 375 mg/kg (Hartley and Kidd, 1987).

Uses: Herbicide used for postemergence control of annual and perennial broadleaf weeds in fruits, vegetables, turfs, and ornamentals.

DALAPON-SODIUM

Synonyms: Basfapon F; Dalapon; Dalapon sodium salt; **2,2-Dichloropropionic acid sodium salt;** α,α-Dichloropropionic acid sodium salt; 2,2-DPA; Dowpon; Gramevin; Radapon; Sodium dalapon; Sodium 2,2-dichloropropanoate; Sodium α,α-dichloropropionate; Unipon.

Structure:

$$CH_3 \overset{\displaystyle Cl}{\underset{\displaystyle Cl}{C}} COONa$$

Designations: CAS: 127-20-8; mf: $C_3H_3Cl_2NaO_2$; fw: 164.95; RTECS: UF1225000.

Properties: Light-colored, very hygroscopic powder. The free acid is a colorless, odorless liquid. Mp: decomposes at 174-176 °C; bp: 185-190 °C (free acid), 98-99 °C at 20 mmHg; ρ: 1.389 at 22.8/4 °C (free acid); fl p: nonflammable; H-t$_{1/2}$:unknown but hydrolyzes readily; K_H: 6 x 10^{-6} atm·m^3/mol (calculated value for the free acid reported by Reinert and Rodgers, 1987); pK_a: 1.84 (free acid); log K_{oc}: 0.37-2.18 (calculated); log K_{ow}: 0.76; S_o (g/kg at 25 °C): acetone (1.4), benzene (0.02), ethanol (185), ethyl ether (0.16), methanol (179); S_w: 450-900 g/L at 25 °C; vp: not applicable.

Transformation Products

Soil: Undergoes dechlorination and the liberation of carbon dioxide in soil. The residual activity is limited to approximately 3-4 mo (Hartley and Kidd, 1987).

Photolytic: Dalapon (free acid) is subject to photodegradation. When an aqueous solution (0.25 M) was irradiated with UV light at 253.7 nm at 49 °C, 70% degraded in 7 hr. Pyruvic acid is formed which is subsequently decarboxylated to acetaldehyde, carbon dioxide and small quantities of 1,1-dichloroethane (2-4%) and a water-insoluble polymer (Kenaga, 1974). The photolysis of aqueous solution of dalapon (free acid) by UV light (λ = 2537 Å) yielded chloride ions, carbon dioxide, carbon monoxide, and methyl chloride at quantum yields of 0.29, 0.10, 0.02 and 0.02, respectively (Baxter and Johnston, 1968).

Chemical/Physical: Slowly reacts with moisture at room temperature forming pyruvic acid (Frank and Demint, 1969; Kenaga, 1974), hydrochloric acid, and sodium chloride (Kenaga, 1974; Wolfe et al., 1990). The reported hydrolysis half-life of dalapon sodium salt at concentrations <1% and temperatures less than 25 °C is several months (Kenaga, 1974). Products reported from the combustion of the free acid (dalapon) at 900 °C include carbon monoxide, carbon dioxide, chlorine, and hydrochloric acid (Kennedy et al., 1972a).

Symptoms of Exposure: Irritating to eyes and skin.

Formulation Types: Granules; wettable powder; water-soluble powder.

Toxicity: LC_{50} (96 hr) for rainbow trout, goldfish, and channel catfish >100 mg/L, carp >500 mg/L, guppies >1,000 mg/L (Hartley and Kidd, 1987), fathead minnow 290 mg/L, and bluegill sunfish 290 mg/L (Verschueren, 1983); LC_{50} (48 hr) for coho salmon 340 mg/L (Verschueren, 1983); acute oral LD_{50} for male and female rats is 9,330 and 7,570 mg/kg, respectively (Hartley and Kidd, 1987).

Uses: Selective systemic herbicide used to control perennial and annual grasses on noncrop land, fruits, vegetables, and some aquatic weeds.

DAZOMET

Synonyms: Basamid; Basamid G; Basamid-granular; Basamid P; Basamid-puder; Carbothialdin; Carbothialdine; Crag 974; Crag fungicide 974; Crag nematocide; Crag 85W; Dazomet-powder BASF; Dimethylformocarbothialdine; 3,5-Dimethyl-tetrahydro-1,3,5-thiadiazine-2-thione; 3,5-Dimethyl-1,2,3,5-tetrahydro-1,3,5-thiadiazinethione-2; 3,5-Dimethyltetrahydro-1,3,5-2H-thiadiazine-2-thione; 3,5-Dimethyl-1,3,5,2H-tetrahydrothiadiazine-2-thione; 3,5-Dimethyltetrahydro-2H-1,3,5-thiadiazine-2-thione; Dimethyl-2H-1,3,5-tetrahydrothiadiazine-2-thione; 3,5-Dimethyl-2-thionotetrahydro-1,3,5-thiadiazine; DMTT; Fennosan B 100; Micofume; Mylon; Mylone; Mylone 85; NA 521; Nalcon 243; Nefusan; Prezervit; Stauffer N 521; Tetrahydro-2H-3,5-dimethyl-1,3,5-thiadiazine-2-thione; **Tetrahydro-3,5-dimethyl-2H-1,3,5-thiadiazine-2-thione**; Thiazon; Thiazone; 2-Thio-3,5-dimethyltetrahydro-1,3,5-thiadiazine; Tiazon; Troysan 142; UC 974.

Structure:

Designations: CAS: 533-74-4; mf: $C_5H_{10}N_2S_2$; fw: 162.28; RTECS: XI2800000.

Properties: Colorless to white, almost odorless crystalline solid. Mp: decomposes at 104-105 °C; ρ: 1.37 at room temperature; fl p: nonflammable; K_H: 2.0 x 10^{-10} atm·m^3/mol at 20 °C (approximate - calculated from water solubility and vapor pressure); log K_{oc}: 0.48 (calculated); log K_{ow}: 0.15; S_o (g/kg at 25 °C): acetone (173), benzene (51), chloroform (391), cyclohexane (400), ethanol (15), ethyl ether (6); S_w: 3 g/kg at 20 °C, 1.2 g/kg at 30 °C; vp: 2.78 x 10^{-6} mmHg at 20 °C.

Transformation Products

Soil: Soil metabolites include formaldehyde, hydrogen sulfide, methylamine, and methyl(methylaminomethyl)dithiocarbamic acid (Hartley and Kidd, 1987) which further decomposes to methyl isothiocyanate (Ashton and Monaco, 1991; Cremlyn, 1991; Hartley and Kidd, 1987). The rate of decomposition is dependent upon the soil type, temperature, and humidity (Cremlym, 1991).

Chemical/Physical: Hydrolyzes in acidic solutions forming carbon disulfide, methylamine, and formaldehyde (Hartley and Kidd, 1987; Humburg et al., 1989). These compounds are probably formed following the decomposition of dazomet with alcohol and water (Windholz et al., 1983). Emits toxic fumes of nitrogen and sulfur oxides when heated to decomposition (Sax and Lewis, 1987).

Symptoms of Exposure: Skin and eye irritant. Symptoms of ingestion include nausea, irritation of the gastrointestinal tract, vomiting, cramps, and diarrhea.

Formulation Types: Dustable powder; granules; wettable powder.

Toxicity: Toxic to fish (Hartley and Kidd, 1987); acute oral LD_{50} for male albino mice 650 mg/kg (Ashton and Monaco, 1991); acute oral LD_{50} for male and females rats is 500 and 400 mg/kg, respectively (Verschueren, 1983), 320 mg/kg (RTECS, 1985).

Uses: Soil fungicide; nematocide; herbicide; insecticide; soil sterilant.

p,p′-DDD

Synonyms: 1,1-Bis(4-chlorophenyl)-2,2-dichloroethane; 1,1-Bis(p-chlorophenyl)-2,2-dichloroethane; 2,2-Bis(4-chlorophenyl)-1,1-dichloroethane; 2,2-Bis(p-chlorophenyl)-1,1-dichloroethane; DDD; 4,4′-DDD; 1,1-Dichloro-2,2-bis(p-chlorophenyl)ethane; 1,1-Dichloro-2,2-di(4-chlorophenyl)ethane; 1,1-Dichloro-2,2-di-(p-chlorophenyl)ethane; Dichlorodiphenyldichloroethane; 4,4′-Dichlorodiphenyldichloroethane; p,p′-Dichlorodiphenyldichloroethane; 1,1′-(2,2-**Dichloroethylidene)bis(4-chlorobenzene)**; Dilene; ENT 4225; ME 1700; NA 2761; NCI-C00475; RCRA waste number U060; Rhothane; Rhothane D-3; Rothane; TDE; 4,4′-TDE; p,p′-TDE; Tetrachlorodiphenylethane.

Structure:

Designations: CAS: 72-54-8; DOT: 2761; mf: $C_{14}H_{10}Cl_4$; fw: 320.05; RTECS: KI0700000.

Properties: White crystalline solid. Mp: 88-90 °C, 109-112 °C; bp: 193 °C; ρ: 1.476 at 20/4 °C; fl p: not pertinent; H-$t_{1/2}$: 28 yr at 25 °C and pH 7; K_H: 2.16 x 10^{-5} atm·m^3/mol at 25 °C (approximate - calculated from water solubility and vapor pressure); log K_{oc}: 5.38; log K_{ow}: 5.061-6.217; S_o (mg/mL at 22 °C): acetone (50-100), dimethylsulfoxide (50-100), 95% ethanol (10-50); S_w: 20-90 μg/L at 25 °C; vap d: 17.2 ng/L at 30 °C; vp: 4.68 x 10^{-6} mmHg at 25 °C.

Transformation Products
Biological: Incubation of p,p′-DDD with hematin and ammonia gave 4,4′-dichlorobenzophenone, 1-chloro-2,2-bis(p-chlorophenyl)ethylene, and bis(p-chlorophenyl)acetic acid methyl ester (Quirke et al., 1979).

Soil: Fries (1972) reported that p,p′-DDD, a major biodegradation product of p,p′-DDT, was further degraded by *Aerobacter aerogenes* under aerobic conditions to 1-chloro-2,2-bis(p-chlorophenyl)ethylene, 1-chloro-2,2-bis(p-chlorophenyl)-ethane, and 1,1-bis(p-chlorophenyl)ethylene. Under anaerobic conditions, however, four additional compounds were identified: bis(p-chlorophenyl)acetic acid, p,p′-dichlorodiphenylmethane, p,p′-dichlorobenzhydrol, and p,p′-dichlorobenzophenone (Fries, 1972). Under reducing conditions, indigenous microbes in Lake Michigan sediments degraded DDD to 2,2-bis(p-chlorophenyl)ethane and 2,2-bis(p-chlorophenyl)ethanol (Leland et al., 1973).

Chemical/Physical: Based on a reported hydrolysis half-life of 28 yr at pH 7 (Ellington et al., 1988), chemical hydrolysis is not expected to be an environmentally relevant fate process (Lyman et al., 1982).

Symptoms of Exposure: Contact with eyes causes irritation. Ingestion causes vomiting.

Formulation Types: Wettable powders; dusts; emulsions.

Toxicity: EC_{50} (96 hr) for catfish <2.6 mg/L; LC_{50} (24 hr) for brown shrimp 6.8 µg/L (salt water); acute oral LD_{50} for rats 3,400 mg/kg (Verschueren, 1983), 113 mg/kg (RTECS, 1985).

Uses: Contact control of leaf rollers and other insects on vegetables and tobacco.

p,p'-DDE

Synonyms: 2,2-Bis(4-chlorophenyl)-1,1-dichloroethene; 2,2-Bis(p-chlorophenyl)-1,1-dichloroethene; 1,1-Bis(4-chlorophenyl)-2,2-dichloroethylene; 1,1-Bis(p-chlorophenyl)-2,2-dichloroethylene; DDE; 4,4'-DDE; DDT dehydrochloride; 1,1-Dichloro-2,2-bis(p-chlorophenyl)ethylene; Dichlorodiphenyldichloroethylene; p,p'-Dichlorodiphenyldichloroethylene; **1,1'-(Dichloroethenylidene)bis(4-chlorobenzene)**; NCI-C00555.

Structure:

Designations: CAS: 72-55-9; DOT: 2761; mf: $C_{14}H_8Cl_4$; fw: 319.03; RTECS: KV9450000.

Properties: Solid. Mp: 88-90 °C; K_H: 1.22 x 10^{-3} atm·m^3/mol at 23 °C; log K_{oc}: 5.386; log K_{ow}: 5.69-6.956; P-t$_{1/2}$: 206.66 hr (absorbance λ = 248.5 nm, concentration on glass plates = 6.7 μg/cm^2); S_o: soluble in cod liver oil (245.5, 269.2, and 398.1 mM at 4, 12, and 20 °C, respectively), n-octanol (44.0 and 63.7 g/L at 4 and 20 °C, respectively), triolein (73.1, 96.3, and 105.6 g/L, at 4, 12, and 20 °C, respectively); S_w: 40 and 65 μg/L at 20 and 24 °C, respectively; vap d: 109 ng/L at 30 °C; vp: 1.57 x 10^{-5} mmHg at 25 °C.

Transformation Products

Photolytic: When an aqueous solution of p,p'-DDE (0.004 μM) in natural water samples collected from California and Hawaii were irradiated (maximum λ = 240 nm) for 120 hr, 62% was photooxidized to p,p'-dichlorobenzophenone (Ross and Crosby, 1985). In an air-saturated distilled water medium irradiated with monochromic light (λ = 313 nm), p,p'-DDE degraded to p,p'-dichlorobenzophenone, 1,1-bis(p-chlorophenyl)-2-chloroethylene (DDMU), and 1-(4-chlorophenyl)-1-(2,4-dichlorophenyl)-2-chloroethylene (o-chloro DDMU). Identical photoproducts also were observed using tap water containing Mississippi River sediments (Miller and Zepp, 1972). The photolysis half-life under sunlight irradiation was reported to be 1.5 d (Mansour et al., 1989).

Chemical/Physical: May degrade to bis(chlorophenyl)acetic acid (DDA) and hydrochloric acid in water (Verschueren, 1983) or oxidize to p,p'-dichlorobenzophenone using UV light as a catalyst (U.S. Department of Health and Human Services, 1989).

124

Toxicity: Acute oral LD_{50} for rats 880 mg/kg (RTECS, 1985).

Uses: Military product; chemical research. Degradation product of *p,p'*-DDT.

p,p'-DDT

Synonyms: Agritan; Anofex; Arkotine; Azotox; 2,2-Bis(4-chlorophenyl)-1,1,1-trichloroethane; 2,2-Bis(*p*-chlorophenyl)-1,1,1-trichloroethane; α,α-Bis(*p*-chlorophenyl)-β,β,β-trichloroethane; 1,1-Bis(*p*-chlorophenyl)-2,2,2-trichloroethane; Bosan Supra; Bovidermol; Chlorophenothan; Chlorophenothane; Chlorophenotoxum; Citox; Clofenotane; DDT; 4,4'-DDT; Dedelo; Deoval; Detox; Detoxan; Dibovan; Dichlorodiphenyltrichloroethane; *p,p'*-Dichlorodiphenyltrichloroethane; 4,4'-Dichlorodiphenyltrichloroethane; Dicophane; Didigam; Didimac; Diphenyltrichloroethane; Dodat; Dykol; ENT 1506; Estonate; Genitox; Gesafid; Gesapon; Gesarex; Gesarol; Guesapon; Gyron; Havero-extra; Ivoran; Ixodex; Kopsol; Mutoxin; NCI-C00464; Neocid; Parachlorocidum; PEB1; Pentachlorin; Pentech; PPzeidan; RCRA waste number U061; Rukseam; Santobane; Trichlorobis(4-chlorophenyl)ethane; Trichlorobis(*p*-chlorophenyl)ethane; 1,1,1-Trichloro-2,2-bis(*p*-chlorophenyl)ethane; 1,1,1-Trichloro-2,2-di(4-chlorophenyl)ethane; 1,1,1-Trichloro-2,2-di(*p*-chlorophenyl)ethane; **1,1'-(2,2,2-Trichloroethylidene)bis(4-chlorobenzene)**; Zeidane; Zerdane.

Structure:

Designations: CAS: 50-29-3; DOT: 2761; mf: $C_{14}H_9Cl_5$; fw: 354.49; RTECS: KJ3325000.

Properties: Odorless to slightly fragrant colorless crystals or white powder. Mp: 108.5 °C; bp: 185 °C at 0.05 mmHg (decomposes); ρ: 1.56 at 15/4 °C; fl p: 72.2-77.2 °C; K_H: 1.29 x 10^{-5} atm·m^3/mol at 23 °C; log K_{oc}: 5.146-6.26; log K_{ow}: 4.89-6.914; P-t$_{1/2}$: 192.31 hr (absorbance λ = 237.5 nm, concentration on glass plates = 6.7 μg/cm^2); S_o: soluble in cyclohexane, morpholine, pyridine, 1,4-dioxane, acetone (580 g/L), benzene (780 g/L), benzyl benzoate (420 g/L), carbon tetrachloride (450 g/L), chlorobenzene (740 g/L), cod liver oil (151.4, 162.2, and 204.2 mM at 4, 12, and 20 °C, respectively), cyclohexanone (1,160 g/L), ethyl ether (280 g/L), gasoline (100 g/L), isopropanol (30 g/L), kerosene (80-100 g/L), methylene chloride (850 g/L at 27 °C), morpholine (750 g/L), *n*-octanol (26.9, 43.6, and 57.5 g/L at 4, 12, and 20 °C, respectively), peanut oil (110 g/L), pine oil (100-160 g/L), tetralin (610 g/L), tributyl phosphate (500 g/L), trichloroethylene (720 g/L at 27 °C), triolein (134.9, 182.0, and 323.6 mM at 4, 12, and 20 °C, respectively), xylene (600 g/L at 27 °C); S_w: 1.2-5.5 μg/L at 25 °C; vap d: 13.6 ng/L at 30 °C; vp: 1.7 x 10^{-10} mmHg at 20 °C.

Soil properties and adsorption data

Soil	K_d (mL/g)	f_{oc} (%)	K_{oc} (mL/g)	pH
Catlin	2,800	2.01	152,000	6.2
Commerce	1,070	0.68	157,000	6.7
Montcalm sandy loam	1,300	1.00	130,000	--
Sims clay	13,700	3.81	360,000	--
Tracy	1,830	1.12	147,000	6.2

Source: McCall et al., 1981; Shin et al., 1970.

Transformation Products
Biological: The white rot fungus *Phanerochaete chrysosporium* degraded *p,p'*-DDT yielding the following metabolites: 1,1-dichloro-2,2-bis(4-chlorophenyl)-ethane (*p,p'*-DDD), 2,2,2-trichloro-1,1-bis(4-chlorophenyl)ethanol (dicofol), 2,2-dichloro-1,1-bis(4-chloro-phenyl)ethanol, and 4,4'-dichlorobenzo-phenone. Mineralization of *p,p'*-DDT by the white rot fungi *Pleurotus ostreatus*, *Phellinus weirri*, and *Polyporus versicolor* was also demonstrated (Bumpus and Aust, 1987). *Aerobacter aerogenes* degraded *p,p'*-DDT under aerobic conditions to *p,p'*-DDD, *p,p'*-DDE, 1-chloro-2,2-bis(*p*-chlorophenyl)ethylene, 1-chloro-2,2-bis-(*p*-chlorophenyl)ethane, and 1,1-bis(*p*-chlorophenyl)ethylene (Fries, 1972). Under anaerobic conditions the same organism produced four additional compounds. These were bis(*p*-chlorophenyl)acetic acid, *p,p'*-dichlorodiphenylmethane, *p,p'*-dichlorobenzhydrol, and *p,p'*-dichlorobenzophenone. Other degradation products of *p,p'*-DDT under aerobic and anaerobic conditions in soils by various cultures not previously mentioned include 1,1-bis(*p*-chlorophenyl)-2,2,2-trichloroethanol (Kelthane) and *p*-chlorobenzoic acid (Fries, 1972). Under aerobic conditions, the amoeba *Acanthamoeba castellanii* (Neff strain ATCC 30.010) degraded *p,p'*-DDT to *p,p'*-DDE, *p,p'*-DDD, and dibenzophenone (Pollero and dePollero, 1978).

Thirty-five microorganisms isolated from marine sediment and marine water samples taken from Hawaii and Houston, TX were capable of degrading *p,p'*-DDT. *p,p'*-DDD was identified as the major metabolite. Minor transformation products included 2,2-bis(*p*-chlorophenyl)ethanol (DDOH), 2,2-bis(*p*-chloro-phenyl)ethane (DDNS), and *p,p'*-DDE (Patil et al., 1972). Similarly, Matsumura et al. (1971) found that *p,p'*-DDT was degraded by numerous aquatic microorganisms isolated from water and silt samples collected from Lake Michigan and its tributaries in Wisconsin. The major metabolites identified were TDE, DDNS, and DDE. *p,p'*-DDT was metabolized by the following microorganisms under laboratory conditions to *p,p'*-DDD: *Actinomycetes* (Chacko et al., 1966), *Escherichia coli* (Langlois, 1967), *Aerobacter aerogenes* (Plimmer et al., 1968; Wedemeyer, 1966), and *Proteus vulgaris* (Wedemeyer, 1966). In addition, *p,p'*-DDT was degraded to *p,p'*-DDD, *p,p'*-DDE, and dicofol by *Trichoderma viride* (Matsumura and Boush, 1968) and to *p,p'*-DDD and *p,p'*-DDE by *Ankistrodemus amalloides* (Neudorf and Khan, 1975).

Jensen et al. (1972) studied the anaerobic degradation of *p,p'*-DDT (100 mg) in 1-L of sewage sludge containing *p,p'*-DDD (4.0%) and *p,p'*-DDE (3.1%) as contaminants. The sludge was incubated at 20 °C for 8 d under a nitrogen atmosphere. The parent compound degraded rapidly (t$_{1/2}$ = 7 hr) forming *p,p'*-DDD, *p,p'*-dichlorodiphenylbenzophenone (DBP), 1,1-bis(*p*-chlorophenyl)-2-chloroethylene (DDMU) and bis(*p*-chlorophenyl)acetonitrile. After 48 hr, the original amount of *p,p'*-DDD added to the sewage sludge had completely reacted. In a similar study, Pfaender and Alexander (1973) observed the cometabolic conversion of DDT (0.005%) in unamended sewage sludge to give DDD, DDE and DBP. When the sewage sludge was amended with glucose (0.10%), the rate of DDD formation was enhanced. However, with the addition of diphenylmethane, the rate of formation of both DDD and DBP was reduced. The diphenylmethane-amended sewage sludge showed the greatest abundance of bacteria capable of cometabolizing DDT, whereas the unamended sewage showed the fewest number of bacteria.

In an *in vitro* fermentation study, rumen microorganisms metabolized both isomers of ^{14}C-labeled DDT (*o,p'*- and *p,p'*-) to the corresponding DDD isomers at a rate of 12%/hr. With *p,p'*-DDT, 11% of the ^{14}C detected was an unidentified polar product associated with microbial and substrate residues (Fries et al., 1969). In another *in vitro* study, extracts of *Hydrogenomonas* sp. cultures degraded DDT to DDD, 1-chloro-2,2-bis(*p*-chlorophenyl)ethane (DDMS), DBP, and several other products under anaerobic conditions. Under aerobic conditions containing whole cells, one of the rings is cleaved and *p*-chlorophenylacetic acid is formed (Pfaender and Alexander, 1972).

Soil: *p,p'*-DDD and *p,p'*-DDE are the major metabolites of *p,p'*-DDT in the environment (Metcalf, 1973). In soils under anaerobic conditions, *p,p'*-DDT is rapidly converted to *p,p'*-DDD via reductive dechlorination (Johnsen, 1976) and very slowly to *p,p'*-DDE under aerobic conditions (Guenzi and Beard, 1967; Kearney and Kaufman, 1976). The aerobic degradation of *p,p'*-DDT under flooded conditions is very slow with *p,p'*-DDE forming as the major metabolite. Dicofol was also detected in minor amounts (Lichtenstein et al., 1971). In addition to *p,p'*-DDD and *p,p'*-DDE, 2,2-bis(*p*-chlorophenyl)acetic acid (DDA), bis(*p*-chlorophenyl)methane (DDM), *p,p'*-dichlorobenzhydrol (DBH), DBP, and *p*-chlorophenylacetic acid (PCPA) were also reported as metabolites of *p,p'*-DDT in soil under aerobic conditions (Subba-Rao and Alexander, 1980).

The anaerobic conversion of *p,p'*-DDT to *p,p'*-DDD in soil was catalyzed by the presence of ground alfalfa or glucose (Burge, 1971). Under flooded conditions, *p,p*-DDT was rapidly converted to TDE via reductive dehalogenation and other metabolites (Castro and Yoshida, 1971; Guenzi and Beard, 1967). Degradation was faster in flooded soil than an upland soil and was faster in soils containing high organic matter (Castro and Yoshida, 1971). Other reported degradation products under aerobic and anaerobic conditions by various soil microbes include 1,1'-bis(*p*-chlorophenyl)-2-chloroethane, 1,1'-bis(*p*-chlorophenyl)-2-hydroxyethane, and *p*-chlorophenyl acetic acid (Kobayashi and Rittman, 1982). It was also reported that *p,p'*-DDE formed by hydrolyzing *p,p'*-DDT (Wolfe et al., 1977). The clay-catalyzed reaction of DDT to form DDE was reported by Lopez-Gonzales and Valenzuela-Calahorro (1970). They observed that DDT adsorbed of sodium bentonite clay surfaces was transformed more rapidly than on the corresponding hydrogen-bentonite clay. In 1 d, *p,p'*-DDT reacted rapidly with reduced hematin forming *p,p'*-DDD and unidentified products (Baxter, 1990). In an Everglades muck, *p,p'*-DDT was slowly converted to *p,p'*-DDD and *p,p'*-DDE (Parr and Smith, 1974). The reported half-life in soil is 3,800 d (Jury et al., 1987).

Oat plants were grown in two soils treated with [^{14}C]*p,p'*-DDT. Most of the residues remained bound to the soil. Metabolites identified were *p,p'*-DDE, *o,p'*-DDT, TDE, DBP, dicofol, and DDA (Fuhremann and Lichtenstein, 1980).

In a 42-d experiment, ^{14}C-labeled *p,p'*-DDT applied to soil-water suspensions under aerobic and anaerobic conditions gave $^{14}CO_2$ yields of 0.8 and 0.7%, respectively (Scheunert et al., 1987).

Photolytic: Photolysis of *p,p'*-DDT in nitrogen-sparged methanol solvent by UV light (λ = 260 nm) produced DDD and DDMU. But photolysis of *p,p'*-DDT at 280 nm in an oxygenated methanol solution yielded a complex mixture containing the methyl ester of 2,2-bis(*p*-chlorophenyl)acetic acid (Plimmer et al., 1970). *p,p'*-DDT in an aqueous solution containing suspended titanium dioxide as a catalyst and irradiated with UV light (λ >340 nm) formed chloride ions. Based on the amount of chloride ions generated, carbon dioxide and hydrochloric acid were reported as the end products (Borello et al., 1989). When *p,p'*-DDT on quartz was subjected to UV radiation (2537 Å) for 2 d, 80% of *p,p'*-DDT degraded to 4,4'-dichlorobenzophenone, 1,1-dichloro-2,2-bis(*p*-chlorophenyl)ethane, and 1,1-dichloro-2,2-bis(*p*-chlorophenyl)ethene. Irradiation of a hexane solution yielded 1,1-dichloro-2,2-bis(*p*-chlorophenyl)ethane and hydrochloric (Mosier et al., 1969). When an aqueous solution containing *p,p'*-DDT was photooxidized by UV light at 90-95 °C, 25, 50, and 75% degraded to carbon dioxide after 25.9, 66.5, and 120.0 hr, respectively (Knoevenagel and Himmelreich, 1976).

Chemical/Physical: In alkaline solutions and temperatures >108.5 °C, *p,p'*-DDT undergoes dehydrochlorination releasing hydrochloric acid to give the icidal *p,p'*-DDE (Hartley and Kidd, 1987; Worthing and Hance, 1991). This reaction is also catalyzed by ferric and aluminum chlorides and UV light (Worthing and Hance, 1991).

Castro (1964) reported that iron(II) porphyrins in dilute aqueous solution was rapidly oxidized by DDT to form the corresponding iron(III) chloride complex (hematin) and DDE, respectively. Incubation of *p,p'*-DDT with hematin and ammonia gave *p,p'*-DDD, *p,p'*-DDE, bis(*p*-chlorophenyl)acetonitrile, 1-chloro-2,2-bis(*p*-chlorophenyl)ethylene, 4,4'-dichlorobenzophenone, and the methyl ester of bis(*p*-chlorophenyl)acetic acid (Quirke et al., 1979).

When *p,p'*-DDT was heated at 900 °C, carbon monoxide, carbon dioxide, chlorine, hydrochloric acid, and other unidentified substances were produced (Kennedy et al., 1972, 1972a). Emits hydrochloric acid and chlorine when incinerated (Sittig, 1985).

Exposure Limits: NIOSH REL: lowest detectable limit; OSHA PEL: TWA 1 mg/m^3; ACGIH TLV: TWA 1 mg/m^3.

Symptoms of Exposure: Paresthesia tongue, lips, face; tremor; apprehension, confusion, malaise; headache; convulsions; paresis of the hands; vomiting; irritates eyes, skin.

Formulation Types: Emulsifiable concentrate; wettable powder; granules; dustable powder; aerosol.

Toxicity: LC_{50} (96 hr) for fathead minnow 19 µg/L, bluegill sunfish 8 µg/L, largemouth bass 2 µg/L, rainbow trout 7 µg/L, brown trout 2 µg/L, coho salmon 4 µg/L, perch 9 µg/L, channel catfish 16 µg/L, black bullhead 5 µg/L, goldfish 21 µg/L, minnow 19 µg/L (Verschueren, 1983); LC_{50} (48 hr) for killifish 2.8 µg/L; EC_{50} for fresh water bass 3.9 and 1.8 µg/L at 24 and 96 hr, respectively; acute oral LD_{50} for rats 113 mg/kg (Hartley and Kidd, 1987), 87 mg/kg (RTECS, 1985).

Uses: Use as an insecticide is now prohibited.

DESMEDIPHAM

Synonyms: Bentanex; Betanal AM; Betanex; EP 475; 3-Ethoxycarbonylamino-phenyl-*N*-phenylcarbamate; **Ethyl 3-phenylcarbamoyloxyphenylcarbamate**; Schering 38107; SN 38107.

Structure:

Designations: CAS: 13684-56-5; mf: $C_{16}H_{16}N_2O_4$; fw: 300.32; RTECS: FD0425000.

Properties: Colorless to pale yellow crystals. Mp: 120 °C; H-t$_{1/2}$: 31 d (pH 5), 14 hr (pH 7), 20 min (pH 9); fl p: 68 °C (open cup); K_H: 1.7 x 10^{-10} atm·m^3/mol at 20 °C (approximate - calculated from water solubility and vapor pressure); log K_{oc}: 3.18 (calculated); log K_{ow}: 3.39 (pH 3.9); S_o (g/L at 20 °C): acetone (400), benzene (1.6), chloroform (80), ethyl acetate (149), *n*-hexane (0.5), methanol (180), toluene (1.2); S_w: 7 mg/L at 20 °C; vp: 3 x 10^{-9} mmHg at 20 °C.

Transformation Products

Soil: Degrades in soil forming the intermediate 3-hydroxycarbanilate (Worthing and Hance, 1991). The reported half-lives in soil are 70 d, 20 hr, and 10 min at pHs 5, 7, and 9, respectively (Worthing and Hance, 1991).

Plant: In sugar beets, *m*-aminophenol and ethyl-*N*-(3-hydroxyphenyl)carbamate were identified as metabolites (Hartley and Kidd, 1987).

Symptoms of Exposure: May cause skin irritation. Symptoms of poisoning include hypoactivity and muscular weakness.

Formulation Types: Emulsifiable concentrate (1.3 and 4 lb/gal); granules (10%).

Toxicity: LC$_{50}$ (96 hr) for rainbow trout 3.8 mg/L and bluegill sunfish 13.4 mg/L (Hartley and Kidd, 1987); acute oral LD$_{50}$ of pure desmedipham and the formulated product for rats is 10,300 and 3,700 mg/kg (Ashton and Monaco, 1991).

Uses: Selective systemic herbicide used to control broadleaf weeds such as buckwheat, chickweed, fiddleneck, kochnia, mustard, pigweed, ragweed, in sugar beets.

DIALIFOS

Synonyms: S-(2-Chloro-1-(1,3-dihydro-1,3-dioxo-2H-isoindol-2-yl)ethyl) O,O-diethyl phosphorodithioate; S-(2-Chloro-1-phthalimidoethyl) O,O-diethyl phosphorodithioate; Dialifor; O,O-Diethyl S-(2-chloro-1-phthalimidoethyl) phosphorodithioate; ENT 27320; Hercules 14503; **Phosphorodithioic acid S-(2-chloro-1-(1,3-dihydro-1,3-dioxo-2H-isoindol-2-yl)ethyl) O,O-diethyl ester;** Phosphorodithioic acid S-(2-chloro-1-phthalimidoethyl) O,O-diethyl ester; Torax.

Structure:

Designations: CAS: 10311-84-9; mf: $C_{14}H_{17}ClNO_4PS_2$; fw: 393.84; RTECS: TD5165000.

Properties: White crystalline solid. Mp: 67-69 °C; H-t$_{1/2}$: 14 hr at 20 °C and pH 7.4, 1.8 hr at 37.5 °C and pH 7.4; K_H: 1.4 x 10^{-6} atm·m^3/mol at 20 °C; log K_{oc}: 4.05 (calculated); log K_{ow}: 4.69; S_o (g/kg at 20 °C): acetone (760), chloroform (620), ethanol (<10), ethyl ether (500), isophorone (400), xylene (570); S_w: 0.18 mg/L at room temperature; vp: 6.2 x 10^{-8} mmHg at 30 °C.

Transformation Products
Chemical/Physical: Though no products were identified, the hydrolysis half-lives at 20 °C were 15 d and 14 hr at pH 6.1 and pH 7.4, respectively (Freed et al., 1979).

Formulation Types: Emulsifiable concentrate.

Toxicity: LC$_{50}$ (96 hr) for rainbow trout 0.55-1.08 mg/L and goldfish 1.80-8.30 mg/L (Hartley and Kidd, 1987); acute oral LD$_{50}$ for male and female rats is 45-53 and 5 mg/kg, respectively (Hartley and Kidd, 1987).

Uses: Nonsystemic insecticide and acaricide.

DIALLATE

Synonyms: Avadex; **Bis(1-methylethyl)carbamothioic acid S-(2,3-dichloro-2-propenyl) ester;** CP 15336; DATC; 2,3-DCDT; Di-allate; Dichloroallyl diisopropylthiocarbamate; S-Dichloroallyl diisopropylthiocarbamate; Dichloroallyl N,N-diisopropylthiolcarbamate; S-2,3-Dichlorodiisopropylthiocarbamate; Pyradex; RCRA waste number U062.

Structure:

$$(CH_3)_2 \diagdown$$
$$\diagup NCOSCH_2CCl = CHCl$$
$$(CH_3)_2 \diagup$$

Designations: CAS: 2303-16-4; mf: $C_{10}H_{17}Cl_2NOS$; fw: 270.21; RTECS: EZ8225000.

Properties: Amber to brown, oily liquid. Mp: 25-30 °C; bp: 108 and 150 °C at 0.25 and 9.0 mmHg, respectively. Decomposes >200 °C; ρ: 1.188 at 25/15.6 °C; fl p: 144 °C (technical), 38 °C (emulsifiable concentrate); H-t$_{1/2}$: 6.6 yr at 25 °C and pH 7; K_H: 2.5 x 10^{-6} atm·m^3/mol at 20-25 °C (approximate - calculated from water solubility and vapor pressure); log K_{oc}: 2.28; log K_{ow}: 3.29 (calculated); S_o at 25 °C: soluble and/or miscible with acetone, benzene, ethanol, ethyl acetate, ethyl ether, n-hexane, n-heptane, kerosene, methanol, 2-propanol, xylene, and many other organic solvents; S_w: 14 mg/L at 25 °C; vap d: 11.04 g/L at 25 °C, 9.36 (air = 1); vp: 1.5 x 10^{-4} mmHg at 20 °C.

Transformation Products

Soil: Soil metabolites include 2,3-dichloroallyl alcohol and 2,3-dichloroallyl mercaptan (Hartley and Kidd, 1987). In an agricultural soil, $^{14}CO_2$ was the only biodegradation identified; however, bound residue and traces of benzene and water-soluble radioactivity were also detected in large amounts (Anderson and Domsch, 1980). The reported half-life in soil is approximately 30 d (Hartley and Kidd, 1987). In four microbially-active agricultural soils, the half-life was 4 wk but in sterilized soil the half-life was 20 wk (Anderson and Domsch, 1976).

Plant: In plants, diallate is metabolized and carbon dioxide is released (Hartley and Kidd, 1987). Diallate undergoes *cis/trans* isomerization and oxidative cleavage when irradiated at 300 nm forming 2,3-dichloroacrolein and 2-chloroacrolein as products (Ruzo and Casida, 1985).

Photolytic: Irradiation of diallate was also conducted in oxygenated chloroform or

133

water until a 10% conversion was obtained. Products formed included acetaldehyde, 2,3-dichloroacrolein, and trace amounts of 2-chloroacrolein (Ruzo and Casida, 1985).

Chemical/Physical: Emits toxic fumes of chlorine, nitrogen and sulfur oxides when heated to decomposition (Sax and Lewis, 1987). Though no products were reported, the calculated hydrolysis half-life at 25 °C and pH 7 is 6.6 yr (Ellington et al., 1988).

Symptoms of Exposure: Eye, skin, and mucous membrane irritant.

Formulation Types: Emulsifiable concentrate (4 lb/gal); granules (10%).

Toxicity: LC_{50} (96 hr) for rainbow trout 7.9 mg/L and bluegill sunfish 5.9 mg/L (Hartley and Kidd, 1987); LC_{50} (96-h for technical grade) for rainbow trout 3.2 mg/L, for bluegill sunfish 2.5 mg/L (Humburg et al., 1989); LC_{50} (48-h for technical grade) for harlequin fish 8.2 mg/L and for *Daphnia magna* 7.5 mg/L (Humburg et al., 1989); acute oral LD_{50} for rats is 395 mg/kg (RTECS, 1985).

Uses: Preemergent, selective herbicide used to control wild oats and blackgrass in barley, corn, flax, lentils, peas, potatoes, soybeans, and sugar beets.

DIAZINON

Synonyms: Alfatox; Basudin; Basudin 10 G; Bazinon; Bazuden; Dazzel; Desapon; Dianon; Diaterr-fos; Diazatol; Diazide; Diethyl 4-(2-isopropyl-6-methylpyrimidinyl)phosphorothioate; *O,O*-Diethyl *O*-(2-isopropyl-4-methyl-6-pyrimidinyl)-phosphorothioate; *O,O*-Diethyl *O*-(2-isopropyl-6-methyl-4-pyrimidinyl)phosphorothioate; *O,O*-Diethyl *O*-(2-isopropyl-4-methyl-6-pyrimimidinyl)thionophosphate; *O,O*-Diethyl 2-isopropyl-4-methylpyrimidinyl-6-thiophosphate; *O,O*-Diethyl *O*-6-methyl-2-isopropyl-4-pyrimidinyl phosphorothioate; ***O,O*-Diethyl *O*-(6-methyl-2-(1-methylethyl)-4-pyrimidinyl) phosphorothioate**; Dimpylate; Dipofene; Dizinon; Dyzol; ENT 19507; G 301; G 24480; Gardentox; Geigy 24480; *O*-2-Isopropyl-4-methylpyrimidinyl-*O,O*-diethyl phosphorothioate; Isopropylmethylpyrimidinyl diethyl thiophosphate; Kayazinon; Kayazol; NA 2763; NCI-C08673; Nedcisol; Neocidol; Nipsan; Nucidol; Sarolex; Spectracide.

Structure:

Designations: CAS: 333-41-5; DOT: 2783; mf: $C_{12}H_{21}N_2O_3PS$; fw: 304.35; RTECS: TF3325000.

Properties: Clear, colorless liquid with a faint ester-like odor. Technical grades are yellow. Mp: decomposes >120 °C; bp: 83-84 °C at 0.0002 mmHg; ρ: 1.116-1.118 at 20/4 °C; fl p: difficult to burn; pK_a: <2.5 (estimated); H-$t_{1/2}$ (20 °C): 11.77 hr (pH 3.1), 185 d (pH 7.4), 136 d (pH 9.0), 6.0 d (pH 10.4); K_H: 1.13 x 10^{-7} atm·m^3/mol at 20 °C (approximate - calculated from water solubility and vapor pressure); log K_{oc}: 3.00-3.27; log K_{ow}: 3.02-3.81; S_o: miscible with acetone, benzene, cyclohexane, dimethyl sulfoxide, ethyl ether, ethanol, ketones, methylene chloride, *n*-octanol, petroleum ether, toluene; S_w: 71.1, 53.5, and 43.7 mg/L at 10, 20, and 30 °C, respectively; vp: 8.47 x 10^{-5} mmHg at 20 °C.

Soil properties and adsorption data

Soil	K_d (mL/g)	f_{oc} (%)	K_{oc} (mL/g)	pH
Batcombe silt loam	4.6	2.05	224	6.1
Kanuma high clay	2.4	1.35	175	5.7

continued

Soil	K_d (mL/g)	f_{oc} (%)	K_{oc} (mL/g)	pH
Rothamsted Farm	3.4	1.51	225	5.1
Sand	15.0	0.81	1,847	7.0
Sandy loam	17.9	1.16	1,543	5.4
Silty loam	8.2	0.81	1,010	7.0
Tsukuba clay loam	13.9	4.24	327	6.5

Source: Briggs, 1981; Kanazawa, 1989; Lord et al., 1980; U.S. Department of Agriculture, 1990.

Transformation Products

Biological: Sethunathan and Yoshida (1973a) isolated a *Flavobacterium* sp. (ATCC 27551) from rice paddy water that metabolized diazinon as the sole carbon source. Diazinon was readily hydrolyzed to 2-isopropyl-4-methyl-6-hydroxypyrimidine under aerobic conditions but less rapidly under anaerobic conditions. This bacterium as well as enrichment cultures isolated from a diazinon-treated rice field mineralized the hydrolysis product to carbon dioxide (Sethunathan and Pathak, 1971; Sethunathan and Yoshida, 1973).

Soil: The reported half-life in soil is 32 d (Jury et al., 1987). Reported half-lives in soil following incubation of 10 ppm diazinon in sterile sand loam, sterile organic soil, nonsterile sandy loam and nonsterile organic soil are 12.5, 6.5, <1, and 2 wk, respectively (Miles et al., 1979). The reported half-life of diazinon in sterile soil at pH 4.7 was 43.8 d (Sethunathan and MacRae, 1969).

Plant: Oxidizes in plants to diazoxon (Ralls et al., 1966; Wolfe et al., 1976) although 2-isopropyl-4-methyl-6-pyrimidin-6-ol was identified in bean plants (Kansouh and Hopkins, 1968) and as a hydrolysis product in soil (Somasundaram et al., 1991) and water (Suffet et al., 1967). Five days after spraying, pyrimidine ring-labeled ^{14}C-diazinon was oxidized to oxodiazinon which was then hydrolyzed to 2-isopropyl-4-methylpyrimidin-6-ol which in turn, was further metabolized to carbon dioxide (Ralls et al., 1966). Diazinon was transformed in field-sprayed kale plants to form hydroxydiazinon {O,O-diethyl-O-[2-(2'-hydroxy-2'-propyl)-4-methyl-6-pyrimidinyl] phosphorothioate} which was not previously reported (Pardue et al., 1970). Hydrolyzes in soil to 2-isopropyl-4-methyl-2-hydroxy-pyrimidine, diethylphosphorothioic acid, carbon dioxide (Bartsch, 1974; Getzin, 1967; Lichtenstein et al., 1968; Sethunathan and Pathak, 1972; Sethunathan and Yoshida, 1969; Somasundaram and Coats, 1991; Wolfe et al., 1976), and tetraethyl-pyrophosphate (Paris and Lewis, 1973).

Photolytic: O,O-diethyl-O-[2-(2'-propyl)-4-methyl-6-pyrimidinyl]phosphorothioate (hydroxydiazinon) was reported as a product of UV irradiation of diazinon (Paris and Lewis, 1973). When diazinon in an aqueous buffer solution (25 °C and pH 7.0) was exposed to filtered UV light (λ >290 nm) for 24 hr, 36% decomposed to 2-isopropyl-6-methylpyrimidin-4-ol. The photolysis half-life for this reaction was calculated to be 15 d (Burkhard and Guth, 1979).

Chemical/Physical: In water, diazinon is hydrolyzed following first-order kinetics to form 2-isopropyl-4-methyl-6-hydroxypyrimidine and diethyl thiophosphoric acid or diethyl phosphoric acid in the pH range 3.1 to 10.4 (Gomaa et al., 1969). Cowart et al. (1971) reported a half-life of approximately 2-3 wk in a neutral solution at room temperature. Chapman and Cole (1982) reported the following hydrolysis half-lives of diazinon in a 1% ethanol/buffered water solution at 25 °C: 14, 54.6, 70, and 54 d at pHs of 5.0, 6.0, 7.0, and 8.0, respectively. Diazoxon was also found in fogwater collected near Parlier, CA (Glotfelty et al., 1990). It was suggested that diazinon was oxidized in the atmosphere during daylight hours prior to its partitioning from the vapor phase into the fog. On 12 January 1986, the distribution of diazinon (1.6 ng/m^3) in the vapor phase, dissolved phase, air particles and water particles were 76.1, 19.8, 3.7, and 0.4%, respectively. For diazoxon (0.82 ng/m^3), the distribution in the vapor phase, dissolved phase, air particles and water particles were 13.4, 81.7, 4.9, and 0.02%, respectively (Glotfelty et al., 1990).

Diazinon begins to decompose at a temperature of 100 °C. During distillation procedures at this temperature, <0.5% is decomposed to 2-isopropyl-4-methyl-6-hydroxypyrimidine. When technical diazinon is dissolved in 20% hydrochloric acid, triethyl thiophosphate is formed (Gysin and Margot, 1958). Above 120 °C (Windholz et al., 1983), diazinon decomposes and emits toxic fumes of phosphorus, nitrogen, and sulfur oxides (Sax and Lewis, 1987; Lewis, 1990).

In excess water under acidic conditions, diazinon is hydrolyzed to 2-isopropyl-4-methyl-6-hydroxypyrimidine and diethylthiophosphoric acid. With insufficient water, tetraethyl monothiopyrophosphate is formed (Sittig, 1985).

Exposure Limits: OSHA PEL: TWA 0.1 mg/m^3; ACGIH TLV: TWA 0.1 mg/m^3.

Symptoms of Exposure: Irritates eyes and skin.

Formulation Types: Emulsifiable concentrate; wettable powder; granules; aerosol; dry seed treatment; capsule suspension, smoke tablet; coating agent; microcapsule suspension.

Toxicity: LC$_{50}$ (96 hr) for rainbow trout 16 mg/L, bluegill sunfish 2.6-3.2 mg/L, carp 7.6-23.4 mg/L (Worthing and Hance, 1991), fathead minnow 2.7-10.0 mg/L

(Verschueren, 1983); LC_{50} (24 hr) for bluegill sunfish 52 ppb, and rainbow trout 380 ppb (Verschueren, 1983); acute oral LD_{50} for rats 240-480 mg/kg (Hartley and Kidd, 1987), 66 mg/kg (RTECS, 1985).

Uses: Nonsystemic contact insecticide used against flies, aphids, and spider mites in soil, fruit, vegetables, and ornamentals.

1,2-DIBROMO-3-CHLOROPROPANE

Synonyms: BBC 12; 1-Chloro-2,3-dibromopropane; 3-Chloro-1,2-dibromo-propane; DBCP; Dibromochloropropane; Fumagon; Fumazone; Fumazone 86; Fumazone 86E; NCI-C00500; Nemabrom; Nemafume; Nemagon; Nemagon 20; Nemagon 20G; Nemagon 90; Nemagon 206; Nemagon soil fumigant; Nemanax; Nemapaz; Nemaset; Nematocide; Nematox; Nemazon; OS 1987; Oxy DBCP; RCRA waste number U066; SD-1897; UN 2872.

Structure:

Designations: CAS: 96-12-8; DOT: 2872; mf: $C_3H_5Br_2Cl$; fw: 236.36; RTECS: TX8750000.

Properties: Colorless to brown liquid with a mildly pungent odor. Mp: 6.1 °C; bp: 196 °C; ρ: 2.05 at 20/4 °C; fl p: 76.7 °C (open cup); H-t$_{1/2}$: 38 ± 4 yr at 25 °C and pH 7; K_H: 2.49 x 10^{-4} atm·m^3/mol at 20 °C (approximate - calculated from water solubility and vapor pressure); log K_{oc}: 1.49-2.16; log K_{ow}: 2.63 (calculated); S_o: miscible with oils, acetone, dichloropropane, dimethylsulfoxide, 95% ethanol, halogenated hydrocarbons, isopropanol, liquid hydrocarbons, and methanol; S_w: 1,230 mg/L at 20 °C; vap d: 9.66 g/L at 25 °C, 8.16 (air = 1); vp: 0.8 mmHg at 20 °C.

Soil properties and adsorption data

Soil	K_d (mL/g)	f_{oc} (%)	K_{oc} (mL/g)	pH	CEC (meq/100 g)
Fresno aquifer solid (53.6 m)	0.06	0.02	305	7.7	--
Fresno aquifer solid (109.7 m)	0.07	0.02	355	7.3	--
Panoche clay loam	0.20	0.65	31	--	22.2
Panoche clay loam	0.39	0.27	144	--	17.2
Panoche clay loam	0.30	0.29	103	--	17.4

Source: Biggar et al., 1984; Deeley et al., 1991.

Transformation Products
Soil: Soil water cultures converted 1,2-dibromo-3-chloropropane to *n*-propanol, bromide, and chloride ions. Precursors to the alcohol formation include allyl

139

chloride and allyl alcohol (Castro and Belser, 1968). The reported half-life in soil is 6 mo (Jury et al., 1987).

Chemical/Physical: Hydrolysis of 1,2-dibromo-3-chloropropane yielded the intermediates 2-bromo-3-chloropropene and 2,3-dibromopropene. Both were readily converted to 2-bromoallyl alcohol (Burlinson et al., 1982). Hydrolysis of 1,2-dibromo-3-chloropropane by alkalies give 2-bromoallyl alcohol (Sittig, 1985). Emits toxic chloride and bromide fumes when heated to decomposition (Lewis, 1990).

Exposure Limits: NIOSH REL: 10 hr TWA 10 ppm; OSHA PEL: TWA 1 ppb.

Formulation Types: Fumigant.

Toxicity: Acute oral LD_{50} for male rats 173 mg/kg (Verschueren, 1983), 170 mg/kg (RTECS, 1985).

Uses: Soil fumigant, nematocide; pesticide.

DI-*n*-BUTYL PHTHALATE

Synonyms: 1,2-Benzenedicarboxylate; **1,2-Benzenedicarboxylic acid dibutyl ester;** *o*-Benzenedicarboxylic acid dibutyl ester; Benzene-*o*-dicarboxylic acid di-*n*-butyl ester; Butyl phthalate; *n*-Butyl phthalate; Celluflex DPB; DBP; Dibutyl-1,2-benzenedicarboxylate; Dibutyl phthalate; Elaol; Hexaplas M/B; Palatinol C; Phthalic acid dibutyl ester; Polycizer DBP; PX 104; RCRA waste number U069; Staflex DBP; Witicizer 300.

Structure:

$$\text{COO(CH}_2)_3\text{CH}_3$$
$$\text{COO(CH}_2)_3\text{CH}_3$$

Designations: CAS: 84-74-2; DOT: 9095; mf: $C_{16}H_{22}O_4$; fw: 278.35; RTECS: TI0875000.

Properties: Colorless, oily, viscous liquid with a mild aromatic or ammoniacal odor. Mp: -35 °C; bp: 340 °C; ρ 1.046 at 20/4 °C; fl p: 157 °C; lel: 0.5% at 235 °C; uel: 2.5% (calculated); K_H: 6.3 x 10^{-5} atm·m^3/mol at 20-25 °C (approximate – calculated from water solubility and vapor pressure); log K_{oc}: 3.14; log K_{ow}: 4.31-4.79; S_o: soluble in ethanol, benzene, ethyl ether; very soluble in acetone; miscible with acetone, dimethylsulfoxide, 95% ethanol, and most organic solvents; S_w: 10.1 mg/L at 20 °C; vap d: 11.38 g/L at 25 °C, 9.61 (air = 1); vp: 10^{-5} mmHg at 25 °C.

Transformation Products

Biological: In anaerobic sludge, di-*n*-butyl phthalate degraded as follows: monobutyl phthalate to phthalic acid to protocatechuic acid followed by ring cleavage and mineralization (Shelton et al., 1984). Engelhardt et al. (1975) reported that a variety of microorganisms were capable of degrading of di-*n*-butyl phthalate and suggested the following degradation scheme: di-*n*-butyl phthalate to mono-*n*-butyl phthalate to phthalic acid to 3,4-dihydroxybenzoic acid and other unidentified products. Di-*n*-butyl phthalate was degraded to benzoic acid by tomato cell suspension cultures (*Lycopericon lycopersicum*) (Pogány et al., 1990).

Soil: Under aerobic conditions using a fresh-water hydrosol, mono-*n*-butyl phthalate and phthalic acid were produced. Under anaerobic conditions, however, phthalic acid was not formed (Verschueren, 1983).

Photolytic: An aqueous solution containing titanium dioxide and subjected to UV radiation (λ >290 nm) produced hydroxyphthalates and dihydroxyphthalates as intermediates (Hustert and Moza, 1988).

Chemical/Physical: Hydrolyzes in water to phthalic acid and *n*-butyl alcohol (Lyman et al., 1982). Pyrolysis of di-*n*-butyl phthalate in the presence of polyvinyl chloride at 600 °C gave the following compounds: indene, methylindene, naphthalene, 1-methylnaphthalene, 2-methylnaphthalene, biphenyl, dimethylnaphthalene, acenaphthene, fluorene, methylacenaphthene, methylfluorene, and six unidentified compounds (Bove and Dalven, 1984).

Exposure Limits: NIOSH REL: IDLH 9,300 mg/m^3; OSHA PEL: TWA 5 mg/m^3; ACGIH TLV: TWA 5 mg/m^3.

Symptoms of Exposure: Irritates nasal passages, stomach, upper respiratory system; light sensitivity.

Formulation Types: Aerosol.

Toxicity: Acute oral LD$_{50}$ for rats 8,000 mg/kg (RTECS, 1985).

Use: Insect repellent.

DICAMBA

Synonyms: Banex; Banvel; Banvel CST; Banvel D; Banvel herbicide; Banvel II herbicide; Banvel 4S; Banvel 4WS; Brush buster; Compound B dicamba; Dianate; Dicambe; 3,6-Dichloro-*o*-anisic acid; **2,5-Dichloro-6-methoxybenzoic acid;** 3,6-Dichloro-2-methoxybenzoic acid; MDBA; Mediben; 2-Methoxy-3,6-dichloro-benzoic acid; NA 2769; Velsicol compound 'R'; Velsicol 58-CS-11.

Structure:

Designations: CAS: 1918-00-9; DOT: 2769; mf: $C_8H_6Cl_2O_3$; fw: 221.04; RTECS: DG7525000.

Properties: Odorless, colorless to pale buff crystals. Mp: 114-116 °C; bp: decomposes >200 °C; ρ: 1.57 at 25/4 °C; fl p: nonflammable; pK_a: 1.90, 1.95; K_H: 1.2 x 10^{-9} atm·m^3/mol at 20-25 °C (approximate - calculated from water solubility and vapor pressure); log K_{oc}: -0.40, 0.34; log K_{ow}: 0.48; S_o (g/L at 25 °C): acetone (810), 1,4-diacetone alcohol (910), cyclohexanone (916), diacetone alcohol (910), 1,4-dioxane (1,180), ethanol (922), heavy aromatic naphthalene solvent (52), methylene chloride (260), toluene (130), xylene (78); S_w: 6.5 g/L at 25 °C; vp: 3.38 x 10^{-5} mmHg at 20 °C.

Soil properties and adsorption data

Soil	K_d (mL/g)	f_{oc} (%)	K_{oc} (mL/g)	pH	CEC (meq/100 g)
Indian Head loam	0.03	2.35	1.28	7.8	--
Melfort loam	0.08	6.05	1.32	5.9	--
Melfort loam	0.30	6.05	4.96	5.9	--
Weyburn Oxbow loam	0.05	3.72	1.34	6.5	--
Weyburn Oxbow loam	0.07	3.72	1.88	6.5	--

Source: Grover, 1977; Grover and Smith, 1974.

Transformation Products

Biological: In a model ecosystem containing sand, water, plants, and biota, dicamba was slowly transformed to 5-hydroxydicamba (10% after 32 d) which slowly underwent decarboxylation (Yu et al., 1975).

Soil: Smith (1974) studied the degradation of ^{14}C-ring- and ^{14}C-carboxyl-labeled dicamba in moist prairie soils at 25 °C. After 4 wk, >50% of the herbicide degraded to the principal degradation products 3,6-dichlorosalicylic acid and carbon dioxide (Smith, 1974). The reported half-life in soil is <14-25 d (Worthing and Hance, 1991).

Plant: Dicamba is hydrolyzed in wheat and Kentucky bluegrass plants to 5-hydroxy-2-methoxy-3,6-dichlorobenzoic acid and 3,6-dichlorosalicylic acid at yields of 90 and 5%, respectively. The remaining 5% was unreacted dicamba (Broadhurst et al., 1966). Dicamba was absorbed from treated soils, translocated in corn plants and then converted to 3,6-dichlorosalicylic acid, *p*-aminobenzoic acid and benzoic acid (Krumzdorf, 1974).

Photolytic: When dicamba on silica gel plates was exposed to UV radiation (λ = 254 nm), it slowly degraded to the 5-hydroxy analog and water solubles (Humburg et al., 1989).

Chemical/Physical: Reacts with alkalies (Hartley and Kidd, 1987), amines, and alkali metals (Worthing and Hance, 1991) forming very water-soluble salts. Emits toxic fumes of chlorine when heated to decomposition (Sax and Lewis, 1987). When dicamba was heated at 900 °C, carbon monoxide, carbon dioxide, chlorine, hydrochloric acid, oxygen, and ammonia were produced (Kennedy et al., 1972, 1972a).

Formulation Types: Granules (10%); water-soluble liquid (1 and 4 lb/gal).

Toxicity: LC_{50} (96 hr) for rainbow trout and bluegill sunfish 135 mg/L (Hartley and Kidd, 1987); LC_{50} (48 hr) for bluegill sunfish 20.0 mg/L (Verschueren, 1983), for *Daphnia magna* 110.7 mg/L, for grass shrimp >100.0 mg/L, for both sheaps-head minnow and fiddler crab >180.0 mg/L (Humburg et al., 1989); acute oral LD_{50} of technical dicamba for rats is 1,700 mg/kg (Ashton and Monaco, 1991), pure dicamba 1,039 mg/kg (RTECS, 1985).

Uses: Selective, systemic preemergence and postemergence herbicide used to control both annual and perennial broadleaf weeds, chickweed, mayweed, and bindweed in cereals and other related crops.

DICHLOBENIL

Synonyms: Barrier 2G; Barrier 50W; Carsoron; Casaron; Casoron; Casoron 133; Casoron G; Casoron G-4; Casoron G-10; Casoron W-50; Code H 133; 2,6-DBN; DCB; Decabane; **2,6-Dichlorobenzonitrile**; Du-sprex; Dyclomec; Dyclomec G2; Dyclomec 4G; H 133; H 1313; NA 2769; NIA 5996; Niagara 5006; Niagara 5996; Norosac; Norosac 4G; Norosac 10G.

Structure:

Designations: CAS: 1194-65-6; DOT: 2769; mf: $C_7H_3Cl_2N$; fw: 172.02; RTECS: DI3500000.

Properties: Colorless, white to off-white crystals with an aromatic odor. Mp: 144-145 °C; bp: 270-270.1 °C; fl p: nonflammable; K_H: 6.6 x 10^{-6} atm·m^3/mol at 20-25 °C (approximate - calculated from water solubility and vapor pressure); log K_{oc}: 2.03-2.32; log K_{ow}: 2.90; S_o (g/L at 8 °C unless otherwise noted): acetone (50), benzene (50), 2-butanone (7 g/L at 15-50 °C), cyclohexanone (70 g/L at 15-20 °C), ethanol (50), furfural (70), methylene chloride (10 g/L at 20 °C), 2-propanol, tetrahydro-furan (90), toluene (40), xylene (50); S_w: 18 mg/L at 20 °C, 25 mg/L at 25 °C; vp: 5.5 x 10^{-4} mmHg at 25 °C, 1.5 x 10^{-2} mmHg at 50 °C, 1.1 mmHg at 100 °C.

Soil properties and adsorption data

Soil	K_d (mL/g)	f_{oc} (%)	K_{oc} (mL/g)	pH
Amenia silt loam	5.0	3.4	147	6.2
Camroden silt loam	6.6	4.8	138	4.6
Colonie loamy fine sand	1.9	0.9	211	5.0
Covington silty clay	6.7	5.3	126	5.7
Croghan loamy fine sand	3.7	2.7	137	5.4
Empeyville stony loam	7.8	5.6	139	4.9
Granby fine sandy loam	5.1	3.0	170	6.8
Howard gravelly loam	4.4	3.1	142	4.6
Lackawanna stony loam	6.0	5.6	107	4.6

continued

Soil	K_d (mL/g)	f_{oc} (%)	K_{oc} (mL/g)	pH
Lima gravelly silt	4.2	2.6	162	6.3
Mardin silt loam	4.5	2.9	155	5.9
Sodus gravelly loam	5.3	3.1	171	4.8
Troy gravelly silt loam	3.9	2.4	163	3.8
Vergennes clay	4.1	2.6	158	4.8
Vergennes silt loam	4.5	2.7	167	4.6
Williamson silt loam	4.3	2.6	165	3.9

Source: Briggs and Dawson, 1970.

Transformation Products
Biological: A cell suspension of *Arthrobacter* sp., isolated from a hydrosol, degraded dichlobenil to 2,6-dichlorobenzamide (71% yield), several unidentified water soluble metabolites (Miyazaki et al., 1975). This microorganism was capable of rapidly degrading dichlobenil in both aerobic sediment-water suspensions and in enrichment cultures (Miyazaki et al., 1975).

Soil: The major soil metabolite is 2,6-dichlorobenzamide which undergoes further degradation to form 2,6-dichlorobenzoic acid ($t_{1/2} \approx$ 1-12 mo) (Hartley and Kidd, 1987). Under field conditions, dichlobenil persists from 2-12 mo (Ashton and Monaco, 1991).

Plant: In plants, dichlobenil is transformed into glucose conjugates, insoluble residues, and hydroxy products that are phytotoxic (Ashton and Monaco, 1991). These may include three phytotoxic compounds, namely 2,6-dichlorobenzonitrile, 3-hydroxy-2,6-dichlorobenzonitrile, and 4-hydroxy-2,6-dichlorobenzonitrile (Duke et al., 1991). Massini (1961) provided some evidence that dichlobenil is metabolized by plants. French dwarf beans, tomatoes, gherkin, and oat plants were all exposed to a saturated atmosphere of dichlobenil at room temperature for 4 d. Most of the herbicide was taken absorbed and translocated by the plants in 3 d. After 6 d of exposure, bean seedlings were analyzed for residues using thin-layer plate chromatography. In addition to dichlobenil, another compound was found but it was not 2,6-dichlorobenzoic acid (Massini, 1963).

Photolytic: When dichlobenil was irradiated in methanol with a 450-W mercury lamp and a Corex filter for 8 hr, *o*-chlorobenzonitrile and benzonitrile formed as the major and minor products, respectively (Plimmer, 1970).

Chemical/Physical: Dichlobenil is hydrolyzed, especially in the presence of alkali, to 2,6-dichlorobenzamide (Briggs and Dawson, 1970; Worthing and Hance, 1991).

Emits toxic fumes of nitrogen oxides and chlorine when heated to decomposition (Sax and Lewis, 1987).

Formulation Types: Granules (4 and 10%).

Toxicity: LC_{50} (48 hr) for guppies >18 mg/L (Worthing and Hance, 1991), *Daphnia magna* 10 mg/L (Sanders, 1970), rainbow trout 22.0 mg/L and bluegill sunfish 20 mg/L (Verschueren, 1983); acute oral LD_{50} for rats 3,100 mg/kg (Ashton and Monaco, 1991), 2,710 mg/kg (RTECS, 1985).

Uses: Soil-applied herbicide used to control many annual and perennial broadleaf weeds.

DICHLONE

Synonyms: Algistat; Compound 604; **2,3-Dichloro-1,4-naphthalenedione**; 2,3-Dichloro-1,4-naphthaquinone; Dichloronaphthoquinone; 2,3-Dichloronaphtho-quinone; 2,3-Dichloro-1,4-naphthoquinone; 2,3-Dichloro-α-naphthoquinone; 2,3-Dichloronaphthoquinone-1,4; ENT 3776; NA 2761; Phygon; Phygon paste; Phygon seed protectant; Phygon XL; Quintar; Quintar 540F; Sanquinon; U.S. Rubber 604; Uniroyal; USR 604.

Structure:

Designations: CAS: 117-80-6; DOT: 2761; mf: $C_{10}H_4Cl_2O_2$; fw: 227.06; RTECS: QL7525000.

Properties: Golden yellow leaflets or crystals. Mp: 193 °C (sublimes >32 °C); bp: 275 °C at 2 mmHg; log K_{oc}: 4.19 (calculated); log K_{ow}: 5.62 (calculated); S_o: moderately soluble in acetone, acetic acid, benzene, *o*-dichlorobenzene (≈ 4 wt %), *N,N*-dimethylformamide, 1,4-dioxane, ethanol, ethylbenzene, ethyl ether, ethyl acetate, toluene, xylene (≈ 4 wt %), and many other organic solvents; S_w: 1.0 mg/L at 25 °C.

Transformation Products

Plant: In plants, dichlone loses both chlorine atoms and are replaced by sulphydryl groups to give a substituted dimercapto compound (Hartley and Kidd, 1987).

Photolytic: Irradiation of dichlone in a variety of organic solvents (benzene, isopropanol, ethanol) using UV light produced a number of dehalogenated compounds. In the absence or presence of oxygen, 2-chloro-1,4-naphthoquinone, 1,4-naphthoquinone, and 1,4-naphthalenediol were produced. Further irradiation in the presence of oxygen yielded phthalic acid and phthalic anhydride as the major products. In a mixture of benzene and isopropanol, dichlone degraded to the minor products: 2-chloro-3-hydroxy-1,4-naphthoquinone, 2-chloro-3-phenoxy-1,4-naphthoquinone, 2,3-dichloro-4-hydroxy-1-keto-2-phenyl-1,2-dihydro-naphthalene, and isopropyl-1-chloro-2,3-dioxo-1-indanecarboxylate (Ide et al., 1979).

Chemical/Physical: Emits toxic fumes of chlorine when heated to decomposition (Sax and Lewis, 1987).

Symptoms of Exposure: Irritates skin, eyes and mucous membranes; central nervous system depressant.

Formulation Types: Wettable powder.

Toxicity: LC_{50} (96 hr) for brown trout 310 µg/L (Hartley and Kidd, 1987); LC_{50} (48 hr) for bluegill sunfish 70 µg/L, largemouth bass 120 µg/L, *Daphnia magna* 25 µg/L (Verschueren, 1983); acute oral LD_{50} for rats 1,300 mg/kg (Hartley and Kidd, 1987), 160 mg/kg (RTECS, 1985).

Uses: Fungicide used on fruits, field crops, and vegetables.

1,2-DICHLOROPROPANE

Synonyms: α,β-Dichloropropane; ENT 15406; NCI-C55141; Propylene chloride; Propylene dichloride; α,β-Propylene dichloride; RCRA waste number U083.

Structure:

$$Cl \quad \overset{Cl}{\underset{}{}}$$

Designations: CAS: 78-87-5; DOT: 1279; mf: $C_3H_6Cl_2$; fw: 112.99; RTECS: TX9625000.

Properties: Colorless, mobile liquid with a sweet, chloroform-like odor. Mp: -100.4 °C, -70 °C; bp: 96.4 °C; ρ: 1.560 at 20/4 °C; fl p: 15.6 °C; lel: 3.4%; uel: 14.5%; H-t$_{1/2}$: 15.8 yr at 25 °C and pH 7; K_H: 2.94 x 10^{-3} atm·m^3/mol at 25 °C; IP: 10.87 eV; log K_{oc}: 1.431, 1.71; log K_{ow}: 2.28; S_o: miscible with acetone, dimethyl-sulfoxide, 95% ethanol, and most other organic solvents; S_w: 2,700 mg/L at 20 °C; vap d: 4.62 g/L at 25 °C, 3.90 (air = 1); vp: 42 mmHg at 20 °C.

Transformation Products

Soil: Boesten et al. (1992) investigated the transformation of ^{14}C-labeled 1,2-dichloropropane under laboratory conditions of three subsoils collected from the Netherlands (Wassenaar low-humic sand, Kibbelveen peat, Noord-Sleen humic sand podsoil). The groundwater saturated soils were incubated in the dark at 9.5-10.5 °C. In the Wassenaar soil, no transformation of 1,2-dichloropropane was observed after 156 d of incubation. After 608 and 712 d, however, more than 90% was degraded to nonhalogenated volatile compounds which were detected in the headspace above the soil. These investigators postulated these compounds could be propylene and *n*-propane in a ratio of 8:1. Degradation of 1,2-dichloropropane in the Kibbelveen peat and Noord-Sleen humic sand podsoil was not observed possibly because the soil redox potentials in both soils (50-180 and 650-670 mV, respectively) were higher than the redox potential in the Wassenaar soil (10-20 mV).

Photolytic: Distilled water irradiated with UV light (λ = 290 nm) yielded the following photolysis products: 2-chloro-1-propanol, allyl chloride, allyl alcohol, and acetone (t$_{1/2}$ in distilled water = 50 min but t$_{1/2}$ in distilled water containing hydrogen peroxide = 30 min) (Milano et al., 1988).

Chemical/Physical: Hydrolysis of 1,2-dichloropropane in distilled water at 25 °C produced 1-chloro-2-propanol and hydrochloric acid (Milano et al., 1988). The calculated hydrolysis half-life at 25 °C and pH 7 is 15.8 yr (Ellington et al., 1988). Ozonolysis yielded carbon dioxide at low ozone concentrations (Medley and Stover, 1983). Emits toxic chloride fumes when heated to decomposition (Lewis, 1990).

Exposure Limits: NIOSH REL: IDLH 2,000 ppm; OSHA PEL: TWA 75 ppm, STEL 110 ppm; ACGIH TLV: TWA 75 ppm, STEL 110 ppm.

Symptoms of Exposure: Eye and skin irritation, drowsiness, light-headedness.

Toxicity: LC_{50} (7 d) for guppies 116 ppm (Verschueren, 1983); acute oral LD_{50} for rats 2,196 mg/kg (RTECS, 1985).

Uses: Soil fumigant for nematodes.

cis-1,3-DICHLOROPROPYLENE

Synonyms: *cis*-1,3-Dichloropropene; *cis*-1,3-Dichloro-1-propene; (Z)-1,3-Dichloropropene; **(Z)-1,3-Dichloro-1-propene**; 1,3-Dichloroprop-1-ene; *cis*-1,3-Dichloro-1-propylene.

Structure:

Cl $\diagup\!\!\!\diagdown$ Cl

Designations: CAS: 10061-01-5; DOT: 2047 (isomeric mixture); mf: $C_3H_4Cl_2$; fw: 110.97; RTECS: UC8325000.

Properties: Colorless to amber-colored liquid with a sweet, penetrating, chloroform-like odor. Mp: -84 °C (isomeric mixture), bp: 104.3 °C; ρ: 1.224 at 20/4 °C; fl p: 35 °C (isomeric mixture); lel: 5.3% (isomeric mixture); uel: 14.5% (isomeric mixture); K_H: 3.55 x 10^{-3} atm·m^3/mol; log K_{oc}: 1.36, 1.68; log K_{ow}: 1.41; S_o: soluble in benzene, chloroform, 95% ethanol, ethyl ether, *n*-octane, toluene; S_w: 2,700 mg/L at 25 °C; vap d: 4.54 g/L at 25 °C, 3.83 (air = 1); vp: 25 mmHg at 20 °C.

Soil properties and adsorption data

Soil	$K_d{}^a$ (mL/g)	f_{oc} (%)	$K_{oc}{}^a$ (mL/g)	pH	CEC (meq/100 g)
Humus sand	14.0	3.17	442	--	--
Humus sand	22.0	3.17	694	--	--
Humus sand	38.0	3.17	1,199	--	--
Peaty sand	47.0	10.39	452	--	--
Peaty sand	78.0	10.39	751	--	--
Peaty sand	130.0	10.39	1,251	--	--

Source: Leistra, 1970. a) The reported K_d and calculated K_{oc} values are expressed on a soil/vapor relationship, i.e., (μg absorbed per g of dry soil)/(μg vapor per mL vapor phase). The author reported an average K_{om} (μg absorbed per g of organic matter)/(μg dissolved per mL water phase) of 14 which is approximately equal to a K_{oc} value of 26.

Transformation Products
Soil: Hydrolyzes in wet soil forming *cis*-3-chloroallyl alcohol (Castro and Belser,

1966). *cis*-1,3-Dichloropropylene was reported to hydrolyze to 3-chloro-2-propen-1-ol and can be biologically oxidized to 3-chloropropenoic acid which is further oxidized to formylacetic acid. Decarboxylation of this compound yields carbon dioxide (Connors et al., 1990).

Chemical/Physical: Hydrolyzes in distilled water at 25 °C forming 2-chloro-3-propenol and hydrochloric acid ($t_{1/2}$ = 1 d) (Milano et al., 1988). Chloroacetaldehyde, formyl chloride, and chloroacetic acid were formed from the ozonation of dichloropropylene at about 23 °C and 730 mmHg. Chloroacetaldehyde and formyl chloride also formed from the reaction of dichloropropylene with hydroxyl radicals (Tuazon et al., 1984). Emits chlorinated acids when incinerated. Incomplete combustion may release toxic phosgene (Sittig, 1985).

Formulation Types: Liquid.

Toxicity: LC_{50} (96 hr) for bluegill sunfish 7.1 mg/L and rainbow trout 7.1 mg/L (Worthing and Hance, 1991); acute oral LD_{50} of the isomeric mixture for male and female rats is 713 and 470 mg/kg, respectively (Verschueren, 1983).

Uses: The isomeric mixture is used as a soil fumigant and a nematocide.

trans-1,3-DICHLOROPROPYLENE

Synonyms: (*E*)-1,3-Dichloropropene; *trans*-1,3-Dichloropropene; (*E*)-1,3-Dichloro-1-propene; *trans*-1,3-Dichloro-1-propene; 1,3-Dichloroprop-1-ene; *trans*-1,3-Dichloro-1-propylene.

Structure:

Designations: CAS: 10061-02-6; DOT: 2047 (isomeric mixture); mf: $C_3H_4Cl_2$; fw: 110.97; RTECS: UC8320000.

Properties: Clear, colorless liquid with a chloroform-like odor. Mp: -84 °C (isomeric mixture), bp: 77 °C at 757 mmHg, 112.1 °C; ρ: 1.1818 at 20/4 °C; fl p: 5.3 °C (isomeric mixture); lel: 5.3% (isomeric mixture); uel: 14.5% (isomeric mixture); K_H: 3.55 x 10^{-3} atm·m^3/mol; log K_{oc}: 1.415, 1.68; log K_{ow}: 1.41; S_o: soluble in acetone, benzene, chloroform, dimethylsulfoxide, 95% ethanol, ethyl ether, *n*-octane, toluene; S_w: 2,800 mg/L at 25 °C; vap d: 4.54 g/L at 25 °C, 3.83 (air = 1); vp: 25 mmHg at 20 °C.

Soil properties and adsorption data

Soil	$K_d{}^a$ (mL/g)	f_{oc} (%)	$K_{oc}{}^a$ (mL/g)	pH	CEC (meq/100 g)
Humus sand	14.0	3.17	442	--	--
Humus sand	22.0	3.17	694	--	--
Humus sand	38.0	3.17	1,199	--	--
Peaty sand	47.0	10.39	452	--	--
Peaty sand	78.0	10.39	751	--	--
Peaty sand	130.0	10.39	1,251	--	--

Source: Leistra, 1970. a) The reported K_d and calculated K_{oc} values are expressed on a soil/vapor relationship, i.e., (μg absorbed per g of dry soil)/(μg vapor per mL vapor phase). The author reported an average K_{om} (μg absorbed per g of organic matter)/(μg dissolved per mL water phase) of 14 which is approximately equal to a K_{oc} value of 24.

Transformation Products
Soil: Hydrolysis in wet soil resulted in the formation of *trans*-3-chloroallyl alcohol

154

(Castro and Belser, 1966). *trans*-1,3-Dichloropropylene was reported to hydrolyze to 3-chloro-2-propen-1-ol and can be biologically oxidized to 3-chloropropenoic acid which is further oxidized to formylacetic acid. Decarboxylation of this compound yields carbon dioxide (Connors et al., 1990).

Chemical/Physical: Hydrolyzes in distilled water at 25 °C forming 2-chloro-3-propenol and hydrochloric acid ($t_{1/2}$ = 2 d) (Milano et al., 1988). Chloroacetaldehyde, formyl chloride, and chloroacetic acid were formed from the ozonation of dichloropropylene at about 23 °C and 730 mmHg. Chloroacetaldehyde and formyl chloride also formed from the reaction of dichloropropylene with hydroxyl radicals (Tuazon et al., 1984). Emits chlorinated acids when incinerated. Incomplete combustion may release toxic phosgene (Sittig, 1985).

Formulation Types: Liquid.

Toxicity: LC_{50} (96 hr) for bluegill sunfish 7.1 mg/L and rainbow trout 7.1 mg/L (Worthing and Hance, 1991); acute oral LD_{50} of the isomeric mixture for male and female rats is 713 and 470 mg/kg, respectively (Verschueren, 1983).

Uses: The isomer mixture is used as a soil fumigant and a nematocide.

DICHLORVOS

Synonyms: Apavap; Astrobot; Atgard; Atgard C; Atgard V; Bay 19149; Benfos; Bibesol; Brevinyl; Brevinyl E50; Canogard; Cekusan; Chlorvinphos; Cyanophos; Cypona; DDVF; DDVP; Dedevap; Deriban; Derribante; Devikol; Dichlorman; 2,2-Dichloroethenyl dimethyl phosphate; 2,2-Dichloroethenyl phosphoric acid dimethyl ester; Dichlorophos; 2,2-Dichlorovinyl dimethyl phosphate; 2,2-Dichlorovinyl dimethyl phosphoric acid ester; Dichlorovos; Dimethyl 2,2-dichloroethenyl phosphate; Dimethyl dichlorovinyl phosphate; Dimethyl 2,2-dichlorovinyl phosphate; *O,O*-Dimethyl *O*-(2,2-dichlorovinyl)phosphate; Divipan; Duo-kill; Duravos; ENT 20738; Equigard; Equigel; Estrosel; Estrosol; Fecama; Fly-die; Fly fighter; Herkal; Herkol; Krecalvin; Lindan; Mafu; Mafu strip; Marvex; Mopari; NA 2783; NCI-C00113; Nerkol; Nogos; Nogos 50; Nogos G; No-pest; No-pest strip; NSC 6738; Nuva; Nuvan; Nuvan 100EC; Oko; OMS 14; **Phosphoric acid 2,2-dichloroethenyl dimethyl ester;** Phosphoric acid 2,2-dichlorovinyl dimethyl ester; Phosvit; SD-1750; Szklarniak; Tap 9VP; Task; Task tabs; Tenac; Tetravos; UDVF; Unifos; Unifos 50 EC; Vapona; Vaponite; Vapora II; Verdican; Verdipor; Vinyl alcohol 2,2-dichlorodimethyl phosphate; Vinylofos; Vinylophos.

Structure:

$$CH_3O \diagdown \overset{\overset{O}{\|}}{\underset{CH_3O \diagup}{P}} - O - CH = CCl_2$$

Designations: CAS: 62-73-7; DOT: 2783; mf: $C_4H_7Cl_2O_4P$; fw: 220.98; RTECS: TC0350000.

Properties: Colorless to yellow liquid with an aromatic odor. Mp: <25 °C; bp: 35, 120, and 140 °C at 0.05, 14, and 20 mmHg, respectively; ρ: 1.44 and 1.415 at 20/4 and 25/4 °C, respectively; fl p: practically nonflammable; H-t$_{1/2}$: 462 min (pH 7), 30 min (pH 8); K_H: 5.0 x 10^{-3} atm·m^3/mol; log K_{oc}: 1.70 (calculated); log K_{ow}: 1.40-2.29; P-t$_{1/2}$: 7.25 hr (absorbance λ = 226.0 nm, concentration on glass plates = 6.7 μg/cm^2); S_o: soluble in glycerol ($\approx$ 5 g/L); miscible with alcohol and most nonpolar solvents; S_w: 16.0 g/L at 20 °C; vap d: 9.03 g/L at 25 °C, 7.63 (air = 1); vp: 0.012 mmHg at 20 °C.

Transformation Products

Biological: When dichlorvos was incubated with sewage sludge for 1 wk at 29 °C, it was converted to dichloroethanol, dichloroacetic acid, ethyl dichloroacetate, and an inorganic phosphate. In addition, dimethyl phosphate formed in the presence or absence of microorganisms (Lieberman and Alexander, 1983).

156

Plant: Metabolites identified in cotton leaves include dimethyl phosphate, phosphoric acid, methyl phosphate, and *O*-demethyl dichlorvos (Bull and Ridgway, 1969).

Chemical/Physical: Releases very toxic fumes of phosphorus oxides and chlorine when heated to decomposition (Sax and Lewis, 1987). Slowly hydrolyzes in water and in acidic media but is more rapidly hydrolyzed under alkaline conditions to dimethyl hydrogen phosphate and dichloroacetaldehyde (Hartley and Kidd, 1987; Worthing and Hance, 1991).

Exposure Limits: NIOSH REL: IDLH 200 mg/m^3; OSHA PEL: TWA 0.1 ppm; ACGIH TLV: TWA 0.1 ppm.

Symptoms of Exposure: Miosis, ache eyes; rhinorrhea; headache; chest, wheezing, laryngeal spasm, salivation; cyanosis; anorexia, nausea, vomiting, diarrhea, sweating; muscle fasiculation, paralysis, ataxia; convulsions; low blood pressure.

Formulation Types: Emulsifiable concentrate; granules; hot and cold fogging concentrates; oil-miscible liquid; aerosol; impregnated strip.

Toxicity: LC$_{50}$ (96 hr) for bluegill sunfish 869 μg/L (Verschueren, 1983); LC$_{50}$ (24 hr) for bluegill sunfish 1.0 mg/L (Hartley and Kidd, 1987); acute oral LD$_{50}$ for rats 56-80 mg/kg (Hartley and Kidd, 1987), 25 mg/kg (RTECS, 1985).

Uses: Insecticide and fumigant used against flies, mosquitoes, and moths.

DICLOFOP-METHYL

Synonyms: 2-(4-(2,4-Dichlorophenoxy)phenoxy)methyl propionoate; 2-(4-(2,4-Dichlorophenoxy)phenoxy)propanoic acid methyl ester; Hoe 23408; Hoe-Grass; Hoelon; Hoelon 3EC; Illoxan; Iloxan; Methyl 2-(4-(2,4-dichlorophenoxy)phenoxy)propanoate.

Structure:

Designations: CAS: 51338-27-3; mf: $C_{16}H_{14}Cl_2O_4$; fw: 341.20; RTECS: UF1180000.

Properties: Colorless and odorless crystals. Mp: 39-41 °C; bp: 175-176 °C at 0.1 mmHg, decomposes at 288.4 °C; ρ: 1.30 at 40/4 °C; H-t$_{1/2}$: 1 d (pH 3), 265 d (pH 5), 29 d (pH 7); log K_{oc}: 4.89; log K_{ow}: 4.58; S_o (kg/L at 20 °C): acetone (2.49), ethanol (0.11), ethyl ether (2.28), xylene (2.53); S_w: 3 g/L at 22 °C; vp: 2.6 x 10^{-6} mmHg at 20 °C.

Transformation Products
Soil: Under aerobic conditions, diclofop-methyl decomposes in soil forming diclofop-acid (Hartley and Kidd, 1987; Smith, 1977, 1979; Humburg et al., 1989) which undergoes further degradation to 4-(2,4-dichlorophenoxy)phenol (Smith, 1979), 4-(2,4-dichlorophenoxy)ethoxybenzene (Smith, 1977, 1979), and hydroxylated free acids (Hartley and Kidd, 1987; Humburg et al., 1989). The half-lives in sandy soils and sandy clay soils were reported to be 10 and 30 d, respectively (Ashton and Monaco, 1991).

Formulation Types: Emulsifiable concentrate (3 lb/gal).

Toxicity: LC$_{50}$ (96 hr) for rainbow trout 350 μg/L (Hartley and Kidd, 1987); acute oral LD$_{50}$ for rats 557-580 (sesame oil) mg/kg (Hartley and Kidd, 1987), 563 mg/kg (RTECS, 1985).

Uses: Postemergence herbicide used to control annual grasses, including wild oats, in flax, lentils, peas, soybeans, wheat, and barley crops.

DICROTOPHOS

Synonyms: Bidirl; Bidrin; C 709; Carbicron; CIBA 709; **(E)-3-(Diethylamino)-1-methyl-3-oxo-1-propenyl dimethyl phosphate**; 3-(Dimethoxyphosphinyloxy)-N,N-dimethyl-cis-crotonamide; 3-(Dimethoxyphosphinyloxy)-N,N-dimethyliso-crotonamide; 3-(Dimethylamino)-1-methyl-3-oxo-1-propenyl dimethyl phosphate; cis-2-Dimethylcarbamoyl-1-methylvinyl dimethyl phosphate; O,O-Dimethyl O-(N,N-dimethylcarbamoyl-1-methylvinyl)phosphate; O,O-Dimethyl-O-(1,4-di-methyl-3-oxo-4-azapent-1-enyl)phosphate; Dimethyl phosphate of 3-hydroxy-N,N-dimethyl-cis-crotonamide; Dimethyl phosphate ester with 3-hydroxy-N,N-dimethyl-cis-crotonamide; Dimethyl phosphate ester with (E)-3-hydroxy-N,N-dimethylcrotonamide; ENT 24482; 3-Hydroxy-N,N-dimethyl-cis-crotonamide dimethyl phosphate; SD 3562; Shell SD-3562.

Structure:

Designations: CAS: 141-66-2; DOT: 2783; mf: $C_8H_{16}NO_5P$; fw: 237.20; RTECS: TC3850000.

Properties: Yellow to brown liquid with a mild ester-like odor. Mp: <25 °C; bp: 130 °C at 0.1 mmHg (pure), 400 °C (technical); ρ: 1.216 at 20/4 °C; fl p: >93.3 °C; H-t$_{1/2}$ (20 °C): 88 d (pH 5), 23 d (pH 9); log K_{oc}: 1.04-2.27; log K_{ow}: -0.50; S_o: slightly soluble in kerosene but miscible with acetone, ethanol, hexylene glycol, isobutyl alcohol, xylene; S_w: miscible; vp: 6.98 x 10^{-5} mmHg at 20 °C.

Soil properties and adsorption data

Soil	K_d (mL/g)	f_{oc} (%)	K_{oc} (mL/g)	pH	CEC (meq/100 g)
Catlin silty loam	0.92	2.32	40	5.7	13.0
Georgia sand	0.07	0.64	11	6.7	2.0
Handford sandy loam	0.40	0.75	53	6.4	7.1
Sharkey clay loam	3.58	1.91	187	5.9	23.5

Source: Lee et al., 1989.

Transformation Products
Chemical/Physical: Emits toxic fumes of phosphorus and nitrogen oxides when

heated to decomposition (Sax and Lewis, 1987; Lewis, 1990). Dicrotophos is hydrolyzed in sodium hydroxide solutions forming dimethylamine. The hydrolysis half-lives at 38 °C and pHs 1.1 and 9.1 are 100 and 50 d, respectively (Sittig, 1985). Lee et al. (1989) reported that the hydrolysis half-lives of dicrotophos in pH 5, 7, and 9 buffer solutions at 25 °C were 117, 72, and 28 d, respectively. *N,N*-Dimethylacetoacetamide and *O*-desmethyldicrotophos were the major products identified.

Soil: The dimethylamino group is converted to an *N*-oxide then to $-CH_2OH$ and aldehyde groups which further degrade via demethylation and hydrolysis (Hartley and Kidd, 1987). Dicrotophos is rapidly degraded under both aerobic and anaerobic conditions forming *N,N*-dimethylacetoacetamide and 3-hydroxy-*N,N*-dimethylbutyramide as the major metabolites. Other metabolites included carbon dioxide and unextractable residues. The half-life of dicrotophos in a Hanford sandy loam soil was 3 d (Lee et al., 1989).

Exposure Limits: OSHA PEL: TWA 0.25 mg/m^3; ACGIH TLV: TWA 0.25 mg/m^3.

Symptoms of Exposure: Headache, anorexia, nausea, vertigo, weakness, abdominal cramps, diarrhea, salivation, lacrimation, ataxia, cyanosis, pulmonary edema, convulsions, coma, shock.

Formulation Types: Emulsifiable concentrate; water soluble concentrate.

Toxicity: LC_{50} (24 hr) for harlequin fish >1,000 mg/L and mosquito fish 200 mg/L (Hartley and Kidd, 1987); acute oral LD_{50} for rats 17-22 mg/kg (Hartley and Kidd, 1987), 13 mg/kg (RTECS, 1985), for male and female rats, 21 and 16 mg/kg, respectively (Windholz et al., 1983).

Uses: Contact and systemic insecticide and acaricide used to control pests on rice, cotton, maize, soybeans, coffee, citrus, and potatoes.

DIELDRIN

Synonyms: Alvit; Compound 497; Dieldrite; Dieldrix; ENT 16225; HEOD; Hexachloroepoxyoctahydro-*endo,exo*-dimethanonaphthalene; 1,2,3,4,10,10-Hexachloro-6,7-epoxy-1,4,4a,5,6,7,8,8a-octahydro-1,4-*endo,exo*-5,8-dimethanonaphthalene; **3,4,5,6,9,9-Hexachloro-1a,2,2a,3,6,6a,7,7a-octahydro-2,7:3,6-dimethanonaphth[2,3-*b*]oxirene**; Illoxol; Insecticide 497; NA 2761; NCI-C00124; Octalox; Panoram D-31; Quintox; RCRA waste number P037.

Structure:

Designations: CAS: 60-57-1; DOT: 2761; mf: $C_{12}H_8Cl_6O$; fw: 380.91; RTECS: IO1750000.

Properties: A stereo-isomer of endrin, dieldrin may be white crystals or pale tan flakes with an odorless to mild chemical odor. Odor threshold: 41 μg/kg. Mp: 175–176 °C; bp: decomposes; ρ: 1.75 at 20/4 °C; fl p: nonflammable; H-t$_{1/2}$: 10.5 yr at 25 °C and pH 7; K_H: 5.8 x 10^{-5} atm·m^3/mol at 25 °C; log K_{oc}: 4.08–4.55; log K_{ow}: 3.692–6.2; P-t$_{1/2}$: 153.84 hr (absorbance λ = 227.0 nm, concentration on glass plates = 6.7 μg/cm^2); S_o (g/L at 20 °C): benzene (400), carbon tetrachloride (380), cod liver oil (57.7, 61.8, 77.8 g/L at 4, 12, and 20 °C, respectively), methylene chloride (480), *n*-octanol (91.2, 39.9, and 40.8 g/L at 4, 12, and 20 °C, respectively), triolein (49.1, 63.2, and 72.6 at 4, 12, and 20 °C, respectively); S_w: 80, 140, and 200 μg/L at 10, 20, and 30 °C, respectively; vap d: 54 ng/L at 20 °C; vp: 1.78 x 10^{-7} mmHg at 20 °C.

Soil properties and adsorption data

Soil	K_d (mL/g)	f_{oc} (%)	K_{oc} (mL/g)	pH	CEC (meq/100 g)
Batcombe silt loam	268	2.10	12,762	6.1	--
Bryce silty clay loam	297	7.50	3,960	--	55.5
Bryce-Swygert silty clay	265	2.70	9,815	--	34.4
Clay loam	195	1.42	14,000	5.9	12.4
Drummer silty clay loam	198	3.10	6,387	--	24.8
Flanagan silty loam	260	3.50	7,429	--	27.7

continued

Soil	K_d (mL/g)	f_{oc} (%)	K_{oc} (mL/g)	pH	CEC (meq/100 g)
Gilford-Hoopeston-Ade sandy loam	147	1.20	12,250	--	7.5
Houghton muck	1,507	16.80	8,970	--	72.4
Plainfield-Bloomfield sand	39	0.40	9,750	--	1.7
Rothamsted Farm	194	1.51	12,847	5.1	--

Source: Briggs, 1981; Felsot and Wilson, 1980; Kishi et al., 1990; Lord et al., 1980.

Transformation Products

Soil/Biological: Identified metabolites of dieldrin from solution cultures containing *Pseudomonas* sp. in soils include aldrin and dihydroxydihydroaldrin. Other unidentified by-products included a ketone, an aldehyde, and an acid (Matsumura et al., 1968; Kearney and Kaufman, 1976). The reported half-life in soil is 868 d (Jury et al., 1987). At least 10 different types of bacteria comprising a mixed anaerobic population degraded dieldrin, via monodechlorination at the methylene bridge carbon, to give *syn-* and *anti*-monodechlorodieldrin. Three isolates, *Clostridium bifermentans*, *Clostridium glycolium*, and *Clostridium* sp., were capable of dieldrin dechlorination but the rate was much lower than that of the mixed population (Maule et al., 1987). A pure culture of the marine alga, namely *Dunaliella* sp, degraded dieldrin to photodieldrin and an unknown metabolite at yields of 8.5 and 3.2%, respectively. Photodieldrin and the diol were also identified as metabolites in field collected samples of marine water, sediments, and associated biological materials (Patil et al., 1972).

Photolytic: Photolysis of an aqueous solution by sunlight for 3 mo resulted in a 70% yield of photodieldrin (Kearney and Kaufman, 1976). A solid film of dieldrin exposed to sunlight for 2 mo resulted in a 25% yield of photodieldrin (Benson, 1971). In addition to sunlight, UV light converts dieldrin to photodieldrin (Georgacakis and Khan, 1971). Solid dieldrin exposed to UV light (λ <300 nm) under a stream of oxygen yielded small amounts of photodieldrin (Gäb et al., 1974). Many other investigators reported photodieldrin as a photolysis product of dieldrin under variety of conditions (Crosby and Moilanen, 1974; Ivie and Casida, 1970, 1971, 1971a; Rosen and Carey, 1968; Rosen et al., 1966; Robinson et al., 1966). After a 1-hr exposure to sunlight, dieldrin was converted to photodieldrin. Photodecomposition was accelerated by a number of photosensitizing agents (Ivie and Casida, 1971a). When an aqueous solution containing dieldrin was photo-oxidized by UV light at 90-95 °C, 25, 50, and 75% degraded to carbon dioxide after 2.9, 4.8, and 12.5 hr, respectively (Knoevenagel and Himmelreich, 1976).

Chemical/Physical: Products reported from the combustion of dieldrin at 900 °C include carbon monoxide, carbon dioxide, hydrochloric acid, chlorine, and unidentified compounds (Kennedy et al., 1972a). Emits very toxic chloride fumes when heated to decomposition (Lewis, 1990).

Exposure Limits: NIOSH REL: 0.15 mg/m^3; OSHA PEL: TWA 0.25 mg/m^3; ACGIH TLV: TWA 0.25 mg/m^3.

Symptoms of Exposure: Headache, dizziness, nausea, vomiting, malaise, sweating, myoclonic limb jerks, clonic and tonic convulsions, coma, respiratory failure.

Formulation Types: Emulsifiable concentrate; wettable powder; granules; dustable powder.

Toxicity: LC_{50} (96 hr) for goldfish 37 μg/L (Hartley and Kidd, 1987), bluegill sunfish 8 μg/L, fathead minnow 16 μg/L (Henderson et al., 1959), rainbow trout 10 μg/L, coho salmon 11 μg/L, chinook 6 μg/L (Katz, 1961), pumpkinseed 6.7 μg/L, channel catfish 4.5 μg/L (Verschueren, 1983); LC_{50} (48 hr) for mosquito fish 8 ppb (Verschueren, 1983); LC_{50} (24 hr) for bluegill sunfish 170 ppb and fathead minnow 24 ppb (Verschueren, 1983); acute oral LD_{50} for rats 37–87 g/kg (Hartley and Kidd, 1987), 38,300 mg/kg (RTECS, 1985).

Use: Insecticide.

DIFENZOQUAT METHYL SULFATE

Synonyms: AC 84777; Avenge; 1,2-Dimethyl-3,5-diphenyl-1*H*-pyrazolium methyl sulfate; Finaven; Mataven; Superaven; Yeh-Yan-Ku.

Structure:

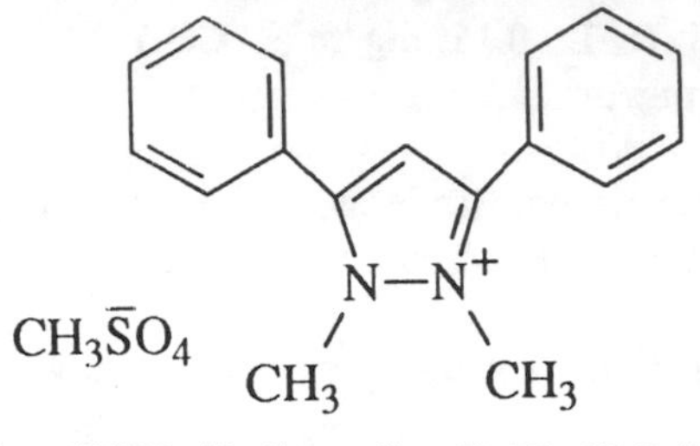

$CH_3\overline{S}O_4$

Designations: CAS: 43222-48-6; mf: $C_{18}H_{20}N_2O_4S$; fw: 360.40; RTECS: UQ9820000.

Properties: Colorless to off-white, odorless, hygroscopic solid. Mp: 156.5-158 °C with decomposition; ρ: 1.13 at 25 °C; fl p: >82 °C (technical grade and formulation); K_H: 5.66 x 10^{-14} atm·m^3/mol at 20-25 °C (approximate - calculated from water solubility and vapor pressure); log K_{oc}: 4.49-5.80; log K_{ow}: 0.65 (pH 5), -0.62 (pH 7), -0.32 (pH 9); S_o (g/L at 25 °C unless otherwise noted): acetone (130 g/L at 20 °C), chlorobenzene (0.4), chloroform (5,000 g/L at 20 °C), dichloroethane (80), methanol (6,200), 2-propanol (0.7), xylene (<0.1 g/L at 20 °C); S_w: 817 g/L at 25 °C; vp: 9.75 x 10^{-8} mmHg at 20 °C.

Soil properties and adsorption data

Soil	K_d (mL/g)	f_{oc} (%)	K_{oc} (mL/g)	pH
Clayey loam	2,680	2.90	92,414	7.7
Sandy loam	181	0.58	31,207	6.9
Sandy clay loam	636	1.80	35,333	6.4
Silty loam	1,093	1.74	62,816	5.2

Source: U.S. Department of Agriculture, 1990.

Transformation Products

Soil: Though no products were reported, the half-life in soil is approximately 3 mo (Hartley and Kidd, 1987).

Photolytic: Degrades photolytically to the volatile monomethyl pyrazole (Hartley and Kidd, 1987).

Chemical/Physical: May react with aluminum releasing hydrogen (Hartley and Kidd, 1987).

Symptoms of Exposure: Do not get in eyes - causes irreversible eye damage.

Formulation Types: Soluble concentrate (aqueous solution for cation = 2 lb/gal); water-soluble powder.

Toxicity: LC_{50} (96 hr) for rainbow trout 694 mg/L and bluegill sunfish 696 mg/L (Worthing and Hance, 1991); acute oral LD_{50} for male rats 270 mg/kg (Ashton and Monaco, 1991).

Uses: Selective postemergence herbicide used to control wild oats in wheat, barley, flax, maize, rye grass, vetches, and rye grass crops.

DIFLUBENZURON

Synonyms: *N*-(((4-Chlorophenyl)amino)carbonyl)-2,6-difluorobenzamide; 1-(4-Chlorophenyl)-3-(2,6-difluorobenzoyl)urea; Dfluron; Dimilin; DU 112307; ENT 29054; OMS 1804; PDD 60401; PH 60-40; Philips-duphar PH 60-40; TH 6040; Thompson–Hayward TH6040.

Structure:

Designations: CAS: 35367-38-5; mf: $C_{14}H_9ClF_2N_2O_2$; fw: 310.69; RTECS: YS6200000.

Properties: Colorless crystals when pure. Technical grade is off-white to yellow crystalline solid. Mp: 230-232 °C (pure), 210-230 °C (technical); bp: decomposes; K_H: 7.3 x 10^{-9} atm·m^3/mol at 20-25 °C (approximate - calculated from water solubility and vapor pressure); log K_{oc}: 3.01 (calculated); log K_{ow}: 3.29 (calculated); S_o (g/L at 25 °C): *N,N*-dimethylformamide (104), 1,4-dioxane (20); S_w: 14 mg/L at 25 °C; vp: 2.5 x 10^{-7} mmHg at 20 °C.

Transformation Products
Soil: Though no products were reported, the half-life in soil is <1 wk (Hartley and Kidd, 1987).

Chemical/Physical: Hydrolyzes in water to 4-chlorophenylurea (Verschueren, 1983).

Formulation Types: Granules; wettable powder.

Toxicity: LC_{50} (96 hr) for rainbow trout 140 mg/L, bluegill sunfish 135 mg/L (Hartley and Kidd, 1987), coho salmon, and juvenile rainbow trout >150 mg/L (Verschueren, 1983); acute oral LD_{50} for mice 4,640 mg/kg (RTECS, 1985).

Uses: Nonsystemic insecticide used to control leaf-eating larvae and leaf miners in forestry, woody ornamentals, and fruit trees.

DIMETHOATE

Synonyms: AC 12880; AC 18682; American Cyanamid 12880; BI 58; BI 58 EC; 8014 Bis HC; Cekuthoate; Chemathoate; CL 12880; Cygon; Cygon 4E; Cygon insecticide; Daphene; De-Fend; Demos-L40; Devigon; Dimate 267; Dimetate; Dimethoate-267; Dimethoat tecvhnisch 95%; Dimethogen; *O,O*-Dimethyl dithiophosphorylacetic acid *N*-monomethylamide salt; *O,O*-Dimethyl *S*-(2-(methylamino)-2-oxoethyl) phosphorodithioate; *O,O*-Dimethyl *S*-(*N*-methylcarbamoylmethyl)dithiophosphate; *O,O*-Dimethyl *S*-methylcarbamoylmethyl phosphorodithioate; *O,O*-Dimethyl *S*-(*N*-methylcarbamoylmethyl) phosphorodithioate; *O,O*-Dimethyl *S*-(*N*-methylcarbamylmethyl) thiothionophosphate; *O,O*-Dimethyl *S*-(*N*-monomethyl)carbamyl methyldithiophosphate; Dimeton; Dimevur; EI-12880; ENT 24650; Experimental insecticide 12880; Ferkethion; Fip; Fortion NM; Fosfamid; Fosfotox; Fosfotox R; Fosfotox R 35; Fostion MM; L 395; Lurgo; *S*-Methylcarbamoylmethyl *O,O*-dimethyl phosphorodithioate; *N*-Monomethylamide of *O,O*-dimethyldithiophosphorylacetic acid; NC 262; NCI-C00135; PEI 35; Perfecthion; Perfekthion; Perfektion; Phosphamid; Phosphamide; Phosphorodithioic acid *O,O*-dimethyl ester, ester with 2-mercapto-*N*-methylacetamide; **Phosphorodithioic acid *O,O*-dimethyl-*S*-(2-(methylamino)-2-oxoethyl) ester**; Racusan; RCRA waste number P044; Rebelate; Rogodial; Rogor; Rogor 40; Rogor L; Rogor 20L; Rogor P; Roxion; Roxion U.A.; Sinoratox; Trimetion.

Structure:

$$CH_3O \diagdown \overset{\overset{S}{\|}}{\underset{CH_3O \diagup}{P}} - SCH_2CONHCH_3$$

Designations: CAS: 60-51-5; DOT: 2783; mf: $C_5H_{12}NO_3PS_2$; fw: 229.30; RTECS: TE1750000.

Properties: Colorless crystals with a mercaptan-like odor. Technical grade (96%) forms white to grayish crystals. Mp: 52-52.5 °C; bp: 117 °C at 0.1 mmHg (technical grade); ρ: 1.281 and 1.277 at 50/4 and 65/4 °C, respectively; fl p: burns readily on contact with flame; H-t$_{1/2}$: 118 hr at 25 °C and pH 7; K_H: 2.63 x 10^{-11} atm·m^3/mol at 20-21 °C (approximate - calculated from water solubility and vapor pressure); log K_{oc}: 0.96; log K_{ow}: 0.508-0.78; P-t$_{1/2}$: 64.10 hr (absorbance λ = 226.5 nm, concentration on glass plates = 6.7 $\mu g/cm^2$); S_o: very soluble in lower alcohols, benzene, carbon tetrachloride, chloroform, ethylbenzene, methyl chloride, methylene chloride, methyl ethyl ketone, toluene, xylene, and many other common organic solvents except saturated hydrocarbons; S_w: 25 g/L at 21 °C; vp: 5.06 x 10^{-6} mmHg at 20 °C.

Soil properties and adsorption data

Soil	K_d (mL/g)	f_{oc} (%)	K_{oc} (mL/g)	pH
Alluvial-Rio Nacimiento	1.51	0.91	166	8.0
Batcombe silt loam	0.20	2.05	10	6.1
Brown Clay-Almanzora Alto	4.21	1.17	359	8.5
Brown Lime-Almanzora Bajo	3.15	1.48	213	8.9
Brown Lime-Los Velez	8.94	2.06	434	8.1
Coarse sand (Jutland, Denmark)	0.08	0.21	38	5.3
Desert-Campo de Tabernas	2.21	0.33	670	7.9
Kanuma high clay	0.50	1.35	36	5.7
Rendzine-Andarax	2.35	0.65	361	8.1
Saline-Campo de Dalías	1.81	1.65	110	8.2
Sandy loam (Jutland, Denmark)	0.05	0.15	33	6.4
Tsukuba clay loam	0.80	4.24	18	6.5
Volcanic-Campo de Nijar	1.06	0.37	286	8.7

Source: Briggs, 1981; Kanazawa, 1989; Kjeldsen et al., 1990; Valverde-García et al., 1988.

Transformation Products

Soil: Duff and Menzer (1973) reported that in moist soils, dimethoate is converted to the oxygen analogue, dimethoate carboxylic acid (dimethoxon) and two unidentified metabolites. The degradation rate of dimethoate in three different soils increased almost twofold with a 10 °C increase in temperature (Kolbe et al., 1991). The reported half-lives of dimethoate in a humus-rich sandy soil, clay loam, and heavy clay soil at 10 and 20 °C are 15.3, 10.3, 15.5 d and 9.7, 4.8, and 8.5 d, respectively. Degradates included dimethoxon (*O,O*-dimethyl-*S*-(*N*-methylcarbamoylmethyl)phosphorothiolate) and unidentified polar compounds (Kolbe et al., 1991).

Plant: In plants, oxidation/hydrolysis leads to the formation of the phosphorothioate. Other hydrolysis products in plants include *O,O*-dimethylphosphorodithioate and *O,O*-dimethylphosphorophosphate which occurs via demethylation and hydrolytic cleavage of the methylamino group (Hartley and Kidd, 1987). In bean plants, dimethoate degraded to *N*-hydroxymethyl dimethoate which further degraded to des-*N*-methyl dimethoate (*N,N*-dimethyl *S*-(carbamoylmethyl) phosphorodithioate). Other metabolites identified include des-*O*-methyl carboxylic acid, dimethoate carboxylic acid, dimethyl phosphorothioic acid, dimethyl phosphorodithioic acid, *N*-hydroxymethyl dimethoate (*O,O*-dimethyl *S*-(*N*-hydroxymethylcarbamoylmethyl) phosphorothioate), an oxygen analog (*O,O*-dimethyl *S*-(*N*-methylcarbamoylmethyl) phosphorothiolate), a des-*N*-methyl oxygen analog (*O,O*-dimethyl *S*-(carbamoylmethyl) phosphorothiolate), *N*-hydroxy-methyl oxygen analog (*O,O*-dimethyl *S*-(*N*-hydroxymethylcarbamoylmethyl) phosphorothiolate), and three unknown substances (Garner and Menzer, 1986; Lucier and Menzer, 1966, 1970). These compounds were not detected on grapes treated with dimethoate 28 d after application (Steller and Brand, 1974).

Chemical/Physical: On heating, dimethoate is converted to the *O,S*-dimethyl analog (Worthing and Hance, 1991). Burns readily in contact with flame releasing toxic fumes of phosphorus, nitrogen, and sulfur oxides (Sax and Lewis, 1987). The calculated hydrolysis half-life at 25 °C and pH 7 is 118 hr (Ellington et al., 1988).

Formulation Types: Emulsifiable concentrate; wettable powder; granules; aerosol; dustable powder.

Toxicity: LC_{50} (96 hr) for mosquito fish 40-60 mg/L (Hartley and Kidd, 1987); LC_{50} (24 hr) for bluegill sunfish 28.0 mg/L and rainbow trout 20.0 mg/L (Verschueren, 1983); acute oral LD_{50} for rats 500-680 mg/kg (Hartley and Kidd, 1987), 250 mg/kg (Windholz et al., 1983).

Uses: Systemic and contact insecticide and acaricide used to control thrips and red spider mites on many agricultural crops, sawflies on apples and plums, wheat bulb and olive flies.

DIMETHYL PHTHALATE

Synonyms: Avolin; **1,2-Benzenedicarboxylic acid dimethyl ester;** Dimethyl-1,2-benzenedicarboxylate; Dimethylbenzeneorthodicarboxylate; DMP; ENT 262; Fermine; Methyl phthalate; Mipax; NTM; Palatinol M; Phthalic acid dimethyl ester; Phthalic acid methyl ester; RCRA waste number U102; Solvanom; Solvarone.

Structure:

Designations: CAS: 131-11-3; mf: $C_{10}H_{10}O_4$; fw: 194.19; RTECS: TI1575000.

Properties: Clear, colorless, odorless, moderately viscous liquid. Technical grades have a slight aromatic odor. Mp: 5.5 °C; bp: 283.8 °C; ρ: 1.1905 at 20/4 °C; fl p: 146 °C; lel: 1.2% at 146 °C; K_H: 4.2 x 10^{-7} atm·m^3/mol; IP: 9.75 eV; log K_{oc}: 0.88-2.28; log K_{ow}: 1.53-2.00; S_o: soluble in acetone, dimethylsulfoxide, benzene, mineral oil 0.34 wt % at 20 °C, miscible with ethanol, ethyl ether, chloroform; S_w: 4,320 mg/L at 25 °C; vap d: 7.94 g/L at 25 °C, 6.70 (air = 1); vp: <10^{-2} mmHg at 20 °C.

Transformation Products

Biological: In anaerobic sludge, degradation occurred as follows: monomethyl phthalate to phthalic acid to protocatechuic acid followed by ring cleavage and mineralization (Shelton et al., 1984).

Photolytic: An aqueous solution containing titanium dioxide and subjected to UV light (λ >290 nm) yielded mono and dihydroxyphthalates as intermediates (Hustert and Moza, 1988).

Chemical/Physical: Hydrolyzes in water forming phthalic acid and methyl alcohol (Wolfe et al., 1980).

Exposure Limits: NIOSH REL: IDLH 9,300 mg/m^3; OSHA PEL: TWA 5 mg/m^3; ACGIH TLV: TWA 5 mg/m^3.

Symptoms of Exposure: Irritates nasal passages, upper respiratory system, stomach; eye pain. Ingestion may cause central nervous system depression.

Formulation Types: Liquid; aerosol.

Toxicity: LC_{50} (8 d) for grass shrimp larvae 100 ppm (Verschueren, 1983); acute oral LD_{50} for rats 6,900 mg/kg (Verschueren, 1983), 6,800 mg/kg (RTECS, 1985).

Use: Insect repellant.

4,6-DINITRO-o-CRESOL

Synonyms: Antinonin; Antinonnon; Arborol; Capsine; Chemsect DNOC; Degrassan; Dekrysil; Detal; Dinitrocresol; Dinitro-o-cresol; 2,4-Dinitro-o-cresol; 3,5-Dinitro-o-cresol; Dinitrodendtroxal; 3,5-Dinitro-2-hydroxytoluene; Dinitrol; Dinitromethyl cyclohexyltrienol; 2,4-Dinitro-2-methylphenol; 2,4-Dinitro-6-methylphenol; 4,6-Dinitro-2-methylphenol; Dinitrosol; Dinoc; Dinurania; DN; DNC; DN-dry mix no. 2; DNOC; Effusan; Effusan 3436; Elgetol; Elgetol 30; Elipol; ENT 154; Extrar; Hedolit; Hedolite; K III; K IV; Kresamone; Krezotol 50; Lipan; **2-Methyl-4,6-dinitrophenol;** 6-Methyl-2,4-dinitrophenol; Nitrador; Nitrofan; Prokarbol; Rafex; Rafex 35; Raphatox; RCRA waste number P047; Sandolin; Sandolin A; Selinon; Sinox; Trifina; Trifocide; Winterwash.

Structure:

Designations: CAS: 534-52-1; DOT: 1598; mf: $C_7H_6N_2O_5$; fw: 198.14; RTECS: GO9625000.

Properties: Yellow, odorless crystals. Mp: 86.5 °C; bp: 312 °C; pK_a: 4.35-4.46; K_H: 1.4 x 10^{-6} atm·m^3/mol at 25 °C; log K_{oc}: 2.64 (calculated); log K_{ow}: 2.12-2.85; S_o (mg/L at 15 °C): methanol (7.33), ethanol (9.12), chloroform (37.2), acetone (100.6); slightly soluble in petroleum ether; S_w: 198 mg/L at 20 °C; vp: 5 x 10^{-5} mmHg at 20 °C.

Transformation Products

Soil/Plant: In plants and soils, the nitro groups reduced to amino groups (Hartley and Kidd, 1987).

Chemical/Physical: 4,6-Dichloro-o-cresol should react with amines and alkali metals forming water-soluble salts which are indicative of phenols (Morrison and Boyd, 1971).

Exposure Limits: NIOSH REL: 10 hr TWA 0.2 mg/m^3, IDLH 5 mg/m^3; OSHA PEL: TWA 0.2 mg/m^3.

Symptoms of Exposure: Sense of well-being, headache, fever, lassitude, profuse sweating, excess thirst, tachycardia, coughing, shortness of breath, coma; eye and skin irritant.

Formulation Types: Emulsifiable concentrate; soluble concentrate; suspension concentrate; wettable powder; liquid cream.

Toxicity: LC_{50} for carp 6-13 mg/L (Hartley and Kidd, 1987); acute oral LD_{50} for rats 25-40 mg/kg (Hartley and Kidd, 1987), 10 mg/kg (RTECS, 1985).

Uses: Dormant ovicidal spray for fruit trees (highly phytotoxic and cannot be used successfully on actively growing plants); herbicide; insecticide.

DINOSEB

Synonyms: Aretit; Basanite; Butaphene; BNP 30; 2-*sec*-Butyl-2,4-dinitrophenol; Caldon; Chemox general; Chemox P.E.; Dinitro; Dinitro-3; Dinitrobutylphenol; 2-Dinitro-6-*sec*-butylphenol; 4,6-Dinitro-2-*sec*-butylphenol; 4,6-Dinitro-*o-sec*-butylphenol; 4,6-Dinitro-2-(1-methyl-*n*-propyl)phenol; DN 289; DNBP; Dnosbp; DNSBP; Dow general; Dow general weed killer; Dow selective weed killer; Elgetol; Elgetol 318; ENT 1122; Gebutox; Hel-fire; Kiloseb; **2-(1-Methylpropyl)-4,6-dinitrophenol**; Nitropone P; Phenotan; Premerge; Premerge 3; RCRA waste number P020; Sinox general; Sparic; Spurge; Subitex; Unicrop DNBP; Vertac dinitro weed killer; Vertac general weed killer; Vertac selective weed killer.

Structure:

Designations: CAS: 88-85-7; mf: $C_{10}H_{12}N_2O_5$; fw: 240.22; RTECS: SJ9800000.

Properties: Dark brown solid or orange liquid with a pungent odor. Mp: 38-42 °C; ρ: 1.2647 at 45/4 °C; fl p: 177 °C; pK_a: 4.62; K_H: 5.0 x 10^{-4} atm·m^3/mol at 20 °C (calculated); log K_{oc}: 2.09, 2.70; log K_{ow}: 2.29; S_o (g/kg): ethanol (480), *n*-heptane (270); S_w: 52 mg/L at 20 °C; vp: 1 mmHg at 151.1 °C.

Transformation Products

Plant: When dinoseb on bean leaves was exposed to sunlight, photodegradation resulted in the formation of persistent, polar compounds. The compounds could not be identified by TLC (Matsuo and Casida, 1970).

Chemical/Physical: Reacts with organic and inorganic bases forming water-soluble salts (Worthing and Hance, 1991). Emits toxic fumes of chlorine when heated to decomposition (Sax and Lewis, 1987).

Symptoms of Exposure: Eye and skin irritant. Symptoms of poisoning include sweating, increased body temperature, excessive fatigue, excessive thirst, and nausea.

Formulation Types: Emulsifiable concentrate; aqueous solution; soluble concentrate; water-in-oil emulsion.

Toxicity: Toxic to fish (Hartley and Kidd, 1987); LC_{100} (24 hr) for goldfish 0.4

174

ppm (Humburg et al., 1989); acute oral LD_{50} for rats 58 mg/kg (Hartley and Kidd, 1987), 25 mg/kg (RTECS, 1985).

Uses: The amine, ammonium salt or acetate ester are used as a contact herbicide for postemergence weed control in cereals, cotton, peas, beans, potatoes, pumpkins, soybeans, and strawberries.

DIPHENAMID

Synonyms: Dif 4; Diamide; *N,N*-Dimethyldiphenylacetamide; *N,N*-Dimethyl-2,2-diphenylacetamide; *N,N*-Dimethyl-α,α-diphenylacetamide; 2,2-Dimethyl-*N,N*-dimethylacetamide; *N,N*-**Dimethyl-α-phenylbenzeneacetamide**; Dimid; Diphenamide; Diphenylamide; Dymid; Enide; Enide 50; Enide 90; FDN; Fenam; L-34314; Lilly 34314; Nor-Am; U 4513; 80W.

Structure:

Designations: CAS: 957-51-7; mf: $C_{16}H_{17}NO$; fw: 239.30; RTECS: AB8050000.

Properties: Colorless to white, nearly odorless crystals. Mp: 134.5-135.5 °C; bp: decomposes at 210 °C; ρ: 1.17 at 23.3/4 °C; fl p: nonflammable; log K_{oc}: 2.31 (calculated); log K_{ow}: 1.92 (calculated); S_o (g/L at 27 °C): acetone (189); *N,N*-dimethylformamide (165), xylene (65); S_w: 260 mg/L at 27 °C; vp: negligible at 20 °C.

Transformation Products
Soil: Degradation of diphenamid in soils was reported to form desmethyldiphenamid (a monodemethylated metabolite of diphenamid) and a bidemethylated product of diphenamide (Somasundaram and Coats, 1991). The persistence of diphenamide under warm-moist soil conditions is 3-6 mo (Ashton and Monaco, 1991).

Plant: Underwent *N*-demethylation in strawberries yielding *N*-methyl-2,2-diphenylacetamide (Golab et al., 1966).

Photolytic: *N*-Methyl-2,2-diphenylacetamide and benzoic acid were reported as major photoproducts following the UV irradiation of diphenamid in distilled water (Cessna and Muir, 1991). Emits toxic fumes of nitrogen oxides when heated to decomposition (Sax and Lewis, 1987).

Formulation Types: Wettable powder (50 and 90%); granules (5%); liquid dispersion (4 lb/gal).

Toxicity: Slightly toxic to fish (Hartley and Kidd, 1987); LC_{50} (48 hr) for *Daphnia magna* 56 mg/L (Sanders, 1970); acute oral LD_{50} for rats 1,050 mg/kg (Hartley and Kidd, 1987), 685 mg/kg (RTECS, 1985).

Uses: Selective preemergence herbicide used to control many broadleaf weeds and most grass weeds in okra, cotton, peanuts tomatoes, sweet potatoes, potatoes, tobacco, fruits, turf, and ornamentals.

DIPROPETRYN

Synonyms: Cotofor; Dipropetryne; 2-Ethylthio-4,6-bis(isopropylamino)-s-triazine; 6-(Ethylthio)-N,N'-bis(1-methylethyl)-1,3,5-triazine-2,4-diamine; GS 16068; Sancap 80W.

Structure:

$$CH_3CH_2S \quad \underset{N}{\overset{N}{\bigcirc}} \quad NHCH(CH_3)_2$$
$$NHCH(CH_3)_2$$

Designations: CAS: 4147-51-7; mf: $C_{11}H_{21}N_5S$; fw: 255.40; RTECS: XY4100000.

Properties: Colorless to white crystalline powder. Mp: 104-106 °C; ρ: 1.120 at 20/4 °C; fl p: nonflammable; H-t$_{1/2}$ (25 °C): 24-28 d (pH 1), >2.5 yr (7 <pH <13); K_H: 1.53 x 10^{-8} atm·m^3/mol at 20 °C (approximate - calculated from water solubility and vapor pressure); log K_{oc}: 2.58-2.95; log K_{ow}: 3.45, 3.81; S_o (g/L at 20 °C): acetone (270), benzene (540), ethanol (180), n-hexane (9), kerosene (10), methanol (190), methylene chloride (300), n-octanol (130), toluene (220), xylene (220); S_w: 16 mg/L at 20 °C; vp: 7.28 x 10^{-7} mmHg at 20 °C.

Soil properties and adsorption data

Soil	K_d (mL/g)	f_{oc} (%)	K_{oc} (mL/g)	pH	CEC (meq/100 g)
Brewer clay loam	18.50	1.61	1,149	5.8	13.5
Cobb sand	1.32	0.34	388	7.3	3.8
Cobb sand + 2% muck	32.50	1.21	2,686	5.3	9.0
Loam	24.60	2.78	885	5.9	--
Port silty clay	8.91	1.04	857	6.3	17.9
Sand	1.50	0.35	431	7.7	--
Sandy loam	5.00	0.70	718	7.6	--
Sandy loam	8.90	2.32	384	7.8	--
Teller fine sandy loam	6.18	0.75	824	5.7	8.6

Source: Murray et al., 1975; U.S. Department of Agriculture, 1990.

Transformation Products

Soil: Degradation of dipropetryn includes dealkylation of the side chain(s), ring opening, and the evolution of carbon dioxide (Hartley and Kidd, 1987). The reported half-life in soil is approximately 100 d (Worthing and Hance, 1991).

Formulation Types: Wettable powder; suspension concentrate.

Toxicity: LC_{50} (96 hr) for rainbow trout 2.7 mg/L and bluegill sunfish 1.6 mg/L (Hartley and Kidd, 1987); acute oral LD_{50} for rats 3,900–5,000 mg/kg (Hartley and Kidd, 1987), 7,144 mg/kg (RTECS, 1985).

Uses: Preemergence herbicide used to control weeds in cotton and melon crops.

DIQUAT

Synonyms: Aquacide; Deiquat; Dextrone; 9,10-Dihydro-8a,10-diazoniaphenanthrene dibromide; Dihydro-8a,10a-diazoniaphenanthrene-(1,1'-ethylene-2,2'-bipyridylium)dibromide; 5,6-Dihydrodipyrido(1,2a;2,1c)pyrazinium dibromide; **6,7-Dihydropyrido[1,2-α:2',1'-c]pyrazinedium dibromide**; 1,1-Ethylene-2,2-bipyridylium dibromide; 1,1'-Ethylene-2,2'-bipyridylium dibromide; FB/2; NA 2781; Ortho; Pathclear; Preeglone; Reglon; Reglone; Weedol; Weedtrine-D.

Structure:

Designations: CAS: 85-00-7; mf: $C_{12}H_{12}Br_2N_2$; fw: 344.06; RTECS: JM5690000.

Properties: Colorless to pale yellow crystals. Mp: decomposes at 320 °C; ρ: 1.22-1.27 at 20/4 °C; fl p: aqueous salt solutions are nonflammable; H-t$_{1/2}$: 74 d under simulated sunlight at pH 7; K_H: <6.3 x 10^{-14} atm·m^3/mol at 20-25 °C (approximate - calculated from water solubility and vapor pressure); log K_{oc}: 0.42 (calculated); log K_{ow}: -4.60; S_o: slightly soluble in alcohols and hydroxylic solvents; S_w: 700 g/L at 25 °C; vp: <9.75 x 10^{-8} mmHg at 20 °C.

Transformation Products

Biological: Under aerobic and anaerobic conditions, the rate of diquat mineralization in eutrophic water and sediments were very low. After 65 d, only 0.88 and 0.21% of the applied amount (5 μg/mL) evolved as carbon dioxide (Simsiman and Chesters, 1976). Diquat is readily mineralized to carbon dioxide in nutrient solutions containing microorganisms. The addition of montmorillonite clay in an amount equal to adsorb one half of the diquat decreased the amount of carbon dioxide by 50%. Additions of kaolinite clay had no effect on the amount of diquat degraded by microorganisms (Weber and Coble, 1968).

Photolytic: Diquat has an absorption maximum of 310 nm (Slade and Smith, 1967). The sunlight irradiation of a diquat solution (0.4 mg/100 mL) yielded 1,2,3,4-tetrahydro-1-oxopyrido[1,2-a]-5-pyrazinium chloride (TOPPS) as the principal metabolite. This compound also formed when diquat was adsorbed on filter paper and silica gel is subjected to sunlight irradiation (Slade and Smith, 1974). When an aqueous diquat solution (5 μg/L) contained in a borosilicate glass beaker was exposed to sunlight, TOPPS, *o*-picolinic acid, and picolinamide formed as major products (Smith and Grove, 1969). When the diquat solution was exposed to

180

sunlight in May and June for 3 wk, 70% was degraded giving rise only to TOPPS and *o*-picolinic acid. The sunlight photolysis half-life of diquat in aqueous solution was estimated to be about 14 d (Smith and Grove, 1969). Extensive photo-degradation was also observed as thin films on leaf surfaces. *o*-Picolinic acid and other products formed in small amounts (Smith and Grove, 1969). Funderburk and Bozarth (1967) reported that dry diquat was decomposed by UV light ($t_{1/2}$ = 48 hr). However in an aqueous solution, degradation was complete after 8 d.

Chemical/Physical: Decomposes at 320 °C (Windholz et al., 1983) emitting toxic fumes of bromides and nitrogen oxides (Lewis, 1990).

Exposure Limits: ACGIH TLV: TWA 0.5 mg/m^3.

Symptoms of Exposure: Eye and skin irritant; nose bleeding if inhaled. Symptoms of poisoning include diarrhea, vomiting, and general malaise.

Formulation Types: Soluble concentrate; gel; aqueous solution (2 lb/gal).

Toxicity: LC_{50} (96 hr) for rainbow trout 21 mg/L, mirror carp 67 mg/L (Worthing and Hance, 1991), fathead minnow 14 mg/L, largemouth bass 7.8 mg/L (Surber and Pickering, 1962), bluegill sunfish 35 mg/L, walleye 2.1 mg/L (Gilderhus, 1967), and striped bass 0.25 ppm (Wellborn, 1969); LC_{50} (96 hr) for northern pike 16 mg/L, rainbow trout 11.2 mg/L (Gilderhus, 1967); acute oral LD_{50} for rats is 230 mg/kg (Ashton and Monaco, 1991), 120 mg/kg (RTECS, 1985).

Uses: Nonselective contact herbicide used to control broadleaf weeds in fruit and vegetable crops.

DISULFOTON

Synonyms: Bay 19639; Bayer 19639; *O,O*-Diethyl *S*-(2-eththioethyl) phosphoro-dithioate; *O,O*-Diethyl *S*-(2-ethylthioethyl) thiothionophosphate; *O,O*-Diethyl *S*-(2-ethylmercaptoethyl) dithiophosphate; *O,O*-Diethyl *S*-(2-(ethylthio)ethyl) phos-**phorodithioate**; Dimaz; Disulfaton; Di-syston; Disystox; Dithiodemeton; Dithio-systox; ENT 23437; *O,O*-Ethyl *S*-2-((ethylthio)ethyl) phosphorodithioate; *S*-2-(Ethylthio)ethyl *O,O*-diethyl ester of phosphorodithioic acid; Frumin AL; Frumin G; M 74; NA 2783; RCRA waste number P039; S 276; Solvirex; Thiodemeton; Thiodemetron.

Structure:

$$CH_3CH_2O \quad \underset{\underset{}{\overset{\|}{P}}}{\overset{S}{}} -SCH_2CH_2SCH_2CH_3$$
$$CH_3CH_2O$$

Designations: CAS: 298-04-4; DOT: 2783; mf: $C_8H_{19}O_2PS_3$; fw: 274.40; RTECS: TD9275000.

Properties: Colorless to yellowish oil with a characteristic odor. Mp: <-25 °C; bp: 128 °C at 0.1 mmHg; ρ: 1.144 at 20/4 °C; H-t$_{1/2}$: 3.04 yr at 20 °C and pH 1-5, 1.2-103 d at 25 °C and pH 7, 7.2 hr at 20 °C and pH 9; K_H: 5.42 x 10^{-6} atm·m^3/mol at 20-22 °C (approximate - calculated from water solubility and vapor pressure); log K_{oc}: 2.76-2.89; log K_{ow}: 3.95, 4.02; S_o: miscible with most solvents; S_w: 12 mg/L at 22 °C; vp: 1.80 x 10^{-4} mmHg at 20 °C.

Soil properties and adsorption data

Soil	K_d (mL/g)	f_{oc} (%)	K_{oc} (mL/g)	pH
Adventurers	134.60	31.00	434	6.9
Bottisham	58.60	8.80	660	7.7
Broadbalk loam	5.80	0.90	644	8.1
Broadbalk loam	21.50	2.70	796	7.8
Elkhorn sandy loam	5.01	0.86	583	6.0
Hugo gravelly sand loam	3.03	0.12	2,525	5.5
Isleham	49.10	7.60	646	7.5
Isleham	55.50	2.80	1,982	6.3
Loam	7.06	1.22	579	6.7
Moulton	21.10	1.70	1,241	8.1
Oakington	16.00	1.80	889	7.2

Soil	K_d (mL/g)	f_{oc} (%)	K_{oc} (mL/g)	pH
Peacock	70.30	11.00	639	7.6
Prickwillow	100.50	15.00	670	5.1
Sand	14.29	2.15	665	6.9
Sandy loam	13.43	1.74	772	5.5
Spinney	95.70	12.00	798	7.2
Stretham	14.70	1.40	1,050	7.5
Sweeney sandy clay loam	11.49	0.65	1,768	6.3
Tierra heavy clay loam	36.86	0.33	11,170	6.2
Wicken	20.30	1.70	1,194	8.0
Woburn	20.00	1.80	1,111	6.5
Woburn	14.90	1.10	1,354	6.8
Woburn	20.50	1.30	1,577	6.8
Worlington	5.30	0.70	757	8.1

Source: Graham-Bryce, 1967; King and McCarty, 1968; U.S. Department of Agriculture, 1990.

Transformation Products

Soil/Plant: Metabolized in soil and plants to the sulfoxide and sulfone (Clapp et al., 1976; Getzin and Shanks, 1970; Metcalf et al., 1957; Worthing and Hance, 1991), the corresponding phosphorothioate analogs, and then to derivatives of O,O-diethyl hydrogen phosphate and 2-ethylthioethyl mercaptan (Worthing and Hance, 1991). Disulfoton is rapidly oxidized in soil to its sulfoxide and sulfone with disulfoton oxon sulfoxide and disulfoton oxon sulfone appearing in small amounts (Szeto et al., 1983). In a Portneuf silt loam soil, the persistence of the sulfoxide and sulfone was 32 and >64 d, respectively (Clapp et al., 1976). Disulfoton was translocated from a sandy loam soil into asparagus tips. Disulfoton sulfoxide, disulfoton sulfone, disulfoton oxon sulfoxide, and disulfoton oxon sulfone were recovered as metabolites (Szeto and Brown, 1982; Szeto et al., 1983). Disulfoton sulfoxide and disulfoton sulfone were also identified in spinach plants 5.5 mo after application (Menzer and Dittman, 1968). In a later study, Menzer et al. (1970) reported that degradation of disulfoton in soil degraded at a higher rate in the winter months than in the summer months. They postulated that soil type, rather than temperature, had greater influence on the rate of decomposition of disulfoton. Soils used in the winter and summer months were an Evesboro loamy sand and Chillum silt loam, respectively (Menzer et al., 1970). The half-life in soil is approximately 5 d (Jury et al., 1987).

Photolytic: Disulfoton was rapidly oxidized to disulfoton sulfoxide and trace amounts (<5% yield) of disulfoton sulfone when sorbed on soil and exposed to sunlight (H-t$_{1/2}$ = 1-4 d) (Gohre and Miller, 1986). The photosensitized oxidation was probably due to the presence of singlet oxygen (Zepp et al., 1981). The degradation rate was higher in soils containing the lowest organic carbon (Gohre and Miller, 1986). When fertilizers containing superphosphate and ammonium nitrate were impregnated with disulfoton, the latter chemically degraded to form disulfoton sulfone and disulfoton sulfoxide (Ibrahim et al., 1969).

Chemical/Physical: Emits toxic fumes of phosphorus and sulfur oxides when heated to decomposition (Sax and Lewis, 1987; Lewis, 1990). The reported half-lives for abiotic hydrolysis, photochemical transformation, and primary degradation in Rhine River water samples were 170, 1,000, and 7-41 d, respectively (Wanner et al., 1989).

Exposure Limits: OSHA PEL: TWA 0.1 mg/m^3; ACGIH TLV: TWA 0.1 mg/m^3.

Symptoms of Exposure: Headache, anorexia, nausea, diarrhea, salivation, lacrimation, sweating, tremor, shortness of breath, cyanosis, fever, pulmonary edema, convulsions, coma, shock.

Formulation Types: Emulsifiable concentrate; granules; dry seed treatment.

Toxicity: LC$_{50}$ (96 hr) for rainbow trout 1.85 mg/L, bluegill sunfish 39 µg/L, goldfish 6.5 mg/L, guppies 0.25 mg/L (Hartley and Kidd, 1987), and fathead minnow 3.70 mg/L and bluegill sunfish 63 µg/L (Pickering et al., 1962); acute oral LD$_{50}$ for rats 2-12 mg/kg (Hartley and Kidd, 1987), male and female rats, 2.3 and 6.8 mg/kg, respectively (Windholz et al., 1983).

Uses: Systemic insecticide and acaricide for control of sucking insects and mites in fruits, vegetables, cotton, and forestry nurseries.

DIURON

Synonyms: AF 101; Cekiuron; Crisuron; Dailon; DCMU; Diater; Dichlorfenidim; 3-(3,4-Dichlorophenol)-1,1-dimethylurea; 3-(3,4-Dichlorophenyl)-1,1-dimethylurea; *N'*-**(3,4-Dichlorophenyl)-*N*,*N*-dimethylurea**; 1,1-Dimethyl-3-(3,4-dichlorophenyl)urea; Di-on; Direx; Direx 4L; Diurex; Diurol; Diuron 4L; Diuron 80W; DMU; Dynex; Farmco diuron; Herbatox; HW 920; Karmex; Karmex diuron herbicide; Karmex DW; Marmer; NA 2767; Sup'r flo; Telvar; Telvar Diuron Weed Killer; Unidron; Urox D; USAF P-7; USAF XR-42; Vonduron.

Structure:

Designations: CAS: 330-54-1; DOT: 2767; mf: $C_9H_{10}Cl_2N_2O$; fw: 233.11; RTECS: YS8925000.

Properties: White, odorless, crystalline solid. Mp: 150-155 °C; bp: 180 °C (decomposes); ρ: 1.385 (calculated); fl p: nonflammable; pK_a: -1 to -2; K_H: 1.46 x 10^9 atm·m^3/mol at 25-30 °C (approximate - calculated from water solubility and vapor pressure); log K_{oc}: 2.21-2.87; log K_{ow}: 1.97-2.81; S_o (ppm at 27 °C): acetone (53,000), benzene (1,200), butyl stearate (1,400), refined cottonseed oil (900); S_w: 40 ppm at 20 °C; vp: 2 x 10^{-7} mmHg at 30 °C.

Soil properties and adsorption data

Soil	K_d (mL/g)	f_{oc} (%)	K_{oc} (mL/g)	pH	CEC (meq/100 g)
Aguadilla loamy sand	4.73	1.44	328	7.4	10.0
Aguirre clay loam	4.63	1.84	252	9.0	13.8
Alonso clay	4.63	1.84	252	5.1	13.8
Altura loam	5.36	2.13	252	8.0	27.6
Ascalon sandy clay loam	1.33	0.85	156	7.3	12.7
Asquith sandy loam	6.90	1.02	676	7.5	--
Adkins loamy sand	1.91	0.40	657	7.3	--
Barnes clay loam	8.51	3.98	214	7.4	33.8
Baymón sandy clay loam	2.29	0.98	234	4.7	5.0
Begbroke sandy loam	2.70	1.11	243	7.1	--

continued

Soil	K_d (mL/g)	f_{oc} (%)	K_{oc} (mL/g)	pH	CEC (meq/100 g)
Beltsville silty loam	1.67	1.40	119	4.3	4.2
Benevola clay	1.49	1.30	115	7.6	20.1
Benevola silty clay	3.62	2.70	134	7.7	19.5
Berkley clay	1.25	0.99	126	7.3	34.4
Berkley silty clay	5.87	4.63	127	7.1	33.7
Bosket silty loam	1.25	0.57	219	5.8	8.4
Boxworth clay	15.00	2.08	721	7.9	22.0
Bridgets silty loam	12.00	3.09	388	8.0	24.0
Catalina clay	1.86	1.09	171	4.7	11.8
Catlin	4.70	2.01	419	6.2	--
Cataño sand	3.90	1.21	322	7.9	6.9
Cayaguá sandy loam	2.96	1.15	257	5.2	7.3
Cecil sandy loam	2.40	0.40	600	5.8	--
Cecil sandy clay	1.58	1.09	145	5.3	3.6
Chester loam	2.81	1.67	168	4.9	5.2
Chillum silty loam	4.62	2.54	182	4.6	7.6
Christiana loam	0.88	0.57	155	4.4	5.6
Cialitos clay	12.40	2.82	440	5.4	18.6
Coloso clay loam	12.16	2.13	571	5.7	23.0
Commerce	2.20	0.68	325	6.7	--
Coto clay	9.36	1.84	509	7.7	14.0
Crosby silty loam	4.26	1.90	224	4.8	11.5
Dundee silty clay loam	2.25	0.96	234	5.0	18.1
Fe clay loam	5.86	1.96	299	7.5	27.6
Fortuna silty clay loam	13.40	1.90	705	5.4	23.3
Fraternidad clay	15.89	2.92	544	5.9	58.0
Fraternidad clay	69.60	1.21	575	6.3	36.0
Garland clay	1.09	0.65	169	7.7	23.2
Great Horse sandy loam	75.00	12.00	625	6.3	18.0
Guánica clay	9.39	2.77	339	8.1	52.1
Hagerstown silty clay loam	2.14	1.30	165	7.5	8.8
Hagerstown silty clay loam	4.26	2.48	172	5.5	12.5
Humata silty clay loam	2.13	0.98	217	4.5	10.1
Indian Head clay loam	13.30	2.34	568	7.8	--
Iredell clay	0.95	6.17	15	5.6	20.9
Irdell silty loam	6.90	3.04	227	5.4	17.0
Josefa silty loam	12.06	1.90	635	6.0	16.8
Juncos silty clay	10.10	1.55	652	6.2	13.4
Keyport silt loam	4.00	1.21	328	5.4	--

Soil	K_d (mL/g)	f_{oc} (%)	K_{oc} (mL/g)	pH	CEC (meq/100 g)
Kirton sandy loam	10.00	1.50	667	7.6	13.0
Lakeland sandy loam	1.10	1.88	58	6.2	2.9
Liscombe sandy loam	25.00	3.45	725	6.2	13.0
Mabí clay	14.46	2.82	513	7.0	55.2
Mabí clay loam	10.33	1.38	749	5.7	31.0
Melfort loam	83.30	6.05	1,377	5.9	--
Mercedita silty clay	5.46	2.19	249	8.1	19.9
Moca clay	12.03	1.90	633	5.8	31.0
Monona clay loam	14.30	1.67	856	6.5	21.2
Múcara loam	6.53	3.06	213	5.8	19.6
Nipe clay loam	15.10	1.15	1,313	5.7	11.9
Ontario clay	0.21	0.86	24	5.9	8.4
Ooster silty loam	2.14	1.31	164	4.7	6.8
Pantura sandy loam	4.73	2.02	234	5.7	7.7
Regina heavy clay	13.40	2.39	561	7.7	--
Río Piedras silty clay	3.63	2.02	180	4.9	11.5
Rosemaunde silty clay loam	10.20	1.76	580	6.7	14.0
Ruston sandy loam	1.49	1.05	142	5.1	3.4
San Antón loam	15.80	1.55	1,019	6.7	26.1
Semiahmoo mucky peat	244.34	27.8	879	5.4	--
Sharkey clay	6.36	2.25	283	6.2	40.2
Sterling clay loam	1.41	0.94	150	7.7	22.5
Talante sandy loam	2.36	0.80	295	5.1	4.0
Terrington silty loam	14.00	1.54	909	8.0	15.0
Thurlow clay loam	2.14	1.25	171	7.7	21.6
Tifton sandy loam	0.62	0.56	110	4.9	2.4
Toa loam	3.56	1.15	310	5.3	13.0
Toa sandy loam	1.26	0.34	371	6.0	8.0
Toledo silty clay	5.64	2.80	201	5.5	29.8
Toll Farm HP	53.00	11.70	453	7.4	41.0
Tracy	4.70	1.12	419	6.2	--
Trawscoed silty clay loam	16.00	3.69	434	6.2	12.0
Tripp loam	1.17	0.86	136	7.6	14.7
Truckton sandy loam	0.56	0.25	222	7.0	4.4
Valentine loamy fine sand	6.50	0.80	813	6.6	10.1
Vega Alta sandy loam	6.30	2.02	312	5.0	5.6
Via loam	5.13	1.32	389	5.1	39.9

continued

Soil	K_d (mL/g)	f_{oc} (%)	K_{oc} (mL/g)	pH	CEC (meq/100 g)
Webster silty clay loam	24.40	3.34	733	7.3	22.0
Weed Res. sandy loam	13.60	1.93	705	7.1	11.0
Wehadkee silty loam	1.41	1.11	127	5.6	10.2
Weyburn Oxbow loam	26.90	3.72	723	6.5	--

Source: Grover, 1975; Grover and Hance, 1969; Hance, 1965; Harris and Sheets, 1965; Liu et al., 1970; Madhun et al., 1986; Majka and Lavy, 1977; McCall et al., 1981; Nkedi-Kizza et al., 1983; Rhodes et al., 1970.

Transformation Products

Biological: Degradation of radiolabeled diuron (20 ppm) was not observed after 2 wk of culturing with *Fusarium* and two unidentified microorganisms (Lopez and Kirkwood, 1974). After 80 d, only 3.5% of the applied amount evolved as $^{14}CO_2$ (Lopez and Kirkwood, 1974). 3,4-Dichloroaniline was reported as a minor degradation product of diuron in water (Drinking Water Health Advisory, 1989) and soils (Duke et al., 1991). Under aerobic conditions, mixed cultures isolated from pond water and sediment degraded diuron (10 μg/mL) to CPDU, 3,4-dichloroaniline, and 3-(3,4-dichlorophenyl)-1-methylurea, carbon dioxide, and a monodemethylated product (Ellis and Camper, 1982). The extent of biodegradation varied with time, glycerol concentration, and microbial population. The degradation half-life was <70 d at 30 °C (Ellis and Camper, 1982).

Soil: Incubation of diuron in soils releases carbon dioxide (Madhun and Freed, 1987). The rate of carbon dioxide formation nearly tripled when the soil temperature was increased from 25 to 35 °C. Reported half-lives in an Adkins loamy sand are 705, 414, and 225 d at 25, 30, and 35 °C, respectively. However, in a Semiahoo mucky peat, the half-lives were considerable higher: 3,991, 2,164, and 1,165 d at 25, 30, and 35 °C, respectively (Madhun and Freed, 1987). Under aerobic conditions, biologically active, organic-rich, diuron-treated pond sediment (40 μg/mL) converted diuron exclusively to 3-(3-chlorophenyl)-1,1-dimethylurea (CPDU) (Attaway et al., 1982, 1982a; Stepp et al., 1985). At 25 and 30 °C, 90% degradation was observed after 55 and 17 d, respectively (Attaway, 1982a).

Photolytic: Tanaka et al. (1985) studied the photolysis of diuron (40 mg/L) in aqueous solution using UV light (λ = 300 nm) or sunlight. After 25 d of exposure to sunlight, diuron degraded to 2,2,3'-trichloro-2,4'-di-*N*,*N*-dimethylurea biphenyl (yield = 1.3%) and hydrogen chloride (Tanaka et al., 1985).

Chemical/Physical: Diuron decomposes at 180 to 190 °C releasing dimethylamine and 3,4-dichlorophenylisocyanate. Dimethylamine and 3,4-dichloroaniline are produced when hydrolyzed or when acids or bases are added at elevated temperatures (Sittig, 1985). The hydrolysis half-life of diuron in a 0.5 N sodium hydroxide solution at 20 °C is 150 d (El-Dib and Aly, 1976). When diuron was pyrolyzed in a helium atmosphere between 400 and 1,000 °C, the following products were identified: dimethylamine, chlorobenzene, 1,2-dichlorobenzene, benzonitrile, a trichlorobenzene, aniline, 4-chloroaniline, 3,4-dichlorophenyl isocyanate, bis(1,3-(3,4-dichlorophenyl)urea), 3,4-dichloroaniline, and monuron [3-(4-chlorophenyl)-1,1-dimethylurea] (Gomez et al., 1982). Products reported from the combustion of diuron at 900 °C include carbon monoxide, carbon dioxide, chlorine, nitrogen oxides, and hydrochloric acid (Kennedy et al., 1972a).

Symptoms of Exposure: May irritate eyes, skin, nose, and throat.

Exposure Limits: OSHA PEL: TWA 10 mg/m^3.

Formulation Types: Liquid concentrate (4 lb/gal); wettable powder (80%).

Toxicity: LC_{50} (96 hr) for rainbow trout 5.6 mg/L, bluegill sunfish 5.9 mg/L, and guppies 25 mg/L (Hartley and Kidd, 1987); LC_{50} (48 hr) for bluegill sunfish 7.4 ppm, rainbow trout 4.3 ppm, and coho salmon 16.0 mg/L (Verschueren, 1983); acute oral LD_{50} for rats 3,400 mg/kg (Hartley and Kidd, 1987), 1,017 mg/kg (RTECS, 1985), 437 mg/kg (Windholz et al., 1983).

Uses: Preemergence herbicide used in soils to control germinating broadleaf grasses and weeds in crops such as apples, cotton, grapes, pears, pineapple, and alfalfa; sugar cane flowering depressant.

α-ENDOSULFAN

Synonyms: Benzoepin; Beosit; Bio 5462; Chlorthiepin; Crisulfan; Cyclodan; Endocel; Endosol; Endosulfan; Endosulfan I; Endosulphan; ENT 23979; FMC 5462; 1,2,3,7,7-Hexachlorobicyclo[2.2.1]-2-heptene-5,6-bisoxymethylene sulfite; α,β-1,2,3,7,7-Hexachlorobicyclo[2.2.1]-2-heptene-5,6-bisoxymethylene sulfite; Hexachlorohexahydromethano-2,4,3-benzodioxathiepin-3-oxide; (3α,5aβ,6α,9α,9aβ)-6,7,8,9,10,10-Hexachloro-1,5,5a,6,9,9a-hexahydro-6,9-methano-2,4,3-benzo-dioxathiepin-3-oxide; 1,4,5,6,7,7-Hexachloro-5-norborene-2,3-dimethanol cyclic sulfite; Hildan; Hoe 2671; Insectophene; Kop-thiodan; Malix; NCI-C00566; NIA 5462; Niagara 5462; OMS-570; RCRA waste number P050; Thifor; Thimul; Thiodan; Thiofor; Thiomul; Thionex; Thiosulfan; Tionel; Tiovel.

Structure:

Designations: CAS: 959-98-8; DOT: 2761; mf: $C_9H_6Cl_6O_3S$; fw: 406.92; RTECS: RB9275000.

Properties: Colorless to brown crystals with a sulfur dioxide odor. Mp: 106 °C, 70-100 °C (technical grade containing both α and β isomers); ρ: 1.745 at 20/4 °C; H-t$_{1/2}$: 218 hr at 25 °C and pH 7; K_H: 1.01 x 10^{-4} atm·m^3/mol at 25 °C; log K_{oc}: 3.31 (calculated); log K_{ow}: 3.55; S_o: soluble in acetone, benzene, ethyl ether, 95% ethanol, toluene, xylene, and most other organic solvents; S_w: 530 ppb at 25 °C; vp: 10^{-5} mmHg at 25 °C.

Transformation Products

Plant: Endosulfan sulfate was formed when endosulfan was translocated from the leaves to roots in both bean and sugar beet plants (Beard and Ware, 1969). In tobacco leaves, α-endosulfan is hydrolyzed to endosulfandiol (Chopra and Mahfouz, 1977). Stewart and Cairns (1974) reported the metabolite endosulfan sulfate was identified in potato peels and pulp at concentrations of 0.3 and 0.03 ppm, respectively. They also reported that the half-life for the conversion of α-endosulfan to β-endosulfan was 60 d.

Soil: Metabolites of endosulfan identified in soils were: endosulfandiol, endosulfanhydroxy ether, endosulfan lactone, and endosulfan sulfate (Dreher and Podratzki, 1988; Martens, 1977). These compounds, as well as endosulfan ether, were also reported as metabolites identified in aquatic systems (Day, 1991).

Endosulfan sulfate was the major biodegradation product in soils under aerobic, anaerobic, and flooded conditions (Martens, 1977). In flooded soils, endolactone was detected only once whereas endodiol and endohydroxy ether were identified in all soils under these conditions. Under anaerobic conditions, endodiol formed in low amounts in two soils (Martens, 1977). Indigenous microorganisms obtained from a sandy loam degraded α-endosulfan to endosulfandiol. This diol was converted to endosulfan α-hydroxy ether and trace amounts of endosulfan ether and both were degraded to endosulfan lactone (Miles and Moy, 1979).

Surface Water: Endosulfan sulfate was also identified as a metabolite in a survey of 11 agricultural watersheds located in southern Ontario, Canada (Frank et al., 1982).

Photolytic: Thin films of endosulfan on glass and irradiated by UV light (λ >300 nm) produced endosulfandiol with minor amounts of endosulfan ether, lactone, α-hydroxyether, and other unidentified compounds (Archer et al., 1972). When an aqueous solution containing endosulfan was photooxidized by UV light at 90-95 °C, 25, 50, and 75% degraded to carbon dioxide after 5.0, 9.5, and 31.0 hr, respectively (Knoevenagel and Himmelreich, 1976).

Chemical/Physical: Undergoes slow hydrolysis forming the endosulfandiol and sulfur dioxide (Worthing and Hance, 1991). Emits toxic fumes of chlorides and sulfur oxides when heated to decomposition (Lewis, 1990).

Exposure Limits: ACGIH TLV: TWA 0.1 mg/m^3 (isomeric mixture).

Formulation Types: Emulsifiable concentrate; wettable powder; granules; dustable powder; smoke tablet.

Toxicity: LC_{50} (96 hr) for golden orfe 2 μg/L (Hartley and Kidd, 1987), rainbow trout 0.3 μg/L, and white sucker 3.0 μg/L (Verschueren, 1983).

Use: Insecticide for vegetable crops.

β-ENDOSULFAN

Synonyms: Benzoepin; Beosit; Bio 5462; Chlorthiepin; Crisulfan; Cyclodan; Endocel; Endosol; Endosulfan; Endosulfan II; Endosulphan; ENT 23979; FMC 5462; 1,2,3,7,7-Hexachlorobicyclo[2.2.1]-2-heptene-5,6-bisoxymethylene sulfite; α,β-1,2,3,7,7-Hexachlorobicyclo[2.2.1]-2-heptene-5,6-bisoxymethylene sulfite; Hexachlorohexahydromethano-2,4,3-benzodioxathiepin-3-oxide; **(3α,5aα,6β,9β,9aα)-6,7,8,9,10,10-Hexachloro-1,5,5a,6,9,9a-hexahydro-6,9-methano-2,4,3-benzodioxathiepin-3-oxide**; 1,4,5,6,7,7-Hexachloro-5-nor-borene-2,3-dimethanol cyclic sulfite; Hildan; Hoe 2671; Insectophene; KOP-thiodan; Malix; NCI-C00566; NIA 5462; Niagara 5462; OMS-570; RCRA waste number P050; Thifor; Thimul; Thiomul; Thiodan; Thiofor; Thionex; Thiosulfan; Tionel; Tiovel.

Structure:

Designations: CAS: 33213-65-9; DOT: 2761; mf: $C_9H_6Cl_6O_3S$; fw: 406.92; RTECS: RB9275000.

Properties: Colorless to brown crystals with a sulfur dioxide odor. Mp: 207-209 °C; ρ: 1.745 at 20/20 °C; H-t$_{1/2}$: 187 hr at 25 °C and pH 7; K_H: 1.91 x 10^{-5} atm·m^3/mol at 25 °C (approximate - calculated from water solubility and vapor pressure); log K_{oc}: 3.37 (calculated); log K_{ow}: 3.62; S_o: soluble in acetone, benzene, ethyl ether, 95% ethanol, toluene, xylene, and most other organic solvents; S_w: 280 ppb at 25 °C; vp: 10^{-5} mmHg at 25 °C.

Transformation Products

Soil: Metabolites of endosulfan identified in soils were: endosulfandiol, endosulfanhydroxy ether, endosulfan lactone, and endosulfan sulfate (Dreher and Podratzki, 1988; Martens, 1977). These compounds, as well as endosulfan ether, were also reported as metabolites identified in aquatic systems (Day, 1991). In soils under aerobic conditions, β-endosulfan is converted to the corresponding alcohol and ether (Perscheid et al., 1973). Endosulfan sulfate was the major biodegradation product in soils under aerobic, anaerobic, and flooded conditions (Martens, 1977). In flooded soils, endolactone was detected only once whereas endodiol and endohydroxy ether were identified in all soils under these conditions. Under anaerobic conditions, endodiol formed in low amounts in two soils (Martens, 1977). Indigenous microorganisms obtained from a sandy loam degraded β-endosulfan to endosulfan diol. This diol was converted to endosulfan α-hydroxy ether and trace amounts of endosulfan ether and both were degraded to endosulfan lactone (Miles and Moy, 1979).

Plant: In addition, endosulfan sulfate was formed when endosulfan was translocated from the leaves to roots in both bean and sugar beet plants (Beard and Ware, 1969). In tobacco leaves, β-endosulfan hydrolyzed into endosulfandiol (Chopra and Mahfouz, 1977). Stewart and Cairns (1974) reported the metabolite endosulfan sulfate was identified in potato peels and pulp at concentrations of 0.3 and 0.03 ppm, respectively. They also reported that the half-life for the oxidative conversion of β-endosulfan to endosulfan sulfate was 800 d.

Surface Water: Endosulfan sulfate was also identified as a metabolite in a survey of 11 agricultural watersheds located in southern Ontario, Canada (Frank et al., 1982).

Photolytic: Thin films of endosulfan on glass and irradiated by UV light (λ >300 nm) produced endosulfan diol with minor amounts of endosulfan ether, lactone, α-hydroxyether, and other unidentified compounds (Archer et al., 1972). Gaseous β-endosulfan subjected to UV light (λ >300 nm) produced endosulfan ether, endosulfan diol, endosulfan sulfate, endosulfan lactone, α-endosulfan, and a dechlorinated ether (Schumacher et al., 1974). Irradiation of β-endosulfan in *n*-hexane by UV light produced the photoisomer α-endosulfan (Putnam et al., 1975). When an aqueous solution containing endosulfan was photooxidized by UV light at 90-95 °C, 25, 50, and 75% degraded to carbon dioxide after 5.0, 9.5, and 31.0 hr, respectively (Knoevenagel and Himmelreich, 1976).

Chemical/Physical: Endosulfan detected in Little Miami River, OH was readily hydrolyzed and tentatively identified as endosulfan diol (Eichelberger and Lichtenberg, 1971). Undergoes slow hydrolysis forming the endosulfan diol and sulfur dioxide (Worthing and Hance, 1991). Emits toxic fumes of chlorides and sulfur oxides when heated to decomposition (Lewis, 1990).

Exposure Limits: ACGIH TLV: TWA 0.1 mg/m^3 (isomeric mixture).

Formulation Types: Emulsifiable concentrate; wettable powder; granules; dustable powder; smoke tablet.

Toxicity: LC_{50} (96 hr) for golden orfe 2 µg/L (Hartley and Kidd, 1987), rainbow trout 0.3 µg/L, and white sucker 3.0 µg/L (Verschueren, 1983).

Use: Insecticide for vegetable crops.

ENDOSULFAN SULFATE

Synonyms: 6,7,8,9,10,10-Hexachloro-1,5,5a,6,9,9a-hexahydro-3,3-dioxide; 6,9-Methano-2,4,3-benzodioxathiepin.

Structure:

Designations: CAS: 1031-07-8; DOT: 2761; mf: $C_9H_6Cl_6O_4S$; fw: 422.92.

Properties: Solid. Mp: 198-201 °C; log K_{oc}: 3.37 (calculated); log K_{ow}: 3.66; S_w: 117 ppb.

Transformation Products

Soil: A mixed culture of soil microorganisms biodegraded endosulfan sulfate to endosulfan ether, endosulfan-α-hydroxy ether, and endosulfan lactone (Verschueren, 1983). Indigenous microorganisms obtained from a sandy loam degraded endosulfan sulfate (a metabolite of α- and β-endosulfan) to endosulfan diol. This diol was converted to endosulfan α-hydroxy ether and trace amounts of endosulfan ether and both were degraded to endosulfan lactone (Miles and Moy, 1979).

Plant: In tobacco leaves, endosulfan sulfate was converted to α-endosulfan which subsequently hydrolyzed into endosulfandiol (Chopra and Mahfouz, 1977).

Uses: Not known. Compound is described here because it is a degradate of endosulfan, a widely used insecticide.

ENDOTHALL

Synonyms: Accelerate; Aquathol; Des-i-cate; 3,6-*endo*-Epoxy-1,2-cyclohexane-dicarboxylic acid; Endothal; 3,6-Endooxohexahydrophthalic acid; 3,6-Epoxy-cyclohexane-1,2-dicarboxylic acid; Hydout; Hydrothal; Hydrothal-47; Hydrothal-191; **7-Oxabicyclo[2.2.1]heptane-2,3-dicarboxylic acid;** Pennout; RCRA waste number P088; Ripenthal; Triendothal.

Structure:

Designations: CAS: 145-73-3; mf: $C_8H_{10}O_5$; fw: 186.16; RTECS: RN7875000.

Properties: Odorless, colorless crystals (monohydrate). Mp: 144 °C (mono-hydrate); ρ: 1.431 at 20/4 °C; fl p: nonflammable; pK_1 = 3.4, pK_2 = 6.7; log K_{oc}: 2.04, 2.14; log K_{ow}: -0.89 (calculated); S_o (g/kg at 20 °C): acetone (70), benzene (0.1), 1,4-dioxane (76), ethyl ether (1), methanol (280), 2-propanol (17); S_w: 100 g/kg at 25 °C; vp: negligible at 20 °C.

Soil properties and adsorption data

Soil	K_d (mL/g)	f_{oc} (%)	K_{oc} (mL/g)	pH
Pat Mayse Lake sediments	0.94	0.68	138	--
Pond sediments	1.40	1.29	110	--

Source: Reinert and Rogers, 1984.

Transformation Products

Biological: Incubation of ^{14}C-ring labeled endothall (10 μg/mL) by *Arthrobacter* sp., which was isolated from pond water and a hydrosol, in aerobic sediment-water suspensions revealed that after 30 d, 40% evolved as [14]CO_2. Glutamic acid was the major transformation product. Minor products were alanine, citric and aspartic acids, and unidentified products, some of which were tentatively identified as phosphate esters (Sikka and Saxena, 1973).

Chemical/Physical: Reacts with bases forming water-soluble salts. Above 90 °C, endothall is slowly converted to the anhydride (Hartley and Kidd, 1987; Windholz et al., 1983) and water (Humburg et al., 1989).

Symptoms of Exposure: Strong irritant to eyes, nose, throat, and skin. Ingestion may cause vomiting and diarrhea.

Formulation Types: Granules; soluble concentrate.

Toxicity: LC_{50} (96 hr) using disodium salt: bluegill sunfish 125 mg/L, chinook 136 mg/L, fathead minnow 110 mg/L, largemouth bass 120 mg/L (Verschueren, 1983); acute oral LD_{50} for rats 38-51 mg/kg (free acid) (Hartley and Kidd, 1987), 182-197 mg/kg (sodium salt) (Verschueren, 1983).

Uses: Preemergence and postemergence herbicide for control of broadleaf weeds and annual grass in vegetable crops. The disodium and dipotassium salts are used as defoliants and herbicides.

ENDRIN

Synonyms: Compound 269; Endrex; ENT 17251; Experimental insecticide 269; Hexachloroepoxyoctahydro-*endo,endo*-dimethanonaphthalene; 1,2,3,4,10,10-Hexachloro-6,7-epoxy-1,4,4a,5,6,7,8,8a-octahydro-*endo,endo*-1,4:5,8-dimethanonaphthalene; **3,4,5,6,9,9-Hexachloro-1a,2,2a,3,6,6a,7,7a-octahydro-2,7:3,6-dimethanonaphth[2,3-*b*]oxirene**; Hexadrin; Isodrin epoxide; Mendrin; NA 2761; NCI-C00157; Nendrin; RCRA waste number P051.

Structure:

Designations: CAS: 72-20-8; DOT: 2761; mf: $C_{12}H_8Cl_6O$; fw: 380.92; RTECS: IO1575000.

Properties: White, odorless, crystalline solid when pure; light tan color with faint chemical odor for technical grade. Mp: 200 °C; bp: 245 °C (decomposes); ρ: 1.70 at 25/4 °C (pure), 1.65 at 25/4 °C (technical); fl p: >26.6 °C (xylene solution); lel: 1.1% in xylene; uel: 7.0% in xylene; K_H: 5.0 x 10^{-7} atm·m^3/mol; log K_{ow}: 3.209-5.339; S_o (g/L at 25 °C): acetone (170), benzene (138), carbon tetrachloride (33), cod liver oil (76.0, 83.3, and 91.4 g/L at 4, 12, and 20 °C, respectively), *n*-hexane (71), *n*-octanol (36.4, 38.1, and 43.7 g/L at 4, 12, and 20 °C, respectively), triolein (61.8, 72.6, and 87.3 g/L at 4, 12, and 20 °C, respectively), xylene (183); S_w: 220-260 ppb at 25 °C; vp: 7 x 10^{-7} mmHg at 25 °C.

Transformation Products

Biological: Algae isolated from a stagnant fish pond degraded 24.4% of the applied endrin to ketoendrin (Patil et al., 1972).

Soil: Microbial degradation of endrin in soil formed several ketones and aldehydes of which *keto*-endrin was the only metabolite identified (Kearney and Kaufman, 1976). In eight Indian rice soils, endrin degraded rapidly to low concentrations after 55 d. Degradation was highest in a pokkali soil and lowest in a sandy soil (Gowda and Sethunathan, 1976).

Plant: In plants, endrin is converted to the corresponding sulfate (Hartley and Kidd, 1987).

Photolytic: Photolysis of thin films of solid endrin using UV light (λ = 254 nm) produced δ-ketoendrin, endrin aldehyde, and other compounds (Rosen et al., 1966). Endrin exposed to a hot California sun for 17 d completely isomerized to δ-ketoendrin or 1,8-exo-9,10,11,11-hexachlorocyclo[6.2.1.1^{3,6}.0^{2,7}.0^{4,10}]dodecan-5-one (Burton and Pollard, 1974). Irradiation of endrin by UV light (λ = 253.7 nm and 300 nm) or by natural sunlight in cyclohexane and n-hexane solution resulted in an 80% yield of 1,8-exo-9,11,11-pentachloropentacyclo[6.2.1.1^{3,6}.0^{2,7}.0^{4,10}]dodecan-5-one (Zabik et al., 1971). When an aqueous solution containing endrin was photooxidized by UV light at 90-95 °C, 25, 50, and 75% degraded to carbon dioxide after 15.0, 41.0, and 172.0 hr, respectively (Knoevenagel and Himmelreich, 1976).

Exposure Limits: NIOSH REL: IDLH 200 mg/m^3; OSHA PEL: TWA 0.1 mg/m^3; ACGIH TLV: TWA 0.1 mg/m^3.

Symptoms of Exposure: Epileptiform convulsions, stupor, headache, dizziness, abdominal discomfort, nausea, vomiting, insomnia, aggressive confusion, lethargy, weakness, anorexia.

Formulation Types: Emulsifiable concentrate; wettable powder; granules; dustable powder.

Toxicity: LC$_{50}$ (96 hr) for bluegill sunfish 0.6 μg/L, fathead minnow 1.0 μg/L (Henderson et al., 1959), rainbow trout 0.6 μg/L, coho salmon 0.5 μg/L, and chinook 1.2 μg/L (Katz, 1961); acute oral LD$_{50}$ for rats 7-15 mg/kg (Hartley and Kidd, 1987), 3 mg/kg (RTECS, 1985), male and female rats 18 and 7.5 mg/kg, respectively (Windholz et al., 1983).

Use: Insecticide.

EPN

Synonyms: ENT 17798; EPN 300; Ethoxy-4-nitrophenoxy phenylphosphine sulfide; Ethyl *p*-nitrophenyl benzenethionophosphate; Ethyl *p*-nitrophenyl benzenethiophosphonate; Ethyl *p*-nitrophenyl ester; *O*-Ethyl *O*-4-nitrophenyl phenylphosphonothioate; Ethyl *p*-nitrophenyl phenylphosphonothioate; *O*-Ethyl *O*-*p*-nitrophenyl phenylphosphonothioate; Ethyl *p*-nitrophenyl thionobenzene-phosphate; Ethyl *p*-nitrophenyl thionobenzenephosphonate; *O*-Ethyl phenyl *p*-nitrophenyl phenylphosphorothioate; *O*-Ethyl phenyl *p*-nitrophenyl thiophos-phonate; Phenylphosphonothioic acid *O*-ethyl *O*-*p*-nitrophenyl ester; **Phosphono-thioic acid *O,O*-diethyl *O*-(3,5,6-trichloro-2-pyridinyl) ester;** Pin; Santox.

Structure:

Designations: CAS: 2104-64-5; DOT: 2783; mf: $C_{14}H_{14}NO_4PS$; fw: 323.31; RTECS: TB1925000.

Properties: Yellow solid or crystals to brown liquid with an aromatic odor. Mp: 36 °C; bp: 215 °C at 5 mmHg; ρ: 1.268 and 1.5978 at 25/4 and 30/4 °C, respectively; log K_{oc}: 3.12; log K_{ow}: 3.85, 5.07; S_o: miscible with acetone, benzene, methanol, isopropanol, toluene, xylene, and many other aromatic solvents; vp: 0.126 mPa at 25 °C.

Soil properties and adsorption data

Soil	K_d (mL/g)	f_{oc} (%)	K_{oc} (mL/g)	pH
Kanuma high clay	7.8	1.35	706	5.7
Tsukuba clay loam	60.7	4.24	1,997	6.5

Source: Kanazawa, 1989.

Transformation Products

Soil: Though no products were reported, the half-life in soil is 15-30 d (Hartley and Kidd, 1987).

Chemical/Physical: On heating, EPN is converted to the *S*-ethyl isomer (Worthing and Hance, 1991). Releases toxic fumes of phosphorus, nitrogen, and sulfur oxides when heated to decomposition (Sax and Lewis, 1987; Lewis, 1990). Rapidly hydrolyzed in alkaline solutions to *p*-nitrophenol, alcohol, and benzene thiophosphoric acid (Sittig, 1985).

Exposure Limits: OSHA PEL: TWA 0.5 mg/m^3; ACGIH TLV: TWA 0.5 mg/m^3.

Symptoms of Exposure: Miosis, irritates eyes; rhinorrhea; headache; tight chest, wheezing, laryngeal spasm; salivation; cyanosis; anorexia, nausea, abdominal cramps, diarrhea; paralysis convulsions; low blood pressure.

Formulation Types: Emulsifiable concentrate; wettable powder; granules.

Toxicity: LC$_{50}$ for rainbow trout 0.21 mg/L (Worthing and Hance, 1991), bluegill sunfish 100 μg/L (Sanders and Cope, 1968), fathead minnow 110 mg/L (Solon and Nair, 1970); acute oral LD$_{50}$ for male and female rats, 36 and 7.7 mg/kg, respectively (Windholz et al., 1983), 7 mg/kg (RTECS, 1985).

Uses: Insecticide; acaricide.

EPTC

Synonyms: **Dipropylcarbamothioic acid *S*-ethyl ester;** *N,N*-Dipropylthiocarbamic acid *S*-ethyl ester; Eptam; *S*-Ethyl dipropylcarbamothioate; *S*-Ethyl dipropylthiocarbamate; *S*-Ethyl di-*n*-propylthiocarbamate; *S*-Ethyl-*N,N*-di-*n*-propylthiolcarbamate; FDA 1541; R 1608.

Structure:

$$CH_3CH_2S - \overset{\overset{\displaystyle O}{\|}}{C} - N \overset{\diagup CH_2CH_2CH_3}{\diagdown CH_2CH_2CH_3}$$

Designations: CAS: 759-94-4; mf: $C_9H_{19}NOS$; fw: 189.32; RTECS: FA4550000.

Properties: Colorless to light yellow liquid with an amine-like odor. Mp: <25 °C; bp: 127 °C at 20 mmHg, 235 °C (extrapolated); ρ: 0.960 at 25/25 °, 0.9546 at 30/4 °C; fl p: 116 °C (open cup); K_H: 1.0 x 10^{-5} atm·m^3/mol at 20-25 °C (approximate - calculated from water solubility and vapor pressure); log K_{oc}: 2.38; log K_{ow}: 3.20; S_o: miscible with most organic solvents, e.g., acetone, benzene, ethanol, ethylbenzene, isopropyl alcohol, kerosene, methanol, methyl isobutyl ketone, toluene, xylene; S_w: 375 mg/L at 25 °C; vap d: 7.74 g/L at 25 °C, 6.56 (air = 1); vp: 3.4 x 10^{-2} mmHg at 20 °C.

Transformation Products

Soil: EPTC is rapidly degraded by soil microbes yielding carbon dioxide, mercaptan, and amino residues (Hartley and Kidd, 1987). EPTC partially degraded in both sterile and nonsterile clay soils. Mineralization was not observed since carbon dioxide was not detected (MacRae and Alexander, 1965). EPTC sulfoxide was also reported as a metabolite identified in soil (Somasundaram and Coats, 1991) and in corn plants (Casida et al., 1974). The rapid formation of carbon dioxide was also observed from the microbial degradation of EPTC by a microbial metabolite isolated from Jimtown loam soil and designated JE1 (Dick et al., 1990). These researchers proposed that EPTC hydroxylated at the α-propyl carbon forming the unstable α-hydroxypropyl EPTC which degrades to *N*-depropyl EPTC and propionaldehyde. Metabolization of *N*-depropyl EPTC yields *s*-ethyl formic acid and propylamine. Demethylation of *s*-ethyl formic acid gives *s*-methyl formic acid. Propylamine and *s*-methyl formic acid probably degrades to ammonia and methyl mercaptan, respectively, and carbon dioxide (Dick et al., 1990). The reported half-life in soil is 30 d (Jury et al., 1987). However, Rajagopal et al. (1989) reported that the persistence of EPTC in soil ranged from less than 4-6 wk when applied at recommended rates.

Plant: EPTC is rapidly metabolized by plants to carbon dioxide and naturally occurring plants constituents (Humburg et al., 1989).

Chemical/Physical: Emits toxic fumes of phosphorus and sulfur oxides when heated to decomposition (Sax and Lewis, 1987).

Formulation Types: Granules (10%); emulsifiable concentrate (7 lb/gal).

Toxicity: LC_{50} (48 hr) for bluegill sunfish 27 mg/L and rainbow trout 19 mg/L (Hartley and Kidd, 1987); LC_{50} (24 hr) for blue crab 10 mg/L (Humburg et al., 1989); acute oral LD_{50} for male rats and male mice is 1,700 and 3,200 mg/kg, respectively (Ashton and Monaco, 1991), 1,325 mg/kg (RTECS, 1985), 1,631 mg/kg (Windholz et al., 1983).

Uses: Selective systemic herbicide used for preemergence control of perennial and annual grasses such as johnsongrass, nutgrass, and quackgrass. Also for control of some broadleaf weeds such as chickweed, henbit, lambsquarters, pigweed, and purslane. EPTC is also used in vegetable crops, alfalfa, cotton, flax, pineapple, almonds, and walnuts.

ESFENVALERATE

Synonyms: $(S-(R^*,R^*))$-Cyano(3-phenoxyphenyl)methyl 4-chloro-α-(1-methyl-ethyl)benzeneacetate.

Structure:

Designations: CAS: 66230-04-4; mf: $C_{25}H_{22}ClNO_3$; fw: 419.90.

Properties: Brown solid. Mp: 59.0-60.2 °C; bp: 151-157 °C; ρ: 1.26 at 26/4 °C; K_H: 9.26 x 10^{-7} atm·m^3/mol at 25 °C (approximate – calculated from water solubility and vapor pressure); log K_{oc}: 3.93 (calculated); log K_{ow}: 6.22; S_o: very soluble in acetone, acetonitrile, chloroform, ethyl acetate, N,N-dimethylformamide, dimethyl sulf-oxide, 4-methyl-2-pentanone, xylene; S_w: 0.3 mg/L at 25 °C; vp: 5.03 x 10^{-7} mmHg at 25 °C.

Transformation Products
Chemical/Physical: May hydrolyze in aqueous solutions forming acetic acid and other compounds.

Formulation Types: Emulsifiable concentrate; suspension concentrate.

Toxicity: LC_{50} (96 hr) for fathead minnows 0.69 μg/L (Worthing and Hance, 1991); acute oral LD_{50} for rats 75 mg/kg (Hartley and Kidd, 1987).

Use: Insecticide.

ETHEPHON

Synonyms: Amchem 68-250; Bromoflor; Camposan; CEP; Cepha; 2-CEPA; Cepha 10LS; Cerone; Chlorethephon; 2-Chloroethanephosphonic acid; **(2-Chloroethyl)-phosphonic acid**; Ethel; Etheverse; Ethrel; Flordimex; Florel; G 996; Kamposan; Prep; Roll-fruct; Tomathrel.

Structure:

$$ClCH_2CH_2PO(OH)_2$$

Designations: CAS: 16672-87-0; mf: $C_2H_6ClO_3P$; fw: 144.50; RTECS: SZ7100000.

Properties: Grayish-white waxy solid. Needles from benzene are very hygroscopic. Mp: 74-75 °C; ρ: 1.2-1.3; fl p: nonflammable; H-t½: 24 hr at 33 °C and pH 7; K_H: 1.11 x 10^{-11} atm·m^3/mol at 20-23 °C (approximate - calculated from water solubility and vapor pressure); log K_{oc}: 0.29 (calculated); log K_{ow}: -0.22; pK_a: <7.0; S_o: very soluble in acetone, ethanol, ethylene glycol, methanol, n-hexane, methylene chloride, slightly soluble in benzene and toluene; S_w: 1,240 g/L at 23 °C; vp: 7.50 x 10^{-5} mmHg at 20 °C.

Transformation Products
Soil: Degrades rapidly in soil to phosphoric acid, ethylene, and chloride ions (Hartley and Kidd, 1987) and naturally occurring substances (Humburg et al., 1989).

Chemical/Physical: In an aqueous solution at pH 3.5, ethepon begins to hydrolyze, releasing ethylene (Windholz et al., 1983).

Symptoms of Exposure: Irritates eyes and skin.

Formulation Types: Emulsifiable concentrate; soluble concentrate.

Toxicity: LC_{50} (96 hr) for rainbow trout 254 mg/L and bluegill sunfish 222 mg/L (Worthing and Hance, 1991); acute oral LD_{50} for rats 4,229 mg/kg (24% solution in propylene glycol) (Hartley and Kidd, 1987), 3,400 mg/kg (RTECS, 1985).

Uses: Accelerates the preharvest ripening of fruits and vegetables.

ETHIOFENCARB

Synonyms: Bay-Hox-1901; Croneton; Ethiophencarp; 2-Ethylmercapotomethyl-phenyl-*N*-methylcarbamate; 2-((Ethylthio)methyl)phenol methylcarbamate; **2-Ethylthiomethylphenyl methylcarbamate**; 2-Ethylthiomethylphenyl-*N*-methyl-carbamate; α-Ethylthio-*o*-tolyl methylcarbamate; HOX 1901.

Structure:

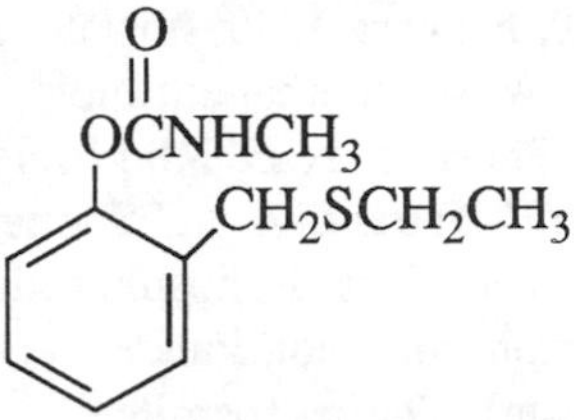

$$O$$
$$\|$$
OCNHCH$_3$
CH$_2$SCH$_2$CH$_3$

Designations: CAS: 29973-13-5; mf: $C_{11}H_{15}NO_2S$; fw: 225.31; RTECS: FC2826000.

Properties: Colorless crystals. Mp: 33.4 °C; bp: decomposes; ρ: 1.1473 at 20/4 °C; H-t$_{1/2}$ (isopropanol/water = 1:1 at 37-40 °C): 330 d (pH 2), 450 hr (pH 7), 5 min (pH 11.4); K_H: 5.3 x 10^{-10} atm·m^3/mol at 20 °C (approximate - calculated from water solubility and vapor pressure); log K_{oc}: 1.84 (calculated); log K_{ow}: 0.98 (calculated); S_o: >600 g/kg in methylene chloride, 2-propanol, toluene; S_w: 1.9 g/L at 20 °C; vp: 3.38 x 10^{-6} mmHg at 20 °C.

Transformation Products
Plant: Degrades in plants to the sulfone and sulfoxide (Hartley and Kidd, 1987).

Formulation Types: Granules, emulsifiable concentrate.

Toxicity: LC$_{50}$ (96 hr) for carp 10-20 mg/L, golden orfe 8-10 mg/L, goldfish 20-40 mg/L, and rudd 10-20 mg/L (Hartley and Kidd, 1987); acute oral LD$_{50}$ for rats 411-499 mg/kg (Hartley and Kidd, 1987), 200 mg/kg (RTECS, 1985).

Uses: Systemic insecticide used to control aphids on fruit crops.

ETHION

Synonyms: AC 3422; Bis(S-(dimethoxyphosphinothioyl)mercapto)methane; Bladan; Diethion; Embathion; ENT 24105; Ethanox; Ethiol; Ethodan; Ethyl methylene phosphorodithioate; FMC 1240; Fosfono 50; Hylemox; Itopaz; Kwit; Methane-dithiol-S,S-diester with O,O-diethyl phosphorodithioate; S,S'-Methylene bis(O,O-diethyl phosphorodithioate); S,S'-Methylene O,O,O',O'-tetraethyl phosphoro-dithioate; NA 2783; NIA 1240; Niagara 1240; Nialate; Phosphorodithioic acid O,O-diethyl ester, S,S-diester with methanedithiol; O,O,O,O-Tetraethyl S,S'-methylenebisdithiophosphate; Phosphotox E; Rhodiacide; Rhodocide; Rodocid; RP 8167; Soprathion; O,O,O',O'-Tetraethyl S,S'-methylenebisphosphordithioate; O,O,O',O'-Tetraethyl S,S'-methylenebisphosphorodithioate; O,O,O',O'-Tetra-ethyl S,S'-methylenebisphosphorothiolothionate; O,O,O',O'-Tetraethyl S,S'-methylene di(phosphorodithioate); Vegfru fosmite.

Structure:

$$CH_3CH_2O \diagdown \underset{\underset{P}{\diagup}}{\overset{\overset{S}{\|}}{}} - SCH_2S - \underset{\underset{\diagdown}{OCH_2CH_3}}{\overset{\overset{S}{\|}}{P}} \diagup OCH_2CH_3$$

$$CH_3CH_2O$$

Designations: CAS: 563-12-2; DOT: 2783; mf: $C_9H_{22}O_4P_2S_4$; fw: 384.48; RTECS: TE4550000.

Properties: Colorless to amber-colored liquid with a disagreeable odor. Mp: -15 to -12 °C; ρ: 1.220 at 20/4 °C; H-t$_{1/2}$: 5 to 63 d at pH 6 and 25 °C; K_H: 3.79 x 10^{-7} atm·m^3/mol at 25 °C; log K_{oc}: 3.54-4.34; log K_{ow}: 4.28, 5.07; P-t$_{1/2}$: 85.47 hr (absorbance λ = 232.5 nm, concentration on glass plates = 6.7 $\mu g/cm^2$); S_o: soluble in most organic solvents including acetone, chloroform, xylene and kerosene + 1% benzene; S_w: 570, 680, and 760 $\mu g/L$ at 10, 20, and 30 °C, respectively; vap d: 15.71 g/L at 25 °C, 13.27 (air = 1); vp: 1.50 x 10^{-6} mmHg at 25 °C.

Soil properties and adsorption data

Soil	K_d (mL/g)	f_{oc} (%)	K_{oc} (mL/g)	pH
Sand	167	0.75	22,149	6.2
Sandy loam	215	1.74	12,356	7.0
Silty loam	105	1.80	5,833	7.1
Silty loam	47	1.33	3,534	7.5

Source: U.S. Department of Agriculture, 1990.

Transformation Products
Chemical/Physical: Emits toxic fumes of phosphorus and sulfur oxides when heated to decomposition (Sax and Lewis, 1987; Lewis, 1990). Though no products were identified, the hydrolysis half-lives of ethion in water at 25 °C and pHs of 4.5, 5.0, 6.0, 7.0, and 8.0 were 99, 63, 58, 25, and 8.4 wk, respectively (Chapman and Cole, 1982).

Exposure Limits: OSHA PEL: TWA 0.4 mg/m^3; ACGIH TLV: TWA 0.4 mg/m^3.

Symptoms of Exposure: Headache, anorexia, nausea, weakness, dizziness, blurred vision, salivation, lacrimation, sweating, shortness of breath, ataxia, fever, cyanosis, pulmonary edema, convulsions, shock, heart block, respiratory failure.

Formulation Types: Emulsifiable concentrate; wettable powder; granules; dustable powder; seed treatment.

Toxicity: LC$_{50}$ (96 hr) for bluegill sunfish 220 μg/L, largemouth bass 150 μg/L, rainbow trout 560 μg/L, cutthroat trout 720 μg/L, channel catfish 7.50 mg/L (Verschueren, 1983); acute oral LD$_{50}$ for rats 208 mg/kg (pure) and 96 mg/kg (technical) (Hartley and Kidd, 1987), 13 mg/kg (RTECS, 1985), male and female rats, 65 and 27 mg/kg, respectively (Windholz et al., 1983).

Uses: Nonsystemic insecticide and acaricide used on apples.

ETHOPROP

Synonyms: ENT 27318; Ethoprophos; *O*-Ethyl *S,S*-dipropyl phosphorodithioate; Jolt; Mobil V-C 9-104; Mocap; **Phosphorodithioic acid *O*-ethyl *S,S*-dipropyl ester;** Prophos; V-C 9-104; V-C chemical V-C 9-104; Virginia-Carolina VC 9-104.

Structure:

$$CH_3CH_2CH_2S \diagdown \quad \overset{\displaystyle O}{\underset{\displaystyle \diagup}{\overset{\displaystyle \|}{P}}} - OCH_2CH_3$$

$$CH_3CH_2CH_2S \diagup$$

Designations: CAS: 13194-48-4; DOT: 2784; mf: $C_8H_{19}O_2PS_2$; fw: 242.33; RTECS: TE4025000.

Properties: Clear, pale yellow liquid. Mp: 20 °C; bp: 86 °C at 0.2 mmHg; ρ: 1.094 at 20/4 °C; K_H: 1.59 x 10^{-7} atm·m^3/mol at 20-25 °C (approximate - calculated from water solubility and vapor pressure); log K_{oc}: 1.82-2.27; log K_{ow}: 3.59 (21 °C); S_o: miscible with acetone, *n*-hexane, xylene; S_w: 700 mg/L at 20 °C; vap d: 9.90 g/L at 25 °C, 8.39 (air = 1); vp: 3.49 x 10^{-4} mmHg at 20 °C.

Soil properties and adsorption data

Soil	K_d (mL/g)	f_{oc} (%)	K_{oc} (mL/g)	pH
Loamy sand	2.44	3.71	66	7.0
Loamy sand	1.08	0.58	186	7.2
Loamy sand	1.24	1.10	113	5.3
Sandy loam	1.61	1.86	87	5.7
Silty loam	2.10	1.33	158	5.6

Source: U.S. Department of Agriculture, 1990.

Transformation Products

Soil: Though no products were identified, the reported half-life in soil in humus-containing soil (pH 4.5) and a sandy loam (pH 7.2-7.3) are 87 and 14-28 d, respectively (Hartley and Kidd, 1987).

Chemical/Physical: Emits toxic fumes of phosphorus and sulfur oxides when heated to decomposition (Sax and Lewis, 1987).

Symptoms of Exposure: Tightness across the chest, nausea, salivation, vomiting, abdominal cramps, diarrhea, abnormal heart rates, arm, and leg weakness, constriction of pupils, involuntary urination.

Formulation Types: Emulsifiable concentrate; granules.

Toxicity: LC_{50} (96 hr) for rainbow trout 13.8 mg/L, bluegill sunfish 2.1 mg/L, and goldfish 13.6 mg/L (Hartley and Kidd, 1987); acute oral LD_{50} for rats 262 mg/kg (Hartley and Kidd, 1987), 34 mg/kg (RTECS, 1985).

Uses: Nonsystemic, nonfumigant nematocide and soil insecticide for control of insects in ornamentals, potatoes, sweet potatoes, tomatoes, strawberries bananas, pineapples, sugar cane, turf, and many other crops.

ETHYLENE DIBROMIDE

Synonyms: Acetylene dibromide; Bromofume; Celmide; DBE; Dibromoethane; **1,2-Dibromoethane;** *sym*-Dibromoethane; α,β-Dibromoethane; Dowfume 40; Dowfume EDB; Dowfume W-8; Dowfume W-85; Dowfume W-90; Dowfume W-100; EDB; EDB-85; E-D-BEE; ENT 15349; Ethylene bromide; Ethylene bromide glycol dibromide; 1,2-Ethylene dibromide; Fumo-gas; Glycol bromide; Glycol dibromide; Iscobrome D; Kopfume; NCI-C00522; Nephis; Pestmaster; Pestmaster EDB-85; RCRA waste number U067; Soilbrom-40; Soilbrom-85; Soilbrom-90; Soilbrom-90EC; Soilbrom-100; Soilbrome-85; Soilfume; UN 1605; Unifume.

Structure:

$$Br\ CH_2\ CH_2\ Br$$

Designations: CAS: 106-93-4; DOT: 1605; mf: $C_2H_4Br_2$; fw: 187.86; RTECS: KH9275000.

Properties: Clear, colorless liquid with a sweet, chloroform-like odor. Mp: 9.8 °C; bp: 131.3 °C; ρ: 2.1792 at 20/4 °C; fl p: nonflammable; H-t$_{1/2}$: 8 yr at 25 °C and pH 7; K_H: 7.06 x 10^{-4} atm·m^3/mol at 25 °C; IP: 9.45 eV; log K_{oc}: 1.56-2.21; log K_{ow}: 1.76; S_o: soluble in acetone, alcohol, benzene, dimethylsulfoxide, ether, and most organic solvents; S_w: 4,321 mg/L at 20 °C; vap d: 7.68 g/L at 25 °C, 6.49 (air = 1); vp: 11 mmHg at 20 °C.

Transformation Products

Biological: A mutant of strain *Acinetobacter* sp. GJ70 isolated from activated sludge degraded ethylene dibromide to ethylene glycol and bromide ions (Janssen et al., 1987). When *Methanococcus thermolithotrophicus*, *Methanococcus deltae*, and *Methanobacterium thermoautotrophicum* were grown with H$_2$-carbon dioxide in the presence of ethylene dibromide, methane, and ethylene were produced (Belay and Daniels, 1987).

Soil: In both soils and water, chemical and biological mediated reactions can transform ethylene dibromide in the presence of hydrogen sulfides to ethyl mercaptan and other sulfur-containing compounds (Alexander, 1981). Complete biodegradation by soil cultures resulted in the formation of ethylene and bromide ions (Castro and Belser, 1968). In a shallow aquifer material, ethylene dibromide aerobically degraded to carbon dioxide, microbial biomass, and nonvolatile water-soluble compound(s) (Pignatello, 1987).

Chemical/Physical: Hydrolyzes in water to ethylene glycol and bromoethanol (Leinster et al., 1978). Dehydrobromination of ethylene dibromide to vinyl bromide was observed in various aqueous buffer solutions (pH 7-11) over the temperature range of 45 to 90 °C. The estimated half-life for this reaction at 25 °C and pH 7 was 2.5 yr (Vogel and Reinhard, 1986). In an aqueous phosphate buffer solution (0.05 M) containing hydrogen sulfide ion (HS⁻), ethylene dibromide was transformed into 1,2-dithioethane and vinyl bromide. The hydrolysis half-lives for solutions with and without sulfides present ranged from 37-70 d and 0.8-4.6 yr, respectively (Barbash and Reinhard, 1989).

Anticipated products from the reaction of ethylene dibromide with ozone or hydroxyl radicals in the atmosphere are bromoacetaldehyde, formaldehyde, bromoformaldehyde, and Br• (Cupitt, 1980). Emits toxic bromide fumes when heated to decomposition (Lewis, 1990).

Exposure Limits: NIOSH REL: TWA 0.045 ppm, 15-min C 0.13 ppm; OSHA PEL: TWA 20 ppm, C 30 ppm, 5-min peak 50 ppm; ACGIH TLV: suspected human carcinogen.

Symptoms of Exposure: Irritation of the respiratory system, eyes; dermatitis with vesiculation.

Formulation Types: Fumigant.

Toxicity: LC_{50} (48 hr) for bluegill sunfish 18 mg/L (Davis and Hardcastle, 1959); acute oral LD_{50} for male and female rats is 148 and 117 mg/kg, respectively (Verschueren, 1983), 108 mg/kg (RTECS, 1985).

Uses: Grain and fruit fumigant; insecticide.

FENAMIPHOS

Synonyms: Bay 68138; ENT 27572; **Ethyl 3-methyl-4-(methylthio)phenyl (1-methylethyl)phosphoramidate;** Ethyl-4-methylthio-*m*-tolyl isopropyl phosphoramidate; Isopropylamino-*o*-ethyl-(4-methylmercapto)-3-methylphenyl)phosphate; 1-(Methylethyl)ethyl 3-methyl-4-(methylthio)phenyl phosphoramidate; Phenamiphos.

Structure:

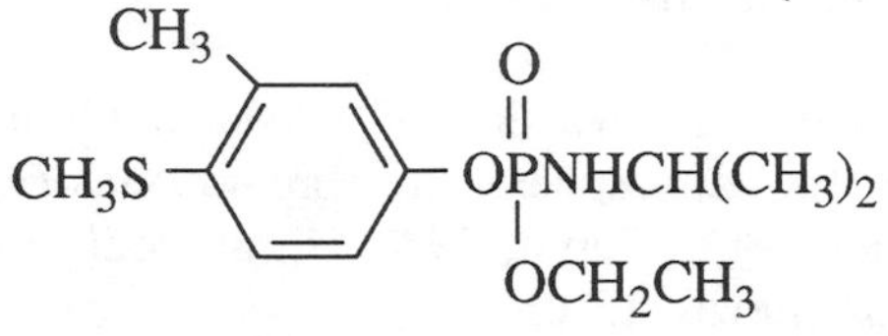

Designations: CAS: 22224-92-6; mf: $C_{13}H_{22}NO_3PS$; fw: 303.40; RTECS: TB3675000.

Properties: Colorless solid. Mp: 49.2 °C; ρ: 1.15 at 20/4 °C; pK_a: 10.5 at 25 °C; K_H: 9.5 x 10^{-10} atm·m^3/mol at 30 °C (approximate - calculated from water solubility and vapor pressure); log K_{oc}: 2.27-3.20; log K_{ow}: 3.23, 3.25; S_o: miscible with acetone, dimethylsulfoxide, 95% ethanol, and many other common organic solvents; S_w: 306, 329, and 419 mg/L at 10, 20, and 30 °C, respectively; vp: 9.98 x 10^{-7} mmHg at 30 °C.

Soil properties and adsorption data

Soil	K_d (mL/g)	f_{oc} (%)	K_{oc} (mL/g)	pH
Arredondo sand	1.18	0.80	148	6.8
Batcombe silt loam	6.74	2.05	329	6.1
Cecil sandy loam	1.77	0.90	197	5.6
Clayey loam	5.78	2.90	199	7.9
Clayey loam	4.59	0.29	1,583	6.0
Loam	9.62	2.32	415	7.3
Loamy sand	3.05	1.62	188	6.6
Rothamsted Farm	4.99	1.51	330	5.1
Sand	1.18	0.46	254	6.8
Sandy loam	1.77	0.52	339	5.6
Webster silty clay loam	9.62	3.97	249	7.3

Source: Bilkert and Rao, 1985; Briggs, 1981; Lord et al., 1980; U.S. Department of Agriculture, 1990.

212

Transformation Products

Soil: Oxidizes in soil to the corresponding sulfone and sulfoxide (Lee et al., 1986). Fenamiphos rapidly degraded in Arredondo soil to fenamiphos sulfone and at the same time to the corresponding phenol. The half-life in this soil is 38-67 d (Ou and Rao, 1986).

Chemical/Physical: Emits toxic fumes of phosphorus, nitrogen, and sulfur oxides when heated to decomposition (Sax and Lewis, 1987; Lewis, 1990).

Exposure Limits: OSHA PEL: TWA 0.1 mg/m^3; ACGIH TLV: TWA 0.1 mg/m^3.

Formulation Types: Emulsifiable concentrate; granules.

Toxicity: LC$_{50}$ (96 hr) for rainbow trout 72.1 μg/L, bluegill sunfish 9.6 μg/L, and goldfish 3,200 μg/L (Hartley and Kidd, 1987); acute oral LD$_{50}$ for male and female rats is 15.3 and 19.4 mg/kg, respectively (Hartley and Kidd, 1987), 8 mg/kg (RTECS, 1985).

Use: Nematocide.

FENBUTATIN OXIDE

Synonyms: Bendex; Bis(tris(β,β-dimethylphenethyl)tin)oxide; Bis(tris(2-methyl-2-phenylpropyl)tin)oxide; Di(tri-(2,2-dimethyl-2-phenylpropyl)tin)oxide; ENT 27738; Hexakis(β,β-dimethylphenethyl)distannoxane; **Hexakis(2-methyl-2-phenylpropyl)distannoxane;** SD 14114; Shell SD-14114; Torque; Vendex.

Structure:

Designations: CAS: 13356-08-6; mf: $C_{60}H_{78}OSn_2$; fw: 1052.70; RTECS: JN8770000.

Properties: Colorless to white crystals. Mp: 138-139 °C; log K_{oc}: 4.91 (calculated); log K_{ow}: 5.10; S_o (g/L): acetone (6), benzene (140), methylene chloride (380); S_w: 5 μg/L at 23 °C.

Transformation Products
Chemical/Physical: Reacts with moisture forming tris(2-methyl-2-phenylpropyl)tin hydroxide (Worthing and Hance, 1991).

Formulation Types: Suspension concentrate; wettable powder.

Toxicity: LC_{50} (48 hr) for rainbow trout 0.27 mg/L (Hartley and Kidd, 1987); acute oral LD_{50} for rats 2,631 mg/kg (Hartley and Kidd, 1987).

Use: Acaricide.

FENSULFOTHION

Synonyms: Bay 25141; Bayer 25141; Bayer S 767; Chemagro 25,141; Dasanit; *O,O*-Diethyl *O*-4-methylsulphinylphenyl phosphorothioate; *O,O*-Diethyl *O-p*-methylsulphinylphenyl phosphorothioate; *O,O*-Diethyl *O-p*-methylsulphinylphenyl thiophosphate; DMSP; ENT 24945; OMS 37; **Phosphorothioic acid *O,O*-diethyl *O*-(*p*-(methylsulfinyl)phenyl) ester;** Terracur P.

Structure:

$$CH_3CH_2O\diagdown \overset{\displaystyle S}{\underset{\displaystyle \diagup}{\overset{\|}{P}}} - O - \underset{\bigcirc}{} - \overset{\displaystyle O}{\underset{\|}{S}}CH_3$$

CH₃CH₂O — P(=S) — O — (ring) — S(=O)CH₃

CH₃CH₂O

Designations: CAS: 115-90-2; DOT: 2765; mf: $C_{11}H_{17}O_4PS_2$; fw: 308.35; RTECS: TF3850000.

Properties: Yellowish-brown oil. Mp: <25 °C; bp: 138-141 °C at 0.01 mmHg; ρ: 1.202 at 20/4 °C; log K_{oc}: 1.89 (calculated); log K_{ow}: 2.23; S_o: miscible with most solvents except aliphatics; S_w: 1.54 g/L at 25 °C; vap d: 12.60 g/L at 25 °C, 9.02 (air = 1).

Transformation Products

Soil: In soils, the bacterium *Klebsiella pneumoniae* degraded fensulfothion to fensulfothion sulfide (Timms and MacRae, 1982, 1983).

Plant: Readily oxidized in plants to the corresponding sulfone (Hartley and Kidd, 1987).

Chemical/Physical: Emits toxic fumes of phosphorus and sulfur oxides when heated to decomposition (Sax and Lewis, 1987; Lewis, 1990). Isomerizes readily to the *O,S*-diethyl isomer (Worthing and Hance, 1991). The hydrolysis half-life of fensulfothion in water at 25 °C at a pH range of 4.5 to 8.0 was 58-87 d (Chapman and Cole, 1982).

Exposure Limits: ACGIH TLV: TWA 0.1 mg/m³.

Formulation Types: Emulsifiable concentrate; dustable powder; granules; wettable powder.

Toxicity: LC_{50} (96 hr) for bluegill sunfish 0.12 mg/L, golden orfe 6.8 mg/L, and rainbow trout 8.8 mg/L (Hartley and Kidd, 1987); acute oral LD_{50} for male and female rats is 10.5 and 2.2 mg/kg, respectively (Hartley and Kidd, 1987), 2 mg/kg (RTECS, 1985).

Uses: Nematocide and pesticide used to control free-living, cyst-forming, and root-knot nematotodes and soil insects in vegetable and fruit crops.

FENTHION

Synonyms: B 29493; Bay 29493; Baycid; Bayer 9007; Bayer 24493; Bayer S 1752; Baytex; *O,O*-Dimethyl *O*-4-(methylmercapto)-3-methylphenyl phosphoro- thioate; *O,O*-Dimethyl *O*-4-(methylmercapto)-3-methylphenyl thiophosphate; *O,O*-Dimethyl *O*-(3-methyl-4-methylmercaptophenyl) phosphorothioate; *O,O*- Dimethyl *O*-(3-methyl-4-methylthiophenyl) phosphorothioate; *O,O*-Dimethyl *O*- (4-methylthio-3-methylphenyl) phosphorothioate; *O,O*-Dimethyl *O*-(4-methyl- thio-*m*-tolyl) phosphorothioate; DMTP; ENT 25540; Entex; Lebaycid; Mercaptophos; 4-Methylmercapto-3-methylphenyl dimethyl thiophosphate; MPP; NCI-C08651; OMS 2; **Phosphorothioic acid *O,O*-dimethyl *O*-(3-methyl-4- (methylthio)phenyl) ester;** Queletox; S 1752; Spottan; Talodex; Tiguvon.

Structure:

Designations: CAS: 55-38-9; DOT: 2784; mf: $C_{10}H_{15}O_3PS_2$; fw: 278.33; RTECS: TF9625000.

Properties: Colorless to amber liquid with a garlic odor. Mp: 7.0 °C; bp: 87 °C at 0.01 mmHg; ρ: 1.250 at 20/4 °C; K_H: 5.49 x 10^{-6} atm·m^3/mol at 25 °C; log K_{oc}: 0.89-1.58; log K_{ow}: 4.09, 4.84; P-t½: 55.83 hr (absorbance λ = 268 nm, concen- tration on glass plates = 6.7 μg/cm^2); S_o: miscible with methylene chloride, 2- propanol; S_w: 6.4, 9.3, and 11.3 mg/L at 10, 20, and 30 °C, respectively; vap d: 11.37 g/L at 25 °C, 9.64 (air = 1); vp: 3.0 x 10^{-5} mmHg at 20 °C.

Soil properties and adsorption data

Soil	K_d (mL/g)	f_{oc} (%)	K_{oc} (mL/g)	pH
Sand	36.2	2.15	1,684	6.9
Sandy loam	7.7	0.81	948	7.7
Sandy loam	38.0	1.74	2,184	5.5
Silty loam	19.8	1.22	1,623	6.7
Silty loam	12.4	1.16	1,069	6.3

Source: U.S. Department of Agriculture, 1990.

Transformation Products
Plant: In plants, fenthion oxidizes to the mesulfenfos and sulfone which further

degrades to the sulfone phosphate before undergoing hydrolysis (Hartley and Kidd, 1987).

Photolytic: Fenthion was oxidized to the corresponding sulfoxide and trace amounts (<5% yield) of sulfone when sorbed on soil and exposed to sunlight. The photosensitized oxidation was probably due to the presence of singlet oxygen. The degradation rate was higher in soils containing the lowest organic carbon (Gohre and Miller, 1986).

Chemical/Physical: Emits very toxic fumes of phosphorus and sulfur oxides when heated to decomposition (Sax and Lewis, 1987; Lewis, 1990). Hydrolyzes in water forming *O,O*-dimethyl-*O*-(4-(methylthio)-*m*-tolyl) phosphate (bayoxon) and 3-methyl-4-methylthiophenol (Suffet et al., 1967).

Exposure Limits: OSHA PEL: TWA 0.2 mg/m^3; ACGIH TLV: TWA 0.2 mg/m^3.

Formulation Types: Emulsifiable concentrate; wettable powder; granules; dustable powder; fogging concentrate.

Toxicity: LC_{50} (96 hr) for fathead minnow 2.44 mg/L, largemouth bass 1.54 mg/L, brown trout 1.33 mg/L, coho salmon 1.32 mg/L, perch 1.65 mg/L, channel catfish 1.68 mg/L, black bullhead 1.62 mg/L, rainbow trout 0.93 mg/L, bluegill sunfish 1.4 mg/L, goldfish 3.4 mg/L, perch 1.7 mg/L, and carp 1.2 mg/L (Macek and McAllister, 1970); LC_{50} (48 hr) for goldfish 1.9 mg/L (Hartley and Kidd, 1987); acute oral LD_{50} for male and female rats is 375 and 290 mg/kg, respectively (Hartley and Kidd, 1987), 180 mg/kg (RTECS, 1985).

Uses: Insecticide and acaricide.

FENVALERATE

Synonyms: Belmark; α-Cyano-3-phenoxybenzyl-2-(4-chlorophenyl)-3-methyl-butyrate; **Cyano(3-phenoxyphenyl)methyl 4-chloro-α-(1-methylethyl)benzene-acetate;** Phenoxybenzyl-2-(4-chlorophenyl)isovalerate; S 5602; Sanmarton; SD 43775; Sumicidin; Sumifly; Sumipower; WL 43775.

Structure:

Designations: CAS: 51630-58-1; mf: $C_{25}H_{22}ClNO_3$; fw: 419.92; RTECS: CY1576350.

Properties: Clear yellow or brown, viscous liquid. Mp: <23 °C; ρ: 1.175 at 25/25 °C; K_H: 1.5 x 10^{-7} atm·m^3/mol at 20-25 °C (approximate - calculated from water solubility and vapor pressure); log K_{oc}: 3.64 (calculated); log K_{ow}: 4.09-6.25; P-t½: 168.85 hr (absorbance λ = 228.0 nm, concentration on glass plates = 6.7 μg/cm^2); S_o: miscible with acetone, chloroform, cyclohexanone, ethanol, xylene; S_w: <1 mg/L at 20 °C, 24 μg/L in seawater; vap d: 17.16 g/L at 25 °C, 14.54 (air = 1); vp: 2.78 x 10^{-7} mmHg at 25 °C.

Transformation Products
Surface Water: In an estuary, the half-life of fenvalerate was 27-42 d (Schimmel et al., 1983).

Chemical/Physical: Undergoes hydrolysis at the ester bond (Hartley and Kidd, 1987). Decomposes gradually at 150-300 °C (Windholz et al., 1983) probably releasing toxic fumes of nitrogen and chlorine.

Formulation Types: Emulsifiable concentrate; suspension concentrate.

Toxicity: LC_{50} (96 hr) for rainbow trout 3.6 μg/L (Worthing and Hance, 1991), estuarine mysid 0.008 μg/L, pink shrimp 0.84 μg/L, sheepshead minnow 5.0 μg/L, Atlantic silverside 0.31 μg/L, striped mullet 0.58 μg/L, Gulf toadfish 5.4 μg/L (Schimmel et al., 1983); LC_{50} (48 hr) for carp <100 μg/L (Hartley and Kidd, 1987); LC_{50} (24 hr) for rainbow trout 76.0 ppb (technical) and 21.0 ppb (formulated product) (Coats and O'Donnell-Jeffrey, 1979); acute oral LD_{50} for rats 451 mg/kg (dimethylsulfoxide) (Hartley and Kidd, 1987).

Use: Insecticide used against a wide variety of pests.

FERBAM

Synonyms: Aafertis; Bercema Fertam 50; Carbamate; Dimethylcarbamodithioc acid iron complex; Dimethylcarbamodithioc acid iron(3+) salt; Dimethyldithiocarbamic acid iron salt; Dimethyldithiocarbamic acid iron(3+) salt; ENT 14689; Ferbam 50; Ferbame; Ferbam, iron salt; Ferbeck; Ferberk; Fermate; Fermate ferbam fungicide; Fermocide; Ferradow; Ferric dimethyldithiocarbamate; Fuklasin; Fuklasin ultra; Hexaferb; Hokmate; Iron flowable; Iron tris(dimethyldithiocarbamate); Karbam black; Trifungol; **Tris(dimethylcarbamodithioato-*S,S'*)iron**; Tris(dimethyldithiocarbamato)iron; Vancide FE95.

Structure:

$$[(CH_3)_2NCS_2]_3Fe$$

Designations: CAS: 14484-64-1; mf: $C_9H_{18}FeN_3S_6$; fw: 416.50; RTECS: NO8750000.

Properties: Black, fluffy powder. Mp: >180 °C with decomposition; pK_a: unknown but pH of saturated solution is 5.0; log K_{oc}: 0.83 (calculated); log K_{ow}: -1.00 (calculated); S_o: soluble in acetone, acetonitrile, benzene, carbon tetrachloride, chloroform, ethanol, methanol, *n*-propanol, pyridine, toluene, xylene, and many other organic solvents; S_w: 120-130 g/L at 20 °C (pH of solution = 5.0); vp: negligible at room temperature.

Transformation Products

Plant: Decomposes in plants to ethylene thiourea, ethylene thiuram monosulfide, ethylene thiuram disulfide, and sulfur (Hartley and Kidd, 1987).

Chemical/Physical: Hydrolyzes in acidic media releasing carbon disulfide. Decomposes in water forming ethylene thiourea (Hartley and Kidd, 1987). Melts and decomposes >180 °C (Windholz et al., 1983) emitting toxic fumes of nitrogen and sulfur oxides (Lewis, 1990; Sax and Lewis, 1987).

Exposure Limits: OSHA PEL: TWA 15 mg/m^3; ACGIH TLV: TWA 10 mg/m^3.

Symptoms of Exposure: May cause irritation of skin and mucous membranes, and renal damage.

Formulation Types: Wettable powder.

220

Toxicity: Moderately toxic to fish (Hartley and Kidd, 1987); acute oral LD_{50} for rats >17,000 mg/kg (Hartley and Kidd, 1987), 4,000 mg/kg (RTECS, 1985).

Uses: Nonphytotoxic fungicide used to control scab on fruits and other crops.

FLUCYTHRINATE

Synonyms: AC 222705; (RS)-Cyano-(3-phenoxyphenyl)methyl (S)-4-(difluoro-methoxy)phenyl)-α-(1-methylethyl)benzeneacetate; 4-(Difluoromethoxy)-α-(1-methylethyl)benzeneacetic acid cyano(3-phenoxyphenyl)methyl ester; Cybolt; Pay-off.

Structure:

Designations: CAS: 70124-77-5; mf: $C_{26}H_{23}F_2NO_4$; fw: 451.48; RTECS: CY1578620.

Properties: Dark amber viscous liquid with an ester-like odor. Mp: <25 °C; bp: 108 °C at 0.35 mmHg; ρ: 1.189 at 22/4 °C; H-t½ (27 °C): 40 d (pH 3), 52 d (pH 5), 6.3 d (pH 9); K_H: 8.08 x 10^{-2} atm·m³/mol at 21-25 °C (approximate - calculated from water solubility and vapor pressure); log K_{oc}: 3.81 (calculated); log K_{ow}: 4.70; S_o (g/L at 21 °C): acetone (820), n-hexane (90), 2-propanol (780), xylene (1,810); S_w: 0.5 mg/L at 21 °C; vp: 6.80 x 10^{-2} mmHg at 25 °C.

Transformation Products
Surface Water: The half-life of flucythrinate in an estuarine environment is 34 d (Schimmel et al., 1983).

Chemical/Physical: May hydrolyze in aqueous solutions forming acetic acid and other compounds.

Formulation Types: Emulsifiable concentrate; water-dispersible granules.

Toxicity: LC_{50} (96 hr) for rainbow trout 0.32 µg/L, bluegill sunfish 0.71 µg/L, channel catfish 0.51 µg/L, sheepshead minnow 1.6 µg/L (Hartley and Kidd, 1987), estuarine mysid 0.008 µg/L, pink shrimp 0.22 µg/L, and sheapshead minnow 1.1 µg/L (Schimmel et al., 1983); acute oral LD_{50} for male and female rats is 81 and 67 mg/kg, respectively (Hartley and Kidd, 1987).

Use: Nonsystemic insecticide.

FLUOMETURON

Synonyms: C 2059; CIBA 2059; Cotoran; Cotoran multi 50WP; Cottonex; 1,1-Dimethyl-3-(3-trifluoromethylphenyl)urea; **N,N-Dimethyl-N'-(3-(trifluoro-methyl)phenyl)urea**; 1,1-Dimethyl-3-(α,α,α-trifluoro-m-tolyl)urea; Herbicide C-2,059; Lanex; Meturon; Meturon 4L; NCI-C08695; Pakhtaran; N-(m-Trifluoro-methylphenyl)-N',N'-dimethylurea; N-(3-Trifluoromethylphenyl)-N',N'-di-methylurea; 3-(m-Trifluoromethylphenyl)-1,1-dimethylurea.

Structure:

Designations: CAS: 2164-17-2; mf: $C_{10}H_{11}F_3N_2O$; fw: 232.21; RTECS: YT1575000.

Properties: Odorless, white, crystalline powder. Mp: 163-164.5 °C, 155 °C (technical - 95%); ρ: 1.39 at 20/4 °C; fl p: nonflammable; H-t$_{1/2}$ (20 °C): 1.6 yr (pH 1), 2.4 yr (pH 5), 2.8 yr (pH 9); K_H: <2.79 x 10^{-6} atm·m³/mol at 20-25 °C (approximate - calculated from water solubility and vapor pressure); log K_{oc}: 1.46-2.08; log K_{ow}: 2.23, 2.38; P-t$_{1/2}$: 1.2 d at 23 °C; S_o (g/L at 20 °C): acetone (105), n-hexane (170), methanol (110), methylene chloride (23), and n-octanol (220); S_w: 80 mg/L at 25 °C; vp: 5 x 10^{-7} mmHg at 20 °C.

Soil properties and adsorption data

Soil	K_d (mL/g)	f_{oc} (%)	K_{oc} (mL/g)	pH
Loam	1.64	2.78	59	5.9
Sand	0.15	0.52	29	6.5
Sandy loam	0.74	0.70	106	7.6
Sandy loam	2.81	2.32	121	7.8

Source: U.S. Department of Agriculture, 1990.

Transformation Products
Soil: In soils, fluometuron rapidly degrades (t$_{1/2}$ ≈ 30 d) to carbon dioxide, polar and nonextractable compounds (Hartley and Kidd, 1987; Humburg et al., 1989).

Plant: In plants, fluometuron degrades to a demethylated intermediate which subsequently is degraded to the aniline moiety (possibly m-trifluoromethylaniline)

(Hartley and Kidd, 1987; Humburg et al., 1989). Duke et al. (1991) reported that fluometuron degrades in plants via the following degradative pathway: fluometuron to N-methyl-N'-(3-(trifluoromethyl)phenyl)urea which undergoes demethylation to 3-(trifluoromethyl)phenylurea followed by deamination and elimination of the ketone group to form 3-trifluoromethylaniline.

Chemical/Physical: Emits toxic fumes of nitrogen oxides and fluorine when heated to decomposition (Sax and Lewis, 1987).

Symptoms of Exposure: Dust may cause eye irritation.

Formulation Types: Suspension concentrate (4 lb/gal); wettable powder (80%).

Toxicity: LC_{50} (96 hr) for rainbow trout 47 mg/L, bluegill sunfish 96 mg/L, catfish 55 mg/L, and crucian carp 170 mg/L (Hartley and Kidd, 1987); acute oral LD_{50} of the 80% formulation for rats is 1,800 mg/kg (Ashton and Monaco, 1991), 6,416 mg/kg (RTECS, 1985).

Uses: Herbicide used to control many annual broadleaf weeds in sugarcane and cotton.

FONOFOS

Synonyms: Difonate; Dyfonate; Dyphonate; ENT 25796; *O*-Ethyl *S*-phenyl ethyl-dithiophosphonate; *O*-Ethyl *S*-phenyl ethylphosphonodithioate; **Ethylphos-phonodithioic acid *O*-ethyl *S*-phenyl ester**; Fonophos; N-2790; Stauffer NA 2790.

Structure:

$$CH_3CH_2-P\underset{S}{\overset{\underset{\|}{S}}{\diagdown}}\overset{O-CH_2CH_3}{\diagup}$$

Designations: CAS: 944-22-9; mf: $C_{10}H_{15}OPS_2$; fw: 246.32; RTECS: TA5950000.

Properties: Colorless to pale yellow liquid with an aromatic odor. Mp: <25 °C; bp: 130 °C at 0.1 mmHg; ρ: 1.154 at 20/20 °C; H-t½ (40 °C): 101 d (pH 4), 74-127 (pH 7, buffer dependent), 1.8 d (pH 10); K_d: 15.3 (loam soil); K_H: 5.2 x 10^{-6} atm·m^3/mol at 25 °C (approximate - calculated from water solubility and vapor pressure); log K_{oc}: 3.03 (calculated); log K_{ow}: 3.89, 3.90; S_o: miscible with many organic solvents; S_w: 13 mg/L at room temperature; vp: 2.1 x 10^{-4} mmHg at 25 °C.

Transformation Products
Plant: In plants, fonofos is oxidized to the phosphonothioate (Hartley and Kidd, 1987). Oat plants were grown in two soils treated with [^{14}C]fonofos. Most of the residues remained bound to the soil. Less than 2% of the applied [^{14}C]fonofos was recovered from the oat leaves. Metabolites identified in both soils and leaves were methyl phenyl sulfone, 2-, 3-, and 4-hydroxymethyl phenyl sulfone, thiophenol, diphenyl disulfide, and fonofos oxon (Fuhremann and Lichtenstein, 1980; Lichtenstein et al., 1982).

Chemical/Physical: Emits toxic nitrogen and phosphorus oxide fumes when heated to decomposition (Sax and Lewis, 1987; Lewis, 1990).

Exposure Limits: ACGIH TLV: 0.1 mg/m^3.

Formulation Types: Granules; emulsifiable concentrate.

Toxicity: LC$_{50}$ (24 hr) for bluegill sunfish 45 μg/L and rainbow trout 110 μg/L (Verschueren, 1983); acute oral LD$_{50}$ for rats 8-17.5 mg/kg (Hartley and Kidd, 1987), 3 mg/kg (RTECS, 1985).

Uses: Soil insecticide used to control rootworms, wireworms, crickets and similar crop pests in vegetables, sorghum, ornamentals, cereals, maize, vines, olives, sugar beet, sugar cane, potatoes, groundnuts, tobacco, turf, and fruit crops.

FORMALDEHYDE

Synonyms: BFV; FA; Fannoform; Formalin; Formalin 40; Formalith; Formic aldehyde; Formol; Fyde; HOCH; Ivalon; Karsan; Lysoform; Methanal; Methyl aldehyde; Methylene glycol; Methylene oxide; Morbicid; NCI-C02799; Oxomethane; Oxymethylene; Paraform; Polyoxymethylene glycols; RCRA waste number U122; Superlysoform; UN 1198; UN 2209.

Structure:

HCHO

Designations: CAS: 50-00-0; DOT: 1198; mf: CH_2O; fw: 30.03; RTECS: LP8925000.

Properties: Clear, colorless liquid with a pungent, suffocating odor and burning taste. Mp: -92 °C; bp: -21 °C, 98-99 °C (40% aqueous solution); ρ: 0.815 at 20/4 °C, 1.081-1.085 at 25/25 °C (37% aqueous solution); fl p: 50 °C (15% methanol-free); lel: 7.0%; uel: 73.0%; K_H: 1.67 x 10^{-7} atm·m^3/mol; IP: 10.88 eV; log K_{oc}: 0.56 (calculated); log K_{ow}: 0.00; S_o: soluble in acetone (>100 mg/mL), benzene, dimethylsulfoxide (>100 mg/mL), 95% ethanol (>100 mg/mL), ethyl ether; S_w: miscible at 25 °C; vap d: 1.23 g/L at 25 °C, 1.067 (air = 1); vp: 760 mmHg at -19.5 °C.

Transformation Products

Biological: Biodegradation products reported include formic acid and ethanol each of which can further degrade to carbon dioxide (Verschueren, 1983).

Photolytic: Major products reported from the photooxidation of formaldehyde with nitrogen oxides are carbon monoxide, carbon dioxide, and hydrogen peroxide (Altshuller, 1983). In synthetic air, photolysis of formaldehyde gave hydrochloric acid and carbon monoxide (Su et al., 1979). Photooxidation of formaldehyde in the absence of nitrogen oxides in air (λ = 2900-3500 Å) gave hydrogen peroxide, alkylhydroperoxides, carbon monoxide, and lower molecular weight aldehydes. In the presence of nitrogen oxides, photooxidation products reported include ozone, hydrogen peroxide, and peroxyacyl nitrates (Kopczynski et al., 1974). Irradiation of gaseous formaldehyde containing an excess of nitrogen dioxide over chlorine yielded ozone, carbon monoxide, nitrogen pentoxide, nitryl chloride, nitric acid, and hydrochloric acid. Peroxynitric acid was the major photolysis product when chlorine exceeded nitrogen dioxide concentrations (Hanst and Gay, 1977).

Chemical/Physical: Oxidizes in air to formic acid (Hartley and Kidd, 1987). Trioxymethylene may precipitate under cold temperatures (Sax, 1984). Polymerizes easily (Windholz et al., 1983). Anticipated products from the reaction of formaldehyde with ozone or hydroxyl radicals in air are carbon monoxide and carbon dioxide (Cupitt, 1980). Reacts with hydrochloric acid in moist air forming bis(chloromethyl)ether. This compound may also form from an acidic solution containing chloride ion and formaldehyde (Frankel et al., 1974). May polymerize in an aqueous solution to trioxymethylene (Hartley and Kidd, 1987).

Exposure Limits: NIOSH REL: 15-min C 0.1 ppm; OSHA PEL: TWA 3 ppm, C 5 ppm, 30-min C 10 ppm; ACGIH TLV: TWA 1 ppm, STEL 2 ppm.

Symptoms of Exposure: Irritates eyes, nose, throat; lacrimation; cough, bronchospasm; pulmonary irritation; dermatitis, nausea, vomiting; loss of consciousness.

Formulation Types: Aqueous solutions.

Toxicity: Toxic to fish (Hartley and Kidd, 1987); acute oral LD_{50} for rats 550-800 mg/kg (Hartley and Kidd, 1987).

Uses: Fungicide; bactericide.

FOSAMINE-AMMONIUM

Synonyms: **Ammonium salt, ammonium ethyl(aminocarbonyl)phosphonate;** Ammonium ethyl carbamoylphosphonate solution; DPX 1108; Krenite; Krenite brush control agent.

Structure:

$$C_2H_5 - O - \overset{\overset{\displaystyle O}{\|}}{\underset{\underset{\displaystyle O^- \ \overset{+}{N}H_4}{|}}{P}} - CONH_2$$

Designations: CAS: 25954-13-6; mf: $C_3H_{11}N_2O_4P$; fw: 170.10; RTECS: BQ4112000.

Properties: Colorless to white crystals. Mp: 175 °C; ρ: 1.33; fl p: nonflammable; K_H: 4.98 x 10^{-13} atm·m³/mol at 20-25 °C (approximate - calculated from water solubility and vapor pressure); log K_{oc}: 0.20 (calculated); log K_{ow}: -2.90; S_o (g/kg at 25 °C): acetone (0.3), benzene (0.4), chloroform (0.04), N,N-dimethylformamide (1.4), ethanol (12.0), n-hexane (0.2), methanol (158.0); S_w: 1,790 g/L at 25 °C; vp: 4 x 10^{-6} mmHg at 20 °C.

Transformation Products
Soil: Fosamine is rapidly degraded to carbon dioxide by microorganisms in soil (Humburg et al., 1989). The reported half-life in soil is approximately 7-10 d (Hartley and Kidd, 1987; Worthing and Hance, 1991).

Plant: Degrades rapidly in plants with a half-life of 2-3 wk (Humburg et al., 1989; Ashton and Monaco, 1991).

Symptoms of Exposure: May irritate eyes, nose, throat, and skin.

Formulation Types: Soluble concentrate (4 lb/gal).

Toxicity: LC_{50} (96 hr) for bluegill sunfish 0.67 g/L, rainbow trout, and fathead minnows >1 g/L (Hartley and Kidd, 1987); acute oral LD_{50} for rats is 24.4 g/kg (Ashton and Monaco, 1991).

Uses: Nonselective contact herbicide used to control many woody and brush species on noncrop land.

FOSETYL-ALUMINUM

Synonyms: Aliette; **Aluminum tris(*O*-ethyl phosphonate)**; Efosite-AL; Epal; Fosetyl AL; LS 74783; Phosethyl; Phosethyl AL; RP 32545.

Structure:

$$\left[CH_3CH_2O - \overset{\displaystyle O}{\underset{\displaystyle O}{\overset{\|}{\underset{|}{P}}}} - H \right]_3^{-} \quad Al^{3+}$$

Designations: CAS: 39148-24-8; mf: $C_6H_{18}AlO_9P_3$; fw: 354.10; RTECS: SZ9640000.

Properties: Colorless powder. Mp: decomposes >200 °C; log K_{oc}: 2.49; log K_{ow}: -2.70 (pH 4); S_o (mg/L): acetone (13), acetonitrile (5), methanol (920); S_w: 120 g/L at room temperature; vp: 9.75 x 10^{-8} mmHg at 25 °C.

Transformation Products
Plant: Felsot and Pedersen (1991) reported that fosetyl-aluminum degrades in plants forming phosphonic acid which ionizes to the dianion phosphonate, HPO_3^{-2}.

Formulation Types: Wettable powder.

Toxicity: LC_{50} (96 hr) for rainbow trout 428 mg/L (Hartley and Kidd, 1987); acute oral LD_{50} for rats 5,800 mg/kg (Hartley and Kidd, 1987), 5,400 mg/kg (RTECS, 1985).

Use: Fungicide.

GLYPHOSATE

Synonyms: Mon 0573; *N*-(Phosphonomethyl)glycine.

Structure:

$$HOOCCH_2NHCH_2 - \overset{\displaystyle O}{\overset{\displaystyle \|}{P}} - OH$$
$$|$$
$$OH$$

Designations: CAS: 1071-83-6; mf: $C_3H_8NO_5P$; fw: 169.08; RTECS: MC1075000.

Properties: Colorless to white, odorless crystals or powdery solid. Mp: 230 °C (decomposes); ρ: 1.74 g/mL; fl p: nonflammable (water-based formulations only); pK_1: 2.32, $pK_2 = 5.86$, $pK_3 = 10.86$; K_H: 1.39 x 10^{-10} atm·m³/mol at 25 °C (approximate - calculated from water solubility and vapor pressure); log K_{oc}: 3.43-3.69; log K_{ow}: -1.60; S_o: insoluble in most organic solvents; S_w: 12 g/L at 25 °C; vp: 7.50 x 10^{-6} mmHg at 25 °C.

Soil properties and adsorption data

Soil	K_d (mL/g)	f_{oc} (%)	K_{oc} (mL/g)	pH
Dupo silt loam	33	--	--	--
Drummer silty clay	324	--	--	--
Houston clay loam	76	1.56	4,871	7.5
Muskingum silt loam	56	1.64	3,414	5.8
Sassafras sandy loam	33	1.24	2,661	5.6
Spinks loamy sand	660	--	--	--

Source: Glass, 1987; U.S. Department of Agriculture, 1990.

Transformation Products

Soil: Degrades microbially in soil releasing phosphoric acid, *N*-nitrosoglyphosate (Newton et al., 1984), ammonia (Cremlyn, 1991), *N,N*-dimethylphosphinic acid, *N*-methylphosphinic acid, aminoacetic acid (glycine), *N*-methylaminoacetic acid (sarcosine), hydroxymethylphosphonic acid (Duke et al., 1991), aminomethylphosphonic acid (Duke et al., 1991; Hoagland, 1980; Muir, 1991; Rueppel et al., 1977), and carbon dioxide (Cremlyn, 1991). *N*-Nitrosoglyphosate also formed from the nitrosation of glyphosate in soil solutions containing nitrite ions (Young and Kahn, 1978). The reported half-life of glyphosate in soil is <60 d (Hartley and Kidd, 1987).

Plant: In a forest brush field ecosystem, the half-life of glyphosate in foliage and litter ranged from 10.4 to 26.6 d (Newton et al., 1984).

Photolytic: When an aqueous solution of glyphosate (1 ppm) was exposed to outdoor sunlight for 9 wk (from August 12 thru October 15, 1983), aminomethyl-phosphonic acid and ammonia formed as major and minor photoproducts, respectively (Lund-Høie and Friestad, 1986). More than 90% degradation was observed after only 4 wk of exposure. Photodegradation was also observed when an aqueous solution was exposed indoors to UV light (λ = 254 nm). The reported half-lives of this reaction at starting concentrations of 1.0 and 2,000 ppm were 4 d and 3-4 wk, respectively. When aqueous solutions were exposed indoors to sodium light (λ = 550-650 nm) and mercury light (λ = 400-600 nm), no photodegradation occurred (Lund-Høie and Friestad, 1986).

Chemical/Physical: Under laboratory conditions, the half-life of glyphosate in natural waters was 7-10 wk (Muir, 1991). A 1% aqueous solution has a pH of 2.5 (Keith and Walters, 1992). This suggests glyphosate will react with alkalies and amines forming water soluble salts. Decomposes at 230 °C (Windholz et al., 1983) probably emitting toxic fumes of phosphorous and nitrogen oxides.

Formulation Types: Soluble concentrate (free acid 3 lb/gal, isopropylamine salt 4 lb/gal); water-soluble powder.

Toxicity: LC_{50} (96 hr) for rainbow trout 86 mg/L, bluegill sunfish 120 mg/L (Hartley and Kidd, 1987), for harlequin fish 168 ppm, for Atlantic oyster >10 mg/L, for shrimp 281 ppm, and fiddler crab 934 mg/L (Humburg et al., 1989); LC_{50} (48 hr) for *Daphnia magna* 780 mg/L (Humburg et al., 1989); acute oral LD_{50} of pure glyphosate and Roundup formulation for rats is 5,600 and 5,400 mg/kg, respectively (Ashton and Monaco, 1991), 470 mg/kg (RTECS, 1985).

Uses: Nonselective, postemergence, broad spectrum herbicide used to control annual and perennial grasses, sedges, broadleaf, and emerged aquatic weeds. This herbicide is also used to control insects on fruit trees.

HEPTACHLOR

Synonyms: Aahepta; Agroceres; Basaklor; 3-Chlorochlordene; Drinox; Drinox H-34; E 3314; ENT 15152; GPKh; H 34; Heptachlorane; 3,4,5,6,7,8,8-Heptachlorodicyclopentadiene; 3,4,5,6,7,8,8a-Heptachlorodicyclopentadiene; 1(3a),4,5,6,7,8,8-Heptachloro-3a(1),4,7,7a-tetrahydro-4,7-methanoindene; 1,4,5,6,7,8,8-Heptachloro-3a,4,7,7a-tetrahydro-4,7-methanoindene; **1,4,5,6,7,8,8-Heptachloro-3a,4,7,7a-tetrahydro-4,7-methanol-1*H*-indene**; 1,4,5,6,7,8,8-Heptachloro-3a,4,7,7a-tetrahydro-4,7-*endo*-methanoindene; 1,4,5,6,7,8,8a-Heptachloro-3a,4,7,7a-tetrahydro-4,7-methanoindene; 1,4,5,6,7,8,8-Heptachloro-3a,4,7,7a-tetrahydro-4,7-methyleneindene; 1,4,5,6,7,10,10-Heptachloro-4,7,8,9-tetrahydro-4,7-methyleneindene; 1,4,5,6,7,10,10-Heptachloro-4,7,8,9-tetrahydro-4,7-*endo*-methyleneindene; 3,4,5,6,7,8,8a-Heptachloro-α-dicyclopentadiene; Heptadichlorocyclopentadiene; Heptagran; Heptagranox; Heptamak; Heptamul; Heptasol; Heptox; NA 2761; NCI-C00180; Soleptax; RCRA waste number P059; Rhodiachlor; Velsicol 104; Velsicol heptachlor.

Structure:

Designations: CAS: 76-44-8; DOT: 2761; mf: $C_{10}H_5Cl_7$; fw: 373.32; RTECS: PC0700000.

Properties: White to light tan, waxy solid or crystals with a camphor-like odor. Mp: 95-96 °C (pure), 46-74 °C (technical); bp: 135-145 °C at 1-1.5 mmHg, decomposes at 760 mmHg; ρ: 1.66 at 20/4 °C; K_H: 2.3 x 10^{-3} atm·m^3/mol; log K_{oc}: 4.38; log K_{ow}: 4.40-5.5; S_o (g/L at 27 °C unless noted otherwise): acetone (750), benzene (1,060), carbon tetrachloride (1,120), cod liver oil (93.8, 115.4, 178.7 g/L at 4, 12, and 20 °C, respectively), cyclohexane (1,190), cyclohexanone (1,190), deodorized kerosene (263), kerosene (1,890), alcohol (45), *n*-octanol (67.9, 81.7, and 87.5 g/L at 4, 12, and 20 °C, respectively), triolein (83.6, 129.4, and 132.4 g/L at 4, 12, and 20 °C, respectively), xylene (1,020); S_w: 180 ppb at 25 °C; vp: 3 x 10^{-4} mmHg at 20 °C.

Transformation Products

Biological: In a model ecosystem containing plankton, *Daphnia magna*, mosquito larva (*Culex pipiens quinquefasciatus*), fish (*Cambusia affinis*), alga (*Oedogonium cardiacum*), and snail (*Physa* sp.), heptachlor degraded to 1-hydroxychlordene, 1-

hydroxy-2,3-epoxychlordene, hydroxychlordene epoxide, heptachlor epoxide, and five unidentified compounds (Lu et al., 1975).

Soil: Many soil microorganisms were found to oxidize heptachlor to heptachlor epoxide (Miles et al., 1969). In addition, hydrolysis produced hydroxychlordene with subsequent epoxidation yielding 1-hydroxy-2,3-epoxychlordene (Kearney and Kaufman, 1976). Heptachlor reacted with reduced hematin forming chlordene which decomposed to hexachlorocyclopentadiene and cyclopentadiene (Baxter, 1990). The reported half-life in soil is 9-10 mo (Hartley and Kidd, 1987).

Photolytic: Sunlight and UV light converts heptachlor to photoheptachlor (Georgacakis and Khan, 1971). Eichelberger and Lichtenberg (1971) reported heptachlor (10 μg/L) in river water, kept in a sealed jar under sunlight and fluorescent light, was completely converted to 1-hydroxychlordene. Under the same conditions, but in distilled water, 1-hydroxychlordene and heptachlor epoxide formed in yields of 60 and 40%, respectively (Eichelberger and Lichtenberg (1971). The photolysis of heptachlor in various organic solvents afforded different photoproducts (McGuire et al., 1972). Photolysis at 253.7 nm in hydrocarbon solvents yields two olefinic monodechlorination isomers: 1,4,5,7,8,8-hexachloro-3a,4,7,7a-tetrahydro-4,7-methanoindene and 1,4,6,7,8,8-hexachloro-3a,4,7,7a-tetrahydro-4,7-methanoindene. Irradiation at 300 nm in acetone, 1,2,3,6,9,10,10-heptachloropentacyclo[5.3.0.0^{2,5}.0^{3,9}.0^{4,8}]decane is the only product formed. This compound and a C-1 cyclohexyl adduct are formed when heptachlor in a cyclohexane/acetone solvent system is irradiated at 300 nm (McGuire et al., 1972).

Chemical/Physical: Slowly releases hydrogen chloride in aqueous media (Hartley and Kidd, 1987). The hydrolysis half-lives of heptachlor in a 1% ethanol/water phosphate-buffered solution at 25 °C and pHs of 4.5, 5.0, 6.0, 7.0, and 8.0 were 5.39, 4.34, 4.48, 4.48, and 3.01 d, respectively (Chapman and Cole, 1982). Chemical degradation of heptachlor give heptachlor epoxide (Newland et al., 1969). Heptachlor degraded in aqueous saturated calcium hypochlorite solution to 1-hydroxychlordene. Although further degradation occurred, no other metabolites were identified (Kaneda et al., 1974). Emits toxic chloride fumes when heated to decomposition (Lewis, 1990).

Exposure Limits: NIOSH REL: IDLH 100 mg/m^3; OSHA PEL: TWA 0.5 mg/m^3; ACGIH TLV: TWA 0.5 mg/m^3.

Symptoms of Exposure: In animals: tremors, convulsions, liver damage.

Formulation Types: Emulsifiable concentrate; wettable powder; granules; dustable powder; seed treatment.

Toxicity: LC_{50} (96 hr) for rainbow trout 7 μg/L, bluegill sunfish 26 μg/L, fathead minnow 78–130 μg/L (Hartley and Kidd, 1987), sheapshead minnow 2.7–8.8 μg/L, and marine pin perch 0.20–4.4 μg/L (Verschueren, 1983); LC_{50} (24 hr) for sheapshead minnow 1.22–4.3 μg/L (Verschueren, 1983); acute oral LD_{50} for rats 147–220 mg/kg (Hartley and Kidd, 1987), 40 mg/kg (RTECS, 1985).

Use: Insecticide for termite control.

HEPTACHLOR EPOXIDE

Synonyms: ENT 25584; Epoxy heptachlor; HCE; 1,4,5,6,7,8,8-Heptachloro-2,3-epoxy-2,3,3a,4,7,7a-hexahydro-4,7-methanoindene; 1,2,3,4,5,6,7,8,8-Heptachloro-2,3-epoxy-3a,4,7,7a-tetrahydro-4,7-methanoindene; **5a,6,6a-Hexahydro-2,5-methano-2H-indeno[1,2-b]oxirene**; 2,3,4,5,6,7,7-Heptachloro-1a,1b,5,5a,6,6a-hexahydro-2,5-methano-2H-oxireno[a]indene; Velsicol 53-CS-17.

Structure:

Designations: CAS: 1024-57-3; DOT: 2761; mf: $C_{10}H_5Cl_7O$; fw: 389.32; RTECS: PB9450000.

Properties: Liquid. Mp: 157-160 °C; K_H: 3.2 x 10^{-5} atm·m^3/mol at 25 °C; log K_{oc}: 4.32 (calculated); log K_{ow}: 3.65, 5.40; S_o: cod liver oil (60.3 g/L at 4 °C), n-octanol (33.9, 30.3, and 46.8 g/L at 4, 12, and 20 °C, respectively), triolein (75.9 and 77.7 g/L at 12 and 20 °C, respectively); S_w: 275 μg/L at 25 °C; vp: 2.6 x 10^{-6} mmHg at 20 °C.

Transformation Products

Biological: In a model ecosystem containing plankton, *Daphnia magna*, mosquito larva (*Culex pipiens quinquefasciatus*), fish (*Cambusia affinis*), alga (*Oedogonium cardiacum*), and snail (*Physa* sp.), heptachlor epoxide degraded to hydroxychlordene epoxide (Lu et al., 1975).

Photolytic: Irradiation of heptachlor epoxide by a 450-W high-pressure mercury lamp gave two half-cage isomers, each containing a ketone functional group (Ivie et al., 1972).

Toxicity: Acute oral LD_{50} for rats 47 mg/kg (RTECS, 1985).

Uses: Not known. A degradation product of heptachlor.

HEXACHLOROBENZENE

Synonyms: Amatin; Anticarie; Bunt-cure; Bunt-no-more; Co-op hexa; Granox NM; HCB; Hexa C.B.; Julin's carbon chloride; No bunt; No bunt 40; No bunt 80; No bunt liquid; Pentachlorophenyl chloride; Perchlorobenzene; Phenyl perchloryl; RCRA waste number U127; Sanocide; Smut-go; Snieciotox; UN 2729.

Structure:

Designations: CAS: 118-74-1; DOT: 2729; mf: C_6Cl_6; fw: 284.78; RTECS: DA2975000.

Properties: White monoclinic crystals or crystalline solid. Mp: 227-230 °C; bp: 323-326 °C (sublimes); ρ: 2.049 and 2.044 at 20/4 and 23/4 °C, respectively; fl p: 242 °C; H-t$_{1/2}$: none observed after 13 d at 85 °C and pHs of 3, 7, and 11; K_H: 7.1 x 10^{-3} atm·m^3/mol at 20 °C; log K_{oc}: 2.56-4.54; log K_{ow}: 3.93-6.42; S_o: soluble in acetone (1-5 mg/mL), benzene, carbon disulfide, chloroform, dimethylsulfoxide (<1 mg/mL), 95% ethanol (<1 mg/mL), ethyl ether, chloroform, cod liver oil (3.1, 4.2, and 5.8 g/L at 4, 12, and 20 °C, respectively), n-octanol (2.4, 3.1, and 3.9 g/L at 4, 12, and 20 °C, respectively), triolein (3.4, 4.7, and 7.3 g/L at 4, 12, and 20 °C, respectively), and many other organic solvents; S_w: 40 µg/L at 20 °C; vp: 1.089 x 10^{-5} mmHg at 20 °C.

Soil properties and adsorption data

Soil	K_d (mL/g)	f_{oc} (%)	K_{oc} (mL/g)	pH
Rothamsted Farm	462.8	1.51	30,649	5.1

Source: Lord et al., 1980.

Transformation Products
Biological: Reductive monodechlorination occurred in an anaerobic sewage sludge yielding principally 1,3,5-trichlorobenzene. Other compounds identified included pentachlorobenzene, 1,2,3,5-tetrachlorobenzene, and dichlorobenzenes (Fathepure et al., 1988). In activated sludge, only 1.5% of the applied hexachlorobenzene mineralized to carbon dioxide after 5 d (Freitag et al., 1985). In a 5-d experiment, ^{14}C-labeled hexachlorobenzene applied to soil-water suspensions under aerobic and

anaerobic conditions gave $^{14}CO_2$ yields of 0.4 and 0.2%, respectively (Scheunert et al., 1987).

Photolytic: Solid hexachlorobenzene exposed to artificial sunlight for 5 mo photolyzed at a very slow rate with no decomposition products identified (Plimmer and Klingebiel, 1976). The sunlight irradiation of hexachlorobenzene (20 g) in a 100 mL borosilicate glass-stoppered Erlenmeyer flask for 56 d yielded 64 ppm pentachlorobiphenyl (Uyeta et al., 1976). A carbon dioxide yield <0.1% was observed when hexachlorobenzene adsorbed on silica gel was irradiated with light (λ >290 nm) for 17 hr (Freitag et al., 1985). Irradiation (λ ≥285 nm) of hexachlorobenzene (1.1-1.2 mM/L) in an acetonitrile-water mixture containing acetone (concentration = 0.553 mM/L) as a sensitizer gave the following products (% yield): pentachlorobenzene (71.0), 1,2,3,4-tetrachlorobenzene (0.6), 1,2,3,5-tetrachlorobenzene (2.2), and 1,2,4,5-tetrachlorobenzene (3.7) (Choudhry and Hutzinger, 1984). Without acetone, the identified photolysis products (% yield) included 1,2,3,4,5-pentachlorobenzene (76.8), 1,2,3,5-tetrachlorobenzene (1.2), 1,2,4,5-tetrachlorobenzene (1.7), and 1,2,4-trichlorobenzene (0.2) (Choudhry and Hutzinger, 1984). In another study, the irradiation (λ = 290-310 nm) of hexa-chlorobenzene in aqueous solution gave only pentachlorobenzene and possibly pentachlorophenol as the transformation products. The photolysis rate increased with the addition of naturally occurring substances (tryptophan and pond proteins) and abiotic sensitizers (diphenylamine and skatole) (Hirsch and Hutzinger, 1989).

Chemical/Physical: Hydrolysis is not considered an environmentally relevant process. No hydrolysis was observed in water at 85 °C and pHs 2, 7, and 11 (Ellington et al., 1988). Emits toxic chloride fumes when heated to decomposition (Lewis, 1990).

Formulation Types: Dry seed treatment.

Toxicity: LC_{50} for five freshwater species 0.05-0.2 mg/L (Hartley and Kidd, 1987); acute oral LD_{50} for rats 10,000 mg/kg (RTECS, 1985).

Use: Seed fungicide.

HEXAZINONE

Synonyms: 3-Cyclohexyl-6-(dimethylamino)-1-methyl-1,3,5-triazine-2,4(1*H*,3*H*)-dione; 3-Cyclohexyl-6-(dimethylamino)-1-methyl-*s*-triazine-2,4(1*H*,3*H*)-dione; DPX 3674; Velpar; Velpar K; Velpar L; Velpar RP; Velpar ULW; Velpar weed killer.

Structure:

$$CH_3$$

(CH₃)₂N ... N ... O

N ... N

O

Designations: CAS: 51235-04-2; mf: $C_{12}H_{20}N_4O_2$; fw: 252.30; RTECS: XY7850000.

Properties: Colorless to white, odorless crystals. Mp: 115-117 °C; bp: decomposes; ρ: 1.25; fl p: nonflammable; log K_{oc}: 1.30-1.43; log K_{ow}: 1.05; S_o (g/kg at 25 °C): acetone (790), benzene (940), chloroform (3,880), *N,N*-dimethylformamide (836), *n*-hexane (3), methanol (2,650), toluene (386); S_w: 33 g/kg at 25 °C; vp: 2 x 10^{-7} mmHg at 20 °C (extrapolated), 6.4 x 10^{-5} mmHg at 86 °C.

Soil properties and adsorption data

Soil	K_d (mL/g)	f_{oc} (%)	K_{oc} (mL/g)	pH	CEC (meq/100 g)
Fallsington sandy loam	0.2	0.81	25	5.6	4.8
Flanagan silt loam	1.0	2.33	173	5.0	23.4

Source: Rhodes, 1980a.

Transformation Products

Soil/Plant: Degrades microbially in soil and natural waters releasing carbon dioxide (Hartley and Kidd, 1987). The reported half-life in soil is 1-6 mo (Hartley and Kidd, 1987). Rhodes (1980a) found that the persistence of hexazinone varied from 4 wk in a Delaware sandy loam to 24 wk in a Mississippi silt loam. Hexazinone is subject to microbial degradation (Rhodes, 1980a). Metabolites identified in soils, alfalfa, and/or sugarcane include 3-(4-hydroxycyclohexyl)-6-(dimethylamino)-1-methyl-1,3,5-triazine-2,4(1*H*,3*H*)-dione, 3-cyclohexyl-6-(methylamino)-1-methyl-1,3,5-triazine-2,4(1*H*,3*H*)-dione (Holt, 1981; Rhodes, 1980a; Roy et al., 1989), 3-(4-hydroxycyclohexyl)-6-(methylamino)-1-methyl-1,3,5-triazine-2,4(1*H*,3*H*)dione, 3-cyclohexyl-1-methyl-1,3,5-triazine-2,4,6(1*H*,3*H*,5*H*)trione, 3-(4-hydroxycyclohexyl)-1-methyl-1,3,5-triazine-2,4,6(1*H*,3*H*,5*H*)trione, 3-

cyclohexyl-6-amino-1-methyl-1,3,5-triazine-2,4(1H,3H)dione, and 3-cyclohexyl-6-(methylamino)-1,3,5-triazine-2,4(1H,3H)-dione (Holt, 1981; Rhodes, 1980a). The half-lives in a sandy soil and clay were 1 and 6 mo, respectively (Rhodes, 1980a).

Photolytic: Photodegradation products identified in aqueous hexazinone solutions following exposure to UV light (λ = 300-400 nm) were 3-(4-hydroxycyclohexyl)-6-(dimethylamino)-1-methyl-1,3,5-triazine-2,4(1H,3H)-dione,   3-cyclohexyl-6-(methylamino)-1-methyl-1,3,5-triazine-2,4(1H,3H)dione,   and   3-cyclohexyl-6-(dimethylamino)-1,3,5-triazine-2,4(1H,3H)dione (Rhodes, 1980).

Symptoms of Exposure: May irritate eyes, nose, skin, and throat.

Formulation Types: Soluble concentrate; water-soluble powder (90%); granules; miscible liquid (2 lb/gal).

Toxicity: LC_{50} (96 hr) for rainbow trout 320-420 mg/L, fathead minnow 274 mg/L, and bluegill sunfish 370-420 mg/L (Hartley and Kidd, 1987); acute oral LD_{50} for rats is 1,700 mg/kg (Ashton and Monaco, 1991).

Uses: Preemergence or postemergence herbicide used to control many annual grasses and broadleaf weeds in noncropped land and certain crops such as alfalfa, blueberries, coffee, pecans, sugarcane.

HEXYTHIAZOX

Synonyms: *trans*-5-(4-Chlorophenyl)-*N*-cyclohexyl-4-methyl-2-oxo-3-thia-zolidinecarboxamide.

Structure:

Designations: CAS: 78587-05-0; mf: $C_{17}H_{21}ClN_2O_2S$; fw: 352.90.

Properties: Colorless crystals. Mp: 108–108.5 °C; H-t$_{1/2}$ (aqueous solution in sunlight): 16.7 d; K_H: 2.4 x 10^{-8} atm·m^3/mol at 20 °C (approximate - calculated from water solubility and vapor pressure); log K_{oc}: 3.81 (calculated); log K_{ow}: 2.53; S_o (g/L at 20 °C): acetone (160), acetonitrile (28.6), chloroform (1,380), *n*-hexane (3.9), methanol (206), xylene (362); S_w: 0.5 mg/L at 20 °C; vp: 2.55 x 10^{-8} mmHg at 20 °C.

Transformation Products
Soil: Though no products were reported, the half-life in a clay loam at 15 °C is 8 d (Worthing and Hance, 1991).

Formulation Types: Emulsifiable concentrate; wettable powder.

Toxicity: LC$_{50}$ (96 hr) for rainbow trout >300 mg/L and bluegill sunfish 11.6 mg/L (Hartley and Kidd, 1987); LC$_{50}$ (48 hr) for carp 3.7 mg/L (Worthing and Hance, 1991); acute oral LD$_{50}$ for mice and rats >5,000 mg/kg (Hartley and Kidd, 1987).

Use: Nonsystemic acaricide.

IPRODIONE

Synonyms: Chipco 26019; 3-(3,5-Dichlorophenyl)-*N*-(1-methylethyl)-2,4-dioxo-1-imidazolidinecarboxamide; Glycophen; Glycophene; 1-Isopropyl carbamoyl-3-(3,5-dichlorophenyl)hydantoin; LFA 2043; MRC 910; Promidione; Rop 500 F; Rovral; RP 26019.

Structure:

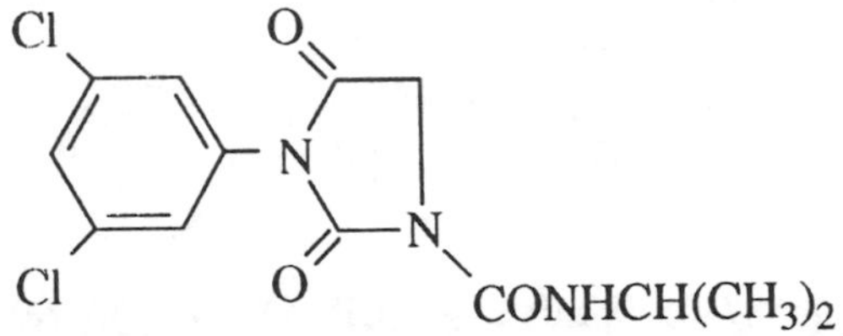

Designations: CAS: 36734-19-7; mf: $C_{13}H_{13}Cl_2N_3O_3$; fw: 330.17; RTECS; NI8870000.

Properties: Colorless, odorless crystals. Mp: 136 °C; K_H: 3.3 x 10^{-8} atm·m^3/mol at 20 °C (approximate - calculated from water solubility and vapor pressure); log K_{oc}: 1.48; log K_{ow}: 3.10; S_o (g/L at 20 °C): acetone (300), ethanol (25), methanol (25), methylene chloride (500), *N,N*-dimethylformamide (500); S_w: 13 mg/L at 20 °C; vp: <10^{-6} mmHg at 20 °C.

Soil properties and adsorption data

Soil	K_d (mL/g)	f_{oc} (%)	K_{oc} (mL/g)	pH
Loam	30.4	2.67	1,139	6.0

Source: U.S. Department of Agriculture, 1990.

Transformation Products

Soil: Readily degrades in soil ($t_{1/2}$ = 20-160 d) releasing carbon dioxide (Hartley and Kidd, 1987; Worthing and Hance, 1991).

Plant: Metabolized in plants to 3,5-dichloroaniline (Hartley and Kidd, 1987).

Chemical/Physical: In an aqueous solution at pH 8.7, iprodione hydrolyzed to *N*-(3,5-dichloroanilinocarbonyl)-*N*-(isopropylaminocarbonyl)glycine (Belafdal et al., 1986). Gomez et al. (1982) studied the pyrolysis of iprodione in an helium atmosphere at 400-1,000 °C. Decomposition began at 300 °C producing isopropyl isocyanate and 3-(3,5-dichlorophenyl)hydantoin. Above 600 °C, the hydantoin ring began to decompose forming the following products: 3-chloroaniline, 3,5-

242

dichloroaniline, chlorinated benzenes, and benzonitrile. From 800 to 1,000 °C, the hydantoin ring was completed destroyed which led to the formation of aryl isocyanates, anilines, and the corresponding diarylureas, namely 3-(3,5-dichloro-phenyl)urea and 1-(3-chlorophenyl)-3-(3,5-dichlorophenyl)urea (Gomez et al., 1982).

Formulation Types: Suspension concentrate; hot fogging concentrate; wettable powder; dustable powder.

Toxicity: LC_{50} (96 hr) for rainbow trout 6.7 mg/L and bluegill sunfish 2.25 mg/L (Hartley and Kidd, 1987); acute oral LD_{50} for rats 3,500 mg/kg (Hartley and Kidd, 1987), 4,400 mg/kg (RTECS, 1985).

Use: Fungicide.

ISOFENPHOS

Synonyms: Amaze; Bay 92114; Bay-sra-12869; 2-((Ethoxy((1-methylethyl)-amino)phosphinothioyl)oxy)benzoic acid 1-methylethyl ester; *O*-Ethyl *O*-(2-iso-propoxycarbonyl)phenyl isopropylphosphoramidothioate; Isophenphos; Isopropyl salicylate *O*-ester with *O*-ethylisopropylphosphoramidothioate; 1-Methylethyl-2-((ethoxy((1-methylethyl)amino)phosphinothioyl)oxy)benzoate; Oftanol; SD 40; SRA 12869.

Structure:

Designations: CAS: 25311-71-1; mf: $C_{15}H_{24}NO_4PS$; fw: 345.40; RTECS: VO4395500.

Properties: Colorless oil. Mp: -12 °C; bp: 120 °C at 0.01 mmHg; ρ: 1.134 at 20/4 °C; H-t$_{1/2}$ (37 °C): 263 d (pH 2), 525 d (pH 7), 52 hr (pH 11.5); K_H: 7.6 x 10^{-6} atm·m^3/mol at 20 °C; log K_{oc}: 2.73, 2.79; log K_{ow}: 3.30, 4.12; S_o: soluble in acetone, benzene, cyclohexanone, ethanol, ethyl ether, methylene chloride; S_w: 23.8 mg/L at 20 °C; vap d: 14.12 g/L at 25 °C, 11.97 (air = 1); vp: 4 x 10^{-6} mmHg at 20 °C.

Soil properties and adsorption data

Soil	K_d (mL/g)	f_{oc} (%)	K_{oc} (mL/g)	pH
Sandy loam	10.3	1.68	613	2.5
Silty loam	5.6	1.04	538	2.6

Source: U.S. Department of Agriculture, 1990.

Transformation Products

Soil: Rapidly degraded by microbes in soils forming isofenphos oxon (Abou-Assaf et al., 1986; Somasundarum et al., 1989), isopropyl salicylate, and carbon dioxide (Somasundaram et al., 1989). A pure culture of *Arthrobacter* sp. was capable of degrading isofenphos at different soil concentrations (10, 50, and 100 ppm) in less than 6 hr. In previously treated soils, isofenphos could be mineralized to carbon dioxide by indigenous microorganisms (Racke and Coats, 1987). Hydrolyzes in soil to salicylic acid (Somasundaram et al., 1991).

Photolytic: Irradiation of an isofenphos (500 mg) in hexane and methanol (100 mL) using a high pressure mercury lamp (λ = 254-360 nm) for 24 hr yielded the following products: isofenphos oxon, and *O*-ethyl hydrogen-*N*-isopropylphosphoroamidothioate and its methylated product, and *O*-ethyl hydrogen-*N*-isopropylphosphoramidate and its methylated product. Photolysis products identified following the irradiation (λ = 253 nm) of a sandy loam soil for 2 d were isofenphos oxon, isopropyl salicylic acid, and *O*-ethyl hydrogen-*N*-isopropylphosphoramidate. When isofenphos was uniformly coated on borosilicate glass and irradiated using a low pressure mercury lamp for 5 d, the following products formed: isofenphos oxon, isopropyl salicylic acid, and *O*-ethyl hydrogen-*N*-isopropylphosphoramidothioate (Dureja et al., 1989).

Formulation Types: Emulsifiable concentrate; wettable powder; granules; seed treatment.

Toxicity: LC_{50} (96 hr) for carp 2-4 mg/L, orfe 1-2 mg/L, goldfish 2 mg/L, and rudd 1 mg/L (Hartley and Kidd, 1987); acute oral LD_{50} for rats 28.0-38.7 mg/kg (Hartley and Kidd, 1987).

Uses: Soil insecticide used to control leaf eating and soil dwelling pests in vegetables, fruits, turf, and field crops.

KEPONE

Synonyms: Chlordecone; CIBA 8514; Compound 1189; 1,2,3,5,6,7,8,9,10,10-Decachloro[5.2.1.0^{2,6}.0^{3,9}.0^{5,8}]decano-4-one; Decachloroketone; Decachloro-1,3,4-metheno-2H-cyclobuta[cd]pentalen-2-one; Decachlorooctahydrokepone-2-one; Decachlorooctahydro-1,3,4-metheno-2H-cyclobuta[cd]pentalen-2-one; **1,1a,3,3a,4,5,5a,5b,6-Decachlorooctahydro-1,3,4-metheno-2H-cyclobuta-[cd]pentalen-2-one;** Decachloropentacyclo[5.2.1.0^{2,6}.0^{3,9}.0^{5,8}]decan-3-one; Decachloropentacyclo[5.2.1.0^{2,6}.0^{4,10}.0^{5,9}]decan-3-one; Decachlorotetra-cyclodecanone; Decachlorotetrahydro-4,7-methanoindeneone; ENT 16391; GC-1189; General chemicals 1189; Merex; NA 2761; NCI-C00191; RCRA waste number U142.

Structure:

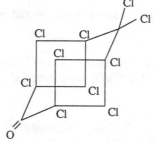

Designations: CAS: 143-50-0; DOT: 2761; mf: $C_{10}Cl_{10}O$; fw: 490.68; RTECS: PC8575000.

Properties: Odorless, white to tan crystals. Mp: sublimes; bp: 350 °C (decomposes); K_H: 3.11 x 10^{-2} atm·m^3/mol at 24-25 °C (approximate - calculated from water solubility and vapor pressure); log K_{oc}: 4.74 (calculated); log K_{ow}: 4.07 (calculated); S_o: soluble in acetic acid (10-50 mg/mL at 20 °C), benzene, corn oil, dimethyl-sulfoxide (>100 mg/mL at 20 °C), 95% ethanol (>100 mg/mL at 20 °C), ketones, light petroleum, petroleum, fats; S_w: 7,600 μg/L at 24 °C (note: solubility increases with pH); vp: 2.25 x 10^{-7} mmHg at 25 °C.

Transformation Products

Photolytic: Kepone-contaminated soils from a site in Hopewell, VA were analyzed by GC/MS. 8-Chloro and 9-chloro homologs identified suggested these were photodegradation products of kepone (Borsetti and Roach, 1978). Products identified from the photolysis of kepone in cyclohexane were 1,2,3,4,6,7,9,10,10-nonachloro-5,5-dihydroxypentacyclo[5.3.0.0^{2,6}.0^{3,9}.0^{4,8}]decane for the hydrate and 1,2,3,4,6,7,9,10,10-nonachloro-5,5-dimethoxypentacyclo[5.3.0.0^{2,6}.0^{3,9}.0^{4,8}]decane (Alley et al., 1974).

Chemical/Physical: Readily reacts with moisture forming hydrates (Hollifield, 1979). Decomposes at 350 °C (Windholz et al., 1983) probably emitting toxic chlorine fumes.

Symptoms of Exposure: May cause tremors.

Toxicity: LC_{50} (96 hr) for sunfish 0.14 ppm, trout 0.02 ppm, bluegill sunfish 0.051, rainbow trout 0.036 ppm (Verschueren, 1983); LC_{50} (48 hr) for sunfish 0.27 ppm, trout 38 ppb (Verschueren, 1983); LC_{50} (24 hr) for bluegill sunfish 257 ppb, trout 66 ppb, bluegill sunfish 257 ppb, and rainbow trout 156 ppb (Verschueren, 1983); acute oral LD_{50} for male and female rats is 132 and 126 mg/kg, respectively (Verschueren, 1983), 95 mg/kg (RTECS, 1985).

Uses: Insecticide; fungicide.

LINDANE

Synonyms: Aalindan; Aficide; Agrisol G-20; Agrocide; Agrocide 2; Agrocide 6G; Agrocide 7; Agrocide III; Agrocide WP; Agronexit; Ameisenatod; Ameisenmittel merck; Aparasin; Aphtiria; Aplidal; Arbitex; BBX; Ben-hex; Bentox 10; Benzenehexachloride; Benzene-γ-hexachloride; γ-Benzenehexachloride; Bexol; BHC; γ-BHC; Celanex; Chloran; Chloresene; Codechine; DBH; Detmol-extrakt; Detox 25; Devoran; Dol granule; ENT 7796; Entomoxan; Exagama; Forlin; Gallogama; Gamacid; Gamaphex; Gamene; Gamiso; Gammahexa; Gammalin; Gammexene; Gammopaz; Gexane; HCCH; HCH; γ-HCH; Heclotox; Hexa; γ-Hexachlor; Hexachloran; γ-Hexachloran; Hexachlorane; γ-Hexachlorane; γ-Hexachlorobenzene; 1,2,3,4,5,6-Hexachlorocyclohexane; 1α,2α,3β,4α,5α,6β-**Hexachlorocyclohexane**; 1,2,3,4,5,6-Hexachloro-γ-cyclohexane; γ-Hexachlorocyclohexane; γ-1,2,3,4,5,6-Hexachlorocyclohexane; Hexatox; Hexaverm; Hexicide; Hexyclan; HGI; Hortex; Inexit; γ-Isomer; Isotox; Jacutin; Kokotine; Kwell; Lendine; Lentox; Lidenal; Lindafor; Lindagam; Lindagrain; Lindagranox; γ-Lindane; Lindapoudre; Lindatox; Lindosep; Lintox; Lorexane; Milbol 49; Mszycol; NA 2761; NCI-C00204; Neo-scabicidol; Nexen FB; Nexit; Nexit-stark; Nexol-E; Nicochloran; Novigam; Omnitox; Ovadziak; Owadziak; Pedraczak; Pflanzol; Quellada; RCRA waste number U129; Silvanol; Spritz-rapidin; Spruehpflanzol; Streunex; Tap 85; TBH; Tri-6; Viton.

Structure:

Designations: CAS: 58-89-9; DOT: 2761; mf: $C_6H_6Cl_6$; fw: 290.83; RTECS: GV4900000.

Properties: Colorless, odorless to slightly musty solid with a bitter taste. Mp: 112.5 °C; bp: 323.4 °C; ρ: 1.5691 and 1.87 at 23.6/4 and 20/4 °C, respectively; fl p: nonflammable; H-t$_{1/2}$: 206 d at 25 °C and pH 7; K_H: 2.43 x 10^{-7} atm·m^3/mol at 23 °C; log K_{oc}: 2.38-3.52; log K_{ow}: 3.20-3.89; P-t$_{1/2}$: 103.95 hr (absorbance $\lambda \leq 300$ nm, concentration on glass plates = 6.7 μg/cm^2); S_o (wt % at 20 °C unless noted otherwise): acetone (30.31), benzene (22.42), chloroform (19.35), cod liver oil (58.0, 69.8, and 82.0 g/L at 4, 12, and 20 °C, respectively), ethyl ether (17.22), ethanol (6.02), n-octanol (29.1, 41.1, and 52.9 g/L at 4, 12, and 20 °C, respectively), triolein (52.9, 68.2, and 78.3 g/L at 4, 12, and 20 °C, respectively); S_w: 7.52 ppm at 25 °C, 17.0 mg/L at 24 °C; vap d: 518 ng/L at 20 °C; vp: 9.4 x 10^{-6} mmHg at 20 °C.

Soil properties and adsorption data

Soil	K_d (mL/g)	f_{oc} (%)	K_{oc} (mL/g)	pH	CEC (meq/100 g)
Brookston sandy loam	22.7	2.10	1,081	--	--
Catlin	10.9	2.01	44	6.20	--
Clay loam	9.8	1.42	690	5.91	12.4
Clay loam	79.4	10.40	760	4.89	35.0
Commerce	49.5	0.68	7,300	6.70	--
Fox sandy loam	17.3	1.84	940	--	--
Honeywood loam	20.4	1.67	1,222	--	--
Kanuma high clay	1.1	1.35	85	5.70	--
Light clay	20.0	1.51	1,300	5.18	13.2
Light clay	37.4	3.23	1,200	5.26	28.3
Sandy loam	75.8	7.91	960	5.51	26.3
Tracy	10.1	1.12	901	6.20	--
Tsukuba clay loam	20.0	4.24	398	6.50	--
Udalf	7.7	0.76	1,010	7.45	--

Source: Kanazawa, 1989; Kay and Elrick, 1967; Kishi et al., 1990; McCall et al., 1981; Rippen et al., 1982.

Transformation Products
Biological: In a laboratory experiment, a strain of *Pseudomonas putida* culture transformed lindane to γ-TCCH, γ-PCCH, and α-BHC (Benezet and Matsumura, 1973). γ-TCCH was also reported as a product of lindane degradation by *Clostridium sphenoides* (Heritage and MacRae, 1977, 1977a; MacRae et al., 1969), an anaerobic bacterium isolated from flooded soils (MacRae et al., 1969; Sethunathan and Yoshida, 1973a). Evidence suggests that degradation of lindane in anaerobic cultures or flooded soils amended with the lindane occurs via reductive dehalogenation producing chlorine-free volatile metabolites (Sethunathan and Yoshida, 1973a).

Soil: Lindane degraded rapidly in flooded rice soils (Raghu and MacRae, 1966). In moist soils, lindane biodegraded to γ-pentachlorocyclohexene (Elsner et al., 1972; Fuhremann and Lichtenstein, 1980; Kearney and Kaufman, 1976). Under anaerobic conditions, degradation by soil bacteria yielded γ-3,4,5,6-tetrachloro-1-cyclohexane and α-BHC (Kobayashi and Rittman, 1982). Other reported biodegradation products include penta- and tetrachloro-1-cyclohexanes and penta- and tetrachlorobenzenes (Moore and Ramamoorthy, 1984). Incubation of lindane for 6 wk in a sandy loam soil under flooded conditions yielded γ-3,4,5,6-tetrachlorocyclohexane, γ-2,3,4,5,6-pentachlorocyclohex-1-ene, and small amounts of 1,2,4-trichlorobenzene, 1,2,3,4-tetrachlorobenzene, 1,2,3,5- and/or 1,2,4,5-

tetrachlorobenzene (Mathur and Saha, 1975). Incubation of lindane in moist soil for 8 wk yielded the following metabolites: γ-3,4,5,6-tetrachlorocyclohexene, γ-1,2,3,4,5-pentachlorocyclohex-1-ene, pentachlorobenzene, 1,2,3,4-tetrachlorobenzene, 1,2,3,5- and/or 1,2,4,5-tetrachlorobenzene, 1,2,4-trichlorobenzene, 1,3,5-trichlorobenzene, m- and/or p-dichlorobenzene (Mathur and Saha, 1977). Microorganisms isolated from a loamy sand soil degraded lindane and some of the metabolites identified were pentachlorobenzene, 1,2,4,5-tetrachlorobenzene, 1,2,3,5-tetrachlorobenzene, γ-2,3,4,5,6-pentachloro-1-cyclohexane (γ-PCCH), γ-3,4,5,6-tetrachloro-1-cyclohexane (γ-TCCH), and β-3,4,5,6-tetrachloro-1-cyclohexane (β-TCCH) (Tu, 1976). γ-PCCH was also reported as a metabolite of lindane in an Ontario soil that was pretreated with p,p'-DDT, dieldrin, lindane, and heptachlor (Yule et al., 1967). The reported half-life in soil is 266 d (Jury et al., 1987).

Indigenous microbes in soil partially degraded lindane to carbon dioxide (MacRae et al., 1967). In a 42 d experiment, ^{14}C-labeled lindane applied to soil-water suspensions under aerobic and anaerobic conditions gave $^{14}CO_2$ yields of 1.9 and 3.0%, respectively (Scheunert et al., 1987).

Plant: Lindane appeared to be metabolized by several grasses to β-BHC (Steinwandter, 1978). Oat plants were grown in two soils treated with [^{14}C]lindane. 2,4,5-Trichlorophenol and possibly γ-PCCH were identified in soils but no other compounds other than lindane were identified in the oat roots or tops (Fuhremann and Lichtenstein, 1980).

Surface Water: Lindane degraded in simulated lake impoundments under both aerobic (15%) and anaerobic (90%) conditions. Lindane degraded primarily to α-BHC with trace amounts of δ-BHC (Newland et al., 1969).

Photolytic: Photolysis of lindane in aqueous solutions gives β-BHC (U.S. Department of Health and Human Services, 1989). When an aqueous solution containing lindane was photooxidized by UV light at 90-95 °C, 25, 50, and 75% degraded to carbon dioxide after 3.0, 17.4, and 45.8 hr, respectively (Knoevenagel and Himmelreich, 1976). Lindane undergoes dehydrochlorination in alkalies (Hartley and Kidd, 1987). Though no products were reported, the calculated hydrolysis half-life at 25 °C and pH 7 is 206 d (Ellington et al., 1988).

Chemical/Physical: In weakly basic media, lindane undergoes dehydrochlorination to give 1,2,4-trichlorobenzene (Cremlyn, 1991) and hydrochloric acid. Three molecules of the acid are produced for every molecule of lindane that reacts.

Exposure Limits: NIOSH REL: IDLH 1,000 mg/m³; OSHA PEL: TWA 0.5 mg/m³; ACGIH TLV: TWA 0.5 mg/m³.

Symptoms of Exposure: Irritates eyes, nose, throat, and skin; headache, nausea, clonic convulsions, respiratory problems, cyanosis, aplastic anemia, and muscle spasms.

Formulation Types: Emulsifiable concentrate; wettable powder; fumigant; suspension concentrate; granules; dustable powder.

Toxicity: LC_{50} (48 hr) for guppies 0.16-0.3 mg/L (Hartley and Kidd, 1987); LC_{50} (96 hr) for bullhead 64 μg/L, carp 90 μg/L, catfish 44 μg/L, coho salmon 41 μg/L, goldfish 131 μg/L, perch 68 μg/L, bass 32 μg/L, minnow 87 μg/L, bluegill sunfish 68 μg/L, and rainbow trout 27 μg/L (Macek and McAllister, 1970); acute oral LD_{50} for rats 88-91 mg/kg (Reuber, 1979), 76 mg/kg (RTECS, 1985).

Uses: Pesticide and insecticide.

LINURON

Synonyms: Afalon; Afalon inuron; Aphalon; Cephalon; 3-(3,4-Dichlorophenyl)-1-methoxymethylurea; 3-(3,4-Dichlorophenyl)-1-methoxy-1-methylurea; *N'*-**(3,4-Dichlorophenyl)-*N*-methoxy-*N*-methylurea;** *N*-(3,4-Dichlorophenyl)-*N'*-methyl-*N'*-methoxyurea; Du Pont 326; Du Pont herbicide 326; Garnitan; Herbicide 326; Hoe 2810; Linex 4L; Linorox; Linurex; Linuron 4L; Lorex; Lorox; Lorox DF; Lorox L; Lorox linuron weed killer; Methoxydiuron; 1-Methoxy-1-methyl-3-(3,4-dichlorophenyl)urea; Premalin; Sarclex; Scarclex; Sinuron.

Structure:

Designations: CAS: 330-55-2; mf: $C_9H_{10}Cl_2N_2O_2$; fw: 249.10; RTECS: YS9100000.

Properties: Colorless to white, odorless crystals or crystalline solid. Mp: 93-94 °C; fl p: nonflammable; K_H: 6.1 x 10^{-8} atm·m^3/mol at 20-25 °C (approximate - calculated from water solubility and vapor pressure); log K_{oc}: 2.70-2.78; log K_{ow}: 2.19, 3.00; S_o (g/kg at 25 °C): acetone (500), benzene (150), ethanol (150), *n*-heptane (150), xylene (130); S_w: 75-81 mg/L at 25 °C; vp: 1.5 x 10^{-5} mmHg at 20 °C.

Soil properties and adsorption data

Soil	K_d (mL/g)	f_{oc} (%)	K_{oc} (mL/g)	pH	Salinity	TOC (mg/L)	CEC[a]
Annandale	11.39	1.70	670	5.8	--	--	11.3
Annandale	12.00	1.70	706	5.9	--	--	11.3
Asquith sandy loam	6.90	1.02	676	7.5	--	--	--
Begbroke sandy loam	10.80	1.11	973	7.1	--	--	--
Bermudian	13.70	1.60	856	6.0	--	--	13.2
Bermudian	14.01	1.60	876	6.0	--	--	13.2
Collington	14.18	2.60	545	5.0	--	--	12.8
Collington	15.05	2.60	579	4.7	--	--	12.8
Colts Neck	1.49	1.20	124	4.2	--	--	7.7
Colts Neck	12.00	1.20	1,000	4.2	--	--	7.7
Dark sandy loam	17.00	12.00	142	6.3	--	--	18.0
Dutchess	11.74	2.90	405	5.5	--	--	12.7
Dutchess	15.42	2.90	532	5.3	--	--	12.7

Soil	K_d (mL/g)	f_{oc} (%)	K_{oc} (mL/g)	pH	Salinity	TOC (mg/L)	CEC[a]
Great House sandy loam	73.00	12.00	608	6.3	--	--	18.0
Indian Head clay loam	17.80	2.34	761	7.8	--	--	--
Lakewood	10.82	0.50	2,163	3.5	--	--	1.8
Lakewood	13.39	0.50	2,678	3.6	--	--	1.8
Melfort loam	97.70	6.05	1,615	5.9	--	--	--
Patuxent River, MD	--	--	6,760	--	13.5	49.0	--
Patuxent River, MD	--	--	6,210	--	14.5	52.5	--
Regina heavy clay	18.20	2.39	762	7.7	--	--	--
Rosemaunde sandy clay loam	35.00	1.76	1,989	6.7	--	--	14.0
Sandy loam	11.40	1.93	591	7.1	--	--	11.0
Sassafras	12.00	2.00	600	5.2	--	--	7.7
Sassafras	10.29	2.00	514	5.2	--	--	7.7
Squires	9.79	1.70	576	6.5	--	--	7.0
Squires	11.39	1.70	670	6.6	--	--	7.0
Toll Farm HP	63.00	11.70	538	7.4	--	--	41.0
Trawscoed silty clay loam	50.00	3.69	1,355	6.2	--	--	12.0
Washington	13.39	2.40	558	6.0	--	--	11.2
Washington	13.39	2.40	558	6.2	--	--	11.2
Weed RES. sandy loam	47.00	1.93	2,435	7.1	--	--	11.0
Weyburn Oxbow loam	19.10	3.72	513	6.5	--	--	--
Whippany	8.31	1.90	437	5.6	--	--	9.4
Whippany	8.90	1.90	479	5.6	--	--	9.4

Source: Grover, 1975; Grover and Hance, 1969; Hance, 1965; MacNamara and Toth, 1970. a) meq/100 g.

Transformation Products
Soil: Linuron degrades in soil forming the common metabolite 3,4-dichloro-aniline (Duke et al., 1991). Under aerobic conditions, linuron in a biologically active, organic-rich, pond sediment was converted to the intermediate 3-(3-chlorophenyl)-1-methoxymethylurea. This compound was further degraded but the compounds were not identified (Stepp et al., 1985). Linuron was degraded by the microorganism *Bacillus sphaericus* in soil forming *N,O*-dimethylhydroxyl-amine and carbon dioxide (Engelhardt et al., 1972). Only 1 ppm 3,4-dichloro-aniline was identified in soils after incubation of soils containing 500 ppm linuron (Belasco and Pease, 1969). The reported half-life in soil is approximately 2-5 mo (Hartley and Kidd, 1987). Boerner (1967) reported that the soil microorganism

Aspergillus niger degraded linuron to phenylmethylurea, phenylmethoxyurea, chloroaniline, ammonia, and carbon dioxide. In an earlier study, Boerner (1965) found 3,4-chloroaniline as a soil metabolite but the microorganism(s) were not identified.

Plant: Undergoes demethylation and demethoxylation in plants (Hartley and Kidd, 1987). Metabolites identified in carrots 117 d after treatment were 3,4-dichlorophenylurea, 3-(3,4-dichlorophenyl)-1-methylurea, and 3,4-dichloro-aniline. About 87% of the linuron remained unreacted (Loekke, 1974).

Photolytic: When an aqueous solution of linuron was exposed to summer sunlight for 2 mo, the photoproducts 3-(3-chloro-4-hydroxyphenyl)-1-methoxy-1-methylurea, 3,4-dichlorophenylurea and 3-(3,4-dichlorophenyl)-1-methylurea formed at yields of 13, 10, and 2%, respectively. The photolysis half-life of this reaction was approximately 97 d (Rosen et al., 1969). In a more recent study, Tanaka et al. (1985) studied the photolysis of linuron (75 mg/L) in aqueous solution using UV light ($\lambda = 300$ nm) or sunlight. After 24 d of exposure to sunlight, linuron degraded to a trichlorinated biphenyl (1% yield) with the concomitant loss of hydrogen chloride (Tanaka et al., 1985).

Chemical/Physical: Emits toxic fumes of nitrogen oxides and chlorine when heated to decomposition (Sax and Lewis, 1987). Linuron can be hydrolyzed to an aromatic amine by refluxing in an alkaline medium (Humburg et al., 1989). The hydrolysis half-life of linuron in 0.5 N sodium hydroxide solution at 20 °C is 1 d (El-Dib and Aly, 1976).

Symptoms of Exposure: May irritate eyes, nose, skin, and throat.

Formulation Types: Suspension concentrate; emulsifiable concentrate; flowable powder (50%); liquid concentrate (4 lb/gal); wettable powder.

Toxicity: LC_{50} (96 hr) for rainbow trout and bluegill sunfish 16 mg/L (Worthing and Hance, 1991); acute oral LD_{50} for rats is approximately 1,500 mg/kg (Ashton and Monaco, 1991), 1,146 mg/kg (RTECS, 1985).

Uses: Selective preemergence and postemergence herbicide used on a wide variety of food crops to control many annual broadleaf and grass weeds.

MALATHION

Synonyms: American Cyanamid 4049; *S*-1,2-Bis(carbethoxy)ethyl-*O,O*-dimethyl dithiophosphate; *S*-1,2-Bis(ethoxycarbonyl)ethyl-*O,O*-dimethyl phosphorodithioate; *S*-1,2-Bis(ethoxycarbonyl)ethyl-*O,O*-dimethyl thiophosphate; Calmathion; Carbethoxy malathion; Carbetovur; Carbetox; Carbofos; Carbophos; Celthion; Chemathion; Cimexan; Compound 4049; Cythion; Detmol MA; Detmol MA 96%; *S*-1,2-Dicarbethoxyethyl-*O,O*-dimethyl dithiophosphate; Dicarboethoxyethyl-*O,O*-dimethyl phosphorodithioate; 1,2-Di(ethoxycarbonyl)ethyl-*O,O*-dimethyl phosphorodithioate; *S*-1,2-Di(ethoxycarbonyl)ethyl dimethyl phosphorothiolothionate; Diethyl (dimethoxyphosphinothioylthio) butanedioate; Diethyl (dimethoxyphosphinothioylthio) succinate; Diethyl mercaptosuccinate, *O,O*-dimethyl phosphorodithioate; Diethyl mercaptosuccinate, *O,O*-dimethyl thiophosphate; Diethyl mercaptosuccinic acid *O,O*-dimethyl phosphorodithioate; **((Dimethoxy-phosphinothioyl)thio)butanedioic acid diethyl ester;** *O,O*-Dimethyl-*S*-1,2-bis(ethoxycarbonyl)ethyldithiophosphate; *O,O*-Dimethyl-*S*-(1,2-dicarbethoxy-ethyl)dithiophosphate; *O,O*-Dimethyl-*S*-(1,2-dicarbethoxyethyl)phosphoro-dithioate; *O,O*-Dimethyl-*S*-(1,2-dicarbethoxyethyl)thiothionophosphate; *O,O*-Dimethyl-*S*-1,2-di(ethoxycarbamyl)ethyl phosphorodithioate; *O,O*-Dimethyldi-thiophosphate dimethylmercaptosuccinate; EL 4049; Emmatos; Emmatos extra; ENT 17034; Ethiolacar; Etiol; Experimental insecticide 4049; Extermathion; Formal; Forthion; Fosfothion; Fosfotion; Four thousand forty-nine; Fyfanon; Hilthion; Hilthion 25WDP; Insecticide 4049; Karbofos; Kop-thion; Kypfos; Malacide; Malafor; Malakill; Malagran; Malamar; Malamar 50; Malaphele; Malaphos; Malasol; Malaspray; Malathion E50; Malathion LV concentrate; Malathion ULV concentrate; Malathiozoo; Malathon; Malathyl LV concentrate & ULV concentrate; Malatol; Malatox; Maldison; Malmed; Malphos; Maltox; Maltox MLT; Mercaptosuccinic acid diethyl ester; Mercaptothion; MLT; Moscardia; NA 2783; NCI-C00215; Oleophosphothion; Orthomalathion; Phosphothion; Prioderm; Sadofos; Sadophos; SF 60; Siptox I; Sumitox; Tak; TM-4049; Vegfru malatox; Vetiol; Zithiol.

Structure:

$$CH_3O \diagdown \overset{\overset{\textstyle S}{\|}}{P} - SCHCOOCH_2CH_3$$
$$CH_3O \diagup \qquad CH_2COOCCH_2CH_3$$

Designations: CAS: 121-75-5; DOT: 2783; mf: $C_{10}H_{19}O_6PS_2$; fw: 330.36; RTECS: WM8400000.

Properties: Yellow to dark brown liquid with a garlic (technical) or mercaptan-

255

like odor. Mp: 2.85 °C; bp: 156-157 °C at 0.7 mmHg (decomposes); ρ: 1.23 at 25/4 °C; fl p: >162.8 °C (open cup); H-t$_{1/2}$: $\approx$ 9 d at pH 6; K$_H$: 4.89 x 10^{-9} atm·m^3/mol at 25 °C; log K$_{oc}$: 2.61; log K$_{ow}$: 2.36-2.89; P-t$_{1/2}$: 51.28 hr (absorbance λ = 210.0 nm, concentration on glass plates = 6.7 μg/cm^2), 15 hr in Suwannee River, FL at pH 4.7 under September sunlight (latitude 34° N); S$_o$: miscible with alcohols, chloroform, dimethylsulfoxide, esters, ethers, n-hexane, ketones, vegetable oils, aromatic and alkylated aromatic hydrocarbons; S$_w$: 141, 145, and 164 mg/L at 10, 20, and 30 °C, respectively; vap d: 13.50 g/L at 25 °C, 11.40 (air = 1); vp: 1.25 x 10^{-6} mmHg at 20 °C.

Soil properties and adsorption data

Soil	K$_d$ (mL/g)	f$_{oc}$ (%)	K$_{oc}$ (mL/g)	pH	CEC (meq/100 g)
Annandale	33.48	1.70	1,969	6.2	11.3
Annandale	76.96	1.70	4,527	6.1	11.3
Bermudian	40.63	1.60	2,540	6.4	13.2
Bermudian	66.92	1.60	4,183	6.4	13.2
Coarse sand (Jutland, Denmark)	0.45	0.21	214	5.3	2.2
Collington	36.51	2.60	1,404	5.7	12.8
Collington	56.67	2.60	2,180	5.6	12.8
Colts Neck	6.94	1.20	579	4.5	7.7
Colts Neck	8.86	1.20	739	4.8	7.7
Dutchess	47.14	2.90	1,626	5.7	12.7
Dutchess	66.92	2.90	2,308	5.8	12.7
Lakewood	4.23	0.50	847	4.7	1.8
Lakewood	4.29	0.50	857	4.5	1.8
Sandy loam (Jutland, Denmark)	0.75	0.15	500	6.4	7.3
Sassafras	24.48	2.00	1,224	5.3	7.7
Sassafras	26.70	2.00	1,335	5.1	7.7
Squires	36.51	1.70	2,148	6.4	7.0
Squires	64.07	1.70	3,769	6.6	7.0
Washington	16.67	2.40	694	6.0	11.2
Whippany	13.26	1.90	698	5.7	9.4
Whippany	16.67	1.90	877	5.7	9.4
Washington	34.44	2.40	1,435	6.1	11.2

Source: Kjeldsen et al., 1990; MacNamara and Toth, 1970.

Transformation Products

Biological: Walker (1976) reported that 97% of malathion added to both sterile and nonsterile estuarine water was degraded after incubation in the dark for 18 d. Complete degradation was obtained after 25 d. Degradation of malathion in organic-rich soils was 3-6 times higher than in soils not containing organic matter. The half-life in an organic-rich soil was about 1 d (Gibson and Burns, 1977). Malathion was degraded by soil microcosms isolated from an agricultural area on Kauai, HI. Degradation half-lives in the laboratory and field experiments were 8.2 and 2 hr, respectively. Dimethyl phosphorodithioic acid and diethyl fumarate were identified as degradation products (Miles and Takashima, 1991). Mostafa et al. (1972) found the soil fungi *Penicillium notatum*, *Aspergillus niger*, *Rhizoctonia solani*, *Rhizobium trifolii*, and *Rhizobium leguminosarum* converted malathion to the following metabolites: malathion diacid, dimethyl phosphorothioate, dimethyl phosphorodithioate, dimethyl phosphate, monomethyl phosphate, and thiophosphates. Malathion also degraded in groundwater and seawater but at a slower rate ($t_{1/2}$ = 4.7 d). Microorganisms isolated from papermill effluents were responsible for the formation of malathion monocarboxylic acid (Singh and Seth, 1989).

Paris et al. (1975) isolated a heterogenous bacterial population that was capable of degrading low concentrations of malathion to β-malathion monoacid. About 1% of the original malathion concentration degraded to malathion dicarboxylic acid, *O,O*-dimethyl phosphorodithioic acid, and diethyl maleate (Paris et al., 1975).

Soil: In soil, malathion was degraded by *Arthrobacter* sp. to malathion monoacid, malathion dicarboxylic acid, potassium dimethylphosphorothioate, and potassium dimethylphosphorodithioate (Walker and Stojanovic, 1973). Chen et al. (1969) reported that the microbial conversion of malathion to malathion monoacid was a result of demethylation of the *O*-methyl group. Malathion was converted by unidentified microorganisms in soil to thiomalic acid, dimethylthiophosphoric acid, and diethylthiomaleate (Konrad et al., 1969).

Plant: When malathion on ladino clover seeds (10.9 ppm) were exposed to UV light (2537 Å) for 168 hr, malathion was the only residue detected. It was reported that 66.1% of the applied amount was lost due to volatilization (Archer, 1971).

Photolytic: When malathion was exposed to UV light, malathion monoacid, malathion diacid *O,O*-diethyl phosphorothioic acid, dimethyl phosphate, and phosphoric acid were formed (Mosher and Kadoum, 1972).

Chemical/Physical: Hydrolyzes in water forming *cis*-diethyl fumarate, *trans*-diethyl fumarate (Suffet et al., 1967), thiomalic acid, and dimethyl thiophosphate (Mulla et al., 1981). Day (1991) reported that the hydrolysis products are dependent upon pH. In basic solutions, malathion hydrolyzes to diethyl fumarate and dimethyl phosphorodithioic acid (Bender, 1969; Day, 1991). Dimethyl phosphorothionic acid and 2-mercaptodiethyl succinate formed in acidic solutions (Day, 1991). The reported hydrolysis half-lives at pH 7.4 at 20 and 37.5 °C were 10.5 and 1.3 d, respectively. At 20 °C and pH 6.1, the hydrolysis half-life is 120 d (Freed et al., 1979). Konrad et al. (1969) reported that after 7 d at pHs 9.0 and 11.0, 25 and 100% of the malathion was hydrolyzed. Hydrolysis of malathion in acidic and alkaline (0.5 M sodium hydroxide) conditions gives $(CH_3O)_2P(S)Na$ and $(CH_3O)_2P(S)OH$ (Sittig, 1985). Malaoxon and phosphoric acid were reported as ozonation products of malathion in drinking water (Richard and Bréner, 1984).

At 87 °C and pH 2.5, malathion degraded in water to malathion α-monoacid and malathion β-monoacid. From the extrapolated acid degradation constant at 27 °C, the half-life was calculated to be >4 yr (Wolfe et al., 1977a). Under alkaline conditions (pH 8 and 27 °C), malathion degraded in water to malathion monoacid, diethyl fumarate, ethyl hydrogen fumarate, and O,O-dimethyl phosphorodithioic acid. At pH 8, the reported half-lives at 0, 27, and 40 °C are 40 d, 36, and 1 hr, respectively (Wolfe et al., 1977a). However, under acidic conditions, it was reported that malathion degraded into diethyl thiomalate and O,O-dimethyl phosphorothionic acid (Wolfe et al., 1977a).

Emits toxic fumes of nitrogen and phosphorus oxides when heated to decomposition (Lewis, 1990). Products reported from the combustion of malathion at 900 °C include carbon monoxide, carbon dioxide, chlorine, sulfur oxides, nitrogen oxides, hydrogen sulfide, and oxygen (Kennedy et al., 1972a).

Exposure Limits: NIOSH REL: 10 hr TWA 15 mg/m^3, IDLH 5,000 mg/m^3; OSHA PEL: TWA 10 mg/m^3 (total dust), 5 mg/m^3 (respirable fraction); ACGIH TLV: TWA 10 mg/m^3.

Symptoms of Exposure: Miosis; eye, skin irritation; rhinorrhea; headache; tight chest, wheezing, laryngeal spasm; salivation; anorexia, nausea, vomiting, abdominal cramps; diarrhea, ataxia.

Formulation Types: Emulsifiable concentrate; wettable powder; aerosol; dustable powder; oil solutions.

Toxicity: LC$_{50}$ (96 hr) for bluegill sunfish 100 μg/L (Hartley and Kidd, 1987), largemouth bass 285 μg/L (Worthing and Hance, 1991), coho salmon 100 μg/L, brown trout 200 μg/L, channel catfish 9.0 mg/L, channel black bullhead 12.9 mg/L, fathead minnow 8.7 mg/L, rainbow trout 170 μg/L, perch 260 μg/L (Macek and McAllister, 1970), green sunfish 120 μg/L (Verschueren, 1983); LC$_{50}$ (24 hr) for bluegill sunfish 120 ppb and rainbow trout 100 ppb (Verschueren, 1983); acute oral LD$_{50}$ for rats 2,800 mg/kg (Hartley and Kidd, 1987), 370 mg/kg (RTECS, 1985).

Uses: Insecticide for control of sucking and chewing insects and spider mites on vegetables, fruits, ornamentals, field crops, greenhouses, gardens, and forestry.

MALEIC HYDRAZIDE

Synonyms: Burtolin; Chemform; De-cut; 1,2-Dihydroxypyridazine-3,6-dione; **1,2-Dihydro-3,6-pyradizinedione;** 1,2-Dihydro-3,6-pyridizinedione; Drexel-super P; ENT 18870; Fair-2; Fair-30; Fair-plus; Fair PS; 6-Hydroxy-2*H*-pyridazin-3-one; 6-Hydroxy-3(2*H*)-pyridazinone; KMH; MAH; Maintain 3; Maleic acid hydrazide; Maleic hydrazide 30%; Malein 30; *N,N*-Maleohydrazine; Malzid; Mazide; MH; MH 30; MH 40; MH 36 Bayer; RCRA waste number U148; Regulox; Regulox W; Regulox 50 W; Retard; Royal MH-30; Royal Slo-Gro; Slo-Grow; Sprout/Off; Sprout-Stop; Stunt-Man; Sucker-Stuff; Super-De-Sprout; Super Sprout Stop; Super Sucker-Stuff; Super Sucker-Stuff HC; 1,2,3,6-Tetrahydro-3,6-dioxopyridazine; Vondalhyde; Vondrax.

Structure:

Designations: CAS: 123-33-1; mf: $C_4H_4N_2O_2$; fw: 112.10; RTECS: UR5950000.

Properties: Colorless to white, odorless crystals. Mp: 292-298 °C; bp: decomposes at 260 °C; ρ: 1.60 at 25/4 °C; pK_a: 5.62 at 20 °C (monobasic acid); log K_{oc}: 1.56 (calculated); log K_{ow}: -1.96; S_o (g/kg at 25 °C): *N,N*-dimethylformamide (24), acetone, ethanol, and xylene (<1); S_w: 6 g/kg at 25 °C.

Transformation Products

Soil: Though no products were reported, the half-life in soil was reported to be 2-8 wk (Hartley and Kidd, 1987).

Plant: Major plant metabolites include fumaric, succinic, and maleic acids (Hartley and Kidd, 1987).

Chemical/Physical: Reacts with alkalies and amines forming water-soluble salts (Hartley and Kidd, 1987). Decomposed by oxidizing acids releasing nitrogen (Worthing and Hance, 1991). Decomposes at 260 °C (Windholz et al., 1983) releasing toxic fumes of nitrogen oxides (Sax and Lewis, 1987).

Formulation Types: Soluble concentrate; water-soluble granules.

Toxicity: LC_{50} (96 hr) for bluegill sunfish 1,608 mg/L, rainbow trout 1,435 mg/L (Hartley and Kidd, 1987), harlequin fish 125 mg/L (Verschueren, 1983); acute oral

LD_{50} for rats >5,000 mg/kg (free acid) (Hartley and Kidd, 1987), 3,800 mg/kg (RTECS, 1985).

Uses: Growth regulator used to control the sprouting of potatoes and onions and to prevent sucker development on tobacco, fruits, ornamentals, vines, field crops and in forestry. Also used to control insects in warehouses, storerooms, empty sacks, and in animal and poultry houses.

MANCOZEB

Synonyms: Carbamic acid ethylenebis(dithio-, manganese zinc complex); Carmazine; Dithane M 45; Dithane S 60; Dithane SPC; Dithane ultra; ((1,2-Ethanediylbis(carbamodithioato))(2-))manganese mixture with ((1,2-ethane-diylbis(carbamodithioato))(2-))zinc; F 2966; Fore; Green-daisen M; Karamate; Mancofol; Maneb-zinc; Manoseb; Manzate 200; Manzeb; Manzin; Manzin 80; Nemispor; Policar MZ; Policar S; Triziman; Triziman D; Zimanat; Zimaneb; Zimman-dithane; Vondozeb.

Structure:

$$[-\underset{\underset{S}{\parallel}}{S}CNHCH_2CH_2NH\underset{\underset{S}{\parallel}}{C}SMn-]_x (Zn^{2+})_y$$

Designations: CAS: 8018-01-7; mf: $[C_4H_6N_2S_2Mn]_xZn_y$; fw: variable; RTECS: ZB3200000.

Properties: Grayish-yellow powder. Mp: 192-194 °C (decomposes); fl p: 137.8 °C (open cup); H-$t_{1/2}$ (20 °C): 20 d (pH 5), 17 hr (pH 7), 34 hr (pH 9); log K_{oc}: 2.93-3.21 (calculated); log K_{ow}: 3.12-3.70 (calculated); S_o: slightly soluble in polar solvents; S_w: 6-20 mg/L.

Transformation Products
Plant: Undergoes metabolism in plants to form ethyl thiourea, ethylene thiuram disulfide, thiuram monosulfide, and sulfur (Hartley and Kidd, 1987).

Chemical/Physical: Decomposes in acids releasing carbon disulfide. In oxygenated waters, mancozeb degrades to ethylene thiuram monosulfide, ethylene diisocyanate, ethylene thiourea, ethylenediamine, and sulfur (Worthing and Hance, 1991).

Formulation Types: Suspension concentrate; wettable powder; dry seed treatment; dustable powder.

Toxicity: LC_{50} (48 hr) for carp 4.0 mg/L (Hartley and Kidd, 1987), catfish 5.2 mg/L, goldfish 9.0 mg/L (21 °C), and rainbow trout 2.2 mg/L (17 °C) (Worthing and Hance, 1991); acute oral LD_{50} for rats >8,000 mg/kg (Hartley and Kidd, 1987), mouse 600 mg/kg (RTECS, 1985).

Use: Fungicide.

MCPA

Synonyms: Agritox; Agroxon; Agroxone; Anicon Kombi; Anicon M; BH MCPA; Bordermaster; Brominal M & Plus; B-Selektonon M; Chiptox; 4-Chloro-*o*-cresol-xyacetic acid; **4-Chloro-2-methylphenoxyacetic acid;** 4-Chloro-*o*-toloxyacetic acid; Chwastox; Cornox; Cornox-M; Ded-weed; Dicopur-M; Dicotex; Dikotes; Dikotex; Dow MCP amine weed killer; Emcepan; Empal; Hedapur M 52; Hederax M; Hedonal M; Herbicide M; Hormotuho; Hornotuho; Kilsem; 4K-2M; Krezone; Legumex DB; Leuna M; Leyspray; Linormone; M 40; 2M-4C; 2M-4CH; MCP; 2,4-MCPA; Mephanac; Metaxon; Methoxone; 2-Methyl-4-chlorophenoxyacetic acid; 2M-4KH; Netazol; Okultin M; Phenoxylene 50; Phenoxylene Plus; Phenoxylene Super; Raphone; Razol dock killer; Rhomenc; Rhomene; Rhonox; Shamrox; Seppic MMD; Soviet technical herbicide 2M-4C; Trasan; U 46; U 46 M-fluid; Ustinex; Vacate; Vesakontuho MCPA; Verdone; Weedar; Weedar MCPA concentrate; Weedone; Weedone MCPA ester; Weed-rhap; Zelan.

Structure:

OCH$_2$COOH

Cl (H$_3$

Cl

Designations: CAS: 94-74-6; mf: C$_9$H$_9$ClO$_3$; fw: 200.63; RTECS: AG1575000.

Properties: Colorless to light-brown solid. Mp: 120 °C (pure), 99-107 °C (technical); ρ: 1.56 at 25/15.5 °C; pK$_a$: 3.05-3.13; log K$_{oc}$: 2.03-2.07 (calculated); log K$_{ow}$: 1.37-1.43 (calculated); S$_o$ (g/L at 25 °C): ethanol (1,530), ethyl ether (770), *n*-heptane (5), toluene (62), xylene (49); S$_w$: 730-825 mg/L at 25 °C, 270 g/L (sodium salt); vp: 1.5 x 10^{-6} mmHg at 20 °C.

Transformation Products

Biological: Cell-free extracts isolated from *Pseudomonas* sp. in a basal salt medium degraded MCPA to 4-chloro-*o*-cresol and glyoxylic acid (Gamar and Gaunt, 1971).

Soil: Residual activity in soil is limited to approximately 3-4 mo (Hartley and Kidd, 1987).

Plant: The penetration, translocation, and metabolism of radiolabeled MCPA in a cornland weed (*Galium aparine*) was studied by Leafe (1962). Carbon dioxide was identified as a metabolite but this could only account 7% of the applied MCPA. Though no additional compounds were identified, it was postulated that MCPA was detoxified in the weed via loss of both carbon atoms of the sidechain (Leafe, 1962).

Photolytic: When MCPA in dilute aqueous solution was exposed to summer sunlight or an indoor photoreactor (λ >290 nm), 2-methyl-4-chlorophenol formed as the major product as well as *o*-cresol and 4-chloro-2-formylphenol (Soderquist and Crosby, 1975). Clapés et al. (1986) studied the photodecomposition of aqueous solution of MCPA (120 ppm, pH 5.4, 25 °C) in a photoreactor equipped with a high pressure mercury lamp. After three minutes of irradiation, 4-chloro-2-methylphenol formed as an intermediate which degraded to 2-methylphenol. Both compounds were not detected after six minutes of irradiation; however, 1,4-dihydroxy-2-methylbenzene and 2-methyl-2,5-cyclohexadiene-1,4-dione formed as major and minor photodecomposition products, respectively. The same experiment was conducted using simulated sunlight (λ <300 nm) in the presence of riboflavin, a known photosensitizer. 4-Chloro-2-methylphenol and 4-chloro-2-methylbenzyl formate formed as major and minor photoproducts, respectively (Clapés et al., 1986). Ozone degraded MCPA in dilute aqueous solution with and without UV light (λ >300 nm) (Benoit-Guyod et al., 1986).

Chemical/Physical: Reacts with alkalies forming water soluble salts (Hartley and Kidd, 1987). Ozonolysis of MCPA in the dark yielded the following benzenoid intermediates: 4-chloro-2-methylphenol, its formate ester, 5-chlorosalicy-aldehyde, 5-chlorosalicyclic acid, and 5-chloro-3-methylbenzene-1,2-diol. Ozonolysis occurred much more rapidly in the presence of UV light. Based upon the intermediate compounds formed, the suggested degradative pathway in the dark included ring hydroxylation followed by cleavage of the ozone molecule and under irradiation, oxidation of the sidechains by hydroxyl radicals (Benoit-Guyod et al., 1986).

Formulation Types: Aqueous solutions; emulsifiable concentrate; soluble concentrate; water-soluble powder.

Toxicity: LC_{50} (48 hr) for bluegill sunfish 100 ppm (Verschueren, 1983); LC_{50} (24 hr) for bluegill sunfish 1.5 mg/L and rainbow trout 117 mg/L (Worthing and Hance, 1991); acute oral LD_{50} for rats is 700 mg/kg (RTECS, 1985).

Uses: Systemic postemergence herbicide used to control annual and perennial weeds in cereals, rice, flax, vines, peas, potatoes, asparagus, grassland and turf.

MEFLUIDIDE

Synonyms: N-(2,4-Dimethyl-5-(((trifluoromethyl)sulfonyl)amino)phenyl)-acetamide; Embark; MBR 12325; Mowchem; 5'-(1,1,1-Trifluoromethanesulphonamido)acet-2',4'-xylidide; VEL 3973; Vistar; Vistar herbicide.

Structure:

Designations: CAS: 53780-34-0; mf: $C_{11}H_{13}F_3N_2O_3S$; fw: 310.29; RTECS: AE2460000.

Properties: Colorless, odorless crystals. Mp: 183-185 °C; fl p: nonflammable (water formulation); pK_a: 4.6; K_H: 1.7 x 10^{-7} atm·m^3/mol at 23-25 °C (approximate - calculated from water solubility and vapor pressure); log K_{oc}: 2.40 (calculated); log K_{ow}: 2.08 (calculated); S_o (g/L at 23 °C): acetone (350), acetonitrile (64), benzene (0.31), ethyl acetate (50), ethyl ether (3.9), methanol (310), methylene chloride (2.1), n-octanol (17), petroleum benzene (0.002), xylene (0.12); S_w: 180 mg/L at 23 °C; vp: <7.5 x 10^{-5} mmHg at 25 °C.

Transformation Products
Soil: Rapidly degrades in soil forming 5-amino-2,4-dimethyltrifluoromethane sulfone anilide as the major metabolite ($t_{1/2}$ <1 wk) (Hartley and Kidd, 1987).

Symptoms of Exposure: Mild eye irritant.

Formulation Types: Aqueous solution.

Toxicity: LC_{50} (96 hr) for bluegill sunfish and rainbow trout >100 mg/L (Hartley and Kidd, 1987); LC_{50} (4 d) for bluegill sunfish 1,600 ppm and trout >1,200 ppm (Humburg et al., 1989); acute oral LD_{50} for rats >4,000 mg/kg (Hartley and Kidd, 1987).

Uses: Plant growth regulator and herbicide used in lawns, turf, grassland, industrial areas, and areas where grass cutting is difficult such as embankments. Also used to enhance sucrose content in sugar cane.

METALAXYL

Synonyms: Apron 2E; CGA 117; CGA 48988; N-(2,6-Dimethylphenyl)-N-methoxyacetyl)alanine methyl ester; N-(2,6-Dimethylphenyl)-N-methoxyacetyl)-DL-alanine methyl ester; Metalaxil; **Methyl N-(2,6-dimethylphenyl)-N-(methoxyacetyl)-DL-alaninate**; Ridomil; Ridomil E; Subdue; Subdue 2E; Subdue 5SP.

Structure:

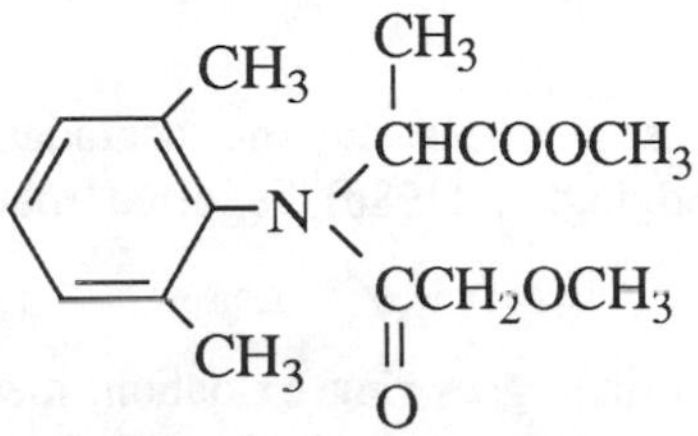

Designations: CAS: 57837-19-1; mf: $C_{15}H_{21}NO_4$; fw: 279.30; RTECS: AY6910000.

Properties: Colorless crystals. Mp: 71.8–72.3 °C; ρ: 1.21 at 20/4 °C; H-t$_{1/2}$ (20 °C): >200 d (pH 1), 115 d (pH 9), 12 d (pH 10); K_H: 1.1 x 10^{-10} atm·m^3/mol at 20 °C (approximate - calculated from water solubility and vapor pressure); log K_{oc}: 1.53–1.84; log K_{ow}: 1.52; S_o (g/L at 20 °C): benzene (550), methanol (650), methylene chloride (750), n-octanol (130), 2-propanol (270); S_w: 7.1 g/L at 20 °C; vp: 2.20 x 10^{-6} mmHg at 20 °C.

Soil properties and adsorption data

Soil	K_d (mL/g)	f_{oc} (%)	K_{oc} (mL/g)	pH
Loamy sand	0.43	1.28	34	7.8
Sand	0.48	0.70	69	6.3
Soil #1 (0–30 cm)	3.47	1.46	238	6.0
Soil #1 (30–50 cm)	17.84	0.93	1,918	6.1
Soil #1 (50–91 cm)	11.93	1.24	962	6.1
Soil #1 (91–135 cm)	2.17	0.70	310	6.4
Soil #1 (135–165 cm)	1.51	0.17	889	6.5
Soil #1 (>165 cm)	1.64	0.07	2,343	6.7
Soil #2 (0–27 cm)	3.06	1.50	204	5.9
Soil #2 (27–53 cm)	3.15	0.16	1,971	6.3
Soil #2 (53–61 cm)	4.89	0.14	3,490	6.4
Soil #2 (61–81 cm)	8.53	0.28	3,046	6.4
Soil #2 (81–157 cm)	13.30	0.50	2,660	6.0
Soil #2 (>157 cm)	13.95	1.30	1,073	5.9

continued

Soil	K_d (mL/g)	f_{oc} (%)	K_{oc} (mL/g)	pH
Sandy loam	1.40	3.25	43	6.7
Silty loam	0.87	2.09	42	6.1

Source: Sukop and Cogger, 1992; U.S. Department of Agriculture, 1990.

Transformation Products
Soil: Little information is available on the degradation of metalaxyl in soil; however, Sharom and Edgington (1986) reported metalaxyl acid as a possible metabolite.

Plant: In plants, metalaxyl undergoes ring oxidation, methyl ester hydrolysis, ether cleavage, ring methyl hydroxylation, and N-dealkylation (Owen and Donzel, 1986). Metalaxyl acid was identified as a hydrolysis product in both sunflower leaves and lettuce treated with the fungicide. Metalaxyl acid was further metabolized to form a conjugated derivative of metalaxyl (Businelli et al., 1984).

Formulation Types: Emulsifiable concentrate; wettable powder; granules; seed treatment.

Toxicity: LC_{50} (96 hr) for rainbow trout, carp, and bluegill sunfish >100 mg/L (Hartley and Kidd, 1987); acute oral LD_{50} for rats 669 mg/kg (RTECS, 1985).

Use: Fungicide.

METHAMIDOPHOS

Synonyms: Acephate-met; Bay 71628; Bayer 71268; Chevron 9006; Chevron ortho 9006; *O,S*-Dimethyl ester amide of aminothioate; *O,S*-Dimethyl phosphoramidothioate; ENT 27396; Hamidop; Metamidofos estrella; Monitor; MTD; NSC 190987; Ortho 9006; **Phosphoramidothioic acid *O,S*-dimethyl ester;** Pillaron; SRA 5172; Tahmabon; Tamaron.

Structure:

$$CH_3O \diagdown \overset{\displaystyle O}{\underset{\displaystyle CH_3S \diagup}{P}} - NH_2$$

Designations: CAS: 10265-92-6; mf: $C_2H_8NO_2PS$; fw: 141.13; RTECS: TB4970000.

Properties: Colorless crystals. Mp: 44.5 °C; ρ: 1.31 at 44.5/4 °C; H-t$_{1/2}$: 120 hr at 37 °C and pH 9, 140 hr at 40 °C and pH 2; log K_{oc}: 0.70; log K_{ow}: -0.79; S_o: soluble in ethanol, 2-propanol, methylene chloride; S_w: miscible; vp: 3.2 x 10^{-5} mmHg at 20 °C.

Soil properties and adsorption data

Soil	K_d (mL/g)	f_{oc} (%)	K_{oc} (mL/g)	pH
Silty loam	0.12	2.32	5	6.1

Source: U.S. Department of Agriculture, 1990.

Transformation Products
Chemical/Physical: Emits toxic fumes of phosphorus, nitrogen, and sulfur oxides when heated to decomposition (Sax and Lewis, 1987).

Formulation Types: Emulsifiable concentrate; soluble concentrate.

Toxicity: LC$_{50}$ (96 hr) for rainbow trout 51 mg/L, guppies 46 mg/L, carp and goldfish approximately 100 mg/L (Hartley and Kidd, 1987); acute oral LD$_{50}$ for rats approximately 30 mg/kg (Hartley and Kidd, 1987), 7,500 µg/kg (RTECS, 1985).

Uses: Insecticide and acaricide.

METHIDATHION

Synonyms: Ciba-Geigy CS 13005; *S*-(2-Dihydro-5-methoxy-2-oxo-1,3,4-thia-diazol-3-methyl)dimethyl phosphorothiolothionate; *S*-2-Dihydro-5-methoxy-2-oxo-1,3,4-thiadiazol-3-ylmethyl *O,O*-dimethyl phosphorodithionate; *O,O*-Dimethyl *S*-(5-methoxy-1,3,4-thiadiazolinyl-3-methyl) dithiophosphate; *O,O*-Dimethyl *S*-(2-methoxy-1,3,4-thiadiazol-5(4*H*)-onyl-4-methyl) phosphoro-dithioate; DMTP; ENT 27193; Fisons NC 2964; Geigy 13005; Geigy GS-13005; GS 13005; Methidathion 50S; *S*-((5-Methoxy-2-oxo-1,3,4-thiadiazol-3(2*H*)-yl)methyl) *O,O*-dimethyl phosphorodithioate; **Phosphorodithioic acid *O,O*-dimethyl ester, *S*-ester with 4-(mercaptomethyl)-2-methoxy-Δ^2-1,3,4-thiadiazolin-5-one**; Somonil; Surpracide; Ultracide.

Structure:

Designations: CAS: 950-37-8; mf: $C_6H_{11}N_2O_4PS_3$; fw: 302.31; RTECS: TE2100000.

Properties: Colorless crystals. Mp: 39-40 °C; ρ: 1.495 at 20/4 °C; H-t$_{1/2}$: 30 min at 25 °C and pH 13; K_H: 2.2 x 10^{-9} atm·m^3/mol at 20 °C; log K_{oc}: 2.29-2.76; log K_{ow}: 2.22, 2.42; S_o (g/kg at 20 °C): acetone (690), cyclohexanone (850), ethanol (260), xylene (600); S_w: 250 mg/L at 20 °C, 240 mg/L at 25 °C; vp: 1.40 x 10^{-6} mmHg at 20 °C.

Soil properties and adsorption data

Soil	K_d (mL/g)	f_{oc} (%)	K_{oc} (mL/g)	pH
Sand	2.48	1.28	194	7.8
Sandy loam	14.83	3.25	456	6.7
Silty loam	4.53	2.09	217	6.1

Source: U.S. Department of Agriculture, 1990.

Transformation Products

Photolytic: When methidathion in an aqueous buffer solution (25 °C and pH 7.0) was exposed to filtered UV light (λ >290 nm) for 24 hr, 17% decomposed to 5-methoxy-3*H*-1,3,4-thiadiazol-2-one. At 50 °C, 56% was degraded after 24 hr. Degradation occurred via hydrolysis of the thiol bond of the phosphorodithioic ester. Under acidic and alkaline conditions, hydrolytic cleavage occurred at the C–S and P–S bonds, respectively (Burkhard and Guth, 1979).

Chemical/Physical: Emits toxic fumes of phosphorus, nitrogen, and sulfur oxides when heated to decomposition (Sax and Lewis, 1987). Methidathion oxon was also found in fogwater collected near Parlier, CA (Glotfelty et al., 1990). It was suggested that methidathion was oxidized in the atmosphere during daylight hours prior to its partitioning from the vapor phase into the fog. On 12 January 1986, the distribution of parathion (0.45 ng/m^3) in the vapor phase, dissolved phase, air particles and water particles were 57.5, 25.4, 16.8, and 0.3%, respectively. For methidathion oxon (0.84 ng/m^3), the distribution in the vapor phase, dissolved phase, air particles and water particles were <7.1, 20.8, 78.6, and 0.1%, respectively (Glotfelty et al., 1990).

Symptoms of Exposure: Eye irritant.

Formulation Types: Emulsifiable concentrate; wettable powder.

Toxicity: LC$_{50}$ (96 hr) for rainbow trout 10 μg/L and bluegill sunfish 2 μg/L (Hartley and Kidd, 1987); acute oral LD$_{50}$ for rats 25-54 mg/kg (Hartley and Kidd, 1987), 20 mg/kg (RTECS, 1985).

Uses: Insecticide and acaricide.

METHIOCARB

Synonyms: B 37344; Bay 9026; Bay 37344; Bayer 37344; **3,5-Dimethyl-4-(methyl-thio)phenyl methylcarbamate**; 3,5-Dimethyl-4-methylthiophenyl *N*-methyl-carbamate; Draza; ENT 25726; H 321; Mercaptodimethur; Mesurol; Methyl carbamic acid 4-(methylthio)-3,5-xylyl ester; 4-Methylmercapto-3,5-dimethyl-phenyl *N*-methyl carbamate; 4-Methylthio-3,5-dimethylphenyl methyl carbamate; 4-Methylthio-3,5-xylyl isomethylcarbamate; Metmercapturon; NA 2757; OMS 93.

Structure:

$$CH_3S \overset{\displaystyle CH_3}{\underset{\displaystyle CH_3}{\diamond}} - OCNHCH_3 \quad (\overset{O}{\|})$$

Designations: CAS: 2032-65-7; DOT: 2757; mf: $C_{11}H_{15}NO_2S$; fw: 225.31; RTECS: FC5775000.

Properties: White crystalline powder. Mp: 121.5 °C; H-t$_{1/2}$ (20 °C): >1 yr (pH 4), <35 d (pH 7), 6 hr (pH 9); log K_{oc}: 2.71; log K_{ow}: 2.97, 3.04; S_o (g/L at 20 °C): methylene chloride (500), 2-propanol (80), toluene (50-100); S_w: 10-30 mg/L at 20 °C.

Soil properties and adsorption data

Soil	K_d (mL/g)	f_{oc} (%)	K_{oc} (mL/g)	pH
Batcombe silt loam	--	0.63-1.46	208	6.7-7.5
Sandy loam	12.6	2.44	516	5.9

Source: Briggs, 1981; U.S. Department of Agriculture, 1990.

Transformation Products

Soil: Methiocarb was oxidized, probably by singlet oxygen, to the corresponding sulfoxide and trace amounts (<5% yield) of sulfone when sorbed on soil and exposed to sunlight. The photosensitized oxidation was faster in soils containing the lowest organic carbon content (Gohre and Miller, 1986).

Plant: On and/or in bean plants, the methylthio group is rapidly oxidized to the sulfoxide and sulfone (Abdel-Wahab et al., 1966) followed by hydrolysis yielding the corresponding thiophenol, methylsulfoxide phenol, and methylsulphonyl phenol (Hartley and Kidd, 1987).

Chemical/Physical: Emits toxic fumes of nitrogen and sulfur oxides when heated to decomposition (Sax and Lewis, 1987).

Formulation Types: Granular bait; seed treatment; wettable powder; dustable powder.

Toxicity: LC_{50} (96 hr) for rainbow trout 0.64 mg/L, bluegill sunfish 0.21-0.75 mg/L, and golden orfe 3.8 mg/L; acute oral LD_{50} for rats 100-130 mg/kg (Hartley and Kidd, 1987), 15 mg/kg (RTECS, 1985).

Uses: Insecticide, acaricide, and bird repellent.

METHOMYL

Synonyms: Du Pont 1179; Du Pont insecticide 1179; ENT 27341; IN 1179; Insecticide 1179; Lannate; Lannate L; Mesomile; *N*-(((Methylamino)carbonyl)-oxy)ethanimidothioate; **N-(((Methylamino)carbonyl)oxy)ethanimidothioic acid methyl ester**; *N*-((Methylcarbamoyl)oxy)thioacetimidic acid methyl ester; Methyl-*N*-(((methylamino)carbonyl)oxy)ethanimidothioate; Methyl-*N*-((methylcarbamoyl)oxy)thioacetimidate; *S*-Methyl *N*-(methylcarbamoyloxy)thioacetimidate; Methyl *O*-(methylcarbamoyl)thioacethohydroxamate; Nu-bait II; Nudrin; RCRA waste number P066; SD 14999; WL 18236.

Structure:

$$CH_3\underset{\underset{SCH_3}{|}}{C} = NO\overset{\overset{O}{||}}{C}NHCH_3$$

Designations: CAS: 16752-77-5; mf: $C_5H_{10}N_2O_2S$; fw: 162.20; RTECS: AK2975000.

Properties: Colorless crystals with a faint sulfur-like odor. Mp: 78-79 °C; ρ: 1.2946 at 24/4 °C; H-t$_{1/2}$: 262 d at 25 °C and pH 7; K_H: 6.4 x 10^{-10} atm·m^3/mol at 25 °C (approximate - calculated from water solubility and vapor pressure); log K_{oc}: 1.86, 2.20; log K_{ow}: 0.13, 1.08; P-t$_{1/2}$: 48.41 hr (absorbance λ = 257.5 nm, concentration on glass plates = 6.7 μg/cm^2); S_o (wt %): soluble in acetone (73), ethanol (42), methanol (100), 2-propanol (22), and many other organic solvents; S_w: 57.9 g/kg at 25 °C; vp: 4.99 x 10^{-5} mmHg at 25 °C.

Transformation Products

Chemical/Physical: Emits toxic fumes of nitrogen and sulfur oxides when heated to decomposition (Sax and Lewis, 1987; Lewis, 1990). Though no products were reported, the calculated hydrolysis half-life at 25 °C and pH 7 is 262 d (Ellington et al., 1988). The hydrolysis half-lives of methomyl in a sterile 1% ethanol/water solution at 25 °C and pHs of 4.5, 6.0, 7.0, and 8.0 were 56, 54, 38, and 20 wk, respectively (Chapman and Cole, 1982). In both soils and water, chemical and biological mediated reactions can transform methomyl into two compounds - one a nitrile and the other a mercaptan (Alexander, 1981).

Exposure Limits: OSHA PEL: TWA 2.5 mg/m^3; ACGIH TLV: TWA 2.5 mg/m^3.

Formulation Types: Emulsifiable concentrate; wettable powder; soluble concentrate.

Toxicity: LC_{50} (96 hr) for rainbow trout 3.4 mg/L and bluegill sunfish 0.9 mg/L (Hartley and Kidd, 1987); acute oral LD_{50} for male and female rats is 17 (RTECS, 1985) and 24 mg/kg (Hartley and Kidd, 1987), respectively.

Use: Insecticide.

METHOXYCHLOR

Synonyms: 2,2-Bis(*p*-anisyl)-1,1,1-trichloroethane; 1,1-Bis(*p*-ethoxyphenyl)-2,2,2-trichloroethane; 2,2-Bis(*p*-methoxyphenyl)-1,1,1-trichloroethane; Chemform; 2,2-Di-*p*-anisyl-1,1,1-trichloroethane; Dimethoxy-DDT; *p,p'*-Dimethoxydiphenyltrichloroethane; Dimethoxy-DT; 2,2-Di(*p*-methoxyphenyl)-1,1,1-trichloroethane; Di(*p*-methoxyphenyl)trichloromethyl methane; DMDT; 4,4'-DMDT; *p,p'*-DMDT; DMTD; ENT 1716; Maralate; Marlate; Marlate 50; Methoxcide; Methoxo; 4,4'-Methoxychlor; *p,p'*-Methoxychlor; Methoxy-DDT; Metox; Moxie; NCI-C00497; RCRA waste number U247; 1,1,1-Trichloro-2,2-bis(*p*-anisyl)ethane; 1,1,1-Trichloro-2,2-bis(*p*-methoxyphenol)ethanol; 1,1,1-Trichloro-2,2-bis(*p*-methoxyphenyl)ethane; 1,1,1-Trichloro-2,2-di(4-methoxyphenyl)ethane; **1,1'-(2,2,2-Trichloroethylidene)bis(4-methoxybenzene).**

Structure:

$$CH_3O \text{—}\langle\text{—}\rangle\text{—}CH\text{—}\langle\text{—}\rangle\text{—}OCH_3$$
$$\underset{CCl_3}{|}$$

Designations: CAS: 72-43-5; DOT: 2761; mf: $C_{16}H_{15}Cl_3O_2$; fw: 345.66; RTECS: KJ3675000.

Properties: White, gray or pale yellow crystals or powder. May be dissolved in an organic solvent or petroleum distillate for application. Pungent to mild fruity odor. May turn color on exposure to light. Mp: 86-88 °C, 89-98 °C; bp: decomposes; ρ: 1.41 at 25/4 °C; H-t$_{1/2}$: 7-18 d in natural surface water samples; log K_{oc}: 4.90, 4.95; log K_{ow}: 3.31-5.08; S_o: soluble in ethanol, chloroform, xylene (440 g/kg at 20 °C); S_w: 40 μg/L at 24 °C.

Soil properties and adsorption data

Soil	K_d (mL/g)	f_{oc} (%)	K_{oc} (mL/g)	pH
Doe Run clay	2,400	3.29	72,948	--
Doe Run coarse silt	2,200	2.78	79,137	--
Doe Run fine silt	2,300	2.89	79,585	--
Doe Run medium silt	1,700	2.34	72,650	--
Hickory Hill clay	1,100	1.20	91,667	--
Hickory Hill coarse silt	2,600	3.27	79,510	--
Hickory Hill fine silt	1,400	1.34	104,407	--
Hickory Hill medium silt	1,800	1.98	90,909	--

Soil	K_d (mL/g)	f_{oc} (%)	K_{oc} (mL/g)	pH
Hickory Hill sand	53	0.13	40,769	--
Oconee River coarse silt	2,500	2.92	85,616	--
Oconee River fine sand	2,100	2.26	92,920	--
Oconee River medium silt	2,000	1.99	100,502	--
Oconee River sand	95	0.57	16,667	--

Source: Karickhoff et al., 1979.

Transformation Products

Biological: Degradation by *Aerobacter aerogenes* under aerobic or anaerobic conditions yielded 1,1-dichloro-2,2-bis(p-methoxyphenyl)ethylene and 1,1-dichloro-2,2-bis(p-methoxyphenyl)ethane (Kobayashi and Rittman, 1982).

Photolytic: In air-saturated distilled water, direct photolysis of methoxychlor (λ >280 nm) produced 1,1-bis(p-methoxyphenyl)-2,2-dichloroethylene (DMDE) which photolyzed to p-methoxybenzaldehyde (estimated $t_{1/2}$ = 4.5 mo) (Zepp et al., 1976). Methoxychlor-DDE and p,p-dimethoxybenzophenone were formed when methoxychlor in water was irradiated by UV light (Paris and Lewis, 1973). Compounds reported from the photolysis of methoxychlor in aqueous, alcoholic solutions were p,p-dimethoxybenzophenone, p-methoxybenzoic acid, and p-methoxyphenol (Wolfe et al., 1976). However, when methoxychlor in milk was irradiated by UV light (λ = 220 and 330 nm), methoxyphenol, methoxychlor-DDE, p,p-dimethoxybenzophenone, and 1,1,4,4-tetrakis(p-methoxyphenyl)-1,2,3-butatriene were formed (Li and Bradley, 1969).

Chemical/Physical: Hydrolysis at common aquatic pHs produced anisoin, anisil, and 2,2-bis(p-methoxyphenyl)-1,1-dichloroethylene (estimated $t_{1/2}$ = 270 d at 25 °C and pH 7.1) (Wolfe et al., 1977). In a model aquatic ecosystem, methoxychlor degraded to ethanol, dihydroxyethane, dihydroxyethylene and unidentified polar metabolites (Metcalf et al., 1971). Kapoor et al. (1970) also studied the biodegradation of methoxychlor in a model ecosystem containing snails, plankton, mosquito larvae, *Daphnia magna*, and mosquito fish (*Gambusia affinis*). The following metabolites were identified: 2-(p-methoxyphenyl)-2-(p-hydroxy-phenyl)-1,1,1-trichloroethane, 2,2-bis(p-hydroxyphenyl)-1,1,1-trichloroethane, 2,2-bis(p-hydroxyphenyl)-1,1,1-trichloroethylene, and polar metabolites (Kapoor et al., 1970).

Decomposes in aqueous alkaline solutions forming diphenylethylene and hydrochloric acid (Hartley and Kidd, 1987). Emits toxic chloride fumes when heated to decomposition (Lewis, 1990).

Exposure Limits: NIOSH REL: IDLH 7,500 mg/m^3; OSHA PEL: TWA 10 mg/m^3 (total dust), 5 mg/m^3 (respirable fraction); ACGIH TLV: TWA 10 mg/m^3.

Symptoms of Exposure: Slightly irritating to skin.

Formulation Types: Aerosol; emulsifiable concentrate; wettable powder; granules; dustable powder.

Toxicity: LC_{50} (96 hr) for fathead minnow 7.5 μg/L, bluegill sunfish 62.0 μg/L, rainbow trout 62.6 μg/L, coho salmon 66.2 μg/L, chinook 27.9 μg/L, perch 20.0 μg/L (Verschueren, 1983); LC_{50} (24 hr) for rainbow trout 52 μg/L and bluegill sunfish 67 μg/L (Worthing and Hance, 1991); acute oral LD_{50} for rats 6,000 mg/kg (Hartley and Kidd, 1987), 5,000 mg/kg (RTECS, 1985).

Uses: Insecticide used to control mosquito larvae, house flies, and other insect pests in field crops, fruits, and vegetables; to control ectoparasites on cattle, sheep, and goats; recommended for use in dairy barns.

METHYL BROMIDE

Synonyms: Brom-o-gas; Brom-o-gaz; **Bromomethane**; Celfume; Dawson 100; Dowfume; Dowfume MC-2; Dowfume MC-2 soil fumigant; Dowfume MC-33; Edco; Embafume; Fumigant-1; Halon 1001; Iscobrome; Kayafume; MB; M-B-C Fumigant; MBX; MEBR; Metafume; Methogas; Monobromomethane; Pestmaster; Profume; R 40B1; RCRA waste number U029; Rotox; Terabol; Terr-o-gas 100; UN 1062; Zytox.

Structure:

$$CH_3Br$$

Designations: CAS: 74-83-9; DOT: 1062; mf: CH_3Br; fw: 94.94; RTECS: PA4900000.

Properties: Colorless liquid or gas with a sweetish odor similar to chloroform at high concentrations. Odorless at low concentrations. Burning taste. Mp: -93.6 °C; bp: 3.55 °C; ρ: 1.732 at 0/4 °C, 1.6755 at 20/4 °C; lel: 10%; uel: 16%; H-t½: 20 d at 25 °C and pH 7: K_H: 3.18 x 10^{-2} atm·m^3/mol; IP: 10.54 eV; log K_{oc}: 1.92 (calculated); log K_{ow}: 1.00-1.19; S_o: soluble in 95% ethanol, ethyl ether, chloroform, carbon disulfide, carbon tetrachloride, benzene, esters, ketones, aromatic and many halogenated hydrocarbons; S_w: 13.200 g/L at 25 °C; vap d: 3.88 g/L at 25 °C, 3.28 (air = 1); vp: 1,420 at 20 °C.

Transformation Products

Chemical/Physical: Hydrolyzes in water forming methanol and hydrobromic acid (Mabey and Mill, 1978). Forms a voluminous crystalline hydrate at 0-5 °C (Keith and Walters, 1992). When methyl bromide was heated to 550 °C in the absence of oxygen, methane, hydrobromic acid, hydrogen, bromine, ethyl bromide, anthracene, pyrene, and free radicals were produced (Chaigneau et al., 1966). Emits toxic bromide fumes when heated to decomposition (Lewis, 1990).

Photolytic: When methyl bromide and bromine gas (concentration = 3%) was irradiated at 1850 Å, methane was produced (Kobrinsky and Martin, 1968).

Exposure Limits: NIOSH REL: lowest feasible limit; OSHA PEL: TWA 5 ppm; ACGIH TLV: TWA 5 ppm.

Symptoms of Exposure: Inhalation causes headache, visual disturbance, vertigo, nausea, vomiting, malaise, hand tremor, convulsions, dyspnea. Eye and skin irritation.

Formulation Types: Vapor.

Toxicity: Acute oral LD_{50} for rats is 100 mg/kg and inhalation LD_{50} for rats is 3,150 mg/L (Ashton and Monaco, 1991).

Uses: Soil, space, and food fumigant; disinfestation of potatoes, tomatoes and other crops.

METHYL ISOTHIOCYANATE

Synonyms: EP 161E; **Isothiocyanatomethane;** Isothiocyanic acid methyl ester; Methyl mustard oil; MIC; MIT; MITC; Morton EP 161E; Mustard oil; Trapex; Trapexide; UN 2477; Vorlex; Vortex; WN 12.

Structure:

$$CH_3NCS$$

Designations: CAS: 556-61-6; DOT: 2477; mf: C_2H_3NS; fw: 73.12; RTECS: PA7625000.

Properties: Colorless crystals with a pungent, horseradish-like odor. Mp: 35–36 °C; bp: 119 °C; ρ: 1.048 at 24/4 °C; fl p: 32 °C; H-t$_{1/2}$ (25 °C): 85 hr (pH 5), 490 hr (pH 7), 110 hr (pH 9); K_H: 2.4 x 10^{-4} atm·m^3/mol at 20 °C (approximate - calculated from water solubility and vapor pressure); log K_{oc}: 1.51 (calculated); log K_{ow}: 1.37; S_o: very soluble in most organic solvents; S_w: 7.6 g/L at 20 °C; vp: 19 mmHg at 20 °C.

Transformation Products
Soil: Though no products were reported, the reported half-life in soil is <14 d (Worthing and Hance, 1991).

Chemical/Physical: Emits toxic fumes of nitrogen and sulfur oxides when heated to decomposition (Sax and Lewis, 1987).

Symptoms of Exposure: Very irritating to eyes, skin, and mucous membranes.

Formulation Types: Emulsifiable concentrate.

Toxicity: LC$_{50}$ (96 hr) for bluegill sunfish 130 μg/L, carp 370 μg/L, and rainbow trout 370 μg/L (Hartley and Kidd, 1987); acute oral LD$_{50}$ for rats 175 mg/kg (Hartley and Kidd, 1987), 97 mg/kg (RTECS, 1985).

Uses: Pesticide and soil fumigant used to control insects, soil fungi, nematodes.

METOLACHLOR

Synonyms: Bicep; CGA 24705; **2-Chloro-*N*-(2-ethyl-6-methylphenyl)-*N*-(2-methoxy-1-methylethyl)acetamide**; 2-Chloro-6'-ethyl-*N*-(2-methoxy-1-methyl-ethyl)acet-*o*-toluidide; α-Chloro-2'-ethyl-6'-methyl-*N*-(1-methyl-2-methoxy-ethyl)acetanilide; 2-Chloro-*N*-(2-ethyl-6-methylphenyl)-*N*-(2-methoxy-1-meth-ylethyl)acetamide; Codal; Cotoran multi; Dual; Dual 8E; Dual 25G; 2-Ethyl-6-methyl-1-*N*-(2-methoxy-1-methylethyl)chloroacetanilide; Metelilachlor; Milocep; Ontrack 8E; Pennant; Pennant 5G; Primagram; Primextra.

Structure:

Designations: CAS: 51218-45-2; mf: $C_{15}H_{22}ClNO_2$; fw: 283.81; RTECS: AN3430000.

Properties: Clear, colorless to tan, odorless liquid. Mp: <25 °C; bp: 100 °C at 0.001 mmHg; ρ: 1.12 at 20/4 °C; fl p: 93.3 °C (Dual 8E formulation); H-t$_{1/2}$ (20 °C): >200 d at 1 ≤pH ≤9; K_H: 9.2 x 10^{-9} atm·m^3/mol at 20 °C (approximate - calculated from water solubility and vapor pressure); log K_{oc}: 2.08-2.49; log K_{ow}: 2.93-3.45; S_o: miscible with acetone, benzene, butyl cellosolve, cyclohexanone, 1,2-dichloroethane, *N,N*-dimethylformamide, ethanol, methanol, methyl cellosolve, methylene chloride, toluene, xylene; S_w: 530 mg/L at 20 °C; vp: 1.3 x 10^{-5} mmHg at 20 °C.

Soil properties and adsorption data

Soil	K_d (mL/g)	f_{oc} (%)	K_{oc} (mL/g)	pH	CEC (meq/100 g)
Appling	1.10	0.81	136	6.8	6.9
Augusta	0.50	0.29	172	5.7	3.2
Cape Fear sandy loam	8.52	5.34	160	--	10.3
Cape Fear sandy loam	10.90	5.05	216	5.1	10.3
Cecil sandy loam	1.00	0.99	101	5.4	3.1
Goldsboro	1.40	0.70	200	5.3	3.3
Lynchburg	2.50	1.45	172	5.5	6.6
Norfolk fine loamy sand	2.20	0.99	222	--	2.3
Norfolk fine loamy sand	0.50	0.29	172	5.4	2.3
Portsmouth	3.30	2.55	129	5.4	10.6

Soil	K_d (mL/g)	f_{oc} (%)	K_{oc} (mL/g)	pH	CEC (meq/100 g)
Rains fine loamy sand	2.40	0.99	242	6.0	7.1
Rains fine loamy sand	3.20	1.45	229	--	7.0
Sand	1.54	1.28	120	7.8	--
Sand	1.69	0.70	243	6.3	--
Sandy loam	10.00	3.25	308	6.7	--
Sandy silt loam	3.18	2.09	152	6.1	--

Source: Kozak et al., 1983; LeBaron et al., 1988; U.S. Department of Agriculture, 1990.

Transformation Products

Soil: The reported half-life of metolachlor in soil is approximately 6 d (Worthing and Hance, 1991). Zimdahl and Clark (1982) reported the half-lives of metolachlor in clay loam soils and sandy loam soils were 15-38 and 33-100 d, respectively. Metolachlor and its degradation products combine with humic acids in soils and small quantities are degraded to carbon dioxide (Ashton and Monaco, 1991). In soil, the fungus *Chaetomium globosum* degraded metolachlor to 2-chloro-*N*-(2-ethyl-6-methylphenyl)acetamide, 2-hydroxy-*N*-(2-methylvinylphenyl)-*N*-(methoxyprop-2-yl)acetamide, 3-hydroxy-8-methyl-*N*-(methoxyprop-2-yl)-2-oxo-1,2,3,4-tetrahydraquinoline, 2-chloro-*N*-(2-ethyl-6-methylphenyl)-*N*-(hydroxyprop-2-yl)acetamide, and the tentatively identified compounds 3-hydroxy-1-isopropyl-8-methyl-2-oxo-1,2,3,4-tetrahydroquinoline, *N*-(methoxyprop-2-yl)-8-methyl-2-oxo-1,2,3,4-tetrahydroquinoline, *N*-(methoxyprop-2-yl)-*N*-(2-methyl-6-vinyl)aniline, and 1-(methoxyprop-2-yl)-7-methyl-2,3-dihydroindole (McGahen and Tiedje, 1978). Metolachlor was transformed by a strain of soil actinomycetes to the following products: 2-chloro-*N*-(2-ethyl-6-methylphenyl)-*N*-(hydroxyprop-2-yl)acetamide, 2-chloro-*N*-2-(1-hydroxyethyl)-6-(methylphenyl)-*N*-(hydroxyprop-2-yl)acetamide, 2-chloro-*N*-(2-ethyl-6-hydroxymethylphenyl)-*N*-(hydroxyprop-2-yl)acetamide, diastereoisomers of 2-chloro-*N*-(2-ethyl-6-hydroxymethylphenyl)-*N*-(methoxyprop-2-yl)acetamide and 2-chloro-*N*-2-(hydroxyethyl)-6-hydroxymethylphenyl)-*N*-(methoxyprop-2-yl)acetamide. These products were formed via hydroxylation of both the *N*-alkyl and alkyl side chains (Krause et al., 1985).

Plant: Metabolizes in plants forming water soluble, polar, nonvolatile products (Hartley and Kidd, 1987).

Symptoms of Exposure: Eye and skin irritant.

Formulation Types: Emulsifiable concentrate (8 lb/gal); granules (5 and 25%).

Toxicity: LC_{50} (96 hr) for rainbow trout 2 mg/L, bluegill sunfish 15 mg/L, carp 4.9 mg/L (Hartley and Kidd, 1987), and channel catfish 4.9 ppm (Humburg et al., 1989); acute oral LD_{50} of technical metolachlor for rats is 2,800 mg/kg (Ashton and Monaco, 1991), pure metolachlor 2,534 mg/kg (RTECS, 1985).

Uses: Preemergence herbicide used to control most annual grasses and many annual weeds in beans, chickpeas, corn, cotton, milo, okra, peanuts, peas, potatoes, safflower, soybeans, and some ornamentals.

METRIBUZIN

Synonyms: 4-Amino-6-(1,1-dimethylethyl)-3-(methylthio)-1,2,4-triazin-5-(4*H*)-one; Bay 61597; Bay dic 1468; Bayer 94337; Bayer 6159H; Bayer 6443H; DIC 1468; Lexone; Lexone DF; Lexone 4L; Preview; Sencor; Sencor 4; Sencor DF; Sencoral; Sencorer; Sencorex.

Structure:

Designations: CAS: 21087-64-9; mf: $C_8H_{14}N_4OS$; fw: 214.28; RTECS: XZ2990000.

Properties: White crystalline solid. Technical grade has a slight sulfur-like odor. Mp: 125-126.5 °C; ρ: 1.31 at 20/4 °C; fl p: nonflammable; pK_a: 1.00; H-t$_{1/2}$: $\approx$ 1 wk in pond water; K_H: 1.2 x 10^{-10} atm·m^3/mol at 20 °C (approximate - calculated from water solubility and vapor pressure); log K_{oc}: 1.81-2.72; log K_{ow}: 1.60, 1.70; S_o (g/kg at 20 °C): acetone (820), benzene (220), *n*-butanol (150), chloroform (850), cyclohexane (1,000), cyclohexanone (1,000), *N,N*-dimethylformamide (1,780), ethanol (190), *n*-hexane (2), kerosene (<10), methanol (450), methylene chloride (333), toluene (120), xylene (90); S_w: 1,050 mg/L at 20 °C; vp: 4.35 x 10^{-7} mmHg at 20 °C.

Soil properties and adsorption data

Soil	K_d (mL/g)	f_{oc} (%)	K_{oc} (mL/g)	pH	CEC (cmol/kg)
Bashaw clay loam	2.13	0.58	367	6.2	35.1
Chehalis sandy loam	2.42	1.39	174	6.0	19.5
Clayey loam	1.90	2.90	66	7.9	--
Clayey loam	1.53	0.29	528	6.0	--
Crooked sandy loam	1.11	0.64	173	8.2	13.7
Loamy sand	1.32	1.62	81	6.6	--
Ontko loam	3.37	3.19	106	6.2	44.2
Woodburn silty clay loam	7.00	1.39	504	4.6	13.7

Source: Peek and Appleby, 1989; U.S. Department of Agriculture, 1990.

Transformation Products

Soil: In soils, metribuzin undergoes deamination and further degradation forming water soluble conjugates (Hartley and Kidd, 1987). The reported half-life in soil is approximately 1-2 mo (Hartley and Kidd, 1987).

Plant: Metribuzin is metabolized in soybean plants to a deaminated diketo derivative which is nonphytoxic (Duke et al., 1991).

Photolytic: The simulated sunlight photolysis (λ >230 nm) as a thin film on silica gel or sand yielded 6-(1,1-dimethylethyl)-3-(methylthio)-1,2,4-triazin-5-(4H)-one and two additional photoproducts. In both of the unnamed photoproducts, the methylthio group in the parent compound is replaced by oxygen and one of the compound also underwent N-deamination (Bartl and Korte, 1975).

Chemical/Physical: Emits toxic fumes of nitrogen and sulfur oxides when heated to decomposition (Lewis, 1990).

Exposure Limits: OSHA PEL: TWA 5 mg/m^3; ACGIH TLV: TWA 5 mg/m^3.

Formulation Types: Water-dispersible granules; emulsifiable concentrate; suspension concentrate (4 lb/gal); flowable powder (75%); wettable powder (50%).

Toxicity: LC_{50} (96 hr) for rainbow trout 64 mg/L and bluegill sunfish 80 mg/L (Worthing and Hance, 1991); acute oral LD_{50} of technical metribuzin for male and female rats is 1,100 and 1,200 mg/kg, respectively (Ashton and Monaco, 1991).

Uses: Selective herbicide used to control many broadleaf weeds and annual grasses in crops such as potatoes and sugar cane.

METSULFURON-METHYL

Synonyms: Ally; Ally 20DF; Escort; 2-(((((4-Methoxy-6-methyl-1,3,5-triazin-2-yl)amino)carbonyl)amino)sulfonyl)benzoic acid.

Structure:

Designations: CAS: 74223-64-6; mf: $C_{14}H_{15}N_5O_6S$; fw: 381.40.

Properties: Colorless crystals with a slight ester-like odor. Mp: 158 °C, decomposes at 172 °C; fl p: nonflammable; pK_a: 3.3 at 25 °C (imide group); H-t$_{1/2}$ (25 °C): 15 hr (pH 2), 33 d (pH 5), >41 d (pH 7-9) and at 45 °C: 2 hr (pH 2), 2 d (pH 5), 33 d (pH 9); K_H: 3.1 x 10^{-9} atm·m^3/mol at pH 7 and 25 °C (approximate - calculated from water solubility and vapor pressure); log K_{oc}: 1.54, 2.31 (Flanagan silt loam); log K_{ow}: 0.00 (pH 5), -1.85 (pH 7); S_o (g/L at 20 °C): acetone (36), ethanol (2.3), n-hexane (0.00079), methanol (7.3), methylene chloride (121), xylene (0.58); S_w (g/L at 25 °C): 0.27 at pH 4.6, 1.1 at pH 5, 1.75 at pH 5.4, 9.5 g/L at pH 7.0; vp: 5.8 x 10^{-5} mmHg at 25 °C.

Transformation Products
Soil/Plant: Hydrolyzes in soil and plants to nontoxic products (Hartley and Kidd, 1987). The half-life in soil varies from 1 wk to 1 mo (Hartley and Kidd, 1987).

Symptoms of Exposure: May irritates eyes, nose, skin, and throat.

Formulation Types: Water-dispersible granules (20 and 60%).

Toxicity: LC_{50} (96 hr) for rainbow trout and bluegill sunfish >125 mg/L (Worthing and Hance, 1991); acute oral LD_{50} for rats is >5,000 mg/kg (Ashton and Monaco, 1991).

Uses: Herbicide used to control broadleaf weeds in barley and wheat.

MEVINPHOS

Synonyms: Apavinphos; 2-Butenoic acid 3-((dimethoxyphosphinyl)oxy)methyl ester; 2-Carbomethoxy-1-methylvinyl dimethyl phosphate; α-Carbomethoxy-1-methylvinyl dimethyl phosphate; 2-Carbomethoxy-1-propen-2-yl dimethyl phosphate; CMDP; Compound 2046; **3-((Dimethoxyphosphinyl)oxy)-2-butenoic acid methyl ester;** O,O-Dimethyl-O-(2-carbomethoxy-1-methylvinyl)phosphate; Dimethyl-1-carbomethoxy-1-propen-2-yl phosphate; O,O-Dimethyl 1-carbomethoxy-1-propen-2-yl phosphate; Dimethyl 2-methoxycarbonyl-1-methylvinyl phosphate; Dimethyl methoxycarbonylpropenyl phosphate; Dimethyl (1-methoxycarboxypropen-2-yl)phosphate; O,O-Dimethyl O-(1-methyl-2-carboxyvinyl)-phosphate; Dimethyl phosphate of methyl-3-hydroxy-cis-crotonate; Duraphos; ENT 22324; Fosdrin; Gesfid; Gestid; 3-Hydroxycrotonic acid methyl ester dimethyl phosphate; Meniphos; Menite; 2-Methoxycarbonyl-1-methylvinyl dimethyl phosphate; cis-2-Methoxycarbonyl-1-methylvinyl dimethyl phosphate; 1-Methoxycarbonyl-1-propen-2-yl dimethyl phosphate; Methyl 3-(dimethoxyphosphinyloxy)crotonate; NA 2783; OS 2046; PD 5; Phosdrin; cis-Phosdrin; Phosfene; Phosphoric acid (1-methoxycarboxypropen-2-yl) dimethyl ester.

Structure:

$$CH_3O\underset{CH_3O}{\overset{O}{\underset{}{\diagdown}}}\overset{\parallel}{P} - O - \overset{\overset{CH_3}{\vert}}{C} = CHCOOCH_3$$

Designations: CAS: 7786-34-7; DOT: 2783; mf: $C_7H_{13}O_6P$; fw: 224.16; RTECS: GQ5250000.

Properties: Colorless to pale yellow liquid. Mp: -56.1 °C; bp: 106-107.5 °C at 1 mmHg; ρ: 1.25 at 20/4 °C; fl p: 79.4 °C (open cup); H-t$_{1/2}$: 120 d (pH 6), 35 d pH 7), 3 d (pH 9), 1.4 hr (pH 11); P-t$_{1/2}$: 23.79 hr (absorbance λ = 230.5 nm, concentration on glass plates = 6.7 μg/cm^2); S_o: miscible with acetone, benzene, carbon tetrachloride, chloroform, ethanol, isopropanol, toluene, and xylene; soluble in carbon disulfide and kerosene (50 g/L); S_w: miscible; vap d: 9.16 g/L at 25 °C, 7.74 (air = 1); vp: 2.2 x 10^{-3} mmHg at 20 °C.

Transformation Products

Plant: In plants, mevinphos is hydrolyzed to phosphoric acid dimethyl ester, phosphoric acid, and other less toxic compounds (Hartley and Kidd, 1987). In 1 d, the compound is almost completely degraded in plants (Cremlyn, 1991). Casida et al. (1956) proposed two degradative pathways of mevinphos in bean plants and cabbage. In the first degradative pathway, cleavage of the vinyl phosphate bond affords methylacetoacetate and acetoacetic acid which may be precursors to the formation of the end products dimethyl phosphoric acid, methanol, acetone, and carbon dioxide. In the other degradative pathway, direct hydrolysis of the carboxylic ester would yield vinyl phosphates as intermediates. The half-life of mevinphos in bean plants was 0.5 d.

Chemical/Physical: The reported hydrolysis half-lives of *cis*-mevinphos and *trans*-mevinphos at pH 11.6 is 1.8 and 3.0 hr, respectively (Casida et al., 1956). Emits toxic phosphorus oxide fumes when heated to decomposition (Lewis, 1990).

Exposure Limits: NIOSH REL: IDLH 40 mg/m^3; ACGIH TWA: TLV 0.01 ppm, STEL 0.03 ppm.

Symptoms of Exposure: Miosis, rhinorrhea, headache, wheezing, laryngeal spasm; salivation, cyanosis; anorexia, nausea, abdominal cramps, diarrhea, paralysis, ataxia, convulsions, low blood pressure; irritates eyes and skin.

Formulation Types: Emulsifiable concentrate; soluble concentrate.

Toxicity: LC_{50} (96 hr) for bluegill sunfish 70 μg/L, largemouth bass 110 μg/L, mummichog 65-300 μg/L, striped killifish 75 μg/L, striped mullet 300 μg/L, Atlantic silverside 320 μg/L, Atlantic eel 65 μg/L, bullhead 74 μg/L, northern puffer 800 μg/L (Verschueren, 1983); LC_{50} (24 hr) for bluegill sunfish 41 ppb and rainbow trout 34 ppb (Verschueren, 1983); acute oral LD_{50} for rats 3-12 mg/kg (Hartley and Kidd, 1987), 1,350 μg/kg mg/kg (RTECS, 1985).

Uses: Contact insecticide and acaricide for control of chewing insects and spider mites in fruits, vegetables, and ornamentals.

MOLINATE

Synonyms: *S*-Ethyl azepane-1-carbothioate; **S-Ethyl hexahydro-1*H*-azepine-1-carbothioate**; *S*-Ethyl *N,N*-hexamethylenethiocarbamate; *S*-Ethyl perhydroazepine-1-thiocarboxylate; Felan; Higalnate; Hydram; Jalan; Molmate; Ordram; Ordram 8E; Ordram 10G; Ordram 15G; R 4572; Stauffer R 4572; Sakkimol; Yalan; Yulan.

Structure:

Designations: CAS: 2212-67-1; mf: $C_9H_{17}NOS$; fw: 187.32; RTECS: CM2625000.

Properties: Clear liquid with an aromatic odor. Mp: <25 °C; bp: 117 °C at 10 mmHg; ρ: 1.0643 at 20/20 °C; fl p: 139 °C (open cup); K_H: 1.6 x 10^{-6} atm·m^3/mol at 25 °C (approximate - calculated from water solubility and vapor pressure); log K_{oc}: 1.93-1.97; log K_{ow}: 2.88; S_o: miscible with most organic solvents; S_w: 880 mg/L at 20 °C; vap d: 7.66 g/L at 25 °C, 6.49 (air = 1); vp: 5.6 x 10^{-3} mmHg at 25 °C.

Soil properties and adsorption data

Soil	K_d (mL/g)	f_{oc} (%)	K_{oc} (mL/g)	pH
Kanuma high clay	1.1	1.35	80	5.7
Tsukuba clay loam	3.7	4.24	89	6.5

Source: Kanazawa, 1989.

Transformation Products

Soil: Hydrolyzes in soil forming ethyl mercaptan, carbon dioxide, and dialkylamine (H-t$_{1/2}$ ≈ 2-5 wk) (Hartley and Kidd, 1987). At recommended rates of application, the half-life of molinate in moist loam soils at 21-27 °C is approximately 3 wk (Humburg et al., 1989). Rajagopal et al. (1989) reported that under flooded conditions, molinate was hydroxylated at the 3- and 4-position with subsequent oxidation forming many compounds including molinate sulfoxide, carboxymethyl molinate, hexahydroazepine-1-carbothioate, 4-hydroxymolinate, 4-hydroxymolinate sulfoxide, hexahydroazepine, *S*-methyl hexahydroazepine-1-carbothioate, 4-ketomolinate, 4-hydroxyhexahydroazepine, 4-hydroxy-*N*-acetyl-hexahydroazepine, carbon dioxide, and bound residues.

Plant: Molinate is rapidly metabolized by plants releasing carbon dioxide and naturally occurring plant constituents (Humburg et al., 1989).

Photolytic: Molinate in a hydrogen peroxide solution (120 μM) was irradiated by UV light (λ = 290 nm) at 23 °C. The major photooxidation products were the two isomers of 2-oxomolinate (20% yield) and *s*-molinate oxide (5% yield) (Draper and Crosby, 1984). Half-lives of 180 and 120 hr were observed using one and two equivalents of hydrogen peroxide, respectively (Draper and Crosby, 1984). Molinate has a UV absorption maximum at 225 nm and no absorption at wavelengths >290 nm. Therefore, molinate is not expected to undergo aqueous photolysis under natural sunlight (λ = 290 nm). In the presence of tryptophan, a naturally occurring photosensitizer, molinate in aqueous solution photodegraded to form 1-((ethylsulfinyl)carbonyl)hexahydro-1*H*-azepine, *S*-ethyl hexahydro-2-oxo-1*H*-azepine-1-carbothioate, and hexamethyleneimine (Soderquist et al., 1977).

Chemical/Physical: Metabolites identified in tap water were molinate sulfoxide, 3- and 4-hydroxymolinate, ketohexamethyleneimine, and 4-ketomolinate (Verschueren, 1983).

Formulation Types: Emulsifiable concentrate (8 lb/gal); granules (10 and 15%).

Toxicity: LC_{50} (96 hr) for bluegill sunfish 29-30 mg/L, goldfish 30 mg/L, mosquito fish 16.4 mg/L, rainbow trout 0.2-1.3 mg/L (Hartley and Kidd, 1987); acute oral LD_{50} of technical molinate for rats and mice is 720 and 795 mg/kg, respectively (Ashton and Monaco, 1991), 501 mg/kg (RTECS, 1985).

Uses: Selective herbicide used to control the germination of annual grasses and broadleaf weeds in rice crops.

MONALIDE

Synonyms: 4'-Chloro-2,2-dimethylvaleranilide; *N*-(**4-Chlorophenyl**)-**2,2-dimeth-ylpentanamide**; *N*-(4-Chlorophenyl)-2,2-dimethylvaleramide; 4'-Chloro-α,α-di-methylvaleranilide; D-90-A; Schering 35830; Potablan; SN 35830.

Structure:

$$Cl-\underset{}{\bigcirc}-NHCO\underset{\underset{CH_3}{|}}{\overset{\overset{CH_3}{|}}{C}}CH_2CH_2CH_3$$

Designations: CAS: 7287-36-7; mf: $C_{13}H_{18}ClNO$; fw: 239.75; RTECS: YV6010000.

Properties: Colorless crystals. Mp: 87-88 °C; H-t$_{1/2}$ at room temperature: 154 d (pH 5), 116 d (pH 8.95); K_H: 2.5 x 10^{-8} atm·m^3/mol at 23-25 °C (approximate - calculated from water solubility and vapor pressure); log K_{oc}: 2.89 (calculated); log K_{ow}: 3.06 (calculated); S_o (g/kg): cyclohexane ($\approx$ 500), petroleum ether (<10), xylene (100); S_w: 22.8 mg/L at 23 °C; vp: 1.8 x 10^{-6} mmHg at 20 °C.

Transformation Products

Soil: Probably degrades via ring hydroxylation and subsequently ring cleavage. Persistence in soil is limited to approximately 6-8 wk (Hartley and Kidd, 1987). Under laboratory conditions, the half-lives in soil were 30, 48, and 59 d at pH 4.85, 5.2, and 10.8, respectively (Worthing and Hance, 1991).

Formulation Types: Emulsifiable concentrate.

Toxicity: LC$_{50}$ for guppy >100 mg/L (Worthing and Hance, 1991); acute oral LD$_{50}$ for rats 2,600 mg/kg (RTECS, 1985).

Uses: Preemergence and postemergence control of broadleaf weeds in parsley, carrots dill, and celery.

MONOCROTOPHOS

Synonyms: Apadrin; Azodrin; Azodrin insecticide; Biloborb; Bilobran; C 1414; CIBA 1414; Crisodrin; 3-(Dimethoxyphosphinyloxy)-*N*-methylisocrotonamide; Dimethyl-1-methyl-2-(methylcarbamoyl)vinyl phosphate; *O,O*-Dimethyl-*O*-(2-*N*-methylcarbamoyl-1-methylvinyl) phosphate; (*E*)-Dimethyl 1-methyl-3-(meth-ylamino)-3-oxo-1-propenyl phosphate; Dimethyl (*E*)-1-methyl-2-(methylcar-bamoyl)vinyl phosphate; Dimethyl phosphate ester of 3-hydroxy-*N*-methyl-*cis*-crotonamide; ENT 27129; Hazodrin; 3-Hydroxy-*N*-methyl-*cis*-crotonamide dimethyl phosphate; *cis*-1-Methyl-2-(methylcarbamoyl) vinyl phosphate; Monocil 40; Monocron; Nuvacron; Pandar; **Phosphoric acid dimethyl (1-methyl-3-(methylamino)-3-oxo-1-propenyl) ester;** Pillardrin; Plantdrin; SD 9129; Shell SD 9129; Susvin.

Structure:

$$CH_3O \diagdown \overset{\displaystyle \overset{O}{\|}}{P} - O \diagdown$$
$$CH_3O \diagup \qquad \qquad C = C \diagup H$$
$$CH_3 \qquad CONHCH_3$$

Designations: CAS: 6923-22-4; mf: $C_7H_{14}NO_5P$; fw: 223.16; RTECS: TC4375000.

Properties: Colorless to reddish-brown, hygroscopic crystals with a mild ester-like odor. Technical grade is a reddish-brown to dark brown semi-solid. Mp: 54–55 °C (pure), 25–35 °C (technical); bp: 125 °C at 5 x 10^{-4} mmHg; ρ: 1.33 at 20/4 °C; H-$t_{1/2}$: 96 d (pH 5), 66 d (pH 7), 17 d (pH 9); K_d: 0.077–0.615; log K_{ow}: -1.97 (calculated); S_o (g/kg): acetone (700), methanol (1,000), methylene chloride (800), *n*-octanol (60), toluene (60); S_w: miscible; vp: 6.75 x 10^{-5} mmHg at 20 °C.

Transformation Products
Plant: Decomposes in plants forming the *N*-hydroxy compound in small amounts (Hartley and Kidd, 1987).

Chemical/Physical: Emits toxic fumes of phosphorus and nitrogen oxides when heated to decomposition (Sax and Lewis, 1987; Lewis, 1990).

Exposure Limits: OSHA PEL: TWA 0.25 mg/m^3; ACGIH TLV: TWA 0.25 mg/m^3.

Formulation Types: Soluble concentrate; granules.

Toxicity: LC$_{50}$ (24 hr) for bluegill sunfish 23 mg/L and rainbow trout 12 mg/L (Worthing and Hance, 1991); LC$_{50}$ (48 hr) for rainbow trout 7 mg/L (Hartley and

Kidd, 1987); acute oral LD_{50} for male and female rats is 18 and 20 mg/kg, respectively (Hartley and Kidd, 1987), 8 mg/kg (RTECS, 1985).

Uses: Systemic insecticide and acaricide used to control pests in cotton, sugarcane, coffee, tobacco, olives, rice, hops, sorghum, maize, deciduous fruits, citrus fruits, potatoes, sugarbeet, tomatoes, soya beans, and ornamentals.

MONURON

Synonyms: Chlorfenidim; 1-(4-Chlorophenyl)-3,3-dimethylurea; 1-(*p*-Chlorophenyl)-3,3-dimethylurea; 3-(4-Chlorophenyl)-1,1-dimethylurea; 3-(*p*-Chlorophenyl)-1,1-dimethylurea; *N'*-**(4-Chlorophenyl)**-*N,N*-**dimethylurea**; *N*-(4-Chlorophenyl)-*N',N'*-dimethylurea; *N'*-(*p*-Chlorophenyl)-*N,N*-dimethylurea; CMU; *N,N*-Dimethyl-*N'*-(4-chlorophenyl)urea; *N,N*-Dimethyl-*N'*-(*p*-chlorophenyl)urea; Karmex; Lirobetarex; Monurex; Monurox; Monuuron; NCI-C02846; Rosuran; Telvar; Telvar Monuron Weed Killer; Urox; USAF P-8; USAF XR-41.

Structure:

$$NHCON(CH_3)_2$$

Cl

Designations: CAS: 150-68-5; mf: $C_9H_{11}ClN_2O$; fw: 198.66; RTECS: YS6300000.

Properties: Colorless to white, odorless crystals. Mp: 174-175 °C; bp: 185-200 °C (decomposes); ρ: 1.27 at 20/4 °C; pK_a: unknown but saturated aqueous solution has a pH of 6.26; K_H: 3.0 x 10^{-8} atm·m^3/mol at 20-25 °C (approximate - calculated from water solubility and vapor pressure); log K_{oc}: 1.99, 2.33; log K_{ow}: 1.46, 2.12; S_o (g/kg at 27 °C): acetone (52), benzene (3); S_w: 230 mg/L at 25 °C (pH = 6.26); vp: 4.5 x 10^{-7} mmHg at 20 °C.

Soil properties and adsorption data

Soil	K_d (mL/g)	f_{oc} (%)	K_{oc} (mL/g)	pH
Cecil loamy sand	0.40	0.41	98	5.8
Keyport silt loam	2.60	1.22	213	5.4

Source: Rhodes et al., 1970.

Transformation Products

Biological: Monuron was mineralized in sewage samples obtained from a water treatment plant in Ithica, NY. (4-Chlorophenyl)urea and 4-chloroaniline were tentatively identified as metabolites (Wang et al., 1985).

Soil/Plant: In soils and plants, monuron is demethylated at the terminal nitrogen atom coupled with ring hydroxylation forming 3-(2-hydroxy-4-chlorophenyl)urea and 3-(3-hydroxy-4-chlorophenyl)urea (Hartley and Kidd, 1987). Wallnöefer et al.

(1973) reported that the soil microorganism *Rhizopus japonicus* degraded monuron 3-(*p*-chlorophenyl)-1-methylurea. However, in the presence of *Pseudomonas* or *Arthrobacter* sp., monuron degraded to 2,4-dichloroaniline, *sym*-bis(3,4-dichlorophenyl)urea and unidentified metabolites (Janko et al., 1970). The reported half-life in soil is 166 d (Jury et al., 1987).

Photolytic: When an aqueous solution of monuron was exposed to sunlight or simulated sunlight, the major degradative pathways were observed to be photooxidation and demethylation of the *N*-methyl groups (Crosby and Tang, 1969; Tanaka et al., 1982a), hydroxylation of the aromatic ring and polymerization. Products identified by TLC and confirmed using IR, mass spectra, and/or chromatographic methods were 3-(*p*-chlorophenyl)-1-formyl-1-methylurea, 1-(*p*-chlorophenyl)-3-methylurea, 3-(4-chloro-2-hydroxyphenyl)-1,1-dimethylurea, 4,4'-dichlorocarbanilide, and the tentatively identified compounds *p'*-chloroformanilide, 1-(*p*-chlorophenyl)-3-formylurea, *p*-chloroaniline (Crosby and Tang, 1969), formaldehyde, formic acid, and carbon dioxide (Tanaka et al., 1982a). A similar study was performed by Rosen et al. (1969). An aqueous solution of monuron was exposed to a late summer sun for 17 d. The solution contained 3-(4-hydroxyphenyl)-1,1-dimethylurea and other unidentified compounds (Rosen et al., 1969). Tanaka et al. (1977) also studied the simulated sunlight photolysis of a saturated aqueous solution (200 ppm) of monuron. The main degradative processes were ring hydroxylation, methyl oxidation, *N*-demethylation, dechlorination, and dimerization. Photoproducts identified by TLC were 3-(4-chlorophenyl)-1-methylurea, 3-(4-chlorophenyl)-1-formyl-1-methylurea, 3-(4-chloro-2-hydroxyphenyl)-1,1-dimethylurea, 3-(4-hydroxyphenyl)-1,1-dimethylurea, 3-(4-hydroxyphenyl)-1-formyl-1-methylurea, 4,4'-dichlorocarbanilide, 3-{4-[*N*-(*N'*,*N'*-dimethylaminocarbonyl)-4'-chloroanilino]phenyl}-1,1'-dimethylurea (dimer), a monodemethylated dimer, a hydroxylated dimer, a dihydroxylated dimer, and a trimer (Tanaka et al., 1977). The rate of photolysis of monuron in aqueous solutions increased with the addition of nonionic surfactants (Tergitol TMN-6, Tergitol TMN-10, Triton X-100, Triton X-405) at concentrations in excess of the critical micelle concentration (Tanaka et al., 1979). The surfactants eliminated ring hydroxylation reactions but enhanced reductive dechlorination reactions with simultaneous formation of biphenyls. Photoproducts included 3-(4-chlorophenyl)-1,1-dimethylurea, 3-(4-chlorophenyl)-1-methylurea (monomethyl monuron), 3-phenyl-1,1-dimethylurea, 3-phenyl-1-methylurea, 3-{4-[*N*-(*N'*,*N'*-dimethylaminocarbonyl)-4'-chloroanilino]phenyl}-1,1-dimethylurea (monuron dimer), 3-{4-[*N*-(*N'*,*N'*-dimethylaminocarbonyl)anilino]phenyl}-1,1'-dimethylurea (fenuron dimer), formaldehyde, and a polymeric material (Tanaka et al., 1979).

Tanaka et al. (1981) studied the photolysis of monuron in dilute aqueous solutions in order to fully characterize a substituted diphenylamine that was observed in an earlier investigation (Tanaka et al., 1977). They identified this compound as an isomeric mixture containing 92% 2-chloro-4',5-bis(N',N'-dimethylureido)biphenyl and 8% 5-chloro-2,4'-bis(N',N'-dimethylureido)biphenyl (Tanaka et al., 1981). Tanaka et al. (1982) undertook a study to identify the several biphenyls formed in earlier photolysis studies (Tanaka et al., 1979, 1981). They identified these compounds as 2,4'-, 3,4'-, and 4,4'-bis(N',N'-dimethylureido)biphenyls (fenuron biphenyls) (Tanaka et al., 1982, 1984). When an aqueous solution containing a surfactant was irradiated by UV light, photodegradation was the major degradative pathway. The compounds isolated were fenuron (3-phenyl-1,1-dimethylurea) (Tanaka et al., 1984) and the three fenuron biphenyls (Tanaka et al., 1982, 1984). In a more recent study, Tanaka et al. (1985) studied the photolysis of linuron (175 mg/L) in aqueous solution using UV light (λ = 300 nm) or sunlight. After 15 d of exposure to sunlight, monuron degraded forming a monochlorinated biphenyl product (1.5% yield) with the concomitant loss of hydrogen chloride (Tanaka et al., 1985). When a methanolic solution containing monuron was irradiated by UV light (λ = 253.7 nm) under aerobic conditions, 3-phenyl-1,1-dimethylurea formed as the major product. Methyl-p-chlorophenylcarbamate also formed but in small quantities (Mazzochi and Rao, 1972).

Chemical/Physical: Hydrolyzes in acidic media forming 4-chloroaniline (Hartley and Kidd, 1987). Under alkaline conditions (0.50 N sodium hydroxide) at 20 °C, the half-life of monuron was 177 d (El-Dib and Aly, 1976). Emits toxic fumes of nitrogen oxides and chlorine when heated to decomposition (Sax and Lewis, 1987).

Symptoms of Exposure: Causes anemia and methemoglobinemia in experimental animals.

Formulation Types: Granules; wettable powder; oil-miscible liquid.

Toxicity: LC_{50} (96 hr) for rainbow trout 76 mg/L; LC_{50} (48 hr) for coho salmon 110 mg/L; acute oral LD_{50} for rats 3,600 mg/kg (Hartley and Kidd, 1987), 1,053 mg/kg (RTECS, 1985).

Uses: Herbicide; sugarcane flowering suppressant.

NALED

Synonyms: Arthodibrom; Bromchlophos; Bromex; Dibrom; 1,2-Dibromo-2,2-di-chloroethyldimethyl phosphate; Dimethyl 1,2-dibromo-2,2-dichloroethyl phosphate; *O,O*-Dimethyl-*O*-(1,2-dibromo-2,2-dichloroethyl)phosphate; *O,O*-Dimethyl *O*-(2,2-dichloro-1,2-dibromoethyl)phosphate; ENT 24988; Hibrom; NA 2783; Ortho 4355; Orthodibrom; Orthodibromo; **Phosphoric acid 1,2-dibromo-2,2-dichloroethyl dimethyl ester;** RE 4355.

Structure:

Designations: CAS: 300-76-5; DOT: 2783; mf: $C_4H_7Br_2Cl_2O_4P$; fw: 380.79; RTECS: TB9450000.

Properties: Colorless to pale yellow liquid or solid with a slight pungent odor. Mp: 26.5-27.5 °C; bp: 110 °C at 0.5 mmHg; ρ: 1.96 at 25/4 °C; S_o: freely soluble in ketone, alcohols, aromatic, and chlorinated hydrocarbons but sparingly soluble in petroleum solvents and mineral oils; S_w: 10 mg/L; vp: 2 x 10^{-3} at 20 °C.

Transformation Products
Chemical/Physical: Completely hydrolyzed in water within 2 d (Windholz et al., 1983). In the presence of metals or reducing agents, naled loses bromine forming dichlorvos (Hartley and Kidd, 1987). Emits toxic fumes of bromines, chlorides, and phosphorus oxides when heated to decomposition (Lewis, 1990).

Exposure Limits: NIOSH REL: IDLH 1,800 mg/m³; OSHA PEL: TWA 3 mg/m³; ACGIH TLV: TWA 3 mg/m³.

Formulation Types: Dustable powder; emulsifiable concentrate.

Toxicity: LC_{50} for bluegill sunfish 180 µg/L, rainbow trout 132 µg/L (Verschueren, 1983); LC_{50} (24 hr) for goldfish 2-4 mg/L (Hartley and Kidd, 1987) and crabs 0.33 mg/L (Worthing and Hance, 1991); acute oral LD_{50} for rats 430 mg/kg (Hartley and Kidd, 1987), 250 mg/kg (RTECS, 1985).

Uses: Insecticide used for control of spider mites, sucking and chewing insects in fruits, vegetables, and ornamentals.

NAPROPAMIDE

Synonyms: Devrinol; Devrinol 2EC; Devrinol 10G; Devrinol 50WP; *N,N*-**Diethyl-2-(1-naphthalenyloxy)propionamide**; (*RS*)-*N,N*-Diethyl-2-(1-naphthyloxy)-propionamide; 2-(α-Napthoxy)-*N,N*-diethylpropionamide; R 7465; R 7475.

Structure:

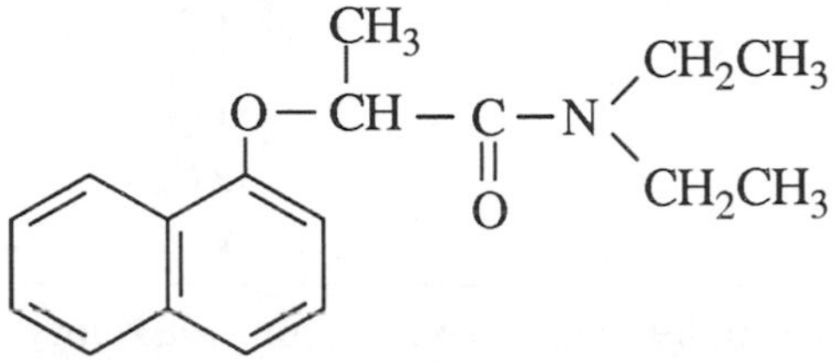

Designations: CAS: 15299-99-7; mf: $C_{17}H_{21}NO_2$; fw: 271.36; RTECS: UE3600000.

Properties: Colorless crystals (pure); brown solid (technical). Mp: 74.8-77.5 °C (pure), 68-70 °C (technical); fl p: 191 °C (open cup); H-t$_{1/2}$: 25.7 min (aqueous solution exposed to artificial sunlight); K_H: 2.9 x 10^{-8} atm·m^3/mol at 20 °C (approximate - calculated from water solubility and vapor pressure); log K_{oc}: 2.83; log K_{ow}: 3.36; S_o (g/L at 20 °C): acetone (>1,000), ethanol (>1,000), *N*-hexane (15), kerosene (62), xylene (505); S_w: 73 mg/L at 20 °C; vp: 4 x 10^{-6} mmHg at 20 °C.

Soil properties and adsorption data

Soil	K_d (mL/g)	f_{oc} (%)	K_{oc} (mL/g)	pH
Bet Dagan I	1.40	0.40	350	7.9
Bet Dagan II	2.96	1.01	293	7.8
Dead Sea sediment	--	2.13	9,769	--
Dead Sea sediment	--	2.13	5,231	2.0
Dead Sea sediment	--	2.13	2,242	6.0
Gilat	1.92	0.55	349	7.8
Kinneret F sediment	--	5.05	414	--
Kinneret F sediment	--	5.05	3,437	2.0
Kinneret F sediment	--	5.05	2,493	6.0
Kinneret G sediment	--	2.84	517	--
Kinneret G sediment	--	2.84	4,121	2.0
Kinneret G sediment	--	2.84	1,883	6.0
Malkiya soil	--	2.21	353	--
Malkiya soil	--	2.21	1,411	3.0
Malkiya soil	--	2.21	760	6.0

continued

Soil	K_d (mL/g)	f_{oc} (%)	K_{oc} (mL/g)	pH
Mivtachim	0.27	0.06	450	8.5
Neve Ya'ar soil	2.94	1.18	249	7.7
Neve Ya'ar soil	--	1.68	196	--
Neve Ya'ar soil	--	1.68	1,819	3.0
Neve Ya'ar soil	--	1.68	760	6.0
Oxford soil	--	9.61	246	--
Tujunga loamy sand	2.01	0.58	346	--

Source: Gerstl and Kliger, 1990; Gerstl and Yaron, 1983; Jury et al., 1986.

Transformation Products
Soil: Degrades slowly in soil to 1-naphthol, 1,4-naphthoquinone, 2-(α-naphth-oxy)-*N*-ethylpropionamide, and 2-(α-naphthoxy)propionamide (Hartley and Kidd, 1987). In moist loam or sandy-loam soils at 70-90 °C, the half-life was 8-12 wk. However, the persistence may be as long as 9 mo under conditions where microbial growth is limited (Ashton and Monaco, 1991).

Plant: Rapidly metabolized in tomatoes and several fruit trees forming the water-soluble hexose conjugates of 4-hydroxynapropamide (Humburg et al., 1989; Ashton and Monaco, 1991).

Photolytic: Under laboratory conditions, irradiation of an aerated aqueous solution with UV light gave 2-hydroxypropananilide, 2-propenanilide, and pyruvinanilide as the major products. Minor photoproducts include 2-(2-naphthoxy)propanoic acid and aniline (Tsao and Eto, 1990). In a deaerated aqueous solution, the photoproducts identified were 2-hydroxypropananilide, 2'-amino-2-hydroxy-propiophenone, 2'-amino-2-naphthoxypropiophenone, and 4'-amino-2-napthoxy-proiophenone (Tsao and Eto, 1990).

Formulation Types: Emulsifiable concentrate (2 lb/gal); granules (10%); wettable powder (50%); suspension concentrate.

Toxicity: LC_{50} (96 hr) for bluegill sunfish 30 mg/L, goldfish >10 mg/L, and rainbow trout 16.6 mg/L (Hartley and Kidd, 1987); acute oral LD_{50} for rats is >5,000 mg/kg (RTECS, 1985).

Uses: Preemergence control of annual grasses and certain broadleaf weeds in mint, orchards, small fruits, tobacco, turf, vegetable crops, and ornamentals.

NAPTALAM

Synonyms: ACP 322; Alanap; Alanap L; Alanape; Alanap 10G AT; Dyanap; Morcran; **2-((1-Naphthalenylamino)carbonyl)benzoic acid;** α-Naphthylphthalamic acid; *N*-1-Naphthylphthalamic acid; Naptalame; Nip-A-Thin; NPA; PA; Peach-Thin; 6Q8; Solo.

Structure:

Designations: CAS: 132-66-1; mf: $C_{18}H_{13}NO_3$; fw: 291.29; RTECS: TH7350000.

Properties: Purple crystals with an unpleasant odor. Mp: 203 °C (pure), 175–185 °C (technical); ρ: 1.362 at 20/4 °C (acid), 1.386 (sodium salt); fl p: nonflammable (water formulation); K_H: <1.9 x 10^{-3} atm·m^3/mol at 20 °C (approximate - calculated from water solubility and vapor pressure); log K_{oc}: 2.37 (calculated); log K_{ow}: 2.04 (calculated); S_o (g/kg at 25 °C): acetone (5), 2-butanone (4), carbon tetrachloride (0.1), *N,N*-dimethylformamide (39), dimethyl sulfoxide (32), methyl ethyl ketone (3.7), 2-propanol (2); S_w: 200 mg/L at 20 °C; vp: <1 mmHg at 20 °C.

Transformation Products
Soil/Plant: Degrades in soils and plants forming 1-naphthylamine and phthalic acid (Hartley and Kidd, 1987; Humburg et al., 1989). Residual activity in soil is limited to approximately 3–4 mo (Hartley and Kidd, 1987).

Chemical/Physical: Forms *N*-(1-naphthyl)phthalimide at elevated temperatures (Worthing and Hance, 1991). Naptalam will precipitate as the free acid in very acidic waters or in extremely hard waters (Humburg et al., 1989).

Formulation Types: Soluble concentrate (sodium salt = 2 lb/gal); granules (sodium salt = 10%); wettable powder.

Toxicity: LC$_{50}$ (96 hr) for bluegill sunfish 354 mg/L and rainbow trout 76 mg/L (Hartley and Kidd, 1987); acute oral LD$_{50}$ of the free acid and sodium salt for rats is >8,200 and 1,800 mg/kg, respectively (Ashton and Monaco, 1991).

Uses: Selective preemergence herbicide used to some grasses and many broadleaf weeds in soybeans, cucurbits, asparagus, groundnuts, potatoes, and established woody ornamentals.

NEBURON

Synonyms: 1-Butyl-3-(3,4-dichlorophenyl)-1-methylurea; *N*-Butyl-*N'*-(3,4-di-chlorophenyl)-*N*-methylurea; Granurex; Kloben; Neburea; Neburex.

Structure:

$$Cl \text{—} \bigcirc \text{—} NHCON \begin{array}{c} CH_3 \\ (CH_2)_3CH_3 \end{array} \quad (Cl)$$

Designations: CAS: 555-37-3; mf: $C_{12}H_{16}Cl_2N_2O$; fw: 275.18; RTECS: YS3810000.

Properties: Colorless crystals. Mp: 101.5-103 °C; log K_{oc}: 3.49; log K_{ow}: 3.80 (calculated); S_o: sparingly soluble in hydrocarbons; S_w: 4.8 mg/L at 24 °C; vp: negligible at room temperature.

Soil properties and adsorption data

Soil	K_d (mL/g)	f_{oc} (%)	K_{oc} (mL/g)	pH	CEC (meq/100 g)
Great House sandy loam	325	12.00	2,708	6.3	18.0
Rosemaunde sandy clay loam	58	1.76	3,296	6.7	14.0
Toll Farm HP	240	11.70	2,051	7.4	41.0
Transcoed silty clay loam	139	3.69	3,767	6.2	12.0
Weed Res. sandy loam	72	1.93	3,731	7.1	11.0

Source: Hance, 1965.

Transformation Products
Soil: In soil, undergoes dealkylation of the terminal nitrogen atom, ring hydroxylation, and degradation to dichlorodihydroxyaniline (Hartley and Kidd, 1987). Residual activity in soil is limited to approximately 3-4 mo (Hartley and Kidd, 1987).

Chemical/Physical: Neburon hydrolyzes under basic conditions forming aniline derivatives. The half-life of neburon in 0.50 N sodium hydroxide at 20 °C is 167 d (El-Dib and Aly, 1976).

Formulation Types: Wettable powder.

Toxicity: LC_{90} (96 hr) for four fish species is 600-900 µg/L (Hartley and Kidd,

1987); acute oral LD$_{50}$ for rats >11,000 mg/kg (Hartley and Kidd, 1987), 1,100 mg/kg (RTECS, 1985).

Uses: Preemergence control of grasses and broadleaf weeds in peas, beans, lucerne, garlic, beet, cereals, beet, strawberries, ornamentals, and forestry.

NITRAPYRIN

Synonyms: 2-Chloro-6-(trichloromethyl)pyridine; Dowco-163; N-serve; N-serve nitrogen stabilizer.

Structure:

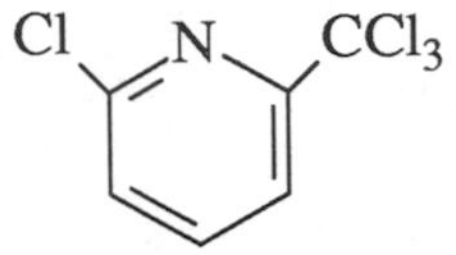

Designations: CAS: 1929-82-4; mf: $C_6H_3Cl_4N$; fw: 230.90; RTECS: US7525000.

Properties: Colorless, crystalline solid. Mp: 62-63 °C; ρ: 1.744 at 20/4 °C (estimated); K_H: 2.13 x 10^{-3} atm·m^3/mol at 22-25 °C (approximate - calculated from water solubility and vapor pressure); log K_{oc}: 2.62-2.68; log K_{ow}: 3.02-3.41; S_o (kg/kg at 20-26 °C): acetone (1.98), ethanol (0.29), methylene chloride (1.85), toluene (1.39), 1,1,1-trichloroethane (0.80), xylene (1.04); S_w: 40 mg/kg at 22 °C; vp: 2.8 x 10^{-3} mmHg at 20 °C.

Soil properties and adsorption data

Soil	K_d (mL/g)	f_{oc} (%)	K_{oc} (mL/g)	pH
Batcombe silt loam	--	0.63-1.46	172	6.7-7.5
British Columbia silty clay loam	4.59	1.97	233	7.6
California clay	2.01	1.04	193	7.7
California loam	8.11	1.80	451	7.8
California loam	44.45	18.68	238	5.9
California sandy loam	1.46	0.46	317	7.3
California sandy loam	0.32	0.17	188	7.5
California silt loam	20.21	6.21	325	5.3
Catlin	9.21	2.01	460	6.2
Commerce	3.01	0.68	440	6.7
Minnesota loam	10.55	7.19	147	8.1
Minnesota loam	9.11	3.25	311	7.8
Texas clay	4.89	1.57	311	6.8
Tracy	5.31	1.12	474	6.2

Source: Briggs, 1981; Goring, 1962; McCall et al., 1981.

Transformation Products
Soil: Hydrolyzes in soil to 6-chloropyridine-2-carboxylic acid (Worthing and

302

Hance, 1991). 6-Chloropicolinic acid and carbon dioxide were reported as biodegradation products (Verschueren, 1983).

Photolytic: Photolysis of nitrapyrin in water yielded 6-chloropicolinic acid, 6-hydroxypicolinic acid, and an unidentified polar material (Verschueren, 1983).

Chemical/Physical: Emits toxic fumes of nitrogen oxides and chlorides when heated to decomposition (Sax and Lewis, 1987; Lewis, 1990).

Exposure Limits: ACGIH TLV: TWA 10 mg/m^3.

Toxicity: LC$_{50}$ for channel catfish 5.8 mg/L (Worthing and Hance, 1991); acute oral LD$_{50}$ for rats 1,230 mg/kg (Verschueren, 1983), 940 mg/kg (RTECS, 1985).

Uses: Bactericide used to inhibit *Nitrosomonas* spp. from oxidizing ammonium ions in soil.

OXADIAZON

Synonyms: 2-*tert*-Butyl-4-(2-2,4-dichloro-5-isopropyloxyphenyl)-1,3,4-oxa-diazolin-5-one; Chipco Ronstar G; Chipco Ronstar 50WP; 3-(2,4-Dichloro-5-(1-methylethoxy)phenyl)-5-(1,1-dimethylethyl)-1,3,4-oxadiazol-2(3*H*)-one; Ronstar; Ronstar 25EC; Ronstar 2G; Ronstar 12L; RP 17623.

Structure:

Designations: CAS: 19666-30-9; mf: $C_{15}H_{18}Cl_2N_2O_3$; fw: 345.22; RTECS: RO0874000.

Properties: Odorless, white crystals. Mp: 88-90 °C; fl p: nonflammable; K_H: <6.5 x 10^{-10} atm·m^3/mol at 20 °C (approximate - calculated from water solubility and vapor pressure); log K_{oc}: 0.70-2.99; log K_{ow}: 4.70; S_o (g/L at 20 °C): acetone (600), benzene (1,000), cyclohexanone (200), ethanol (100); S_w: 700 mg/L at 20 °C; vp: <10^{-6} mmHg at 20 °C.

Soil properties and adsorption data

Soil	K_d (mL/g)	f_{oc} (%)	K_{oc} (mL/g)	pH
Batcombe silt loam	--	0.63-1.46	5	6.7-7.5
Kanuma high clay	5.2	1.35	387	5.7
Tsukuba clay loam	41.3	4.24	973	6.5

Source: Briggs, 1981; Kanazawa, 1989.

Transformation Products
Soil: The reported half-life in soil is approximately 3-6 mo (Hartley and Kidd, 1987). Oxadiazon degraded slowly in both moist and flooded soils. After 25 wk, only 0.1-3.5% degraded to carbon dioxide and 0.5-1.1% as volatile products. Metabolites identified included oxadiazon-phenol, oxadiazon acid, and methoxy-oxadiazon (Ambrosi et al., 1977).

Symptoms of Exposure: May be irritating to skin.

Formulation Types: Wettable powder (50%); granules (5%).

Toxicity: LC_{50} (96 hr) for carp 1.76 mg/L and channel catfish $\geq$15.4 mg/L (Worthing and Hance, 1991); acute oral LD_{50} of technical oxadiazon for rats is >8,000 mg/kg (Ashton and Monaco, 1991), 3,500 mg/kg (RTECS, 1985).

Uses: Preemergence herbicide used for controlling certain annual grasses (e.g., bluegrass, barnyardgrass, crabgrass, goosegrass, sprangletop) and broadleaf weeds (e.g., cudweed, dayflower, filaree, groundsel, jimsonweed, morning-glory, mustards, pigweed, redmaids, smartweed, sowthistle, velvetleaf) in turf, lawns, orchards, and ornamentals.

OXAMYL

Synonyms: D-1410; 2-(Dimethylamino)-N-(((methylamino)carbonyl)oxy)-2-oxo-ethanimidothioic acid methyl ester; 2-Dimethylamino-1-(methylthio)glyoxal O-methylcarbamoylmonoxime; N,N-Dimethyl-α-methylcarbamoyloxyimino-α-(methylthio)acetamide; N',N'-Dimethyl-N-((methylcarbamoyl)oxy)-1-thiooxam-imidic acid methyl ester; DPX 1410; Insecticide-nematocide 1410; **Methyl 2-(dimethylamino)-N-(((methylamino)carbonyl)oxy)-2-oxoethanimidothioate;** Methyl-1-(dimethylcarbamoyl)-N-((methylcarbamoyl)oxy)thioformimidate; S-Methyl 1-(dimethylcarbamoyl)-N-((methylcarbamoyl)oxy)thioformimidate; Methyl-N',N'-dimethyl-N-((methylcarbamoyl)oxy)-1-thiooxamimidate; Thiozamyl; Vydate; Vydate L insecticide/nematocide; Vydate L oxamyl insecticide/nematocide.

Structure:

$$(CH_3)_2N\overset{\overset{\displaystyle O}{\|}}{C} - \underset{\underset{\displaystyle SCH_3}{|}}{C}=N-O\overset{\overset{\displaystyle O}{\|}}{C}NHCH_3$$

Designations: CAS: 23135-22-0; mf: $C_7H_{13}N_3O_3S$; fw: 219.25; RTECS: RP2300000.

Properties: Crystalline solid with a sulfur-like odor. Mp: 100-102 °C which changes to a dimorphic form melting at 108-110 °C; bp: decomposes; ρ: 0.97 at 25/4 °C; K_H: 2.6 x 10^{-6} atm·m^3/mol at 20-25 °C (approximate - calculated from water solubility and vapor pressure); log K_{oc}: -0.70 to 1.40; log K_{ow}: -0.4; P-t$_{1/2}$: 55.38 hr (absorbance λ = 223.0 nm, concentration on glass plates = 6.7 μg/cm^2); S_o (g/L at 25 °C): acetone (670), ethanol (330), methanol (1,440), 2-propanol (110), toluene (10); S_w: 280 g/L at 25 °C; vp: 2.33 x 10^{-4} mmHg at 20 °C.

Soil properties and adsorption data

Soil	K_d (mL/g)	f_{oc} (%)	K_{oc} (mL/g)	pH
Arredondo sand	0.06	0.80	7.5	6.8
Cecil sandy loam	0.05	0.90	5.6	5.6
Cu-montmorillonite	11.50	--	--	--
Devizes	0.12	4.47	2.7	7.8
Mepal	0.62	11.60	5.3	7.2
Pitstone	0.17	2.18	7.8	8.0
Sutton Veany	0.35	9.05	3.9	7.6
Webster silty clay loam	0.40	3.97	10.1	7.3

Soil	K_d (mL/g)	f_{oc} (%)	K_{oc} (mL/g)	pH
Woburn sandy loam	0.02	0.78	2.6	7.0
Woburn sandy loam	0.30	3.43	87.5	6.3
Zn-montmorillonite	6.70	--	--	--

Source: Bilkert and Rao, 1985; Bromilow et al., 1980; Khan and Bansal, 1980.

Transformation Products
Soil: Oxamyl rapidly degraded in a loamy sand and fine sand soil at 25 °C to carbon dioxide and the intermediate methyl N-hydroxy-N,N-dimethyl-1-thio-oxaminidate (Rajagopol et al., 1984). The reported half-life in soil is approximately 1 wk (Worthing and Hance, 1991). Ou and Rao (1986) reported a half-life in soil of 8-50 d. The reported half-lives of oxamyl in Pitstone, Devizes, Sutton Veany, and Mepal soils at 15 °C were reported to be 10.2-13.1, 6.2, 7.1, and 17.8 d, respectively (Bromilow et al., 1980).

Chemical/Physical: Emits toxic fumes of nitrogen and sulfur oxides when heated to decomposition (Sax and Lewis, 1987).

Formulation Types: Soluble concentrate; granules.

Toxicity: LC_{50} (96 hr) for rainbow trout 4.2 mg/L, bluegill sunfish 5.6 mg/L, and goldfish 27.5 mg/L (Hartley and Kidd, 1987); acute oral LD_{50} for rats 5.4 mg/kg (Hartley and Kidd, 1987), 2,500 μg/kg (RTECS, 1985).

Uses: Insecticide, nematocide, acaricide.

OXYDEMETON-METHYL

Synonyms: Bay 21097; Bayer 21097; Demeton-methyl sulfoxide; Demeton-*O*-methyl sulfoxide; Demeton-*S*-methyl sulfoxide; *O,O*-Dimethyl *S*-(2-eththionyl-ethyl) phosphorothioate; Dimethyl *S*-(2-eththionylethyl) thiophosphate; *O,O*-Dimethyl *S*-2-(ethylsulfinyl)ethyl phosphorothioate; *O,O*-Dimethyl *S*-2-(ethyl-sulfinyl)ethyl thiophosphate; *O,O*-Dimethyl *S*-ethylsulphinylethyl phosphoro-thioate; ENT 24964; **S-2-(Ethylsulfinyl)ethyl *O,O*-dimethyl phosphorothioate**; Isomethylsystox sulfoxide; Metaisosystox sulfoxide; Metasystemox; Metasystox-R; Methyl demeton-*O*-sulfoxide; Metilmercaptofosoksid; Oxydemetonmethyl; Phos-phothioic acid *O,O*-dimethyl *S*-2-(ethylsulfinyl)ethyl ester; R 2170.

Structure:

$$CH_3CH_2SCH_2CH_2S \overset{\overset{\displaystyle O}{\|}}{P}(OCH_3)_2$$

Designations: CAS: 301-12-2; mf: $C_6H_{15}O_4PS_2$; fw: 246.29; RTECS: TG1420000.

Properties: Clear, amber-colored liquid. Mp: <-10 °C; bp: >80 °C (decomposes); ρ: 1.289 at 20/4 °C; log K_{oc}: 0.70-1.49; log K_{ow}: -1.97 (calculated); S_o: miscible with many aromatic and halogenated solvents; S_w: miscible; S_w: miscible; vp: 2.85 x 10^{-5} mmHg at 20 °C.

Soil properties and adsorption data

Soil	K_d (mL/g)	f_{oc} (%)	K_{oc} (mL/g)	pH
Sand	0.10	2.15	5	6.9
Sandy loam	0.54	1.74	31	5.5
Silty loam	0.19	1.22	16	6.7

Source: U.S. Department of Agriculture, 1990.

Transformation Products

Plant: In asparagus, oxydemeton-methyl was converted to the corresponding sulfone (Szeto and Brown, 1982).

Soil: The sulfoxide group is oxidized to the sulfone and oxidative and hydrolytic cleavage of the side chain gives dimethylphosphoric and phosphoric acids (Hartley and Kidd, 1987).

Chemical/Physical: Oxydemeton-methyl can be converted to the corresponding sulfone by hydrogen peroxide (Cremlyn, 1991). Emits toxic fumes of phosphorus and sulfur oxides when heated to decomposition (Sax and Lewis, 1987).

Formulation Types: Soluble concentrate; emulsifiable concentrate.

Toxicity: LC_{50} (96 hr) for rainbow trout 4.0 mg/L (Walker, 1964); LC_{50} (24 hr) for rainbow trout and bluegill sunfish 10 mg/L (Hartley and Kidd, 1987); acute oral LD_{50} for rats 65-75 mg/kg (Hartley and Kidd, 1987), 30 mg/kg (RTECS, 1985).

Uses: Systemic and contact insecticide and acaricide used to control spider mites and other insects on vegetables and some ornamentals.

PARATHION

Synonyms: AAT; AATP; AC 3422; ACC 3422; Alkron; Alleron; American Cyanamid 3,422; Aphamite; B 404; Bay E-605; Bladan; Bladan F; Compound 3422; Corothion; Corthion; Corthione; Danthion; DDP; *O,O*-Diethyl-*O*-4-nitrophenyl phosphorothioate; *O,O*-Diethyl *O*-*p*-nitrophenyl phosphorothioate; Diethyl-4-nitrophenyl phosphorothionate; Diethyl-*p*-nitrophenyl thionophosphate; *O,O*-Diethyl-*O*-4-nitrophenyl thionophosphate; *O,O*-Diethyl-*O*-*p*-nitrophenyl thionophosphate; Diethyl-*p*-nitrophenyl thiophosphate; *O,O*-Diethyl-*O*-*p*-nitrophenyl thiophosphate; Diethylparathion; DNTP; DPP; Drexel parathion 8E; E 605; Ecatox; Ekatin WF & WF ULV; Ekatox; ENT 15108; Ethlon; Ethyl parathion; Etilon; Folidol; Folidol E605; Folidol E & E 605; Fosfermo; Fosferno; Fosfex; Fosfive; Fosova; Fostern; Fostox; Gearphos; Genithion; Kolphos; Kypthion; Lethalaire G 54; Lirothion; Murfos; NA 2783; NCI-C00226; Niran; Niran E-4; Nitrostigmine; Nitrostygmine; Niuif-100; Nourithion; Oleofos 20; Oleoparaphene; Oleoparathion; Orthophos; Pac; Panthion; Paradust; Paraflow; Paramar; Paramar 50; Paraphos; Paraspray; Parathene; Parathionethyl; Parawet; Penphos; Pestox plus; Pethion; Phoskil; Phosphemol; Phosphenol; **Phosphorothioic acid *O,O*-diethyl *O*-(4-nitrophenyl) ester;** Phosphostigmine; RB; RCRA waste number P089; Rhodiasol; Rhodiatox; Rhodiatrox; Selephos; Sixty-three special E.C. insecticide; SNP; Soprathion; Stabilized ethyl parathion; Stathion; Strathion; Sulphos; Super rodiatox; T-47; Thiofos; Tiophos; Thiophos 3422; Tox 47; Vapophos; Vitrex.

Structure:

Designations: CAS: 56-38-2; DOT: 2783; mf: $C_{10}H_{14}NO_5PS$; fw: 291.27; RTECS: TF4550000.

Properties: Light yellow to dark brown liquid with a garlic-like odor. Mp: 6.1 °C; bp: 375 °C, 157–162 °C at 0.6 mmHg; ρ: 1.26 at 25/4 °C; fl p: 174 °C; H-t$_{1/2}$: ≈ 3.5 wk at pH 6; K_H: 8.56 x 10^{-8} atm·m^3/mol at 25 °C; log K_{oc}: 2.50–4.20; log K_{ow}: 2.15–3.93; P-t$_{1/2}$: 57.6 hr at 90–95 °C (irradiation of aqueous solution by UV light); S$_o$: freely soluble in lower alcohols, aromatic and saturated hydrocarbons, esters, ethers, ketones and many common other organic solvents but only slightly soluble in petroleum ether and kerosene; S$_w$: 10.3, 12.9, and 15.2 mg/L at 10, 20, and 30 °C, respectively; vap d: 11.91 g/L at 25 °C, 10.06 (air = 1); vp: 4 x 10^{-4} mmHg at 20 °C.

Soil properties and adsorption data

Soil	K_d (mL/g)	f_{oc} (%)	K_{oc} (mL/g)	pH	CEC (meq/100 g)
Amarillo silty loam	5.13	0.29	1,769	7.7	8.00
Clarion soil	33.81	2.64	1,278	5.0	21.02
Elkhorn sandy loam	18.19	0.09	21,151	6.0	--
Harps soil	41.12	3.80	1,083	7.3	37.84
Hugo gravelly sand loam	6.90	0.12	5,750	5.5	--
Katy silty loam	6.05	.58	1,043	5.1	--
Nacogdoches clay	2.26	0.23	983	5.0	14.00
Peat soil	254.68	18.36	1,388	7.0	77.34
Rothamsted Farm	15.68	1.51	1,038	5.1	--
Sarpy fine sandy loam	5.68	0.51	1,114	7.3	5.71
Soil #8	12.30	0.94	1,308	6.3	26.60
Soil #10	7.67	0.44	1,743	6.2	18.60
Soil #11	38.02	1.67	2,276	6.3	42.80
Soil #13	125.90	3.20	3,934	5.2	19.20
Soil #14	457.10	14.28	3,201	3.3	28.90
Soil #15	213.80	4.76	4,492	3.5	21.20
Sweeney sandy clay loam	22.16	0.65	3,409	6.3	--
Thurman loamy fine sand	11.07	1.07	1,034	6.8	6.10
Tierra heavy clay loam	40.56	0.33	12,291	6.2	--

Source: Felsot and Dahm, 1979; King and McCarty, 1968; Lord et al., 1980; Wahid and Sethunathan, 1978.

Transformation Products

Biological: Initial hydrolysis products include diethyl-*O*-thiophosphoric acid, *p*-nitrophenol (Munnecke and Hsieh, 1976; Sethunathan, 1973; Sethunathan et al., 1977; Verschueren, 1983), and the biodegradation products *p*-aminoparathion and *p*-aminophenol (Sethunathan, 1973). Mixed bacterial cultures were capable of growing on technical parathion as the sole carbon and energy source (Munnecke and Hsieh, 1976). Three oxidative pathways were reported. The primary degradative pathway is initial hydrolysis to yield *p*-nitrophenol and diethyl-thiophosphoric acid. The secondary pathway involves the formation of paraoxon (diethyl *p*-nitrophenyl phosphate) which subsequently undergoes hydrolysis to yield *p*-nitrophenol and diethylphosphoric acid. The third degradative pathway involved reduction of parathion under low oxygen conditions to yield *p*-aminoparathion followed by hydrolysis to *p*-aminophenol and diethylphosphoric acid. Other potential degradation products include hydroquinone, 2-hydroxy-hydroquinone, ammonia, and polymeric substances (Munnecke and Hsieh, 1976). The reported half-life in soil is 18 d (Jury et al., 1987).

A *Flavobacterium* sp. (ATCC 27551), isolated from rice paddy water, degraded parathion to *p*-nitrophenol. The microbial hydrolysis half-life of this reaction was <1 hr (Sethunathan and Yoshida, 1973). Sharmila et al. (1989) isolated a *Bacillus* sp. from a laterite soil which degraded parathion in the presence yeast extracts. At yeast concentrations of 0.05%, 0.1 and 0.25%, and 0.5%, parathion degraded via hydrolysis, hydrolysis, and nitro group reduction, exclusively by nitro group reduction, respectively. Aminoparathion and *p*-nitrophenol were the identified metabolites under these conditions (Sharmila et al., 1989).

In both soils and water, chemical and biological mediated reactions transform parathion to paraoxon (Alexander, 1981). Parathion was reported to biologically hydrolyze to *p*-nitrophenol in different soils under flooded conditions (Ferris and Lichtenstein, 1980; Sudhakar-Barik and Sethunathan, 1978).

p-Nitrophenol, paraoxon, and three unidentified metabolites were identified in a model ecosystem containing algae, *Daphnia magna*, fish, mosquito, and snails (Yu and Sanborn, 1975).

Soil: A *Pseudomonas* sp. (ATCC 29354), isolated from parathion-amended treated soil, degraded *p*-nitrophenol to *p*-nitrocatechol which was recalcitrant to further degradation. In an unsterilized soil, however, *p*-nitrocatechol was further degraded to nitrites and other unidentified compounds (Sudhakar-Barik et al., 1978a). *Pseudomonas* sp. and *Bacillus* sp., isolated from a parathion-amended flooded soil, degraded *p*-nitrophenol (parathion hydrolysis product) to nitrite ions (Siddaramappa et al., 1973; Sudhakar-Barik et al., 1976) and carbon dioxide (Sudhakar-Barik et al., 1976). When parathion was equilibrated with aerobic soils, virtually no degradation was observed (Adhya et al., 1981a). However, in flooded (anaerobic) acid sulfate soils or low sulfate soils, aminoparathion and desethyl aminoparathion formed as the major metabolites (Adhya et al., 1981).

p-Nitrophenol was identified as a hydrolysis product in soil (Camper, 1991; Miles et al., 1979; Somasundaram et al., 1991; Suffet et al., 1967). The rate of hydrolysis in soil is accelerated following repeated applications of parathion (Ferris and Lichtenstein, 1980). The reported hydrolysis half-lives at pH 7.4 and 20 and 37.5 °C were 130 and 26.8 d, respectively. At pH 6.1 and 20 °C, the hydrolysis half-life is 170 d (Freed et al., 1979). When equilibrated with a prereduced pokkali soil (acid sulfate), parathion instantaneously degraded to aminoparathion. The quick rate of reaction was reportedly due to soil enzymes and/or other heat labile substances. Desethyl aminoparathion was also identified as a metabolite in two separate studies (Wahid and Sethunathan, 1979; Wahid et al., 1980).

Aminoparathion also was formed when parathion (500 ppm) was incubated in a flooded alluvial soil. The amount of parathion remaining after 6 and 12 d were

43.0 and 0.09%, respectively (Freed et al., 1979). In flooded alluvial soils, parathion degraded to aminoparathion via nitro group reduction. The rate of degradation remained constant despite variations in the redox potential of the soils (Adhya et al., 1981a). In soil, parathion may degrade via two oxidative pathways. The primary pathway is hydrolysis to p-nitrophenol (Sudhakar-Barik et al., 1979) and diethylthiophosphoric acid (Miles et al., 1979). The other pathway involves oxidation to paraoxon but aminoparathion is formed under anaerobic (Miles et al., 1979) and flooded conditions (Sudhakar-Barik et al., 1979). The degradation pathway as well as the rate of degradation of parathion in a flooded soil changed following each successive application of parathion (Sudhakar-Barik et al., 1979). After the first application, nitro group reduction gave aminoparathion as the major product. After the second application, both the hydrolysis product (p-nitrophenol) and aminoparathion were found. Following the third addition of parathion, p-nitrophenol was the only product detected. It was reported that the change from nitro group reduction to hydrolysis occurred as a result of rapid proliferation of parathion-hydrolyzing microorganisms that utilized p-nitrophenol as the carbon source (Sudhakar-Barik et al., 1979).

In a cranberry soil pretreated with p-nitrophenol, parathion was rapidly mineralized to carbon dioxide by indigenous microorganisms (Ferris and Lichtenstein, 1980). The half-lives of parathion (10 ppm) in a nonsterile sandy loam and a nonsterile organic soil were <1 and 1.5 wk, respectively (Miles et al., 1979). Walker (1976) reported that 16-23% of parathion added to both sterile and nonsterile estuarine water was degraded after incubation in the dark for 40 d.

Plant: Oat plants were grown in two soils treated with [^{14}C]parathion. Less than 2% of the applied [^{14}C]parathion was translocated to the oat plant. Metabolites identified in both soils and leaves were paraoxon, aminoparaoxon, aminoparathion, p-nitrophenol, and an aminophenol (Fuhremann and Lichtenstein, 1980).

The following metabolites were identified in a soil-oat system: paraoxon, aminoparathion, p-nitrophenol, and p-aminophenol (Lichtenstein, 1980; Lichtenstein et al., 1982). Mick and Dahm (1970) reported that *Rhizobium* sp. converted 85% ^{14}C-labeled parathion to aminoparathion in 1 d and 10% diethylphosphorothioic acid.

One month after application of [^{14}C]parathion to cotton plants, 6.5-10.5% of the total reactivity was found to be unreacted [^{14}C]parathion. Photo-alteration products identified included S-ethyl parathion, S-phenyl parathion, paraoxon, and p-nitrophenol (Joiner and Baetcke, 1973). Reddy and Sethunathan (1983) studied the mineralization of ring-labeled [2,6-^{14}C]parathion in the rhizosphere of rice seedlings under flooded and nonflooded soil conditions. In unplanted soil, only 5.5% of the ^{14}C in the parathion was evolved as ^{14}CO$_2$ in 15 d under flooded and nonflooded conditions. However, in soils planted with rice, 9.2 and 22.6% of the ^{14}C in the parathion evolved as $^{14}CO_2$ under nonflooded and flooded conditions, respectively (Reddy and Sethunathan, 1983). In an earlier study, the presence of rice straw in a flooded alluvial soil inoculated with an enrichment culture greatly inhibited the hydrolysis of parathion to p-nitrophenol and O,O-diethylphosphorothioic acid. In uninoculated soils, however, rice straw enhanced the degradation of parathion via nitro group reduction to p-aminoparathion and a compound possessing a P=S bond (Sethunathan, 1973).

Photolytic: p-Nitrophenol and paraoxon were formed from the irradiation of parathion in water, aqueous methanol, and aqueous n-propyl alcohol solutions by a low-pressure mercury lamp. Degradation was more rapid in water than in organic solvent/water mixtures with p-nitrophenol forming as the major product (Mansour et al., 1983). Parathion degraded on both glass surfaces and on bean plant leaves. Metabolites reported were paraoxon, p-nitrophenol, and a compound tentatively identified as s-ethyl parathion (El-Refai and Hopkins, 1966). Upon exposure to high intensity UV light, parathion was altered to the following photoproducts: paraoxon, O,S-diethyl O-4-nitrophenyl phosphorothioate, O,O-diethyl S-4-nitrophenyl phosphorothioate, O,O-bis(4-nitrophenyl) O-ethyl phosphorothioate, O,O-bis(4-nitrophenyl) O-ethyl phosphate, O,O-diethyl O-phenyl phosphorothioate, and O,O-diethyl O-phenyl phosphate (Joiner et al., 1971).

When parathion was released in the atmosphere on a sunny day, it was rapidly converted to the photochemical paraoxon (estimated $t_{1/2}$ = two min) (Woodrow et al., 1978). The reaction involving the oxidation of parathion to paraoxon is catalyzed in the presence of UV light, ozone, soil dust, or clay minerals (Spencer et al., 1980). When applied as thin films on leaf surfaces, parathion was converted to paraoxon and p-nitrophenol. The photodegradation half-life was reported to be 88 hr (Hazardous Substances Data Bank, 1989).

When an aqueous solution containing parathion was photooxidized by UV light at 90-95 °C, 25, 50, and 75% degraded to carbon dioxide after 15.1, 57.6, and 148.8 hr, respectively (Knoevenagel and Himmelreich, 1976). p-Nitrophenol and paraoxon were the major products identified following the sunlight irradiation of parathion in distilled water and river water. The photolysis half-life of parathion in river water was 15 hr (Mansour et al., 1989). Kotronarou et al. (1992) studied the degradation of parathion-saturated deionized water solution at 30 °C by ultrasonic irradiation (sonolysis). After 2 hr of sonolysis, all the parathion degraded to the following final end products: sulfate ions, phosphate ions, nitrate ions, hydrogen ions, and carbon dioxide. Precursors/intermediate compounds to the final end products included p-nitrophenol, diethylmonothiophosphoric acid, hydroquinone, benzoquinone, formic acid, oxalic acid, 4-nitrocatechol, nitrite ions, and ethanol (Kotronarou et al., 1992).

Chemical/Physical: Paraoxon was also found in fogwater collected near Parlier, CA (Glotfelty et al., 1990). It was suggested that parathion was oxidized in the atmosphere during daylight hours prior to its partitioning in the fog. On 12 January 1986, the distribution of parathion (9.4 ng/m^3) in the vapor phase, dissolved phase, air particles and water particles were 78, 10, 11, and 0.6%, respectively. For paraoxon (2.3 ng/m^3), the distribution in the vapor phase, dissolved phase, air particles and water particles were 7.8, 35.5, 56.7, and 0.09%, respectively (Glotfelty et al., 1990). The hydrolysis half-lives of parathion in water at 25 °C and pHs of 5.0, 6.0, 7.0, and 8.0 were 43, 33, 24, and 15 wk, respectively (Chapman and Cole, 1982).

Reported ozonation products of parathion in drinking water include sulfuric acid (Richard and Bréner, 1984), paraoxon, 2,4-dinitrophenol, picric, and phosphoric acids (Laplanche and Tonnard, 1984).

At 130 °C, parathion isomerizes to *O,S*-diethyl *O-p*-nitrophenyl phosphorothioate (Hartley and Kidd, 1987; Worthing and Hance, 1991). An 85% yield of this compound was reported when parathion was heated at 150 °C for 24 hr (Wolfe et al., 1976). Emits toxic oxides of nitrogen, sulfur and phosphorus when heated to decomposition (Lewis, 1990).

Exposure Limits: NIOSH REL: 10 hr TWA 0.05 mg/m^3, IDLH 20 mg/m^3; OSHA PEL: TWA 0.1 mg/m^3.

Symptoms of Exposure: Acute effects include miosis, rhinorrhea, headache, bronchoconstriction, chest wheezing, laryngeal spasm, excessive salivation, cyanosis, anorexia, abdominal cramps, diarrhea, sweating, muscle fasiculation and twitching, nausea, weakness, paralysis, ataxia, convulsions, low blood pressure, dermatitis, pupillary constriction.

Formulation Types: Emulsifiable concentrate; wettable powder; granules; dustable powder; aerosol.

Toxicity: LC$_{50}$ (96 hr) for rainbow trout 1.5 mg/L, golden orfe 0.57 mg/L (Hartley and Kidd, 1987), fathead minnow 1.4-2.7 mg/L (Worthing and Hance, 1991), bluegill sunfish 65 μg/L, green sunfish 425 μg/L, and largemouth bass 190 μg/L (Verschueren, 1983); acute oral LD$_{50}$ for rats 3.6-13 mg/kg (Hartley and Kidd, 1987).

Uses: Insecticide and acaricide for control of sucking and chewing insects and mites in fruits, vegetables, ornamentals and field crops.

PEBULATE

Synonyms: Butylethylthiocarbamic acid S-propyl ester; PEBC; S-Propyl butyl-ethylthiocarbamate; Propylethylbutylthiocarbamate; Propyl-N-ethyl-n-butylthio-carbamate; S-(n-Propyl)-N-ethyl-N,N-butylthiocarbamate; n-Propyl-N-ethyl-N-(n-butyl)thiocarbamate; Propyl ethyl-n-butylthiolcarbamate; Propylethylbutyl-thiolcarbamate; Propyl ethyl-n-butylthiolcarbamate; PEBC; R 2061; Stauffer 2061; Tillam; Timmam-6-E.

Structure:

$$CH_3CH_2 \diagdown$$
$$\diagup NCOSCH_2CH_2CH_3$$
$$CH_3CH_2 \diagup$$

Designations: CAS: 1114-71-2; mf: $C_{10}H_{21}NOS$; fw: 203.36; RTECS: EZ0400000.

Properties: Colorless to yellow liquid with an amine-like odor. Mp: <25 °C; bp: 142 °C at 20 mmHg; ρ: 0.9555 at 20/20 °C, 0.9458 at 30/4 °C; fl p: 124 °C; K_H: 1.15 x 10^{-4} atm·m^3/mol at 20 °C (approximate - calculated from water solubility and vapor pressure); log K_{oc}: 2.80; log K_{ow}: 3.84; S_o: miscible with but not limited to acetone, benzene, ethanol, kerosene, methanol, 2-propanol, toluene, xylene; S_w: 60 mg/L at 20 °C; vp: 6.8 x 10^{-2} mmHg at 20 °C.

Transformation Products
Soil: Pebulate rapidly degrades in soil forming carbon dioxide, ethylbutylamine, and the mercaptan (Hartley and Kidd, 1987). The half-life in a moist loam soil at 21-27 °C was reported to be about 2 wk (Humburg et al., 1989).

Plant: Rapidly transformed in plants to carbon dioxide (Worthing and Hance, 1991) and naturally occurring plant constituents (Humburg et al., 1989).

Chemical/Physical: Emits toxic fumes of nitrogen and sulfur oxides when heated to decomposition (Sax and Lewis, 1987).

Symptoms of Exposure: May cause violent vomiting when accompanied by alcohol ingestion.

Formulation Types: Emulsifiable concentrate (6 lb/gal); granules.

Toxicity: LC_{50} (96 hr) for bluegill sunfish and rainbow trout about 7.4 mg/L (Worthing and Hance, 1991); LC_{50} (48 hr) for killifish 7.78 mg/L and silver mullet

6.25 mg/L (Hartley and Kidd, 1987); acute oral LD_{50} for male rats and male mice is 0.9–1.1 and 1.5–1.8 g/kg, respectively (Ashton and Monaco, 1991).

Uses: Selective preemergence herbicide used to control some broadleaf weeds and annual grasses in tomatoes, sugar beet, and tobacco.

PENDIMETHALIN

Synonyms: AC 92553; N-(1-Ethylpropyl)-3,4-dimethyl-2,6-dinitrobenzenamine; N-(1-Ethylpropyl)-2,6-dinitro-3,4-xylidine; Herbadox; Horbadox; Pay-off; Penoxalin; Penoxaline; Penoxyn; Phenoxalin; Prowl; Stomp; Stomp 300D; Stomp 300E; Tendimethalin.

Structure:

Designations: CAS: 40487-42-1; mf: $C_{13}H_{19}N_3O_4$; fw: 281.31; RTECS: BX5470000.

Properties: Orange-yellow crystalline solid with a faint nutty odor. Mp: 54-58 °C; bp: 330 °C; ρ: 1.19 at 25/4 °C; H-t½: <21 d under continuous irradiation; K_H: 8.56 x 10^{-7} atm·m^3/mol at 25 °C; log K_{oc}: 1.48-2.93; log K_{ow}: 5.18; S_o (g/L at 26 °C): acetone (700), dimethyl sulfoxide (214), n-heptane (138), methanol (55), 2-propanol (77), xylene (628); S_w: 0.3 mg/L at 20 °C; vp: 3.0 x 10^{-5} mmHg at 25 °C.

Soil properties and adsorption data

Soil	K_d (mL/g)	f_{oc} (%)	K_{oc} (mL/g)	pH
Loam	301	2.20	13,682	7.0
Loam	854	2.90	29,448	6.5
Sand	30	0.46	6,466	7.6
Sandy loam	110	0.93	11,853	6.4
Silty loam	380	2.73	13,919	7.0

Source: U.S. Department of Agriculture, 1990.

Transformation Products

Biological: In an *in vitro* study, the soil fungi *Fusarium oxysporum* and *Paecilomyces varioti* degraded pendimethalin to N-(1-ethylpropyl)-3,3-dimethyl-2-nitrobenzene-1,6-diamine and 3,4-dimethyl-2,6-dinitroaniline. The latter compound was the only metabolite identified by another soil fungus; namely, *Rhizoctonia bataticola* (Singh and Kulshrestha, 1991).

Soil/Plant: In soil, the 4-methyl group on the benzene ring is oxidized to a hydroxy group which undergoes further oxidation to the carboxylic acid. The major plant metabolite is 4-(1-ethylpropyl)amino-2-methyl-3,5-dinitrobenzyl alcohol (Hartley and Kidd, 1987). Kulshrestha and Singh (1992) investigated the influence of soil moisture and microbial activity on the degradation of pendimethalin in a sandy loam soil. In both nonsterile nonflooded and flooded soils, degradation followed first-order kinetics. The observed half-lives in flooded and nonflooded conditions in nonsterile and sterile soils were 33 and 45, and 52 and 67 d, respectively. In the nonsterile, nonflooded soil, pendimethalin underwent N-dealkylation and reduction of the less hindered nitro group affording N-(1-ethylpropyl)-3,4-dimethyl-2-nitrobenzene-1,6-diamine and 3,4-dimethyl-2,6-dinitroaniline as major products. Under flooded conditions, only one degradation was formed: N-(1-ethylpropyl)-5,6-dimethyl-7-nitrobenzimidazole. Under flooded conditions, pendimethalin was completely degraded after 4 d (Kulshrestha and Singh, 1992).

Photolytic: Irradiation of a sandy loam soil with a high pressure mercury lamp for 48h gave the following products (% yield): 2-amino-6-nitro-N-(1-ethylpropyl)-3,4-xylidine (35), 2,6-dinitro-3,4-xylidine (21), 2,2-diethyl-7-nitro-4,5-dimethyl-3-hydroxy-2,3-dihydrobenzimidazole (13), and 2-ethyl-7-nitro-4,5-dimethyl-3-hydroxy-2,3-dihydrobenzimidazole (8) (Dureja and Walia, 1989).

Symptoms of Exposure: Eye irritant.

Formulation Types: Emulsifiable concentrate (4 lb/gal); wettable powder (50%); granules (2 and 5%).

Toxicity: LC_{50} (96 hr) for rainbow trout 138 μg/L and bluegill sunfish 199 μg/L; acute oral LD_{50} of the technical product for male and female rats is 1,250 and 1,050 mg/kg, respectively (Ashton and Monaco, 1991).

Uses: Herbicide used to control many annual grass and broadleaf weeds in crops such as corn, cotton, peanuts, potatoes, rice, sorghum, sunflower, and tobacco.

PENTACHLOROBENZENE

Synonyms: QCB; RCRA waste number U183.

Structure:

Designations: CAS: 608-93-5; mf: C_6HCl_5; fw: 250.34; RTECS: DA6640000.

Properties: White crystals or needles from alcohol. Mp: 86 °C; bp: 277 °C; ρ: 1.8342 at 16.5/4 °C; H-$t_{1/2}$: >900 yr at 25 °C and pH 7; K_H: 0.0071 atm·m^3/mol at 20 °C; log K_{oc}: 6.3; log K_{ow}: 4.88-5.75; S_o: slightly soluble in benzene, carbon disulfide, chloroform, dimethylsulfoxide (1-10 mg/mL at 22 °C), 95% ethanol (1-10 mg/mL at 22 °C), ethyl ether; S_w: 5.32 μM/L at 25 °C; vp: 6.0 x 10^{-3} mmHg at 20-30 °C.

Transformation Products

Biological: In activated sludge, <0.1% mineralized to carbon dioxide after 5 d (Freitag et al., 1985).

Photolytic: UV irradiation (λ = 2537 Å) of pentachlorobenzene in *n*-hexane solution for 3 hr produced a 50% yield of 1,2,4,5-tetrachlorobenzene and a 13% yield of 1,2,3,5-tetrachlorobenzene (Crosby and Hamadmad, 1971). Irradiation (λ ≥285 nm) of pentachlorobenzene (1.1-1.2 mM/L) in an acetonitrile-water mixture containing acetone (concentration = 0.553 mM/L) as a sensitizer gave the following products (% yield): 1,2,3,4-tetrachlorobenzene (6.6), 1,2,3,5-tetrachlorobenzene (52.8), 1,2,4,5-tetrachlorobenzene (15.1), 1,2,4-trichlorobenzene (1.9), 1,3,5-trichlorobenzene (5.3), 1,3-dichlorobenzene (0.9), 2,2',3,3',4,4',5,6,6'-nonachlorobiphenyl (2.08), 2,2',3,3',4,4',5,5',6-nonachlorobiphenyl (0.34), 2,2',3,3',4,5,5',6,6'-nonachlorobiphenyl (trace), one octachlorobiphenyl (0.53), and one heptachlorobiphenyl (0.49) (Choudhry and Hutzinger, 1984). Without acetone, the identified photolysis products (% yield) included 1,2,3,4-tetrachlorobenzene (3.7), 1,2,3,5-tetrachlorobenzene (13.5), 1,2,4,5-tetrachlorobenzene (2.8), 1,2,4-trichlorobenzene (12.7), 1,3,5-trichlorobenzene (1.0), and 1,4-dichlorobenzene (6.7) (Choudhry and Hutzinger, 1984). A carbon dioxide yield of 2.0% was achieved when pentachlorobenzene adsorbed on silica gel was irradiated with light (λ >290 nm) for 17 hr.

Chemical/Physical: Emits chlorinated acids when incinerated. Incomplete combustion may release toxic phosgene (Sittig, 1985). Though no products were reported, the calculated hydrolysis half-life in water at 25 °C and pH 7 is >900 yr (Ellington et al., 1988).

Toxicity: LC_{50} (14 d) for guppy 178 ppb (Verschueren, 1983); acute oral LD_{50} for rats 1,080 mg/kg (RTECS, 1985).

Use: Agrochemical research.

PENTACHLOROPHENOL

Synonyms: Acutox; Chem-penta; Chem-tol; Chlorophen; Cryptogil OL; Dowcide 7; Dowicide 7; Dowicide EC-7; Dowicide G; Dow pentachlorophenol DP-2 anti-microbial; Durotox; EP 30; Fungifen; Fungol; Glazd penta; Grundier arbezol; Lauxtol; Lauxtol A; Liroprem; Monsanto penta; Moosuran; NCI-C54933; NCI-C55378; NCI-C56655; PCP; Penchlorol; Penta; Pentachlorofenol; Pentachloro-fenolo; Pentachlorophenate; Pentachlorphenol; 2,3,4,5,6-Pentachlorophenol; Pentacon; Penta-kil; Pentasol; Penwar; Peratox; Permacide; Permaguard; Permasan; Permatox DP-2; Permatox Penta; Permite; Priltox; RCRA waste number U242; Santobrite; Santophen; Santophen 20; Sinituho; Term-i-trol; Thompson's wood fix; Weedone; Witophen P.

Structure:

Designations: CAS: 87-86-5; DOT: 2020; mf: C_6HCl_5O; fw: 266.34; RTECS: SM6300000.

Properties: White to dark-colored flakes or beads with a characteristic odor and pungent taste. Very pungent odor when hot. Darkens on exposure to air. Mp: 190–191 °C (anhydrous), 174 °C (hydrous); bp: 310 °C with decomposition; ρ: 1.978 at 22/4 °C; fl p: nonflammable; pK_a: 4.74; K_H: 2.1-3.4 x 10^{-7} atm·m^3/mol; log K_{oc}: 2.47-4.40; log K_{ow}: 3.32-5.86; S_o: very soluble in acetone, carbitol, cellosolve, cod liver oil (184.2, 188.5, and 98.9 g/L at 4, 12, and 20 °C, respectively), ethanol, ethyl ether, n-octanol (266.3, 260.2, and 206.7 g/L at 4, 12, and 20 °C, respectively), and triolein (168.0, 168.0, and 106.0 g/L at 4, 12, and 20 °C, respectively) but slightly soluble in alkanes, carbon tetrachloride, cold petroleum ether, solvents and ligroin; S_w: 14 mg/L at 20 °C at pH 5; vp: 1.7 x 10^{-4} mmHg at 20 °C.

Soil properties and adsorption data

Soil	K_d (mL/g)	f_{oc} (%)	K_{oc} (mL/g)	pH
Aquifer material	200.00	0.84	23,800	--
Coarse sand	1.20	0.21	571	5.3
Lake sediment	3,670.00	9.40	39,000	--
Loamy sand	0.45	0.15	300	6.4

Soil	K_d (mL/g)	f_{oc} (%)	K_{oc} (mL/g)	pH
River sediment	930.00	2.60	35,800	--

Source: Kjeldsen et al., 1990; Schellenberg, 1984.

Transformation Products

Soil: Under anaerobic conditions, pentachlorophenol may undergo sequential dehalogenation to produce tetra-, tri-, di-, and *m*-chlorophenol (Kobayashi and Rittman, 1982). In aerobic and anaerobic soils, pentachloroanisole was the major metabolite along with 2,3,6-trichlorophenol, 2,3,4,5-, and 2,3,5,6-tetrachlorophenol (Murthy et al., 1979).

Weiss et al. (1982) studied the fate of ^{14}C-pentachlorophenol added to flooded rice soil in a plant growth chamber. After one growing season, the following residues were observed (% of applied radioactivity): unidentified unextractable/bound compounds (28.61%), pentachloro-phenol (0.51%), conjugated pentachlorophenol (0.61%), 2,3,6-, 2,4,5-, 2,4,6-, 2,3,4-, 2,3,5-, and 3,4,5-trichlorophenols (1.27%), 2,3,4,5-tetrachlorophenol (0.38%), 2,3,4-, 2,3,6-, 2,4,6-, and 3,4,5-trichloroanisoles (0.08%), 2,3,4,5-tetrachloroanisole (0.02%), pentachloroanisole (0.02%), and unidentified conversion products (including highly polar hydrolyzable and nonhydrolyzable compounds) (4.74%).

Metabolites identified in soil beneath a sawmill environment where pentachlorophenol was used as a wood preservative include pentachloroanisole, 2,3,4,6-tetrachloroanisole, tetrachlorocatechol, tetrachlorohydroquinone, 3,4,5-trichlorocatechol, 2,3,6-trichlorohydroquinone, 3,4,6-trichlorocatechol, and 2,3,4,6-tetrachlorophenol (Knuutinen et al., 1990).

Biological: Under aerobic conditions, microbes in estuarine water partially dechlorinated pentachlorophenol to trichlorophenol (Hwang et al., 1986). The disappearance of pentachlorophenol was studied in four aquaria with and without mud under aerobic and anaerobic conditions. Potential biological and/or chemical products identified include pentachloroanisole, 2,3,4,5-, 2,3,4,6-, and 2,3,5,6-tetrachlorophenol (Boyle et al., 1980). Pentachlorophenol was also subject to methylation by a culture medium containing *Trichoderma viride* affording pentachloroanisole (Cserjesi and Johnson, 1972).

Pentachlorophenol degraded in anaerobic sludge to 3,4,5-trichlorophenol which was further reduced to 3,5-dichlorophenol (Mikesell and Boyd, 1985). In activated sludge, only 0.2% of the applied amount was mineralized to carbon dioxide after 5 d (Freitag et al., 1985).

Photolytic: When pentachlorophenol in distilled water is exposed to UV irradiation, it is photolyzed to give tetrachlorophenols, trichlorophenols, chlorinated dihydroxybenzenes, and dichloromaleic acid (Hwang et al., 1986).

Wood treated with pure pentachlorophenol did not photolyze under natural sunlight or laboratory-induced UV radiation. However, in the presence of an antimicrobial (Dowcide EC-7), pure pentachlorophenol degraded to chlorinated dibenzo-*p*-dioxin. Wood containing composited technical grade pentachlorophenol yielded similar results (Lamparski et al., 1980).

An aqueous solution containing pentachlorophenol and exposed to sunlight or laboratory UV light yielded tetrachlorocatechol, tetrachlororesorcinol, and tetrachlorohydroquinone. These compounds were air-oxidized to chloranil, hydroxyquinones, and 2,3-dichloromaleic acid (DCM). Other compounds identified include a cyclic dichlorodiketone, tetra-, and trichlorophenols (Wong and Crosby, 1981). Photodecomposition of pentachlorophenol was observed when an aqueous solution was exposed to sunlight for 10 d. The violet-colored solution contained 3,4,5-trichloro-6-(2'-hydroxy-3',4',5',6'-tetrachlorophenoxy)-*o*-benzoquinone as the major product. Minor photodecomposition products (% yield) included tetrachlororesorcinol (0.10%), 2,5-dichloro-3-hydroxy-6-penta-chlorophenoxy-*p*-benzoquinone (0.16%), and 3,5-dichloro-2-hydroxy-5-2',4',5',6'-tetrachloro-3-hydroxyphenoxy-*p*-benzoquinone (0.08%) (Plimmer, 1970). UV irradiation (λ = 2537 Å) of pentachlorophenol in a *n*-hexane solution for 32 hr produced a 30% yield of 2,3,5,6-tetrachlorophenol and about a 10% yield of a compound tentatively identified as an isomeric tetrachlorophenol (Crosby and Hamadmad, 1971).

A carbon dioxide yield of 50.0% was achieved when pentachlorophenol adsorbed on silica gel was irradiated with light (λ >290 nm) for 17 hr (Freitag et al., 1985).

When an aqueous solution containing pentachlorophenol (4.5 x 10^{-5} M) and a suspension of titanium dioxide (2 g/L) was irradiated with UV light, carbon dioxide and hydrochloric acid formed in quantitative amounts ($t_{1/2}$ = 9 min at 45-50 °C) (Barbeni et al., 1985). When an aqueous solution containing pentachloro-phenol was photooxidized by UV light at 90-95 °C, 25, 50, and 75% degraded to carbon dioxide after 31.7, 66.0, and 180.7 hr, respectively (Knoevenagel and Himmelreich, 1976). The photolysis half-lives of pentachlorophenol under sunlight irradiation in distilled water and river water were 27 and 53 hr, respectively (Mansour et al., 1989).

Petrier et al. (1992) studied the sonochemical degradation of pentachlorophenol in aqueous solutions saturated with different gases at 24 °C. Ultrasonic irradiation of solutions saturated with air or oxygen resulted in the liberation of chloride ions and mineralization of the parent compound to carbon dioxide. When the solution is saturated with argon, pentachlorophenol completely degraded to carbon monoxide and chloride ions.

Chemical/Physical: Wet oxidation of pentachlorophenol at 320 °C yielded formic and acetic acids (Randall and Knopp, 1980). In a dilute aqueous solution at pH 6.0, pentachlorophenol reacted with excess hypochlorous acid forming 2,3,5,6-tetra-chlorobenzoquinone (chloranil), hexachloro-5-cyclohexadienone, and two other chlorinated compounds at yields of 3, 20, and 27%, respectively (Smith et al., 1976). Reacts with amines and alkali metals forming water-soluble salts (Sanborn et al., 1977). Hexachlorobenzene, octachlorodiphenylene dioxide, and other polymeric compounds were formed when pentachlorophenol was heated to 300 °C for 24 hr (Sanderman et al., 1957). Emits very toxic chloride fumes when heated to decomposition (Lewis, 1990).

Exposure Limits: NIOSH REL: IDLH 150 mg/m^3; OSHA PEL: TWA 0.5 mg/m^3; ACGIH TLV: TWA 0.5 mg/m^3.

Symptoms of Exposure: Ingestion causes fluctuation in blood pressure, respiration; fever; urinary output; motor weakness; convulsions and possibly death.

Formulation Types: Wettable powder; granules; oil-miscible liquid.

Toxicity: LC$_{50}$ (48 hr) for rainbow trout 0.17 mg/L (sodium salt) (Hartley and Kidd, 1987); LC$_{50}$ (24 hr) for goldfish 270 ppb (Verschueren, 1983); acute oral LD$_{50}$ for rats 210 mg/kg (Hartley and Kidd, 1987), 27 mg/kg (RTECS, 1985).

Uses: Insecticide; fungicide; herbicide.

PERMETHRIN

Synonyms: AI3-29158; Ambush; BW-21-Z; 3-(2,2-Dichloroethenyl)-2,2-dimethylcyclopropanecarboxylic acid (3-phenoxyphenyl)methyl ester; Ectiban; Exmin; FMC 33297; FMC 41665; ICI-PP 557; Kestrel; NDRC-143; NIA 33297; Niagara 33297; Outflank; Outflanf-stockade; (3-Phenoxyphenyl)methyl 3-(2,2-dichloroethenyl)-2,2-dimethylcyclopropane carboxylate; Pounce; PP 557; S 3151; SBP-1513; Talcord; WL 43479.

Structure:

Designations: CAS: 52645-53-1; mf: $C_{21}H_{20}Cl_2O_3$; fw: 391.29; RTECS: GZ1255000.

Properties: Colorless crystals to a light yellow, viscous liquid. Technical material contains 60% *trans-* and 40% *cis-* isomers. Mp: 34-39 °C; bp: 220 °C at 0.05 mmHg; ρ: 1.190-1.272 at 20/4 °C; K_H: 4.8 x 10^{-8} atm·m^3/mol at 20 °C (approximate - calculated from water solubility and vapor pressure); log K_{oc}: 1.32-2.79; log K_{ow}: 2.88-6.10; P-t½: 177.51 hr (*cis*), 121.03 (*trans*) (absorbance λ = 227.5 nm, concentration on glass plates = 6.7 μg/cm^2); S_o: soluble in methanol (258 g/kg at 25 °C) and miscible with many organic solvents except ethylene glycol; S_w: 0.2 mg/L at 20 °C; vp: 1.88 x 10^{-8} mmHg at 20 °C.

Soil properties and adsorption data

Soil	K_d (mL/g)	f_{oc} (%)	K_{oc} (mL/g)	pH
Loam	63.3	--	--	--
Loamy sand	20.7	--	--	--
Sandy loam	439.4	--	--	--
Sandy loam	611.0	--	--	--
Silty loam	118.0	--	--	--

Source: U.S. Department of Agriculture, 1990.

Transformation Products
Soil: Permethrin is rapidly degraded in soil by microorganisms via hydrolysis yielding 3-(2,2-dichloroethenyl)-2,2-dimethylcyclopropanecarboxylic acid and 3-phenoxybenzyl alcohol (Kaufman et al., 1981). The reported half-life in soil

containing 1.3-51.3% organic matter and pH 4.2-7.7 is <38 d (Worthing and Hance, 1991).

In lake water, permethrin degraded more rapidly in lake water than in flooded sediment to *trans*- and *cis*-(dichlorovinyl)dimethylcyclopropanecarboxylic acid. It was found that the *cis*-isomer was more stable toward biological and chemical degradation than the *trans*-isomer (Sharom and Solomon, 1981).

Plant: Metabolites identified in cotton leaves included *trans*-hydroxypermethrin, 2'-hydroxypermethrin, 4'-hydroxypermethrin, dichlorovinyl acid, dichlorovinyl acid conjugates, hydroxydichlorovinyl acid, hydroxydichlorovinyl acid conjugates, phenoxybenzyl alcohol and phenoxybenzyl alcohol conjugates. Degradation of permethrin was mainly by ester cleavage and conjugation of the acid and alcohol intermediates (Gaughan and Casida, 1978).

Photolytic: Photolysis of permethrin in aqueous solutions containing various solvents (acetone, hexane, and methanol) under UV light (λ >290 nm) or on soil in sunlight initially resulted in the isomerization of the cyclopropane moiety and ester cleavage. Photolysis products identified included 3-phenoxybenzyldimethyl acrylate, 3-phenoxybenzaldehyde, 3-phenoxybenzoic acid, monochlorovinyl acids, *cis*- and *trans*-dichlorovinyl acids, benzoic acid, 3-hydroxybenzoic acid, 3-hydroxybenzyl alcohol, benzyl alcohol, benzaldehyde, 3-hydroxybenzaldehyde, and 3-hydroxybenzoic acid (Holmstead et al., 1978).

Symptoms of Exposure: Eye and skin irritant.

Formulation Types: Aerosol; emulsifiable concentrate; wettable powder; water-dispersible granules; dustable powder; aerosol; fumigant.

Toxicity: LC_{50} (48 hr) for rainbow trout 5.4 μg/L and bluegill sunfish 1.8 μg/L (Hartley and Kidd, 1987); acute oral LD_{50} for rats approximately 4,000 mg/kg (*cis:trans* = 40:60) (Hartley and Kidd, 1987).

Use: Insecticide.

PHENMEDIPHAM

Synonyms: Beetomax; Beetup; Betanal; Betanal E; Betosip; EP 452; Fender; Fenmedifam; Goliath; Gusto; *m*-Hydroxycarbanilic acid methyl ester *m*-methylcarbanilate; Kemifam; 3-Methoxycarbonylaminophenyl 3-methylcarbanilate; 3-((Methoxycarbonyl)amino)phenyl (3-methylphenyl)carbamate; Methyl *m*-hydroxycarbanilate *m*-methylcarbanilate; Methyl 3-(3-methylcarbaniloyloxy)carbanilate; 3-(Methylphenyl)carbamic acid 3-((methoxycarbonyl)amino)phenyl ester; Methyl 3-(*m*-tolylcarbamoyloxy)phenylcarbamate; Pistol; Protrum K; Schering 38584; SN 4075; SN 38584; Spin-Aid; Vangard.

Structure:

Designations: CAS: 13684-63-4; mf: $C_{16}H_{16}N_2O_4$; fw: 300.32; RTECS: FD9050000.

Properties: Colorless to white crystals or crystalline solid. Mp: 139-142 °C (pure), 143-144 °C (technical); ρ: 0.25-0.30 at 20/4 °C; fl p: 74 °C; H-t$_{1/2}$ (22 °C): 70 d (pH 5), 1 d (pH 7), 10 min (pH 9); K_d: 28-314; K_H: 8.4 x 10^{-11} atm·m^3/mol at 20 °C (approximate - calculated from water solubility and vapor pressure); log K_{oc}: 3.21-3.27 (calculated); log K_{ow}: 3.59 at pH 4; S_o (g/L at 20 °C): acetone ($\approx$ 200), benzene (2.5), chloroform (20), cyclohexanone ($\approx$ 200), *n*-hexane ($\approx$ 0.5), ethanol ($\approx$ 50), methanol (50), methylene chloride (16.7), toluene (0.97), 2,2,4-trimethylpentane (0.16); S_w: 4.7-6.0 mg/L at 20 °C; vp: 10^{-9} mmHg at 20 °C.

Transformation Products

Soil: Phenmedipham degraded in soil forming methyl *N*-(3-hydroxyphenyl)-carbamate and *m*-aminophenol (Hartley and Kidd, 1987). Hydrolysis yields *m*-aminophenol (Rajagopal et al., 1989). The reported half-life in soil is approximately 20 d (Rajagopal et al., 1989) and 26 d (Worthing and Hance, 1991).

Plant: In plants, methyl *N*-(3-hydroxyphenyl)carbamate is the major metabolite (Hartley and Kidd, 1987).

Photolytic: Bussacchini et al. (1985) studied the photolysis (λ = 254 nm) of phenmedipham in ethanol, ethanol/water, and hexane as solvents. In their proposed free radical mechanism, homolysis of the carbon-oxygen bond of the carbamate linkage gave the following photoproducts: 3-(hydroxyphenyl)carbamic acid methyl ester, *m*-toluidine, 2-hydroxy-4-aminomethyl benzoate, 3-hydroxy-

5-aminomethyl benzoate, 2-amino-4-hydroxymethyl benzoate, and 2-amino-6-hydroxymethyl benzoate.

Symptoms of Exposure: Ingestion may cause hyperactivity and muscle spasms.

Formulation Types: Emulsifiable concentrate (1.3 lb/gal).

Toxicity: LC_{50} (96 hr) for bluegill sunfish 3.98 mg/L, rainbow trout 1.4–3.0 mg/L, *Daphnia magna* 3.2 mg/L (Worthing and Hance, 1991), harlequin fish 16.5 mg/L (Hartley and Kidd, 1987); acute oral LD_{50} of pure phenmedipham and the formulated product for rats is 3,700 and 10,300 mg/kg, respectively (Ashton and Monaco, 1991).

Uses: Postemergence herbicide used to control weeds such as chickweed, dogfennel, foxtail, kochia, nightshade, yellow mustard, in strawberries and beet crops, and spinach.

PHORATE

Synonyms: AC 3911; American Cyanamid 3911; *O,O*-Diethyl *S*-ethylmercapto-methyl dithiophosphonate; *O,O*-Diethyl *S*-ethylthiomethyl dithiophosphonate; *O,O*-Diethyl *S*-ethylthiomethyl phosphorodithioate; EI-3911; ENT 24042; Experimental insecticide 3911; Granutox; L 11/6; Phorate-10G; **Phosphorodithioic acid *O,O*-diethyl *S*-((ethylthio)methyl) ester;** Rampart; RCRA waste number P094; Thimet; Timet; Vegfru; Vergfru foratox.

Structure:

$$CH_3CH_2O \diagdown \overset{\displaystyle S}{\underset{\displaystyle }{\overset{\|}{P}}} - SCH_2SCH_2CH_3$$
$$CH_3CH_2O \diagup$$

Designations: CAS: 298-02-2; mf: $C_7H_{17}O_2PS_3$; fw: 260.40; RTECS: TD9450000.

Properties: Clear liquid. Mp: <-15 °C; bp: 118-120 °C at 0.8 mmHg, 125-127 °C at 2.0 mmHg; ρ: 1.156 at 25/4 °C; H-t$_{1/2}$: 96 hr at 25 °C and pH 7; K_H: 6.4 x 10^{-6} atm·m^3/mol at 20-24 °C (approximate - calculated from water solubility and vapor pressure); log K_{oc}: 2.51-2.80; log K_{ow}: 2.91-3.92; S_o: miscible with carbon tetrachloride, di-*n*-butyl phthalate, methylene chloride, vegetable oils, 1,4-dioxane, xylene, and many other organic solvents; S_w: 20 mg/L at 24 °C; vp: 8.4 x 10^{-4} mmHg at 20 °C.

Soil properties and adsorption data

Soil	K_d (mL/g)	f_{oc} (%)	K_{oc} (mL/g)	pH	CEC (meq/100 g)
Batcombe silt loam	4.51	2.05	220	6.10	--
Clarion soil	9.62	2.64	364	5.00	21.02
Elkhorn sandy loam	3.94	0.09	4,581	6.00	--
Harps soil	16.14	3.80	424	7.30	37.84
Hugo gravelly sandy loam	1.96	0.12	1,633	5.50	--
Loam	12.00	2.20	545	7.00	--
Peat soil	73.79	18.36	402	6.98	77.34
Rothamsted Farm	9.83	1.51	651	5.10	--
Sand	1.80	0.58	310	6.00	--
Sandy loam	4.00	0.64	627	6.90	--
Silty loam	5.60	1.74	322	5.20	--
Sarpy fine sandy loam	2.32	0.51	455	7.30	5.71
Sweeney sandy clay loam	13.71	1.65	2,109	6.30	--

Soil	K_d (mL/g)	f_{oc} (%)	K_{oc} (mL/g)	pH	CEC (meq/100 g)
Thurman loamy fine sand	5.48	1.07	514	6.83	6.10
Tierra clay loam	15.50	0.33	4,697	6.20	--

Source: Briggs, 1981; Felsot and Dahm, 1979; Lord et al., 1980; U.S. Department of Agriculture, 1990.

Transformation Products

Biological: [^{14}C]Phorate degraded in a model ecosystem consisting of soil, plants, and water (Lichtenstein et al., 1974). Under both nonpercolating and percolating water conditions, 12% of the applied amount migrated downward as the corresponding sulfone and sulfoxide. Phorate was absorbed in the roots of corn and was transformed primarily to the sulfone with trace amounts of the sulfoxide. Translocation of radioactive insecticide to the leaves was also observed but the major products were identified as phoratoxon sulfone and phoratoxon sulfoxide (Lichtenstein et al., 1974).

Soil: The corresponding sulfoxide and sulfone and their phosphorothioate analogs are major soil metabolites (Hartley and Kidd, 1987; Lichtenstein et al., 1974). The phosphorothioate analogs may hydrolyze forming dithio-, thio-, and orthophosphoric acids (Hartley and Kidd, 1987). Phorate sulfoxide, a microbial metabolite of phorate, was degraded by microbes to phorate oxon (Ahmed and Casida, 1958). Menzer et al. (1970) reported that degradation of disulfoton in soil degraded at a higher rate in the winter months than in the summer months. They postulated that soil type, rather than temperature, had greater influence on the rate of decomposition of disulfoton. Soils used in the winter and summer months were an Evesboro loamy sand and Chillum silt loam, respectively (Menzer et al., 1970). The reported half-life in soil is 82 d (Jury et al., 1987) and 68 d in a sandy soil (Way and Scopes, 1968).

Plant: Oat plants were grown in two soils treated with [^{14}C]phorate. Most of the residues remained bound to the soil. Less than 2% of the applied [^{14}C]phorate was recovered from the oat leaves. The major residues in soil were phorate and the corresponding oxon, sulfone, oxon sulfone, sulfoxide, and oxon sulfoxide (Fuhremann and Lichtenstein, 1980). These compounds were also found in asparagus tissue and soil treated with the insecticide (Szeto and Brown, 1982). In soil and plants, phorate is oxidized to the corresponding sulfone which is further oxidized to the sulfoxide (Getzin and Shanks, 1970). Both metabolites were identified in spinach 5.5 mo after application (Menzer and Dittman, 1968). The half-lives in coastal Bermuda grass and alfalfa are 1.4 (Leuck and Bowman, 1970) and 3.6 d (Dobson et al., 1960), respectively.

Chemical/Physical: Emits toxic fumes of nitrogen and sulfur oxides when heated to decomposition (Lewis, 1990). Though no products were reported, the calculated hydrolysis half-life at 25 °C and pH 7 is 96 hr (Ellington et al., 1988).

Exposure Limits: ACGIH TLV: TWA 0.05 mg/m^3 ppm, STEL 0.2 mg/m^3.

Formulation Types: Emulsifiable concentrate; granules.

Toxicity: LC_{50} (96 hr) for rainbow trout 13 μg/L and channel catfish 280 μg/L (Worthing and Hance, 1991); acute oral LD_{50} for male and female rats is 3.7 and 1.6 mg/kg, respectively (Hartley and Kidd, 1987), 1 mg/kg (RTECS, 1985).

Uses: Systemic insecticide for control of mites, chewing and sucking insects in fruits and vegetables, cotton, and some ornamentals.

PHOSALONE

Synonyms: Azofene; *S*-((3-Benzoxazolinyl-6-chloro-2-oxo)methyl) *O,O*-diethyl phosphorodithioate; Benzphos; Benzophosphate; Chipman 11974; *S*-(6-Chloro-3-(mercaptomethyl)-2-benzoxazolinone) *O,O*-diethyl phosphorothiolothionate; *S*-((6-Chloro-2-oxo-3(2*H*)-benzoxazolyl)methyl) *O,O*-diethyl phosphorodithioate; *O,O*-Diethyl *S*-(6-chlorobenzoxazolinyl-3-methyl) dithiophosphate; *O,O*-Diethyl *S*-((6-chloro-2-oxobenzoxazolin-3-yl)methyl) phosphorodithioate; *O,O*-Diethyl *S*-(6-chloro-2-oxobenzoxazolin-3-yl)methyl phosphorothiolothionate; 3-Diethyldithiophosphorymethyl-6-chlorobenzoxazolone-2; *O,O*-Diethyl phosphorodithioate, *S*-ester with 6-chloro-3-(mercaptomethyl)-2-benzoxazolinone; ENT 27163; Fozalon; NIA 9241; Niagara 9241; NPH-1091; P-974; Phasolon; Phosalon; Phozalon; **Phosphorodithioic acid *S*-((6-chloro-2-oxo-3(2*H*)-benzoxazolyl)-methyl) *O,O*-diethyl ester;** Phosphorodithioic acid *S*-ester of 6-chloro-3-mercaptomethylbenzoxazyol-2-one; Rhodia RP 11974; RP 11974; Rubitox; Zolon; Zolon PM; Zolone.

Structure:

Designations: CAS: 2310-17-0; mf: $C_{12}H_{15}ClNO_4PS_2$; fw: 367.80; RTECS: TD5175000.

Properties: Colorless crystals with a garlic-like odor. Mp: 45-48 °C; K_H: 7.6 x 10^{-8} atm·m^3/mol at 25 °C (approximate - calculated from water solubility and vapor pressure); log K_{oc}: 3.41 (calculated); log K_{ow}: 3.77-4.38; P-t$_{1/2}$: 71.37 hr (absorbance λ = 241.5 nm, concentration on glass plates = 6.7 μg/cm^2); S_o (g/L): acetone (1,000), acetonitrile (1,000), ethanol (200), *n*-hexane (11); S_w: 1.2, 2.6, and 3.7 mg/L at 10, 20, and 30 °C, respectively; vp: 5.03 x 10^{-7} mmHg at 25 °C.

Transformation Products

Plant: Degrades in plants to chlorbenzoxazolone, formaldehyde, and diethyl phosphorodithioate (Hartley and Kidd, 1987).

Soil: Ambrosi et al. (1977a) studied the persistence and metabolism of phosalone in both moist and flooded Matapeake loam and Monmouth fine sandy loam. Phosalone rapidly degraded (t$_{1/2}$ = 3-7 d) but mineralization to carbon dioxide accounted for only 10% of the loss. The primary degradative pathway proceeded by oxidation of phosalone to give phosalone oxon. Subsequent cleavage of the

O,O-diethyl methyl phosphorodithioate linkage gave 6-chloro-2-benzoxazo-linone. Although 2-amino-5-chlorophenol was not detected in this study, they postulated that the condensation of this compound yielded phenoxazinone.

Chemical/Physical: Emits toxic fumes of chlorine, phosphorus, nitrogen, and sulfur oxides when heated to decomposition (Sax and Lewis, 1987).

Formulation Types: Emulsifiable concentrate; wettable powder; dustable powder.

Toxicity: LC_{50} for goldfish 2 mg/L, bluegill sunfish 0.11 mg/L, and rainbow trout 0.3-0.63 mg/L (Worthing and Hance, 1991); acute oral LD_{50} for male and female rats is 120-175 and 135-170 mg/kg, respectively (Hartley and Kidd, 1987), 85 mg/kg (RTECS, 1985).

Uses: Insecticide; acaracide.

PHOSMET

Synonyms: Appa; *N*-Dimethoxyphosphinothioylthiomethyl)phthalimide; *O,O*-Dimethyl phosphorodithoate *S*-ester with *N*-(mercaptomethyl)phthalimide; *O,O*-Dimethyl *S*-phthalimidomethyl phosphorodithioate; *S*-((1,3-Dioxo-2*H*-isoindol-2-yl)methyl) *O,O*-dimethyl phosphorodithioate; ENT 25705; Imidan; Kemolate; **Phosphorodithioic acid** *S*-((1,3-dihydro-1,3-dioxo-2*H*-isoindol-2-yl)methyl) *O,O*-**dimethyl ester;** Phosphorodithioic acid *O,O*-dimethyl ester with *N*-(mercaptomethyl)phthalimide; Phthalimidomethyl-*O,O*-dimethylphosphorodithioate; Phtalofos; PMP; Prolate; R 1504.

Structure:

Designations: CAS: 732-11-6; mf: $C_{11}H_{12}NO_4PS_2$; fw: 317.33; RTECS: TE2275000.

Properties: Colorless crystals. Technical grades are off-white to pale pink crystals with an offensive odor. Mp: 72.0-72.7 °C (pure), 66.5-69.5 °C (technical grade, 95-98% purity); bp: decomposes rapidly >100 °C; H-t$_{1/2}$ (buffered solution at 20 °C): 13 d (pH 4.5), <12 hr (pH 7), <4 hr (pH 8.3); K_H: 9.4 x 10^{-9} atm·m^3/mol at 25-30 °C (approximate - calculated from water solubility and vapor pressure); log K_{oc}: 2.87-2.90 (calculated); log K_{ow}: 2.78-3.04; P-t$_{1/2}$: 53.25 hr (absorbance λ = 243.0 nm, concentration on glass plates = 6.7 $\mu g/cm^2$); S$_o$ (g/L at 25 °C): acetone (650), benzene (600), kerosene (5), methanol (50), 4-methyl-2-pentanone (300), toluene (300), xylene (250); S$_w$: 22-25 mg/L at 25 °C; vp: 4.52 x 10^{-7} mmHg at 30 °C, 10^{-3} mmHg at 50 °C.

Transformation Products

Plant: In plants, phosmet is degraded to nontoxic compounds (Hartley and Kidd, 1987). Dorough et al. (1966) reported the half-life in Bermuda grass was 6.5 d.

Chemical/Physical: Emits toxic fumes of phosphorus, nitrogen, and sulfur oxides when heated to decomposition (Sax and Lewis, 1987). Though no products were identified, the hydrolysis half-lives at 20 °C were 7.0 d and 7.1 hr at pH 6.1 and pH 7.4, respectively. At 37.5 °C and pH 7.4, the hydrolysis is 1.1 hr (Freed et al., 1979).

Formulation Types: Dustable powder; emulsifiable concentrate; wettable powder; soluble concentrate.

Toxicity: LC$_{50}$ (96 hr) for bluegill sunfish 70 μg/L, channel catfish 11,000 μg/L, chinook salmon 150 μg/L, fathead minnow 7,300 μg/L, rainbow trout 560 μg/L, and small mouth bass 150 μg/L (Verschueren, 1983); acute oral LD$_{50}$ for male and female rats is 113 and 160 mg/kg, respectively (Hartley and Kidd, 1987).

Uses: Nonsystemic acaricide and insecticide used on citrus, fruit, grape, and potato crops.

PHOSPHAMIDON

Synonyms: Apamidon; C 570; 2-Chloro-3-(diethylamino)-1-methyl-3-oxo-1-propenyl dimethyl phosphate; 2-Chloro-2-diethylcarbamoyl-1-methylvinyl dimethylphosphate; CIBA 570; Dimecron; Dimecron 100; Dimethyl 2-chloro-2-diethylcarbamoyl-1-methylvinyl phosphate; *O,O*-Dimethyl *O*-(2-chloro-2-(*N,N*-diethylcarbamoyl)-1-methylvinyl) phosphate; Dimethyl phosphate of chloro-*N,N*-diethyl-3-hydroxycrotonamide; Dixon; ENT 25515; Famfos; ML 97; NCI-C00588; OR 1191; **Phosphoric acid 2-chloro-3-(dimethylamino)-1-methyl-3-oxo-1-propenyl dimethyl ester.**

Structure:

$$CH_3O \underset{CH_3O}{\overset{O}{\underset{\diagup}{\overset{\parallel}{\diagdown}}}} P - OC \overset{CH_3}{=} \underset{Cl}{C}CON \overset{CH_2CH_3}{\underset{CH_2CH_3}{\diagup}}$$

Designations: CAS: 13171-21-6; mf: $C_{10}H_{19}ClNO_5P$; fw: 299.69; RTECS: TC2800000.

Properties: Pale yellow, oily liquid with a faint odor. Mp: -45 °C; bp: 120 °C at 0.001 mmHg, 162 °C at 1.5 mmHg; ρ: 1.2132 at 25/4 °C; H-t$_{1/2}$ (20 °C): 60 d (pH 5), 54 d (pH 7), 12 d (pH 9); K_H: 3.6 x 10^{-6} atm·m^3/mol at 20-22 °C (approximate - calculated from water solubility and vapor pressure); log K_{oc}: 0.63-0.98; log K_{ow}: 0.795; S_o: miscible with most organic solvents except saturated hydrocarbons such as *n*-hexane (33.3 g/kg); S_w: miscible at 22 °C; vp: 2.5 x 10^{-5} mmHg at 20 °C.

Soil properties and adsorption data

Soil	K_d (mL/g)	f_{oc} (%)	K_{oc} (mL/g)	pH
Loamy sand	0.034	0.80	4.25	7.4
Sand	0.057	0.60	9.55	6.5
Silty loam	0.078	1.50	5.20	6.2
Silty loam	0.390	5.40	7.22	7.3
Silt	1.360	25.00	5.44	6.9
Silty loam	5.600	1.74	321.84	5.2

Source: U.S. Department of Agriculture, 1990.

Transformation Products
Chemical/Physical: Emits toxic fumes of chlorine, phosphorus, and nitrogen oxides when heated to decomposition (Sax and Lewis, 1987).

Formulation Types: Emulsifiable concentrate; suspension concentrate; soluble concentrate.

Toxicity: LC_{50} (96 hr) for fathead minnow 100.0 mg/L, bluegill sunfish 4.50 mg/L, and channel catfish 70.0 mg/L (Verschueren, 1983); LC_{50} (24 hr) for rainbow trout 5.0 ppm (Verschueren, 1983); acute oral LD_{50} for rats 17.9-30 mg/kg (Hartley and Kidd, 1987), 10,900 μg/kg (RTECS, 1985).

Uses: Insecticide used to control sap-feeding insects and other pests in a wide variety of crops.

PICLORAM

Synonyms: Amdon grazon; 4-Amino-3,5,6-trichloropicolinic acid; 4-Amino-3,5,6-trichloropyridine-2-carboxylic acid; **4-Amino-3,5,6-trichloro-2-pyridine-carboxylic acid;** ATCP; Borolin; Grazon; K-pin; NCI-C00237; Tordon; Tordon 10K; Tordon 101 mixture.

Structure:

Designations: CAS: 1918-02-1; mf: $C_6H_3Cl_3N_2O_2$; fw: 241.48; RTECS: TJ7525000.

Properties: Colorless to white crystals with a chlorine-like odor. Mp: decomposes at 215-219 °C without melting; fl p (open cup): 35 °C (Tordon 101 formulation), 46 °C (Tordon 22K formulation); pK_a: 2.3 (20 °C), 3.6; K_H: 3.4 x 10^{-10} atm·m^3/mol at 25-35 °C (approximate - calculated from water solubility and vapor pressure); log K_{oc}: 1.41; log K_{ow}: 0.30; S_o (g/L at 25 °C): acetone (19.8), acetonitrile (1.6), benzene (0.2), carbon disulfide (<0.05), ethanol (10.5), ethyl ether (1.2), kerosene (0.01), methylene chloride (0.6), 2-propanol (5.5); S_w: 400-430 mg/L at 25 °C; vp: 6.16 x 10^{-7} mmHg at 35 °C, 1.07 x 10^{-7} mmHg at 45 °C.

Soil properties and adsorption data

Soil	K_d (mL/g)	f_{oc} (%)	K_{oc} (mL/g)	pH	CEC (meq/100 g)
Asquith sandy loam	0.030	1.02	2.91	7.5	--
Asquith sandy loam	0.090	1.04	8.65	6.9	--
Ephrata sandy loam	0.070	0.54	12.96	7.1	--
Fiddletown silty loam	0.976	2.42	40.33	5.6	--
Indian Head clay loam	0.240	2.36	10.17	8.1	--
Indian Head loam	0.100	2.35	4.26	7.8	--
Indian Head clay	0.230	2.70	8.52	8.1	--
Kentwood sandy loam	0.118	0.92	12.83	6.4	--
Kinney clay loam	1.600	2.88	55.56	5.0	16.1
Kinney clay loam	4.600	4.27	107.73	5.2	6.5
Kinney clay loam	2.300	1.44	159.72	5.2	8.9
Lacombe loam	0.750	7.20	10.42	7.9	--

continued

Soil	K_d (mL/g)	f_{oc} (%)	K_{oc} (mL/g)	pH	CEC (meq/100 g)
Linne clay loam	0.409	1.38	29.64	7.4	--
Melfort loam	0.490	6.05	8.10	5.9	--
Melfort loam	0.490	6.00	8.17	6.5	--
Minam loam	0.300	2.19	13.70	7.3	24.4
Minam loam	0.600	3.81	15.75	7.0	28.3
Palouse silty loam	0.310	1.38	22.46	5.7	--
Palouse silty loam	0.553	2.07	26.72	5.7	--
Regina heavy clay	0.100	2.40	4.17	8.0	--
Weyburn Oxbow loam	0.240	3.72	6.45	6.5	--
Weyburn Oxbow loam	0.310	3.75	8.27	7.9	--
Weyburn light loam	0.240	2.50	9.60	8.2	--
Woodcock loam	0.300	4.44	6.76	6.1	12.9
Woodcock loam	0.600	2.48	24.19	5.8	3.6
Woodcock sandy loam	0.409	0.92	44.56	5.6	3.2

Source: Farmer and Aochi, 1974; Gaynor and Volk, 1976; Grover, 1971, 1977.

Transformation Products
Soil: Degrades in soil via cleavage of the chlorine atom at the *m*-position to form 4-amino-5,6-dichloro-2-picolinic acid. Replacement of the chlorine at the *m*-position by a hydroxyl group yields 4-amino-3-hydroxy-5,6-dichloropicolinic acid (Hartley and Kidd, 1987). Other soil metabolites reported include carbon dioxide, chloride ions, 4-amino-6-hydroxy-3,5-dichloropicolinic acid (Meikle et al., 1974), 4-amino-3,5-dichloro-6-hydroxypicolinic acid, and 4-amino-3,5,6-trichloropyridine (Goring and Hamaker, 1971).

Plant: Picloram degraded very slowly in cotton plants releasing carbon dioxide (Meikle et al., 1966). Metabolites identified in spring wheat were 4-amino-2,3,5-trichloropyridine, oxalic acid, and 4-amino-3,5-dichloro-6-hydroxypicolinic acid (Redemann et al., 1968; Plimmer, 1970). In soil, 4-amino-3,5-dichloro-6-hydroxypicolinic acid was the only compound positively identified (Redemann et al., 1968).

Photolytic: The sodium salt of picloram in aqueous solution was readily decomposed by UV light (λ = 300-380 nm). Two chloride ions were formed for each molecule of picloram that reacted. It was postulated that degradation proceeded via a free radical and an ionic mechanism (Mosier and Guenzi, 1973). Hall et al. (1968) exposed picloram (0.02 M and pH >7) to UV light (λ = 253.7 nm) at an intensity of 200 μW/cm^2. After 2 d, 20% degraded releasing two chloride ions per molecule, five unidentified products, and the destruction of the pyridine ring. Replication of this experiment in sunlight produced similar, but less effective and variable results (Hall et al., 1968).

Picloram is rapidly degraded in irradiated aqueous solutions (Kearney et al., 1968). Irradiation of picloram in aqueous solution by UV light (2,537 Å) released two chloride ions for each molecule of picloram degraded. Photodecomposition products included acidic compounds and five methylated derivatives (Plimmer, 1970). When picloram in an aqueous solution (25 °C) was exposed by a high intensity monochromatic UV lamp, dechlorination occurred yielding 4-amino-3,5-dichloro-6-hydroxypicolinic acid which underwent decarboxylation to give 4-amino-3,5-dichloropyridin-2-ol. In addition, decarboxylation of picloram yielded 2,3,5-trichloro-4-pyridylamine which may undergo dechlorination to give 4-amino-3,5-dichloro-6-hydroxypicolinic acid (Burkhard and Guth, 1979).

Chemical/Physical: Emits toxic fumes of nitrogen oxides and chlorides when heated to decomposition (Sax and Lewis, 1987; Lewis, 1990). Releases carbon monoxide, carbon dioxide, chlorine, and ammonia when heated to 900 °C (Kennedy, 1972, 1972a). Reacts with alkalies and amines forming water-soluble salts (Hartley and Kidd, 1987).

Exposure Limits: ACGIH TLV: TWA 10 mg/m^3, STEL 20 mg/m^3.

Symptoms of Exposure: Mild eye and skin irritant. Ingestion of large amounts may cause nausea.

Formulation Types: Aqueous solution; granules; pellets; soluble concentrate.

Toxicity: LC_{50} (96 hr) for fathead minnow 55.3 mg/L and rainbow trout 19.3 mg/L (Worthing and Hance, 1991); acute oral LD_{50} for rats is 8,200 mg/kg (Ashton and Monaco, 1991), 2,898 mg/kg (RTECS, 1985).

Uses: Systemic herbicide used to control most broadleaf weeds on grassland and noncrop areas. Use as a pesticide is restricted.

PINDONE

Synonyms: Chemrat; 2-(2,2-Dimethyl-1-oxopropyl)-1*H*-indene-1,3(2*H*)-dione; Pivacin; Pival; 2-Pivaloylindane-1,3-dione; 2-Pivaloyl-1,3-indanedione; Pivalyl; Pivalyl indandione; 2-Pivalyl-1,3-indandione; Pivalyl valone; Tri-Ban; UN 2472.

Structure:

Designations: CAS: 83-26-1; DOT: 2472; mf: $C_{14}H_{14}O_3$; fw: 230.25; RTECS: NK6300000.

Properties: Bright yellow crystals. Mp: 108-110 °C; log K_{oc}: 2.95 (calculated); log K_{ow}: 3.18 (calculated); S_w: 18 ppm at 25 °C.

Exposure Limits: NIOSH REL: IDLH 200 mg/m³; OSHA PEL: TWA 0.1 mg/m³; ACGIH TLV: TWA 0.1 mg/m³.

Symptoms of Exposure: Nosebleed, excess bleeding minor cuts, bruises; smokey urine, black tarry stools; pain abdominal, back.

Formulation Types: Grain bait; tracking powder.

Toxicity: LC_{50} (96 hr) for rainbow trout 0.21 mg/L and bluegill sunfish 1.6 mg/L (Hartley and Kidd, 1987); acute oral LD_{50} for rats 280 mg/kg (RTECS, 1985).

Uses: Insecticide; rodenticide.

PROFENOFOS

Synonyms: *O*-(4-Bromo-2-chlorophenyl) *O*-ethyl *S*-propyl phosphorothioate; CGA 15324; Curacron; Phosphorothioic acid *O*-(4-bromo-2-chlorophenyl)-*O*-ethyl-*S*-propyl ester; Polycron; Selecron.

Structure:

Designations: CAS: 41198-08-7; mf: $C_{11}H_{15}BrClO_3PS$; fw: 373.64; RTECS: TE6975000.

Properties: Pale yellow liquid. Mp: <25 °C; bp: 110 °C at 0.001 mmHg; *ρ*: 1.455 at 20/4 °C; H-t$_{1/2}$ (20 °C): 93 d (pH 5), 14.6 d (pH 7), 5.7 hr (pH 9); K_H: 2.4 x 10^{-7} atm·m^3/mol at 20 °C (approximate - calculated from water solubility and vapor pressure); log K_{oc}: 2.55-3.46; log K_{ow}: 1.9; S_o: miscible with most organic solvents including aromatic and chlorinated hydrocarbons; S_w: 20 mg/L at 20 °C; vp: 9.75 x 10^{-6} mmHg at 20 °C.

Soil properties and adsorption data

Soil	K_d (mL/g)	f_{oc} (%)	K_{oc} (mL/g)	pH
Sand	4.6	1.28	356	7.8
Sand	20.2	0.70	2,902	6.3
Sandy loam	22.2	2.09	1,062	6.1
Sandy-silty loam	55.6	3.25	1,711	6.7
Silty loam	5.6	1.74	322	5.20

Source: U.S. Department of Agriculture, 1990.

Transformation Products

Photolytic: When profenofos in an aqueous buffer solution (pH 7.0) was exposed to filtered UV light (λ >290 nm) for 24 hr at 25 and 50 °C, 29 and 61% respectively decomposed to 4-bromo-2-chlorophenol and 4-bromo-2-chlorophenyl ethyl hydrogen phosphate (Burkhard and Guth, 1979).

Chemical/Physical: Emits toxic fumes of bromine, chlorine phosphorus, and sulfur oxides when heated to decomposition (Sax and Lewis, 1987).

Formulation Types: Water-dispersible granules; granules; suspension concentrate; wettable powder.

Toxicity: LC_{50} (96 hr) for rainbow trout 80 μg/L, bluegill sunfish 300 μg/L, and crucian carp 90 μg/L (Hartley and Kidd, 1987); acute oral LD_{50} for rats 358 mg/kg (Hartley and Kidd, 1987), 400 mg/kg (RTECS, 1985).

Uses: Insecticide; acaricide.

PROMETON

Synonyms: 2,4-Bis(isopropylamino)-6-methoxy-s-triazine; 2,6-Diisopropyl-amino-4-methoxytriazine; *N,N'*-Diisopropyl-6-methoxy-1,3,5-triazine-2,4-diyldiamine; G 31435; Gesafram; Gesafram 50; 2-Methoxy-4,6-bis(isopropyl-amino)-1,3,5-triazine; **6-Methoxy-*N,N'*-bis(1-methylethyl)-1,3,5-triazine-2,4-diamine**; Methoxypropazine; Ontracic 800; Ontrack; Ontrack-WE-2; Pramitol; Primatol; Primatol 25E; Prometone.

Structure:

$$CH_3O \quad N \quad NHCH(CH_3)_2$$
$$N \quad N$$
$$NHCH(CH_3)_2$$

Designations: CAS: 1610-18-0; mf: $C_{10}H_{19}N_5O$; fw: 225.30; RTECS: XY4200000.

Properties: Colorless to white crystals or powder. Mp: 91-92 °C; ρ: 1.088 at 20/4 °C; fl p: nonflammable; pK_a: 4.3 at 21 °C; K_H: 8.9 x 10^{-10} atm·m³/mol at 20 °C (approximate - calculated from water solubility and vapor pressure); log K_{oc}: 1.92-2.24; log K_{ow}: 2.69, 2.99; S_o (g/L): acetone (300), benzene (>250), methanol (600), methylene chloride (350), toluene (250); S_w: 750 mg/L at 20 °C; vp: 2.3 x 10^{-6} mmHg at 20 °C, 7.9 x 10^{-6} mmHg at 30 °C.

Soil properties and adsorption data

Soil	K_d (mL/g)	f_{oc} (%)	K_{oc} (mL/g)	pH	CEC (meq/100 g)
Bates silty loam	0.60	0.80	75	6.5	9.3
Baxter silty clay loam	1.70	1.21	141	6.0	11.2
Chillum silty loam	5.40	2.54	213	4.6	7.6
Clarksville silty clay loam	2.20	0.80	275	5.7	5.7
Cumberland silty loam	0.50	0.69	72	6.4	6.5
Eldon silty loam	1.30	1.73	75	5.9	12.9
Gerald silty loam	6.00	1.55	387	4.7	11.0
Grundy silty clay loam	6.30	2.07	304	5.6	13.5
Hagerstown silty clay loam	4.30	2.48	173	5.5	12.5
Knox silty loam	6.40	1.67	383	5.4	18.8
Lakeland silty loam	1.00	1.90	53	6.2	2.9
Lebanon silty loam	7.80	1.04	750	4.9	7.7

continued

Soil	K_d (mL/g)	f_{oc} (%)	K_{oc} (mL/g)	pH	CEC (meq/100 g)
Lindley loam	4.70	0.86	546	4.7	6.9
Lintonia loamy sand	0.70	0.34	206	5.3	3.2
Loam	2.90	1.68	173	6.9	--
Marian silty loam	14.90	0.80	1,863	4.6	9.9
Marshall silty clay loam	8.80	2.42	364	5.4	21.3
Menfro silty loam	1.20	1.38	87	5.3	9.1
Newtonia silty loam	2.40	0.92	261	5.2	8.8
Oswego silty clay loam	3.40	1.67	204	6.4	21.0
Putnam silty loam	2.80	1.09	257	5.3	12.3
Salix loam	4.60	1.21	380	6.3	17.9
Sandy loam	2.61	1.74	150	6.1	--
Sandy loam	1.20	1.22	98	7.0	--
Sandy clay loam	2.40	2.90	83	7.0	--
Sarpy loam	1.50	0.75	200	7.1	14.3
Sharkey clay	55.20	1.44	3,833	5.0	28.2
Shelby loam	22.30	2.07	1,077	4.3	20.1
Summit silty clay	11.60	2.82	411	4.8	35.1
Union silty loam	6.10	1.04	586	5.4	6.8
Wabash clay	17.00	1.27	1,339	5.7	40.3
Waverley silty loam	3.80	1.15	330	5.6	12.8
Wehadkee silty loam	3.90	1.09	358	5.6	10.2

Source: Harris, 1966; Talbert and Fletchall, 1965; U.S. Department of Agriculture, 1990.

Transformation Products
Soil: Degrades in soil yielding hydroxy metabolites and dealkylation of the sidechains (Hartley and Kidd, 1987).

Photolytic: Pelizzetti et al. (1990) studied the aqueous photocatalytic degradation of prometon and other s-triazines (ppb level) using simulated sunlight (λ >340 nm) and titanium dioxide as a photocatalyst. Prometon rapidly degraded forming cyanuric acid, nitrates, the intermediate tentatively identified as 2,4-diamino-6-hydroxy-N,N'-bis(1-methylethyl)-1,3,5-triazine, and other intermediate compounds similar to those found for atrazine. Mineralization of cyanuric acid to carbon dioxide was not observed (Pelizzetti et al., 1990).

Formulation Types: Emulsifiable concentrate; wettable powder.

Toxicity: LC_{50} (96 hr) for rainbow trout 12 mg/L, bluegill sunfish 40 mg/L, crucian carp 70 mg/L (Worthing and Hance, 1991), and goldfish 8.6 mg/L (Hartley and Kidd, 1987); acute oral LD_{50} of the Prometon EC formulation for rats is 2,300 mg/kg (Ashton and Monaco, 1991), 503 mg/kg (RTECS, 1985).

Uses: Nonselective preemergence and postemergence herbicide used to control most annual and perennial broadleaf weeds, grasses, and brush weeds on noncrop land.

PROMETRYN

Synonyms: 2,4-Bis(isopropylamino)-6-methylmercapto-*s*-triazine; 2,4-Bis-(isopropylamino)-6-methylthio-*s*-triazine; 2,4-Bis(isopropylamino)-6-methylthio-1,3,5-triazine; **N,N'-Bis(1-methylethyl)-6-(methylthio)-1,3,5-triazine-2,4-diamine**; Caparol; Caparol 80W; Cotton-Pro; *N,N'*-Diisopropyl-6-methylthio-1,3,5-triazine-2,4-diyldiamine; G 34161; Gesagard; Merkazin; 2-Methylmercapto-4,6-bis(isopropylamino)-*s*-triazine; 2-Methylthio-4,6-bis(isopropylamino)-*s*-triazine; Polisin; Primatol; Primatol Q; Prometrex; Prometrin; Prometryne; Selektin; Sesagard.

Structure:

$$CH_3S \quad \diagdown \quad N \diagdown \quad NHCH(CH_3)_2$$

$$NHCH(CH_3)_2$$

Designations: CAS: 7287-19-6; mf: $C_{10}H_{19}N_5S$; fw: 241.37; RTECS: XY4390000.

Properties: Colorless to white crystalline solid. Mp: 118-120 °C; ρ: 1.157 at 20/4 °C; fl p: nonflammable; pK_a: 4.05; H-t$_{1/2}$ at 25 °C: 22 d (0.1N hydrochloric acid solution), $\approx$ 500 yr (pH 7), $\approx$ 30 yr (0.01N sodium hydroxide solution); pK_a: 4.05 at 21 °C; K_H: 4.9 x 10^{-9} atm·m^3/mol at 20 °C (approximate - calculated from water solubility and vapor pressure); log K_{oc}: 2.28-2.79; log K_{ow}: 3.34, 3.46; S_o (g/L at 20 °C): acetone (240), *n*-hexane (5.5), methanol (160), methylene chloride (300), toluene (170); S_w: 48 mg/L at 20 °C; vp: 10^{-6} mmHg at 20 °C, 4 x 10^{-6} mmHg at 30 °C.

Soil properties and adsorption data

Soil	K_d (mL/g)	f_{oc} (%)	K_{oc} (mL/g)	pH	CEC (meq/100 g)
Bates silty loam	1.60	0.80	200	6.5	9.3
Baxter silty clay loam	4.30	1.21	355	6.0	11.2
Brewer clay loam	9.95	1.61	618	5.8	13.5
Cape Fear sandy loam	844.00	5.34	15,805	--	10.3
Chillum silt loam	12.90	2.54	508	4.6	7.6
Clarksville silty clay loam	5.10	0.80	638	5.7	5.7
Cobb sand	0.66	0.34	194	7.3	3.8
Cobb sand + 2% muck	28.90	1.21	2,388	5.3	9.0
Cumberland silty loam	1.40	0.69	203	6.4	6.5
Eldon silty loam	3.60	1.73	208	5.9	12.9

Soil	K_d (mL/g)	f_{oc} (%)	K_{oc} (mL/g)	pH	CEC (meq/100 g)
Gerald silty loam	9.40	1.55	606	4.7	11.0
Grundy sandy clay loam	9.20	2.07	444	5.6	13.5
Hagerstown silty clay loam	10.30	2.48	415	5.5	12.5
Kebabib silty loam	9.00	1.04	865	4.9	7.7
Knox silty loam	8.40	1.67	503	5.4	18.8
Lakeland silty loam	1.90	1.90	100	6.2	2.9
Lebanon silty loam	9.00	1.04	865	4.9	7.7
Lindley loam	7.90	0.86	919	4.7	6.9
Lintonia loamy sand	0.90	0.34	265	5.3	3.2
Loam	9.95	1.62	614	5.8	--
Loam	4.90	1.04	471	6.3	--
Marian silty loam	14.20	0.80	1,775	4.6	9.9
Marshall silty clay loam	12.30	2.42	508	5.4	21.3
Menfro silty loam	3.30	1.38	239	5.3	9.1
Newtonia silty loam	3.50	0.92	380	5.2	8.8
Norfolk fine sandy loam	87.00	0.99	8,788	--	2.3
Oswego silty clay loam	5.00	1.67	299	6.4	21.0
Port silty clay	4.90	1.04	471	6.3	17.9
Putnam silty loam	3.80	1.09	349	5.3	12.3
Rains fine sandy loam	139.00	1.45	9,586	--	7.1
Salix loam	6.40	1.21	529	6.3	17.9
Sand	0.66	0.35	190	7.3	--
Sandy loam	3.49	0.75	463	5.7	--
Sarpy loam	2.90	0.75	387	7.1	14.3
Sharkey clay	43.40	1.44	3,014	5.0	28.2
Shelby loam	22.10	2.07	1,068	4.3	20.1
Soil 1	3.10	0.51	608	--	10.6
Soil 2	5.00	1.56	321	--	10.3
Soil 3	5.70	1.76	324	--	22.3
Soil 4	10.40	3.82	272	--	26.9
Soil 5	39.80	13.61	292	--	31.2
Soil 6	65.40	19.98	327	--	83.1
Summit silty clay	17.70	2.82	628	4.8	35.1
Teller fine sandy loam	3.49	0.75	465	8.6	5.7
Union silty loam	8.60	1.04	827	5.4	6.8
Wabash clay	17.30	1.27	1,362	5.7	40.3
Waverley silty loam	5.80	1.15	504	6.4	12.8

continued

Soil	K_d (mL/g)	f_{oc} (%)	K_{oc} (mL/g)	pH	CEC (meq/100 g)
Wehadkee silty loam	0.62	1.09	57	5.6	10.2

Source: Harris, 1966; Kozak et al., 1983; Murray et al., 1975; Talbert and Fletchall, 1965; U.S. Department of Agriculture, 1990; Walker and Crawford, 1970.

Transformation Products

Soil: In soil and plants, the methylthio group is oxidized. The proposed degradative pathway is the formation of the corresponding sulfoxide and sulfone derivatives of prometryn followed by oxidation of the latter to forming hydroxypropazine (Hartley and Kidd, 1987; Kearney and Kaufman, 1976). Cook and Hütter (1982) reported that bacterial cultures degraded prometryne to form the corresponding hydroxy derivative (hydroxyprometryne). ^{14}C-Prometryn was incubated in an organic soil for 1 yr. It was observed that 57.4% of the bound residues were comprised of prometryn (>50%), hydroxypropazine, mono-*n*-dealkylated prometryn, mono-*N*-dealkylated hydroxypropazine, a didealkylated compound [2-(methylthio)-4-amino-6-(isopropylamino)-*s*-triazine], and unidentified methanol soluble products (Khan and Hamilton, 1980; Khan, 1982). The reported half-life in soil is 60 d (Jury et al., 1987). Hydroxyprometryn and ammeline were reported as hydrolysis metabolites in soil (Somasundaram et al., 1991).

Photolytic: When prometryn in aqueous solution was exposed to UV light for 3 hr, the herbicide was completely converted to hydroxypropazine. Irradiation of soil suspensions containing prometryn was found to be more resistant to photodecomposition. About 75% of the applied amount was converted to hydroxypropazine after 72 hr of exposure (Khan, 1982). The UV (λ = 253.7 nm) photolysis of prometryn in water, methanol, ethanol, *n*-butanol, and benzene, yielded 2-methylthio-4,6-bis(isopropylamino)-*s*-triazine. At wavelengths >300 nm, photodegradation was not observed (Pape and Zabik, 1970). Khan and Gamble (1983) also studied the UV irradiation (λ = 253.7 nm) of prometryn in distilled water and dissolved humic substances. In distilled water, 2-hydroxy-4,6-bis(isopropylamino)-*s*-triazine and 4,6-bis(isopropylamino)-*s*-triazine formed as major products. In the presence of humic/fulvic acids, 4-amino-6-(isopropylamino)-*s*-triazine was also formed (Khan and Gamble, 1983). Pelizzetti et al. (1990) studied the aqueous photocatalytic degradation of prometryn and other *s*-triazines (ppb level) using simulated sunlight (λ >340 nm) and titanium dioxide as a photocatalyst. Prometryn rapidly degraded forming cyanuric acid, nitrates, sulfates, the intermediate tentatively identified as 2,4-diamino-6-hydroxy-*N,N'*-bis(1-methylethyl)-1,3,5-triazine, and other intermediate compounds similar to those found for atrazine. It was suggested that the appearance of sulfate ions was due to the attack of the methylthio group at the number two position. Mineralization of cyanuric acid to carbon dioxide was not observed (Pelizzetti et al., 1990).

Symptoms of Exposure: Eye and skin irritant.

Formulation Types: Suspension concentrate (4 lb/gal); wettable powder.

Toxicity: LC_{50} (96 hr) for rainbow trout 2.5 mg/L, bluegill sunfish 10.0 mg/L, goldfish 3.5 mg/L, and carp 8 mg/L (Hartley and Kidd, 1987); acute oral LD_{50} of the 80% formulation for rats is 3,800 mg/kg (Ashton and Monaco, 1991), 2,100 mg/kg (RTECS, 1985).

Uses: Selective herbicide used to control many annual grass and broadleaf weeds in cotton, celery, and peas.

PROPACHLOR

Synonyms: Albrass; Bexton; Bexton 4L; 2-Chloro-N-isopropylacetanilide; α-Chloro-N-isopropylacetanilide; **2-Chloro-N-(1-methylethyl)-N-phenylacet-amide;** CP 31393; N-Isopropyl-2-chloroacetanilide; N-Isopropyl-α-chloro-acetanilide; Niticid; Propachlore; Prolex; Ramrod; Ramrod 65; Satecid.

Structure:

Designations: CAS: 1918-16-7; mf: $C_{11}H_{14}ClNO$; fw: 211.67; RTECS: AE1575000.

Properties: Pale tan solid. Mp: 77 °C (pure), 67-76 °C (technical); bp: 110 °C at 0.03 mmHg, decomposes at 170 °C; ρ: 1.242 at 25/4 °C; fl p: nonflammable; K_H: 1.1 x 10^{-7} atm·m^3/mol at 25 °C (approximate - calculated from water solubility and vapor pressure); log K_{oc}: 2.07-2.11 (calculated); log K_{ow}: 1.61; S_o (g/kg at 25 °C): acetone (448), benzene (737), carbon tetrachloride (174), chloroform (602), ethanol (408), ethyl ether (219), toluene (342), xylene (239); S_w: 613-700 mg/L at 25 °C; vp: 2.25 x 10^{-4} mmHg at 25 °C.

Transformation Products

Plant: In corn seedlings and excised leaves of corn, sorghum, sugarcane, and barley, propachlor was metabolized to at least three water-soluble products. Two of these metabolites were identified as a γ-glutamylcysteine conjugate of propachlor and a glutathione conjugate of propachlor. It was postulated that both compounds were intermediate compounds in corn seedlings since they were not detected 3 d following treatment (Lamoureux et al., 1971).

Photolytic: When propachlor in an aqueous ethanolic solution was irradiated with UV light (λ = 290 nm) for 5 hr, 80% decomposed to the following cyclic photoproducts: N-isopropyloxindole, N-isopropyl-3-hydroxyoxindole and a spiro compound. Irradiation of propachlor in an aqueous solution containing riboflavin as a sensitizer resulted in completed degradation of the parent compound. m-Hydroxypropachlor was the only compound identified in trace amounts which formed via ring hydroxylation (Rejtö et al., 1984). Hydrolyzes under alkaline conditions forming N-isopropylaniline (Sittig, 1985) which is also a product of microbial metabolism (Novick et al., 1986).

Chemical/Physical: Emits toxic fumes of nitrogen oxides and chlorine when heated to decomposition (Sax and Lewis, 1987). Hydrolyzes under alkaline conditions forming *N*-isopropylaniline (Sittig, 1985) which is also a product of microbial metabolism (Novick et al., 1986).

Symptoms of Exposure: May cause allergic skin reactions upon contact.

Formulation Types: Granules (20%); suspension concentrate or flowable liquid (4 lb/gal); wettable powder.

Toxicity: LC_{50} (96 hr) for bluegill sunfish >1.4 mg/L and rainbow trout 0.17 mg/L (Hartley and Kidd, 1987). For the Ramrod flowable formulation the LC_{50} (96 hr) for bluegill sunfish and rainbow trout is 1.6 and 0.42 mg/L, respectively (Humburg et al., 1989); acute oral LD_{50} for rats is 710 mg/kg (RTECS, 1985).

Uses: Selective preemergence herbicide used to control most annual grasses and some broadleaf weeds in brassicas, corn, cotton, flax, leeks, maize, milo, onions, peas, roses, ornamental trees and shrubs, soybeans, sugarcane.

PROPANIL

Synonyms: Bay 30130; Chem-Rice; Crystal Propanil-4; DCPA; *N*-(3,4-**Dichloro-phenyl)propanamide**; *N*-(3,4-Dichlorophenyl)propionamide; Dichloropropion-anilide; 3',4'-Dichloropropionanilide; DPA; Dipram; Erban; Erbanil; Farmco propanil; FW-734; Grascide; Herbax; Herbax technical; Montrose Propanil; Propanex; Propanid; Propanide; Prop-Job; Propionic acid 3,4-dichloroanilide; Riselect; Rogue; Rosanil; S 10165; Stam; Stam F-34; Stam LV 10; Stam M-4; Stampede; Stampede 360; Stampede 3E; Stam Supernox; Strel; Supernox; Surcopur; Surpur; Synpran N; Vertac; Wham EZ.

Structure:

$$NHOCCH_2CH_3$$

(structure of propanil: 3,4-dichlorophenyl ring with Cl, Cl substituents)

Designations: CAS: 709-98-8; mf: $C_9H_9Cl_2NO$; fw: 218.09; RTECS: UE4900000.

Properties: Colorless to light brown to gray-black crystals. Technical grades are light-brown to grayish-black. Mp: 92-93 °C (pure), 88-91 °C (technical); ρ: 1.25 at 25/4 °C; K_H: 3.6 x 10^{-8} atm·m^3/mol at 20 °C (approximate - calculated from water solubility and vapor pressure); log K_{oc}: 2.19; log K_{ow}: 2.03, 2.29; S_o (g/L at 25 °C): acetone (1,700), benzene (70), ethanol (1,100), *n*-hexane (<1), methylene chloride (>200), 2-propanol (>200), toluene (50-100); S_w: 130 mg/L at 20 °C, 225 mg/L at 25 °C; vp: 2 x 10^{-7} mmHg at 20 °C.

Transformation Products

Soil: Propanil degrades in soil forming 3,4-dichloroaniline (Bartha, 1968; Bartha and Pramer, 1970; Duke et al., 1991; Pothuluri et al., 1991) which is further degraded by microbial peroxidases to 3,3',4,4'-tetrachloroazobenzene (Bartha and Pramer, 1967; Bartha et al., 1968), 3,3',4,4'-tetrachloroazooxybenzene (Bartha and Pramer, 1970), 4-(3,4-dichloroanilo)-3,3',4,4'-tetrachloroazobenzene (Linke and Bartha, 1970), and 1,3-bis(3,4-dichlorophenyl)triazine (Plimmer et al., 1970). Under aerobic conditions, propanil in a biologically active, organic-rich pond sediment underwent dechlorination at the *para-* position forming *N*-(3-chloro-phenyl)propanamide (Stepp et al., 1985). Residual activity in soil is limited to approximately 3-4 mo (Hartley and Kidd, 1987).

Plant: In rice plants, propanil is rapidly hydrolyzed via an aryl acylamidase enzyme isolated by Frear and Still (1968) forming the nonphototoxic compounds (Ashton and Monaco, 1991) 3,4-dichloroaniline, propionic acid (Hatzios, 1991;

Matsunaka, 1969; Menn and Still, 1977), and a 3′,4′-dichloroaniline-lignin complex. This complex was identified as a metabolite of N-(3,4-dichlorophenyl)glucosylamine, a 3,4-dichloroaniline saccharide conjugate and a 3,4-dichloroaniline sugar derivative (Yi et al., 1968). In a rice field soil under anaerobic conditions, however, propanil underwent amide hydrolysis and dechlorination at the *para* position forming 3,4-dichloroaniline and *m*-chloroaniline (Pettigrew et al., 1985). In addition, propanil may degrade indirectly via an initial oxidation step resulting in the formation of 3,4-dichlorolacetanilide which is further hydrolyzed to 3,4-dichloroaniline and lactic acid (Hatzios, 1991). In an earlier study, four metabolites were identified in rice plants, two of which were positively identified as 3,4-dichloroaniline and N-(3,4-dichlorophenyl)glucosylamine (Still, 1968).

Photolytic: Photoproducts reported from the sunlight irradiation of propanil (200 mg/L) in distilled water were 3′-hydroxy-4′-chloropropionanilide, 3′-chloro-4′-hydroxypropionanilide, 3′,4′-dihydroxypropionanilide, 3′-chloropropionanilide, 4′-chloropropionanilide, propionanilide, 3,4-dichloroaniline, 3-chloroaniline, propionic acid, propionamide, 3,3′,4,4′-tetrachloroazobenzene, and a dark polymeric humic substance. The photolysis products resulted from the reductive dechlorination, replacement of chlorine substituents by hydroxyl groups, formation of propionamide, hydrolysis of the amide group, and azobenzene formation (Moilanen and Crosby, 1972). Tanaka et al. (1985) studied the photolysis of propanil (100 mg/L) in aqueous solution using UV light (λ = 300 nm) or sunlight. After 26 d of exposure to sunlight, propanil degraded forming a trichlorinated biphenyl product (<1% yield) and hydrogen chloride (Tanaka et al., 1985).

Chemical/Physical: Hydrolyzes in acidic and alkaline media to propionic acid (Worthing and Hance, 1991) and 3,4-dichloroaniline (Sittig, 1985; Worthing and Hance, 1991). The half-life of propanil in a 0.50 N sodium hydroxide solution at 20 °C was determined to be 6.6 d (El-Dib and Aly, 1976).

Symptoms of Exposure: Inhalation of vapors may irritate the nose and throat and cause drowsiness, stupor, and unconsciousness. Irritates the eyes and skin. Ingestion causes dizziness, stupor, drowsiness, fever, and blue lips and fingernails.

Formulation Types: Emulsifiable concentrate.

Toxicity: LC_{50} (48 hr) for carp 13 mg/L, goldfish 14 mg/L, Japanese killifish 14 mg/L, and mosquitofish 11 mg/L (Hartley and Kidd, 1987); acute oral LD_{50} for rats is 1,400 mg/kg (Ashton and Monaco, 1991), 367 mg/kg (RTECS, 1985).

Uses: Selective preemergence and postemergence herbicide used to control many grasses and broadleaf weeds in potatoes, rice, and wheat.

PROPAZINE

Synonyms: 2,4-Bis(isopropylamino)-6-chloro-s-triazine; 2-Chloro-4,6-bis(isopropylamino)-s-triazine; **6-Chloro-N,N'-bis(1-methylethyl)-1,3,5-triazine-2,4-diamine;** 6-Chloro-N,N'-diisopropyl-1,3,5-triazine-2,4-diyldiamine; G 30028; Geigy 30028; Gesamil; Maax; Milogard; Milogard 4L; Milogard 80W; Plantulin; Primatol P; Propasin; Propazin; Prozinex.

Structure:

$(CH_3)_2CHNH$ N   Cl

$NHCH(CH_3)_2$

Designations: CAS: 139-40-2; mf: $C_9H_{16}ClN_5$; fw: 230.09; RTECS: XY5300000.

Properties: Colorless, crystalline solid. Mp: 212 °C; ρ: 1.162 at 20/4 °C; fl p: nonflammable; pK_a: 1.85 at 22 °C; K_H: 9.9 x 10^{-9} atm·m^3/mol at 20 °C (approximate - calculated from water solubility and vapor pressure); log K_{oc}: 1.69-2.56; log K_{ow}: 2.91, 2.94; P-t$_{1/2}$: 108.17 hr (absorbance λ = 222.5 nm, concentration on glass plates = 6.7 μg/cm^2); S_o (g/kg at 20 °C): benzene (6.2), carbon tetrachloride (2.5), ethyl ether (5.0), toluene (6.2); S_w: 8.5 mg/L at 20 °C; vp: 2.9 x 10^{-8} mmHg at 20 °C, 1.6 x 10^{-7} mmHg at 30 °C.

Soil properties and adsorption data

Soil	K_d (mL/g)	f_{oc} (%)	K_{oc} (mL/g)	pH	CEC (meq/100 g)
Bates silty loam	0.70	0.80	88	6.5	9.3
Baxter silty clay loam	1.90	1.21	157	6.0	11.2
Chillum silty loam	4.60	2.54	181	4.6	7.6
Clarksville silty clay loam	2.10	0.80	263	5.7	5.7
Cumberland silty loam	0.70	0.69	101	6.4	6.5
Eldon silty loam	1.80	1.73	104	5.9	12.9
Evouettes	1.15	2.09	55	6.1	--
Gerald silty loam	1.80	1.55	116	4.7	11.0
Grundy silty clay loam	2.80	2.07	135	5.6	13.5
Hagerstown silty clay loam	3.70	2.48	149	5.5	12.5
Hickory Hill silt	11.90	3.27	363	--	--
Knox silty loam	2.70	1.67	162	5.4	18.8
Lakeland loamy sand	0.80	1.90	49	6.2	2.9
Lebanon silty loam	2.00	1.04	192	4.9	7.7

Soil	K_d (mL/g)	f_{oc} (%)	K_{oc} (mL/g)	pH	CEC (meq/100 g)
Lindley loam	2.20	0.86	256	4.7	6.9
Lintonia loamy sand	0.10	0.34	29	5.3	3.2
Marian silty loam	2.10	0.80	263	4.6	9.9
Marshall silty clay loam	3.00	2.42	124	5.4	21.3
Menfro silty loam	1.80	1.38	130	5.3	9.1
Newtonia silty loam	1.40	0.92	152	5.2	8.8
Oswego silty clay loam	1.90	1.67	114	6.4	21.0
Putnam silty loam	1.10	1.09	101	5.9	12.3
Salix loam	1.90	1.21	157	6.3	17.9
Sarpy loam	1.20	0.75	160	7.1	14.3
Sharkey clay	3.00	1.44	208	5.0	28.2
Shelby loam	2.80	2.07	135	4.3	20.1
Soil 1	0.90	0.51	176	--	10.6
Soil 2	3.00	1.56	192	--	10.3
Soil 3	3.00	1.76	170	--	22.3
Soil 4	3.50	3.82	92	--	26.9
Soil 5	19.70	13.61	145	--	31.2
Soil 6	20.50	19.98	103	--	83.1
Summit silty clay	3.40	2.82	121	4.8	35.1
Union silty loam	2.40	1.04	231	5.4	6.8
Vertroz	4.69	3.25	144	6.7	--
Wabash clay	3.10	1.27	244	5.7	40.3
Waverley silty loam	2.00	1.15	174	6.4	12.8
Wehadkee silty loam	1.60	1.09	147	5.6	10.2

Source: Burkhard and Guth, 1981; Harris, 1966; Talbert and Fletchall, 1965; Walker and Crawford, 1970.

Transformation Products

Soil: Undergoes microbial degradation in soil forming hydroxypropazine (Harris, 1967). Dealkylation of both substituted amino groups is presumably followed by ring opening and decomposition (Hartley and Kidd, 1987). Under laboratory conditions, the half-lives of propazine in a Hatzenbühl soil (pH 4.8) and Neuhofen soil (pH 6.5) at 22 °C were 62 and 127 d, respectively (Burkhard and Guth, 1981).

Photolytic: Irradiation of propazine in methanol afforded prometone (2-methoxy-4,6-bis(isopropylamino-*s*-triazine). Photodegradation of propazine in methanol did not occur when irradiated at wavelengths >300 nm (Pape and Zabik, 1970).

Chemical/Physical: In aqueous solutions, propazine is converted exclusively to hydroxypropazine (2-hydroxy-4,6-bisisopropylamino)-s-triazine by UV light (λ = 253.7 nm) (Pape and Zabik, 1970).

Symptoms of Exposure: Eye and skin irritant.

Formulation Types: Water-dispersible granules; wettable powder; liquid (4 lb/gal).

Toxicity: LC_{50} (96 hr) for rainbow trout 17.5 mg/L, bluegill sunfish >100 mg/L, and goldfish >32.0 mg/L (Hartley and Kidd, 1987); acute oral LD_{50} of the 80% formulation for rats is >5,000 mg/kg (Ashton and Monaco, 1991), 3,840 mg/kg (RTECS, 1985).

Uses: Selective preemergence herbicide used to control annual grasses and broadleaf weeds in milo and sweet sorghum.

PROPHAM

Synonyms: Ban-Hoe; Beet-Kleen; Carbanilic acid isopropyl ester; Chem-Hoe; IFC; IFK; INPC; IPC; IsoPPC; Isopropyl carbanilate; Isopropyl carbanilic acid ester; Isopropyl phenylcarbamate; Isopropyl *N*-phenylurethane; *O*-Isopropyl *N*-phenyl carbamate; 1-Methylethyl phenylcarbamate; Ortho grass killer; **Phenylcarbamic acid 1-methylethyl ester;** *N*-Phenyl isopropylcarbamate; Premalox; Profam; Prophame; Triherbide; Triherbide-IPC; Tuberit; Tuberite; USAF D-9; Y 2.

Structure:

$$NHCOOCH(CH_3)_2$$

Designations: CAS: 122-42-9; mf: $C_{10}H_{13}NO_2$; fw: 179.22; RTECS: FD9100000.

Properties: Colorless to light tan crystals. Technical grade is a light tan solid. Mp: 87-88 °C; bp: sublimes but decomposes >150 °C; ρ: ≈ 1.09 at 30/4 °C; fl p: nonflammable; log K_{oc}: 1.71; log K_{ow}: 1.93 (calculated); S_o: soluble in most organic solvents; S_w: 32-250 mg/L at 20-25 °C; vp: sublimes at room temperature.

Transformation Products

Biological: Rajagopal et al. (1989) reported that *Achromobacter* sp. and an *Arthrobacter* sp. utilized propham as a sole carbon source. Metabolites identified were *N*-phenylcarbamic acid, aniline, catechol, monoisopropyl carbonate, 2-propanol, and carbon dioxide (Rajagopal et al., 1989).

Soil: Readily degraded by soil microorganisms forming aniline and carbon dioxide (Humburg et al., 1989). The reported half-life in soil is approximately 15 and 5 d at 16 and 29 °C, respectively (Hartley and Kidd, 1987).

Plant: The major plant metabolite which was identified from soybean plants is isopropyl *N*-2-hydroxycarbanilate (Hartley and Kidd, 1987; Humburg et al., 1989).

Chemical/Physical: Emits toxic fumes of nitrogen oxides when heated to decomposition (Sax and Lewis, 1987).

Formulation Types: Dustable powder; emulsifiable concentrate; granules (15%); suspension concentrate (3 and 4 lb/gal); wettable powder.

Toxicity: LC_{50} (48 hr) for bluegill sunfish 32 mg/L and guppies 35 mg/L (Hartley and Kidd, 1987); acute oral LD_{50} for rats is 9,000 mg/kg (Ashton and Monaco, 1991).

Uses: Preemergence and postemergence herbicide used to control annual grass weeds in peas, beet crops, lucerne, clover, sugar beet, beans, lettuce, flax, safflowers, and lentils.

PROPOXUR

Synonyms: Aprocarb; Arprocarb; Bay 9010; Bay 39007; Bayer 39007; Baygon; Bifex; Blattanex; Boygon; Brygou; Chemagro 90; ENT 25671; o-IMPC; Invisigard; Isocarb; o-Isopropoxyphenyl methylcarbamate; o-Isopropoxyphenyl N-methylcarbamate; 2-(1-Methylethoxy)phenol methylcarbamate; N-Methyl-2-isopropoxyphenylcarbamate; OMS 33; PHC; Propotox M; Propoxure; Propyon; Sendran; Suncide; Tugon fliegenkugel; Unden.

Structure:

OCONHCH$_3$

OCH(CH$_3$)$_2$

Designations: CAS: 114-26-1; mf: C$_{11}$H$_{15}$NO$_3$; fw: 209.25; RTECS: FC3150000.

Properties: White to tan crystalline solid. Mp: 84-87 °C; bp: decomposes; H-t$_{1/2}$: 290 d (pH 7), 17.9 d (pH 8), 40 min at 20 °C and pH 10; K$_H$: 1.3 x 10^{-9} atm·m^3/mol at 20 °C (approximate - calculated from water solubility and vapor pressure); log K$_{oc}$: 0.48-1.97; log K$_{ow}$: 1.45-1.56; S$_o$: soluble in most solvents; S$_w$: 1.74, 1.93, and 2.44 g/L at 10, 20, and 30 °C, respectively; vp: 9.75 x 10^{-6} mmHg at 20 °C.

Soil properties and adsorption data

Soil	K$_d$ (mL/g)	f$_{oc}$ (%)	K$_{oc}$ (mL/g)	pH
Clayey loam	0.27	0.29	93	6.0
Kanuma high clay	0.80	1.35	53	5.7
Loamy sand	0.05	1.62	3	6.6
Sandy loam	0.62	0.81	76	7.7
Silty loam	0.49	1.16	42	6.3
Silty loam	1.12	2.55	44	6.1
Silty loam	0.30	2.96	10	5.1
Tsukuba clay loam	1.70	4.24	41	6.5

Source: Kanazawa, 1989; U.S. Department of Agriculture, 1990.

Transformation Products

Chemical/Physical: Decomposes at elevated temperatures forming methyl isocyanate (Windholz et al., 1983) and nitrogen oxides (Lewis, 1990). Hydrolyzes in water to 1-naphthol and 2-isopropoxyphenol (Miles et al., 1988). Miles et al. (1988)

studied the rate of hydrolysis of propoxur in phosphate-buffered water (0.01 M) at 26 °C with and without a chlorinating agent (10 mg/L hypochlorite solution). The hydrolysis half-lives at pH 7 and 8 with and without chlorine were 3.5 and 10.3 d and 0.05 and 1.2 d, respectively (Miles et al., 1988). The reported hydrolysis half-lives of propoxur in water at pH 8, 9, and 10 were 16.0 d, 1.6 d, and 4.2 hr, respectively (Aly and El-Dib, 1971). In a 0.50 N sodium hydroxide solution at 20 °C, the hydrolysis half-life was reported to be 3.0 d (El-Dib and Aly, 1976).

Exposure Limits: OSHA TWA: 0.5 mg/m^3; ACGIH TLV: TWA 0.5 mg/m^3.

Formulation Types: Emulsifiable concentrate; wettable powder; fumigant; dustable powder; aerosol.

Toxicity: LC_{50} (96 hr) for rainbow trout 3.7 mg/L and bluegill sunfish 6.6 mg/L (Hartley and Kidd, 1987); acute oral LD_{50} for male and female rats is 95 and 104 mg/kg, respectively (Hartley and Kidd, 1987).

Uses: Insecticide used for control of chewing and sucking insects in fruits, vegetables, ornamentals, and forestry.

PROPYZAMIDE

Synonyms: Clanex; **3,5-Dichloro-*N*-(1,1-dimethylpropynyl)benzamide**; 3,5-Di-chloro-*N*-(1,1-dimethyl-2-propynyl)benzamide; Kerb; Kerb 50W; Promamide; Pronamide; RCRA waste number U192; RH-315.

Structure:

Designations: CAS: 23950-58-5; mf: $C_{12}H_{11}Cl_2NO$; fw: 256.13; RTECS: CV3460000.

Properties: Colorless to white powder or crystalline solid with a mild odor. Mp: 155-156 °C; bp: decomposes above mp; fl p: nonflammable; H-t$_{1/2}$: >700 d at 25 °C and pH 7; K_H: 1.9 x 10^{-7} atm·m^3/mol at 20-25 °C (approximate - calculated from water solubility and vapor pressure); log K_{oc}: 2.30-2.34; log K_{ow}: 3.09-3.28; S_o (g/L): 2-butanone (200), cyclohexanone (200), dimethyl sulfoxide (300), methanol (150), methyl ethyl ketone (200), 2-propanol (150); S_w: 15 mg/L at 25 °C; vp: 8.5 x 10^{-5} mmHg at 20 °C.

Transformation Products
Soil: The major soil metabolite is 2-(3,5-dichlorophenyl)-4,4-dimethyl-5-methyleneoxazoline. The half-life in soil is approximately 30 d at 25 °C (Hartley and Kidd, 1987). Residual activity in soil is limited to approximately 2-6 mo (Hartley and Kidd, 1987).

Chemical/Physical: Emits toxic fumes of nitrogen oxides and chlorine when heated to decomposition (Sax and Lewis, 1987). Propyzamide is hydrolyzed to 3,5-dichlorobenzoate by refluxing under strongly acidic conditions (Humburg et al., 1989).

Formulation Types: Granules; suspension concentrate; wettable powder.

Toxicity: LC$_{50}$ (96 hr) for catfish 500 mg/L, goldfish 350 mg/L, guppies 150 mg/L, and rainbow trout 72 mg/L (Hartley and Kidd, 1987); acute oral LD$_{50}$ for male and female rats is 8,350 and 5,620 mg/kg, respectively (Hartley and Kidd, 1987).

Uses: Preemergence or postemergence herbicide used to control many perennial and annual grasses and broadleaf weeds in ornamental trees and shrubs, forestry, fruits, and vegetables.

QUINOMETHIONATE

Synonyms: Bay 36205; Bayer 4964; Bayer 36205; Chinomethionat; Chinomethionate; Cyclic *S,S*-(6-methyl-2,3-quinoxalinediyl) dithiocarbonate; ENT 25606; Erade; Erazidon; Forstan; **6-Methyl-1,3-dithiolo[4,5-*b*]quinoxalin-2-one**; 6-Methyl-2-oxo-1,3-dithiolo[4,5-*b*]quinoxaline; 6-Methyl-2,3-quinoxaline dithiocarbonate; 6-Methyl-2,3-quinoxalinedithiol cyclic carbonate; 6-Methyl-2,3-quinoxalinedithiol cyclic dithiocarbonate; 6-Methyl-2,3-quinoxalinedithiol cyclic *S,S*-dithiocarbonate; 6-Methylquinoxaline-2,3-dithiolcyclocarbonate; *S,S*-(6-Methylquinoxaline-2,3-diyl) dithiocarbonate; Morestan; Morestane; Oxythioquinox; Quinomethoate; SS 2074.

Structure:

Designations: CAS: 2439-01-2; mf: $C_{10}H_6N_2OS_2$; fw: 234.29; RTECS: FG1400000.

Properties: Yellow crystals. Mp: 169.8-170 °C; H-t$_{1/2}$ (22 °C): 10 d (pH 4), 80 hr (pH 7), 225 min (pH 9); K_H: 1.2 x 10^{-9} atm·m^3/mol at 20 °C (approximate - calculated from water solubility and vapor pressure); log K_{oc}: 3.36-3.37; log K_{ow}: 3.78; S_o (g/L at 20 °C): cyclohexanone (18), *N,N*-dimethylformamide (10), petroleum oils (4); S_w: 50 mg/L at 20 °C; vp: 2.03 x 10^{-7} mmHg at 20 °C.

Soil properties and adsorption data

Soil	K_d (mL/g)	f_{oc} (%)	K_{oc} (mL/g)	pH
Sandy loam	37	1.62	2,284	5.1
Silty loam	41	1.74	2,356	5.9

Source: U.S. Department of Agriculture, 1990.

Transformation Products
Chemical/Physical: Reacts with ammonia forming 6-methyl-2,3-quinoxalinedithiol.

Formulation Types: Wettable powder; fumigant; dustable powder.

Toxicity: LC$_{50}$ (96 hr) for rainbow trout 220 μg/L and bluegill sunfish 120 μg/L

(Hartley and Kidd, 1987); acute oral LD_{50} for rats 2,500-3,000 mg/kg (Hartley and Kidd, 1987), 1,100 mg/kg (RTECS, 1985).

Uses: Acaricide; fungicide.

QUIZALOFOP-ETHYL

Synonyms: Assure; (±)-2-(4-((6-Chloro-2-quinoxalinyl)oxy)phenoxy)-propanoic acid ethyl ester.

Structure:

Designations: CAS: 76578-14-8; mf: $C_{19}H_{17}ClN_2O_4$; fw: 372.80.

Properties: Colorless to white crystalline solid with a petroleum-like odor. Mp: 91.7-92.1 °C; bp: 220 °C at 0.2 mmHg (decomposes at 320 °C); ρ: 1.35 at 20/4 °C; fl p: 61 °C; K_H: 4.8 x 10^{-7} atm·m³/mol at 20-25 °C (approximate - calculated from water solubility and vapor pressure); log K_{oc}: ≈ 2.76; log K_{ow}: 4.20; P-t$_{1/2}$: 40 d (sandy loam soil); S_o (g/L at 20 °C): acetone (110), benzene (290), ethanol (9), n-hexane (2.6), xylene (120); S_w: 310 μg/L at 25 °C; vp: 3.0 x 10^{-7} mmHg at 20 °C.

Transformation Products
Plant: Rapidly hydrolyzed in plants to quizalofop and ethanol (Humburg et al., 1989).

Symptoms of Exposure: May irritate eyes, nose, skin, and throat.

Formulation Types: Suspension concentrate; emulsifiable concentrate (0.79 lb/gal).

Toxicity: LC_{50} (96 hr) for rainbow trout 10.7 mg/L and bluegill sunfish 0.46-2.8 mg/L (Worthing and Hance, 1991); acute oral LD_{50} for male and female rats is 1,700 and 1,500 mg/kg, respectively (Ashton and Monaco, 1991).

Uses: Selective postemergence herbicide used to control both annual and perennial grasses in soybeans and cotton.

RONNEL

Synonyms: Dermaphos; Dimethyl trichlorophenyl thiophosphate; *O,O*-Dimethyl-*O*-2,4,5-trichlorophenyl phosphorothioate; *O,O*-Dimethyl *O*-(2,4,5-trichlorophenyl)thiophosphate; Dow ET 14; Dow ET 57; Ectoral; ENT 23284; ET 14; ET 57; Etrolene; Fenchlorfos; Fenchlorophos; Fenchchlorphos; Karlan; Korlan; Korlane; Nanchor; Nanker; Nankor; **Phosphorothioic acid *O,O*-dimethyl *O*-(2,4,5-trichlorophenyl)ester;** Trichlorometafos; Trolen; Trolene; Viozene.

Structure:

$$CH_3O \overset{S}{\underset{CH_3O}{\underset{|}{P}}} - O - \text{(2,4,5-trichlorophenyl)}$$

CH$_3$O — with Cl substituents (2,4,5-trichlorophenyl ring)

Designations: CAS: 299-84-3; DOT: 2922; mf: $C_8H_8Cl_3O_3PS$; fw: 321.57; RTECS: TG0525000.

Properties: White to light brown, waxy solid. Mp: 41 °C; bp: 97 °C at 0.01 mmHg; ρ: 1.48 at 25/4 °C; H-t$_{1/2}$: ≈ 3 d at pH 6; K_H: 8.46 x 10^{-6} atm·m^3/mol at 20-25 °C (approximate - calculated from water solubility and vapor pressure); log K_{oc}: 2.76 (calculated); log K_{ow}: 4.67-5.068; S$_o$: soluble in acetone, benzene, carbon tetrachloride, chloroform, ethyl ether, kerosene, methylene chloride, toluene, xylene, and many other organic solvents; S$_w$: 40 mg/L at 25 °C; vp: 3.37 x 10^{-5} mmHg at 20 °C.

Transformation Products

Chemical/Physical: Though no products were identified, the reported hydrolysis half-life at pH 7.4 and 70 °C using a 1:4 ethanol/water mixture is 10.2-10.4 hr (Freed et al., 1977). Ronnel decomposed at elevated temperatures on five clay surfaces, each treated with hydrogen, calcium, magnesium, aluminum, and iron ions. At temperatures <950 °C (125, 300, and 750 °C), bentonite clays impregnated with technical ronnel (18.6 wt %) decomposed to 2,4,5-trichlorophenol and a rearrangement product tentatively identified as *O*-methyl *S*-methyl-*O*-(2,4,5-trichlorophenyl) phosphorothioate (Rosenfield and Van Valkenburg, 1965). At 950 °C, only the latter product formed. It was postulated that this compound resulted from a acid-catalyzed molecular rearrangement reaction. Ronnel also undergoes base-catalyzed hydrolysis at elevated temperatures. Products include methanol and a new compound that is formed via cleavage of a methyl group from one of the methoxy groups which is then bonded to the sulfur atom (Rosenfield and Van Valkenburg, 1965). Emits very toxic fumes of chlorides, sulfur, and phosphorus oxides when heated to decomposition (Lewis, 1990).

Exposure Limits: NIOSH REL: IDLH 5,000 mg/m^3; OSHA PEL: TWA 10 mg/m^3; ACGIH TLV: TWA 10 mg/m^3.

Symptoms of Exposure: In animals: cholinesterase inhibition; irritates eyes; liver and kidney damage.

Toxicity: LC$_{50}$ (96 hr) for fathead minnow 305 μg/L (Verschueren, 1983); acute oral LD$_{50}$ for rats 1,740 mg/kg (Verschueren, 1983), 625 mg/kg (RTECS, 1985).

Use: Insecticide.

SIDURON

Synonyms: Du Pont herbicide 1318; H 1318; 1-(2-Methylcychohexyl)-3-phenyl-urea; *N*-(2-Methylcyclohexyl)-*N'*-phenylurea; Tupersan.

Structure:

Designations: CAS: 1982-49-6; mf: $C_{14}H_{20}N_2O$; fw: 232.30; RTECS: YT7350000.

Properties: Colorless to white, odorless, crystalline solid. Mp: 133–138 °C; ρ: 1.08 at 25/4 °C; fl p: nonflammable; K_H: 6.8 x 10^{-10} atm·m^3/mol at 25 °C (approximate - calculated from water solubility and vapor pressure); log K_{oc}: 2.88 (estimated); log K_{ow}: 2.70; S_o (g/kg at 25 °C): cellosolve (175), dimethylacetamide (367), *N,N*-dimethylformamide (260), ethanol (160), isophorone (118), methylene chloride (118); S_w: 18 mg/L at 25 °C; vp: 4 x 10^{-8} mmHg at 25 °C.

Transformation Products
Soil: A fungus and two *Pseudomonas* spp. isolated from soil were capable of degrading siduron to form the major metabolites: 1-(4-hydroxy-2-methyl-cyclohexyl)-3-(*p*-hydroxyphenyl)urea, 1-(4-hydroxy-2-methylcyclohexyl)-3-phenylurea, and 1-(4-hydroxyphenyl)-3-(2-methycyclohexyl)urea (Belasco and Langsdorf, 1969).

Symptoms of Exposure: May irritate eyes, nose, throat, and skin.

Formulation Types: Wettable powder (50%).

Toxicity: LC$_{50}$ (48 hr) for carp 18 mg/L and Japanese goldfish 10–40 mg/L (Worthing and Hance, 1991); acute oral LD$_{50}$ for rats is >7,500 mg/kg (Ashton and Monaco, 1991), 5,000 mg/kg (RTECS, 1985).

Uses: Selective preemergence herbicide used to control annual weed grasses and against crabgrass, foxtail, and barnyard grass in turf, cereals, and other food crops.

SIMAZINE

Synonyms: A 2079; Aktinit S; Aquazine; Aquazine 80W; Batazina; 2,4-Bis(ethyl-amino)-6-chloro-s-triazine; Bitemol; Bitemol S 50; Caliper; Caliper 90; CAT; CDT; Cekusan; Cekuzina-S; 2-Chloro-4,6-bis(ethylamino)-1,3,5-triazine; 2-Chloro-4,6-bis(ethylamino)-s-triazine; **6-Chloro-N',N'-diethyl-1,3,5-triazine-2,4-di-amine**; 6-Chloro-N^2,N^4-diethyl-1,3,5-triazine-2,4-diyldiamine; Framed; G 27,692; Geigy 27692; Gesaran; Gesatop; Gesatop 50; H 1803; Herbazin; Herbazin 50; Herbex; Herboxy; Hungazin DT; Premazine; Primatol S; Princep; Princep 4G; Princep 4L; Princep 80W; Printop; Radocon; Radokor; Simadex; Simanex; Simazin; Simazine 80W; Sim-Trol; Symazine; Tafazine; Tafazine 50-W; Taphazine; Triazine A 384; W 6,658; Weedex; Zeapur.

Structure:

Designations: CAS: 122-34-9; mf: $C_7H_{12}ClN_5$; fw: 201.66; RTECS: XY5250000.

Properties: Colorless to white crystalline solid. Mp: 225-227 °C; ρ: 1.203 at 20/4 °C; fl p: nonflammable; pK_a: 1.7 at 21 °C; K_H: 3.4 x 10^{-9} atm·m^3/mol at 20 °C (approximate - calculated from water solubility and vapor pressure); log K_{oc}: 2.14; log K_{ow}: 1.94-2.26; P-t$_{1/2}$: 108.17 hr (absorbance λ = 53.25 nm, concentration on glass plates = 6.7 μg/cm^2); S_o (mg/L at 20-25 °C): chloroform (900), ethyl ether (300), light petroleum (2), methanol (400), n-pentane (3), petroleum ether (2); S_w: 3.5-5 mg/L at 20 °C; vp: 6.1 x 10^{-9} mmHg at 20 °C, 3.6 x 10^{-8} mmHg at 20 °C.

Soil properties and adsorption data

Soil	K_d (mL/g)	f_{oc} (%)	K_{oc} (mL/g)	pH	CEC (meq/100 g)
Aiken loam	2.59	2.94	88	5.2	20.6
Ascalon silty clay loam	0.81	0.85	96	7.3	12.7
Barnes clay loam	3.62	3.98	91	7.4	33.8
Basinger fine sand	0.73	0.61	119	5.8	--
Batcombe silt loam	0.97	2.05	47	6.1	--
Bates sandy loam	1.00	0.80	125	6.5	9.3
Bautista silty loam	0.49	1.03	48	7.2	14.2

continued

Soil	K_d (mL/g)	f_{oc} (%)	K_{oc} (mL/g)	pH	CEC (meq/100 g)
Bautista loamy sand	0.30	0.28	107	6.4	5.5
Baxter silty clay loam	2.30	1.21	190	6.0	11.2
Beltsville silty loam	1.94	1.40	139	4.3	4.2
Benevola clay	0.88	1.30	68	7.6	20.1
Benevola silty clay	1.58	2.70	58	7.7	19.5
Berkley clay	0.81	0.99	82	7.3	34.4
Berkley silty clay	2.94	4.63	63	7.1	33.7
Boca fine sand	0.85	1.67	71	7.1	--
Bosket silty loam	1.76	0.57	308	5.8	8.4
Brentwood clay	3.03	1.67	181	6.1	30.8
Broadbalk 2B	1.50	2.90	52	7.5	--
Broadbalk 3	0.50	1.00	50	7.9	--
Broadbalk 8	0.60	1.20	50	6.6	--
Cajon loam	1.65	1.27	130	6.2	14.6
Cecil silty clay	0.75	1.09	68	5.3	3.6
Chester loam	1.58	1.67	94	4.9	5.2
Chillum silty loam	3.30	2.54	130	4.6	7.6
Chillum silty loam	2.58	2.54	101	4.6	7.6
Chino silty loam	3.60	2.36	152	7.4	14.2
Chobee fine sandy loam	1.06	1.39	76	7.2	--
Christiana loam	1.25	0.57	219	4.4	5.6
Clarksville silty clay loam	1.40	0.80	175	5.7	5.7
Coachella sandy loam	0.54	1.03	52	7.1	11.8
Collombey	0.64	1.28	50	7.8	--
Columbia clay loam	3.23	1.38	234	6.8	26.5
Columbia very fine silty loam	1.99	0.86	231	7.3	15.0
Crosby silty loam	2.81	1.90	148	4.8	11.5
Cumberland silty loam	1.20	0.69	174	6.4	6.5
Diablo sandy clay loam	0.91	0.69	132	6.1	35.8
Dinuba fine sandy loam	0.42	0.46	91	6.2	6.6
Ducor adobe clay loam	1.22	2.13	57	7.2	34.6
Dundee silty clay loam	6.62	0.96	690	5.0	18.1
Eldon silty loam	2.90	1.73	167	5.9	12.9
Elkhorn loamy sand	1.10	1.67	66	6.7	13.2
Elkhorn loamy sand	0.88	1.27	69	7.0	11.6
Evouettes	1.78	2.09	81	6.1	--
Exeter loamy sand	1.07	1.50	71	5.8	14.7
Exeter sandy loam	0.56	1.32	42	6.7	12.8
Fallbrook fine sandy loam	1.41	2.77	51	6.4	19.7

Soil	K_d (mL/g)	f_{oc} (%)	K_{oc} (mL/g)	pH	CEC (meq/100 g)
Garland clay	0.56	0.65	86	7.7	23.2
Gerald silty loam	4.20	1.55	271	4.7	11.0
Greenfield sandy loam	0.93	2.77	34	7.3	15.8
Greenfield sandy loam	0.69	1.38	50	6.5	12.6
Greenfield sandy loam	1.18	2.07	57	6.4	16.4
Grundy silty clay loam	6.50	2.07	314	5.6	13.5
Hagerstown silty clay loam	1.67	2.48	67	5.5	12.5
Hagerstown silty clay loam	1.10	1.30	84	7.5	8.8
Hagerstown silty clay loam	3.30	2.48	133	5.5	12.5
Hanford gravelly sandy loam	1.48	2.82	52	6.3	18.1
Hanford sandy loam	0.69	1.90	36	6.2	12.4
Hanford sandy loam	0.78	1.38	56	6.0	11.0
Hanford sandy loam	0.90	1.55	58	6.4	12.4
Hanford loamy sand	0.59	1.27	46	7.2	9.5
Hanford loamy sand	0.67	1.15	58	5.8	9.6
Hanford loamy sand	0.60	0.86	70	6.5	7.5
Hickory Hill silt	7.04	3.27	215	--	--
Holopaw fine sand	0.29	0.50	58	6.1	--
Hovey adobe loam	2.33	2.25	104	6.6	49.8
Indio loam	0.71	1.32	54	7.2	15.1
Indio loam	0.81	1.21	67	7.2	13.4
Iredell clay	1.68	6.17	26	5.6	20.9
Iredell silty loam	3.48	3.04	114	5.4	17.0
Knox silty loam	5.10	1.67	305	5.4	18.8
Lakeland silty loam	0.68	1.88	36	6.2	2.9
Lakeland silty loam	0.90	1.90	47	6.2	2.9
Las Posas sandy loam	1.20	1.09	110	7.3	11.8
Lebanon silty loam	2.80	1.04	269	4.9	7.7
Lindley loam	2.60	0.86	302	4.7	6.9
Lintonia loamy sand	1.00	0.34	294	5.3	3.2
Manzanita clay loam	1.42	1.21	117	5.9	13.8
Marian silty loam	3.50	0.80	438	4.6	9.9
Marshall silty clay loam	7.20	2.42	298	5.4	21.3
Menfro silty loam	2.50	1.38	181	5.3	9.1
Montalto clay	0.32	0.86	37	5.9	8.4
Newtonia silty loam	3.00	0.92	326	5.2	8.8
Norfolk silty loam	0.21	0.08	260	5.1	0.2

continued

Soil	K_d (mL/g)	f_{oc} (%)	K_{oc} (mL/g)	pH	CEC (meq/100 g)
Ooster silty loam	1.02	1.31	78	4.7	6.3
Oswego silty clay loam	3.90	1.67	234	6.4	21.0
Panoche loam	0.58	0.57	102	7.5	20.2
Park Grass 3A	2.50	5.60	45	7.0	--
Park Grass 3A	1.20	2.40	50	6.8	--
Park Grass 3A	1.80	3.60	50	6.9	--
Park Grass 3D	3.00	5.30	57	5.1	--
Park Grass 3D	1.90	3.10	61	5.1	--
Park Grass 3D	1.50	2.10	71	5.3	--
Park Grass 11/2D	1.40	1.80	78	3.8	--
Park Grass 11/2D	2.90	3.30	88	3.6	--
Park Grass 11/2D	1.19	10.90	109	3.7	--
Park Grass 13A	1.00	2.10	48	6.6	--
Park Grass 13A	1.80	3.50	51	6.5	--
Park Grass 13A	3.10	6.00	52	6.8	--
Park Grass 13D	1.00	2.80	36	4.7	--
Park Grass 13D	0.70	1.80	39	4.8	--
Park Grass 13D	2.00	4.50	44	4.8	--
Park Grass 14A	2.80	5.50	51	6.9	--
Park Grass 14A	1.90	3.30	58	6.9	--
Park Grass 14A	1.50	1.80	83	6.9	--
Pineda fine sand	0.55	1.22	45	7.1	--
Porterville adobe clay	1.27	1.90	67	7.3	37.6
Porterville adobe clay loam	0.75	0.75	100	6.4	18.1
Putnam sandy loam	2.20	1.09	202	5.3	12.3
Ramona sandy loam	0.97	1.78	54	5.9	15.2
Ramona sandy loam	0.62	1.38	45	6.7	10.4
Ramona sandy loam	0.29	0.51	57	6.6	9.5
Rhinebeck silty clay loam	1.97	3.13	63	6.7	--
Rothamsted Farm	0.73	1.51	48	5.1	--
Rincon loam	3.83	1.78	215	5.8	25.2
Riviera fine sand	0.59	0.94	63	6.2	--
Ruston silty loam	0.62	1.05	59	5.1	3.4
Salix loam	3.50	1.21	289	6.3	17.9
San Joaquin clay loam	1.10	1.21	91	7.4	20.8
San Joaquin loam	0.92	1.44	64	5.1	11.2
San Joaquin loam	0.94	1.15	82	6.6	16.7
San Joaquin loam	0.96	1.15	83	5.6	14.0
San Joaquin sandy clay loam	0.95	0.75	127	6.1	16.2

Soil	K_d (mL/g)	f_{oc} (%)	K_{oc} (mL/g)	pH	CEC (meq/100 g)
San Joaquin sandy loam	0.41	0.69	59	7.2	8.6
Sarpy loam	2.00	0.75	267	7.1	14.3
Sharkey clay	5.87	2.25	261	6.2	40.2
Sharkey clay	7.00	1.44	486	5.0	28.2
Shelby loam	5.10	2.07	246	4.3	20.1
Soil #1 (0–30 cm)	3.61	1.46	247	6.0	--
Soil #1 (30–50 cm)	4.50	0.93	484	6.1	--
Soil #1 (50–91 cm)	5.12	1.24	413	6.1	--
Soil #1 (91–135 cm)	2.58	0.70	369	6.4	--
Soil #1 (135–165 cm)	1.94	0.17	1,139	6.5	--
Soil #1 (>165 cm)	2.49	0.07	3,559	6.7	--
Soil #2 (0–27 cm)	5.42	1.50	361	5.9	--
Soil #2 (27–53 cm)	2.70	0.16	1,685	6.3	--
Soil #2 (53–61 cm)	2.37	0.14	1,690	6.4	--
Soil #2 (61–81 cm)	3.13	0.28	1,117	6.4	--
Soil #2 (81–157 cm)	4.66	0.50	931	6.0	--
Soil #2 (>157 cm)	8.53	1.30	656	5.9	--
Sorrento fine sandy loam	1.15	1.09	106	6.9	16.1
Staten peaty loam	0.21	14.00	151	6.9	72.1
Sterling clay loam	0.95	0.94	101	7.7	22.5
Summit sandy clay	7.90	2.82	280	4.8	35.1
Superstition stoney loamy sand	0.31	0.23	135	7.2	5.6
Thurlow clay loam	1.10	1.25	88	7.7	21.6
Tifton loamy sand	0.21	0.56	37	4.9	2.4
Toledo silty clay	3.06	2.80	109	5.5	29.8
Tripp loam	0.88	0.86	103	7.6	14.7
Truckton silty loam	0.49	0.25	198	7.0	4.4
Union silty loam	3.80	1.04	365	5.4	6.8
Valois silty loam	1.97	1.64	120	5.9	--
Vertroz	2.88	3.25	89	6.7	--
Vista loamy sand	0.66	0.86	77	6.5	10.0
Vista sandy loam	1.10	0.86	128	6.4	12.6
Vista sandy loam	0.61	0.75	81	6.2	14.2
Vista sandy loam	0.60	0.69	100	6.8	13.2
Wabash clay	6.00	1.27	472	5.7	40.3
Wabasso sand	0.80	0.61	131	6.6	--
Waverley silty loam	3.10	1.15	270	6.4	12.8

continued

Soil	K_d (mL/g)	f_{oc} (%)	K_{oc} (mL/g)	pH	CEC (meq/100 g)
Wehadkee silty loam	1.25	1.11	113	5.6	10.2
Wehadkee silty loam	2.70	1.09	248	5.6	10.2
Wyman clay loam	2.52	1.78	142	6.4	32.0
Wyman loam	1.78	1.21	147	6.6	15.4
Yolo clay loam	2.37	2.07	114	6.8	31.0
Yolo clay loam	1.15	1.27	91	7.2	27.8
Yolo clay loam	4.95	2.13	232	6.1	31.0
Yolo loam	1.12	1.61	70	5.8	22.1
Yolo loam	1.43	1.78	80	7.2	27.4
Yolo loam	2.05	1.73	118	6.6	17.8
Yolo loam	2.53	1.03	246	7.3	24.0
Yolo sandy loam high fan phase	11.20	2.59	43	6.9	20.4
Yolo silty loam	1.09	2.30	47	7.0	21.2
Yolo sandy loam	0.90	1.32	68	7.0	17.0
Yolo sandy loam	0.96	1.32	73	6.9	17.1

Source: Briggs, 1981; Brown and Flagg, 1981; Burkhard and Guth, 1981; Day et al., 1968; Gamerdinger et al., 1991; Harris, 1966; Harris and Sheets, 1965; Lord et al., 1980; Reddy et al., 1992; Sukop and Cogger, 1992; Talbert and Fletchall, 1965; Williams, 1968.

Transformation Products

Soil: The reported half-life in soil is 75 d (Alva and Singh, 1991). Under laboratory conditions, the half-lives of simazine in a Hatzenbühl soil (pH 4.8) and Neuhofen soil (pH 6.5) at 22 °C were 45 and 100 d, respectively (Burkhard and Guth, 1981).

Plant: Simazine is metabolized by plants to the herbicidally inactive 6-hydroxysimazine which is further degraded via dealkylation of the sidechains and hydrolysis of the amino group releasing carbon dioxide (Castelfranco et al., 1961; Humburg et al., 1989).

Photolytic: Pelizzetti et al. (1990) studied the aqueous photocatalytic degradation of simazine and other s-triazines (ppb level) using simulated sunlight (λ >340 nm) and titanium dioxide as a photocatalyst. Simazine rapidly degraded forming cyanuric acid, nitrates, and other intermediate compounds similar to those found for atrazine. Mineralization of cyanuric acid to carbon dioxide was not observed (Pelizzetti et al., 1990). In aqueous solutions, simazine is converted exclusively to hydroxysimazine by UV light (λ = 253.7 nm). The UV irradiation of methanolic solutions of simazine afforded simetone (2-methoxy-4,6-bis(ethylamino-s-triazine). Photodegradation of simazine in methyl alcohol did not occur when irradiated at wavelengths >300 nm (Pape and Zabik, 1970).

Chemical/Physical: Emits toxic fumes of nitrogen oxides and chlorine when heated to decomposition (Sax and Lewis, 1987). In the presence of hydroxy or perhydroxy radicals generated from Fenton's reagent, simazine undergoes dealkylation to give 2-chloro-4,6-diamino-*s*-triazine as the major product (Kaufman and Kearney, 1970).

Symptoms of Exposure: May cause acute and moderate dermatitis; irritant to eyes, skin, and mucous membranes.

Formulation Types: Emulsifiable concentrate; granules (4%); suspension concentrate; water-dispersible granules (90%); wettable powder (80%).

Toxicity: LC_{50} (96 hr) for bluegill sunfish 90 mg/L, guppies 49 mg/L, rainbow trout and crucian carp >100 mg/L (Hartley and Kidd, 1987); LC_{50} (48 hr) for coho salmon 6.60 mg/L, bluegill sunfish 130 ppm, and rainbow trout 85 ppm (Verschueren, 1983); acute oral LD_{50} of the technical product for rats is >5,000 mg/kg (Ashton and Monaco, 1991), 971 mg/kg (RTECS, 1985).

Uses: Selective preemergence systemic herbicide used to control many broadleaf weeds and annual grasses in deep-rooted fruit and vegetable crops.

STRYCHNINE

Synonyms: Certox; Dolco mouse cereal; Kwik-kil; Mole death; Mole-nots; Mouse-rid; Mouse-tox; Pied piper mouse seed; RCRA waste number P108; Ro-dex; Sanaseed; **Strychnidin-10-one**; Strychnos; UN 1692.

Structure:

Designations: CAS: 57-24-9; DOT: 1692; mf: $C_{21}H_{22}N_2O_2$; fw: 334.42; RTECS: WL2275000.

Properties: Colorless to white, odorless crystals. Very bitter taste. Mp: 286-288 °C; bp: 270 °C at 5 mmHg; ρ: 1.36 at 20/4 °C; pK_a: at 20 °C: pK_1 = 6.0, pK_2 = 11.7; log K_{oc}: 2.45 (calculated); log K_{ow}: 1.93; S_o (g/L): soluble in alcohol (6.67), amyl alcohol (4.55), benzene (5.56), chloroform (200), glycerol (3.13), methanol (3.85), toluene (5); S_w: 0.02 wt % at 20 °C (pH = 9.5).

Transformation Products
Chemical/Physical: Reacts with acids forming water-soluble salts (Worthing and Hance, 1991). Emits toxic nitrogen oxides when heated to decomposition (Lewis, 1990).

Exposure Limits: NIOSH REL: IDLH 3 mg/m³; OSHA PEL: TWA 0.15 mg/m³; ACGIH TLV: TWA 0.15 mg/m³.

Symptoms of Exposure: Stiff neck, muscle fasiculation, restless, apprehension, acuity of perception, reflex excitability, cyanosis, tetanic convulsions, opisthotonos.

Formulation Types: Bait.

Toxicity: Acute oral LD_{50} for rats 16 mg/kg (RTECS, 1985).

Uses: Destroying rodents and predatory animals.

SULFOMETURON-METHYL

Synonyms: Benzoic acid *o*-((3-(4,6-dimethyl-2-pyrimidinyl)ureido)sulfonyl) methyl ester; DPX 5648; 2-(((((4,6-Dimethyl-2-pyrimidinyl)amino)carbonyl)- amino)sulfonyl)benzoic acid methyl ester; Oust.

Structure:

Designations: CAS: 74222-97-2; mf: $C_{15}H_{16}N_4O_5S$; fw: 364.40; RTECS: DG9096550.

Properties: Colorless to white solid. Mp: 203-205 °C; ρ: 1.48; fl p: nonflammable; pK_a: 5.2; H-t$_{1/2}$: 18 d at pH 5; log K_{oc}: 1.85-2.09; log K_{ow}: 1.18 (pH 5), -0.50 (pH 7); S_o (mg/kg at 25 °C): acetone (2,380), acetonitrile (1,530), ethanol (137), ethyl ether (32), xylene (37); S_w at 25 °C: 8-10 mg/L at pH 5, 70-300 mg/L at pH 7; vp: 5.5 x 10^{-16} mmHg at 25 °C.

Soil properties and adsorption data

Soil	K_d (mL/g)	f_{oc} (%)	K_{oc} (mL/g)	pH	CEC (meq/100 g)
Fallsington sandy loam	0.97	0.81	120	5.6	4.8
Flanagan silt loam	2.85	2.33	122	6.5	23.4
Myakka sand	1.00	1.41	71	6.3	3.9

Source: Harvey et al., 1985.

Transformation Products

Soil: In unsterilized soils, 58% of ^{14}C-labeled sulfometuron-methyl degraded after 24 wk. Metabolites identified were 2,3-dihydro-3-oxobenzisosulfonazole (sac- charin), methyl-2-(aminosulfonyl)benzoate, 2-aminosulfonyl benzoic acid, 2- (((aminocarbonyl)amino)sulfonyl)benzoate and [^{14}-C]carbon dioxide. The rate of degradation in aerobic soils was primarily dependent upon pH and soil type (Anderson and Dulka, 1985). The reported half-life in soil was approximately 4 wk (Hartley and Kidd, 1987).

Chemical/Physical: Sulfometuron-methyl is stable in water at pH 7 or 9 but is rapidly hydrolyzed at pH 5.0 (H-t$_{1/2}$ = 14 d) forming methyl 2-(aminosulfonyl)-

benzoate and saccharin. When sulfometuron-methyl in an aqueous solution was exposed to UV light (λ = 300-400 nm), it degraded to the intermediate methyl benzoate which then mineralized to carbon dioxide (Harvey et al., 1985).

Symptoms of Exposure: May irritate eyes, nose, throat, and skin.

Formulation Types: Water-dispersible granules (75%); flowable powder (60%).

Toxicity: LC_{50} (96 hr) for rainbow trout and bluegill sunfish >12.5 mg/L (Worthing and Hance, 1991); acute oral LD_{50} for rats is >5,000 mg/kg (Ashton and Monaco, 1991).

Uses: Preemergence and postemergence herbicide used to control many annual and perennial grasses and broadleaf weeds in noncropland and industrial turf areas.

SULFOTEPP

Synonyms: ASP 47; Bay E-393; Bayer E 393; Bis-*O,O*-diethylphosphorothionic anhydride; Bladafum; Bladafume; Bladafun; Dithio; Dithione; Dithiophos; Dithiophosphoric acid tetraethyl ester; Dithiotep; E 393; ENT 16273; Ethyl thiopyrophosphate; Lethalaire G 57; Pirofos; Plant dithio aerosol; Plantfume 103 smoke generator; Pyrophosphorodithioic acid tetraethyl ester; Pyrophosphorodithioic acid *O,O,O,O*-tetraethyl dithionopyrophosphate; RCRA waste number P109; Sulfatep; Sulfotep; TEDP; TEDTP; Tetraethyl dithionopyrophosphate; Tetraethyl dithiopyrophosphate; *O,O,O,O*-Tetraethyl dithiopyrophosphate; Thiodiphosphoric acid tetraethyl ester; **Thiophosphoric acid tetraethyl ester;** Thiopyrophosphoric acid tetraethyl ester; Thiotepp; UN 1704.

Structure:

$$CH_3CH_2O \underset{CH_3CH_2O}{\overset{\overset{\displaystyle S}{\|}}{\diagdown}} P - O - P \overset{\overset{\displaystyle S}{\|}}{\underset{OCH_2CH_3}{\diagup}} OCH_2CH_3$$

Designations: CAS: 3689-24-5; DOT: 1704; mf: $C_8H_{20}O_5P_2S_2$; fw: 322.30; RTECS: XN4375000.

Properties: Pale yellow liquid with a garlic-like odor. Mp: <25 °C; bp: 136-139 °C at 2 mmHg; ρ: 1.196 at 25/4 °C; K_H: 2.88 x 10^{-6} atm·m^3/mol at 20 °C (approximate - calculated from water solubility and vapor pressure); log K_{oc}: 2.66, 2.82; log K_{ow}: 3.02 (calculated); S_o: miscible with most organic solvents; S_w: 0.0025 wt % at 20 °C; vap d: 13.17 g/L at 25 °C, 11.13 (air = 1); vp: 1.7 x 10^{-4} mmHg at 20 °C.

Soil properties and adsorption data

Soil	K_d (mL/g)	f_{oc} (%)	K_{oc} (mL/g)	pH	CEC (meq/100 g)
Coarse sand (Jutland, Denmark)	0.96	0.21	457	5.3	2.2
Sandy loam (Jutland, Denmark)	1.00	0.15	667	6.4	7.3

Source: Kjeldsen et al., 1990.

Exposure Limits: NIOSH REL: IDLH 35 mg/m^3; ACGIH TLV: TWA 0.2 mg/m^3.

Transformation Products

Soil: Cleavage of the molecule yields diethyl phosphate, monoethyl phosphate, and phosphoric acid (Hartley and Kidd, 1987).

Chemical/Physical: Emits toxic oxides of sulfur and phosphorus oxides when heated to decomposition (Lewis, 1990).

Formulation Types: Fumigant.

Toxicity: LC_{50} (96 hr) for fathead minnow 178 μg/L, bluegill sunfish 1.6 μg/L, and rainbow trout 18 μg/L (Verschueren, 1983); acute oral LD_{50} for rats 5-10 mg/kg (Hartley and Kidd, 1987), 5 mg/kg (RTECS, 1985).

Use: Insecticide.

SULPROFOS

Synonyms: Bay-ntn-9306; Bolstar; *O*-Ethyl *O*-(4-(methylmercapto)phenyl) *S-n*-propylphosphorothionothiolate; *O*-Ethyl *O*-(4-methylthio)phenyl) phosphorodithioic acid *S*-propyl ester; *O*-Ethyl *O*-(4-methylthio)phenyl) *S*-propyl phosphorodithioate; Helothion.

Structure:

Designations: CAS: 35400-43-2; mf: $C_{12}H_{19}O_2PS_3$; fw: 322.45; RTECS: TE4165000.

Properties: Colorless to tan, oily liquid with a phosphorous-like odor. Mp: <25 °C; bp: 155-158 °C at 0.1 mmHg; ρ. 1.20 at 20/4 °C; H-t$_{1/2}$: 52 hr (pH 11.5), 41 d (pH 2.2); K_H: 8.6 x 10^{-7} atm·m^3/mol at 20-25 °C (approximate - calculated from water solubility and vapor pressure); log K_{oc}: 4.10-4.73; log K_{ow}: 5.48; P-t$_{1/2}$: 2 d (thin films exposed to sunlight); S_o (g/kg at 29 °C): cyclohexanone (120), 2-propanol (400-600), toluene (>1,200); S_w: 310 µg/L at 25 °C; vp: 6.30 x 10^{-7} mmHg at 20 °C.

Soil properties and adsorption data

Soil	K_d (mL/g)	f_{oc} (%)	K_{oc} (mL/g)	pH
Clayey loam	155	0.29	53,448	6.0
Loamy sand	204	1.62	12,593	6.6
Silty loam	347	2.90	11,966	7.9

Source: U.S. Department of Agriculture, 1990.

Transformation Products

Photolytic: When sulprofos was exposed to sunlight as deposits on cotton foliage, glass surfaces, and in aqueous solution, the insecticide degraded rapidly (t$_{1/2}$ <2 d). Irradiation of sulprofos in aqueous solution using UV light (λ >290 nm) was also very rapid (t$_{1/2}$ <2 hr). The major degradative pathway involved oxidation of the methylthio sulfur to the corresponding sulfoxide and sulfones, hydrolysis of the *O*-phenyl ester, and oxidative desulfuration of the P=S group. Products included the corresponding sulfone and sulfoxide from the parent compound, an *O*-analog sulfone, a phenol, a phenol sulfone and a phenol sulfoxide, and four unidentified compounds (Ivie and Bull, 1976).

Chemical/Physical: Emits toxic oxides of sulfur and phosphorus when heated to decomposition (Lewis, 1990).

Exposure Limits: OSHA REL: TWA 1 mg/m^3; ACGIH TLV: TWA 1 mg/m^3.

Formulation Types: Emulsifiable concentrate.

Toxicity: LC_{50} (96 hr) for rainbow trout 23 mg/L, bluegill sunfish 11 mg/L, and carp 5.2 mg/L; acute oral LD_{50} for rats 304 mg/kg (Hartley and Kidd, 1987), 65 mg/kg (RTECS, 1985).

Use: Insecticide.

2,4,5-T

Synonyms: Amine 2,4,5-T for rice; BCF-bushkiller; Brush-off 445 low volatile brush killer; Brush-rhap; Brushtox; Dacamine; Debroussaillant concentre; Debroussaillant super concentre; Decamine 4T; Ded-weed brush killer; Ded-weed LV-6 brush-kil and T-5 brush-kil; Dinoxol; Envert-T; Estercide T-2 and T-245; Esteron; Esterone 245; Esteron 245 BE; Esteron brush killer; Farmco fence rider; Fence rider; Forron; Forst U 46; Fortex; Fruitone A; Inverton 245; Line rider; NA 2765; Phortox; RCRA waste number U232; Reddon; Reddox; Spontox; Super D weedone; Tippon; Tormona; Transamine; Tributon; **(2,4,5-Trichlorophenoxy)-acetic acid**; Trinoxol; Trioxon; Trioxone; U 46; Veon; Veon 245; Verton 2T; Visko-rhap low volatile ester; Weddar; Weedone; Weedone 2,4,5-T.

Structure:

OCH$_2$COOH
Cl
Cl
Cl

Designations: CAS: 93-76-5; DOT: 2765; mf: C$_8$H$_5$Cl$_3$O$_3$; fw: 255.48; RTECS: AJ8400000.

Properties: White to pale brown, odorless crystals. Mp: 157-158 °C (pure), 150-151 °C (technical); ρ: 1.80 at 20/20 °C; pK$_a$: 2.80-2.88; K$_H$: 4.87 x 10^{-8} atm·m^3/mol at 20 °C (approximate - calculated from water solubility and vapor pressure); log K$_{oc}$: 1.72, 2.27; log K$_{ow}$: 0.60-3.40; P-t½: 15 d (near-surface waters under a midsummer sun) S$_o$ (g/L at 20-25 °C): 95% ethanol (548.2), ethyl ether (234.3), n-heptane (0.40), methanol (496.0), toluene (7.32), xylene (6.08), benzene, ethylbenzene, xylene, and many other organic solvents; S$_w$: 220 mg/L at 20 °C; vp: 3.75 x 10^{-5} mmHg at 20 °C.

Soil properties and adsorption data

Soil	K$_d$ (mL/g)	f$_{oc}$ (%)	K$_{oc}$ (mL/g)	pH
Ephrata sandy loam	0.31	0.80	38.8	7.5
Glendale silty clay loam	0.49	0.53	92.4	8.5
Ordinance sandy loam	2.40	3.66	65.6	6.6
Palouse silty loam	3.00	2.43	123.5	6.5
Webster sandy clay loam	6.20	3.34	185.6	--

Source: Nkedi-Kizza et al., 1983; O'Connor and Anderson, 1974.

Transformation Products

Biological: 2,4,5-T degraded in anaerobic sludge by reductive dechlorination to 2,4,5-trichlorophenol, 3,4-dichlorophenol, and 4-chlorophenol (Mikesell and Boyd, 1985). An anaerobic methanogenic consortium, growing on 3-chlorobenzoate, metabolized 2,4,5-T to (2,5-dichlorophenoxy)acetic acid at a rate of 1.02 x 10^{-7} M/hr (t$_{1/2}$ = 2 d at 37 °C) (Suflita et al., 1984). Under aerobic conditions, 2,4,5-T degraded to 2,4,5-trichlorophenol and 3,5-dichlorocatechol which may further degrade to 4-chlorocatechol or *cis,cis*-2,4-dichloromuconic acid, 2-chloro-4-carboxymethylenebut-2-enolide, chlorosuccinic acid, and succinic acid (Byast and Hance, 1975).

Soil: 2,4,5-Trichlorophenol and 2,4,5-trichloroanisole were the primary degradation products formed when 2,4,5-T was incubated in soil at 25 °C under aerobic conditions (t$_{1/2}$ = 14 d) (McCall et al., 1981a). Hydrolyzes in soil to 2,4,5-trichlorophenol (Somasundaram et al., 1989, 1991) and 2,4,5-trichloroanisole (Somasundaram et al., 1989). The rate of 2,4,5-T degradation in soil remained unchanged in a soil pretreated with its hydrolysis metabolite (2,4,5-trichlorophenol) (Somasundaram et al., 1989).

Photolytic: When 2,4,5-T (10^{-4} M) in an oxygenated, titanium dioxide (2 g/L) suspension was irradiated by sunlight (λ ≥340 nm), 2,4,5-trichlorophenol, 2,4,5-trichlorophenyl formate and nine chlorinated aromatic hydrocarbons formed as intermediates. Complete mineralization yielded hydrochloric acid, carbon dioxide, and water (Barbeni et al., 1987). Crosby and Wong (1973) studied the photolysis of 2,4,5-T in aqueous solutions (100 mg/L) under alkaline conditions (pH 8) using both outdoor sunlight and indoor irradiation (λ = 300-450 nm). 2,4,5-Trichlorophenol and 2-hydroxy-4,5-dichlorophenoxyacetic acid formed as major products. Minor photodecomposition products included 4,6-dichlororesorcinol, 4-chlororesorcinol, 2,5-dichlorophenol, and a dark polymeric substance. The rate of photolysis increased 11-fold in the presence of sensitizers (acetone or riboflavin) (Crosby and Wong, 1973). The rate of photolysis of 2,4,5-T was also higher in natural waters containing fulvic acids when compared to distilled water. The major photoproduct found in the humic acid-induced reaction was 2,4,5-trichlorophenol. In addition, the presence of ferric ions and/or hydrogen peroxides may contribute to the sunlight-induced photolysis of 2,4,5-T in acidic, weakly absorbing natural waters (Skurlatov et al., 1983).

Chemical/Physical: Carbon dioxide, chloride, dichloromaleic, oxalic, and glycolic acids were reported as ozonation products of 2,4,5-T in water at pH 8 (Struif et al., 1978). Reacts with alkali metals and amines forming water-soluble salts (Worthing and Hance, 1991). When 2,4,5-T was heated at 900 °C, carbon monoxide, carbon dioxide, chlorine, hydrochloric acid, and oxygen were produced (Kennedy et al., 1972, 1972a).

Exposure Limits: NIOSH REL: IDLH 5,000 mg/m^3; OSHA PEL: TWA 10 mg/m^3; ACGIH TLV: TWA 10 mg/m^3.

Symptoms of Exposure: Skin irritation. May also cause eye, nose and throat irritation.

Formulation Types: Emulsifiable concentrate; soluble concentrate.

Toxicity: LC$_{50}$ (96 hr) for rainbow trout 350 mg/L and carp 355 mg/L (Hartley and Kidd, 1987); acute oral LD$_{50}$ for rats 500 mg/kg (Hartley and Kidd, 1987), 300 mg/kg (RTECS, 1985).

Uses: Plant hormone; defoliant; herbicide used to control undesirable brush and woody plants.

TEBUTHIURON

Synonyms: Brulan; Bushwacker; 1-(5-t-Butyl-1,3,4-thiadiazol-2-yl)-1,3-dimethylurea; N-(5-(1,1-Dimethylethyl)-1,3,4-thiadiazol-2-yl)-N,N'-dimethylurea; E 103; EI 103; EL 103; Graslan; Perflan; Perfmid; Preflan; Prefmid; Spike; Tebulan; Tiurolan.

Structure:

$$(CH_3)_3C \diagdown \underset{N \text{———} N}{\overset{S}{\diagup \diagdown}} \overset{\overset{CH_3}{|}}{N}CONHCH_3$$

Designations: CAS: 34014-18-1; mf: $C_9H_{16}N_4OS$; fw: 228.32; RTECS: YS4250000.

Properties: Colorless crystals with a slight pungent odor. Mp: 161.5-164 °C (decomposes); fl p: nonflammable; H-t$_{1/2}$: >64 d at pH 3-9; K_d: 0.11 (sand), 1.82 (clay loam); K_H: 2.5 x 10^{-10} atm·m^3/mol at 20-25 °C (approximate - calculated from water solubility and vapor pressure); log K_{oc}: 2.79; log K_{ow}: 1.79; S_o (g/L at 25 °C): acetone (70), acetonitrile (60), benzene (3.7), chloroform (250), n-hexane (6.1), methanol (170), methyl cellosolve (60); S_w: 2.3-2.5 g/L at 25 °C; vp: 2 x 10^{-6} mmHg at 20 °C.

Transformation Products

Soil: In microbially active soils, tebuthiuron is degraded via demethylation. The average half-life in soil 12-15 mo in areas receiving 60 inches of rainfall a year (Humburg et al., 1989).

Plant: Degrades in plants via N-demethylation and hydroxylation of the t-butyl sidechain (Hartley and Kidd, 1987; Humburg et al., 1989).

Formulation Types: Dry flowable powder (85%), granules; pellets (20 and 40%); wettable powder.

Toxicity: LC$_{50}$ (96 hr) for bluegill sunfish 112 mg/L, rainbow trout 144 mg/L, goldfish and fathead minnow >160 mg/L (Hartley and Kidd, 1987); acute oral LD$_{50}$ for rats is 644 mg/kg (Ashton and Monaco, 1991).

Uses: Nonselective herbicide used to control herbaceous and woody plants on noncrop land.

TERBACIL

Synonyms: 3-*tert*-Butyl-5-chloro-6-methyluracil; 5-Chloro-3-*tert*-butyl-6-methyluracil; **5-Chloro-3-(1,1-dimethylethyl)-6-methyl-2,4(1*H*,3*H*)-pyrimidinedione**; Compound 732; Du Pont 732; Du Pont herbicide 732; Experimental herbicide 732; Sinbar; Turbacil.

Structure:

Designations: CAS: 5902-51-2; mf: $C_9H_{13}ClN_2O_2$; fw: 216.70; RTECS: YQ9360000.

Properties: Colorless to white crystalline solid. Mp: 175-177 °C; bp: sublimes below mp; ρ: 1.34 at 25/25 °C; fl p: nonflammable; K_H: 1.8 x 10^{-10} atm·m^3/mol at 20-25 °C (approximate - calculated from water solubility and vapor pressure); log K_{oc}: $\approx$ 1.74; log K_{ow}: 1.89, 1.90; S_o (g/kg at 25 °C): *n*-butyl acetate (88), cyclohexanone (180), *N,N*-dimethylformamide (252), methyl isobutyl ketone (121), xylene (61); S_w: 710 mg/L at 25 °C; vp: 4.5 x 10^{-7} mmHg at 20 °C.

Soil properties and adsorption data

Soil	K_d (mL/g)	f_{oc} (%)	K_{oc} (mL/g)	pH	CEC (meq/100 g)
Cecil sandy loam	0.15	0.40	38	5.8	--
Cecil sandy loam	0.38	0.90	42	5.6	6.8
Eustis fine sand	0.12	0.50	21	5.6	5.2
Glendale sandy clay loam	0.38	0.56	76	7.4	--
Keyport silt loam	1.70	1.21	140	5.4	--
Webster silty clay loam	2.46	3.87	64	7.3	54.7

Source: Rao and Davidson, 1979; Rhodes et al., 1970; U.S. Department of Agriculture, 1990.

Transformation Products

Plant: The major degradation products of [2-^{14}C]terbacil identified in alfalfa using a mass spectrometer were (% of applied amount): 3-*tert*-butyl-5-chloro-6-hydroxymethyluracil (11.9) and 6-chloro-2,3-dihydro-7-(hydroxymethyl)-3,3-dimethyl-5*H*-oxazolo[3,2-*a*]pyrimidin-5-one (41.2). Two additional compounds tentatively identified by TLC were 3-*tert*-butyl-6-hydroxymethyluracil and 6-

chloro-2,3-dihydro-7-methyl-3,3-dimethyl-5*H*-oxazolo[3,2-*a*]pyrimidin-5-one (Rhodes, 1977).

Photolytic: Acher et al. (1981) studied the dye-sensitized photolysis of aerated aqueous solutions of terbacil over a wide pH range. After a 2-hr exposure to sunlight, terbacil in aqueous solution (pH range 3.0-9.2) in the presence of presence of methylene blue (3 ppm) or riboflavin (10 ppm), photodecomposed to 3-*tert*-5-butyl-5-acetyl-5-hydroxyhydantoin. Deacylation was observed under alkaline conditions (pH 8.0 or 9.2) affording 3-*tert*-5-hydroxyhydantoin. In neutral or acidic conditions (pH 6.8 or 3.0) containing riboflavin, a mono-*N*-dealkylated terbacil dimer and an unidentified water-soluble product formed. Product formation, the relative amounts of products formed, and the rate of photolysis were found to be all dependent upon pH, sensitizer, temperature, and time (Acher et al., 1981).

Symptoms of Exposure: May irritate eyes, nose, throat, and skin.

Formulation Types: Wettable powder (80%).

Toxicity: LC_{50} (48 hr) for pumpkinseed sunfish 86 mg/L and fiddler crab >1,000 mg/L (Worthing and Hance, 1991); acute oral LD_{50} for nonfasted and fasted rats is >5,000 and <7,500 mg/kg, respectively (Ashton and Monaco, 1991).

Uses: Herbicide used to control many annual broadleaf weeds, some perennial weeds and grasses in alfalfa, sugarcane, mints, and certain fruit and nut trees.

TERBUFOS

Synonyms: AC 92100; S-((tert-Butylthio)methyl) O,O-diethyl phosphorodi-thioate; Counter; Counter 15G soil insecticide; Counter 15G soil insecticide-nematocide; S-(((1,1-Dimethylethyl)thio)methyl) O,O-diethyl phosphorodi-thioate; Phosphorodithioic acid S-((tert-butylthio)methyl) O,O-diethyl ester; **Phosphorodithioic acid S-(((1,1-dimethylethyl)thio)methyl) O,O-diethyl ester.**

Structure:

$$(CH_3)_3CSCH_2S - P \overset{\overset{\displaystyle S}{\|}}{\underset{}{}} \begin{array}{l} OCH_2CH_3 \\ \\ OCH_2CH_3 \end{array}$$

Designations: CAS: 13071-79-9; mf: $C_9H_{21}O_2PS_3$; fw: 288.43; RTECS: TD7200000.

Properties: Technical product (85-88%) is clear, colorless to pale yellow liquid with a mercaptan-like odor. Mp: -29.2 °C; bp: 69 °C at 0.01 mmHg (decomposes >120 °C); ρ: 1.105 at 24/4 °C; fl p: 88 °C (open cup); K_H: 2.2 x 10^{-5} atm·m^3/mol at 20-27 °C (approximate - calculated from water solubility and vapor pressure); log K_{oc}: 2.46-3.03; log K_{ow}: 2.22-4.70; S_o: soluble in many organic solvents; S_w: 4.5 mg/L at 27 °C; vap d: 11.79 g/L at 25 °C, 9.99 (air = 1); vp: 2.63 x 10^{-4} mmHg at 20 °C.

Soil properties and adsorption data

Soil	K_d (mL/g)	f_{oc} (%)	K_{oc} (mL/g)	pH	CEC (meq/100 g)
Clarion soil	8.30	2.64	314	5.00	21.02
Harps soil	21.33	3.80	561	7.30	37.84
Peat soil	52.72	18.36	287	6.98	77.34
Sarpy fine sandy loam	3.34	0.51	655	7.30	5.71
Thurman loamy fine sand	11.40	1.07	1,065	6.83	6.10

Source: Felsot and Dahm, 1979.

Transformation Products
Soil: Oxidized in soil to its primary and secondary oxidation products, terbufos sulfoxide and terbufos sulfone (Wei, 1990). Incubation of terbufos (5 µg/g) in a loamy sand containing *Nitrosomonas* sp. and *Nitrobacter* sp. gave terbufos sulfoxide and terbufos sulfone as the primary products. After 2 wk, the sulfoxide increased the bacterial population >55% and the sulfone increased the fungal population at least 66% (Tu, 1980). The half-life in soil is 9-27 d (Worthing and Hance, 1991).

Formulation Types: Granules.

Toxicity: LC_{50} (96 hr) for rainbow trout 10 mg/L and bluegill sunfish 4 mg/L (Hartley and Kidd, 1987); acute oral LD_{50} for rats 1,600 μg/kg (RTECS, 1985).

Use: Soil insecticide.

TERBUTRYN

Synonyms: 2-*tert*-Butylamino-4-ethylamino-6-methylmercapto-*s*-triazine; 2-*tert*-Butylamino-4-ethylamino-6-methylthio-*s*-triazine; Clarosan; *N*-(1,1-Dimethylethyl)-*N'*-ethyl-6-(methylthio)-1,3,5-triazine-2,4-diamine; GS 14260; HS 14260; Igran; Igran 50; Igran 80W; 2-Methylthio-4-ethylamino-6-*tert*-butyl-amino-*s*-triazine; Prebane; Shortstop; Shortstop E; Terbutrex B; Terbutryne.

Structure:

$$CH_3S \quad \underset{N}{\overset{N}{\diagup}} \quad NHC(CH_3)_3$$

$$NHCH_2CH_3$$

Designations: CAS: 886-50-0; mf: $C_{10}H_{19}N_5S$; fw: 241.36; RTECS: XY4725000.

Properties: Colorless crystals or white powder. Mp: 104-105 °C; bp: 154-160 °C at 0.06 mmHg; ρ: 1.115 at 20/4 °C; fl p; nonflammable; pK_a: 4.07; K_H: 1.2 x 10^{-7} atm·m^3/mol at 20 °C (approximate - calculated from water solubility and vapor pressure); log K_{oc}: 3.21-4.07; log K_{ow}: 3.43-3.73; S_o (g/L at 20 °C): acetone (280), *n*-hexane (9), methanol (280), methylene chloride (300), toluene (45); S_w: 25 mg/L at 20 °C; vp: 9.6 x 10^{-6} mmHg at 20 °C.

Soil properties and adsorption data

Soil	K_d (mL/g)	f_{oc} (%)	K_{oc} (mL/g)	pH
Boyce loam	5.10	1.27	402	8.0
Boyce silty loam	13.70	0.41	3,341	9.6
Chehalis sandy loam	21.50	0.98	2,194	5.2
Clayey loam	183.00	1.57	11,656	--
Deschutes sandy loam	6.10	0.51	1,196	5.9
Gila silty clay	3.90	0.63	619	8.0
Loamy sand	2.84	0.17	1,632	7.4
Metolius sandy loam	4.80	0.81	593	7.1
Powder silty loam	7.20	1.67	431	7.7
Quincy sandy loam	1.70	0.28	607	8.0
Sand	49.30	1.80	2,742	5.1
Silty loam	14.10	0.87	1,621	6.7
Woodburn silty loam	14.70	1.32	1,113	5.2

Source: Colbert et al., 1975; U.S. Department of Agriculture, 1990.

Transformation Products
Plant: In plants, the methylthio group is oxidized to hydroxy derivatives and by dealkylation of the sidechains. Residual activity in soil is limited to approximately 3-10 wk (Hartley and Kidd, 1987).

Symptoms of Exposure: Eye and skin irritant.

Formulation Types: Suspension concentrate; wettable powder; granules.

Toxicity: LC_{50} (96 hr) for rainbow trout 3 mg/L, bluegill sunfish 4 mg/L, carp 4 mg/L, and perch 4 mg/L (Hartley and Kidd, 1987); acute oral LD_{50} for rats 2,500 mg/kg (Hartley and Kidd, 1987), 2,045 mg/kg (RTECS, 1985).

Uses: Selective herbicide for control of annual broadleaf and grass weeds in wheat.

TETRAETHYL PYROPHOSPHATE

Synonyms: Bis-*O,O*-diethylphosphoric anhydride; Bladan; **Diphosphoric acid tetraethyl ester;** ENT 18771; Ethyl pyrophosphate; Fosvex; Grisol; Hept; Hexamite; Killax; Kilmite 40; Lethalaire G 52; Lirohex; Mortopal; NA 2783; Nifos; Nifos T; Nifost; Pyrophosphoric acid tetraethyl ester; RCRA waste number P111; TEP; TEPP; Tetraethyl diphosphate; Tetraethyl pyrofosfaat; Tetrastigmine; Tetron; Tetron-100; Vapotone.

Structure:

$$CH_3CH_2O \underset{CH_3CH_2O}{\overset{\overset{\textstyle O}{\|}}{\diagup}} P - O - P \overset{\overset{\textstyle O}{\|}}{\underset{OCH_2CH_3}{\diagdown}} OCH_2CH_3$$

Designations: CAS: 107-49-3; DOT: 2784; mf: $C_8H_{10}O_7P_2$; fw: 290.20; RTECS: UX6825000.

Properties: Colorless to amber liquid with a fruity odor. Hygroscopic. Mp: 0 °C; bp: 135-138 °C at 1 mmHg; ρ: 1.185 at 20/4 °C; H-t$_{1/2}$: 6.8-7.5 hr at 25 °C and pH 7; S_o: miscible with acetone, benzene, carbon tetrachloride, chloroform, ethanol, ethylbenzene, ethylene glycol, glycerol, kerosene, methanol, methyl ethyl ketone, petroleum ether, *n*-propanol, propylene glycol, toluene, xylene, and many other organic solvents; S_w: miscible; vap d: 11.86 g/L at 25 °C, 10.02 (air = 1); vp: 1.55 x 10^{-4} mmHg at 20 °C.

Transformation Products
Chemical/Physical: Though no products were identified, tetraethyl pyrophosphate is quickly hydrolyzed by water. The hydrolysis half-lives at 25 and 38 °C are 6.8 and 3.3 hr, respectively (Sittig, 1985). Decomposes at 170-213 °C releasing large amounts of ethylene (Hartley and Kidd, 1987). Emits toxic phosphorus oxide fumes when heated to decomposition (Lewis, 1990).

Exposure Limits: NIOSH REL: IDLH 10 mg/m³; ACGIH TLV: TWA 0.004 ppm.

Symptoms of Exposure: Eye pain, vision, tears, headache, chest, cyanosis, anorexia, nausea, vomiting, diarrhea, local sweating, weakness, twitch, paralysis, Cheyne-Stokes respiratory, convulsions, low blood pressure.

Formulation Types: Aerosol; emulsifiable concentrate.

Toxicity: LC$_{50}$ (96 hr) for fathead minnow 1.90 mg/L and bluegill sunfish 1.10

mg/L (Verschueren, 1983); acute oral LD_{50} for rats 1.12 mg/kg (Hartley and Kidd, 1987), 500 μg/kg (RTECS, 1985).

Uses: Insecticide for mites and aphids; rodenticide.

THIABENDAZOLE

Synonyms: Apl-Luster; Arbotect; Bovizole; Comfuval; Eprofil; Equizole; Lombristop; Mertec; Mertect; Mertect 160; Metasol TK-100; Mintesol; Mintezol; Minzolum; MK 360; Mycozol; Nemapan; Omnizole; Polival; Storite; TBDZ; TBZ; Tecto; Tecto 60; Tecto RPH; Thiaben; Thiabendazol; Thiabenzazole; Thiabenzol; 2-(Thiazol-4-yl)benzimidazole; 2-(1,3-Thiazol-4-yl)benzimidazole; 2-(4-Thiazol-yl)benzimidazole; **2-(4-Thiazolyl)-1*H*-benzimidazole**; Thibenzol; Thibenzole; Thibenzole ATT; Top Form Wormer.

Structure:

Designations: CAS: 148-79-8; mf: $C_{10}H_7N_3S$; fw: 201.25; RTECS: DE0700000.

Properties: Colorless to pale white, odorless powder. Mp: 304-305 °C; bp: sublimes at 310 °C; log K_{oc}: 2.71 at pH 5-12 (calculated); log K_{ow}: 2.69 at pH 5-12 (calculated); S_o (g/L at 25 °C): acetone (4.2), benzene (0.23), chloroform (80), N,N-dimethylformamide (39), dimethyl sulfoxide (80), ethanol (7.9), ethyl acetate (2.1), methanol (9.3); S_w: <50 mg/L at pH 5-12, ≈ 250 mg/L at pH 3-5, 39.9 g/L at pH 2.2.

Transformation Products

Photolytic: When thin films of thiabendazole on glass plates were exposed to sunlight for 128 d, benzimidazole-2-carboxamide and benzimidazole formed as photolysis products. Both compounds were also formed when aqueous methanolic solutions of thiabendazole were subjected to UV light for 1 hr (Zbozinek, 1984).

Formulation Types: Smoke tablets; suspension concentrate; wettable powder.

Toxicity: Low toxicity to fish (Hartley and Kidd, 1987); acute oral LD_{50} for rats 3,810 mg/kg (Hartley and Kidd, 1987), 3,100 mg/kg (RTECS, 1985).

Uses: Systemic fungicide used for diseases of fruits and vegetables and for control of Dutch elm disease.

THIAMETURON-METHYL

Synonyms: DPX M6316; Harmony; 3-(((((4-Methoxy-6-methyl-1,3,5-triazin-2-yl)amino)-carbonyl)amino)sulfonyl)-2-thiophenecarboxylic acid methyl ester.

Structure:

Designations: CAS: 79277-27-3; mf: $C_{12}H_{13}N_5O_6S_2$; fw: 387.40.

Properties: Colorless to white crystalline solid. Mp: 186 °C; ρ: 1.49; fl p: nonflammable; pK_a: 4.0; log K_{oc}: $\approx$ 1.65; log K_{ow}: -1.60; S_o (mg/L at 25 °C): acetone (11.9), acetonitrile (7.3), ethanol (0.9), ethyl acetate (2.6), n-hexane (<0.1), methanol (2.6), methylene chloride (27.5), xylene (0.2); S_w at 25 °C: 24 mg/L at pH 4, 260 mg/L at pH 5, 2.4 g/L at pH 6; vp: 2.03 x 10^{-6} mmHg at 20 °C.

Transformation Products
Chemical/Physical: May hydrolyze in aqueous solutions forming methyl alcohol and 3-(((((4-Methoxy-6-methyl-1,3,5-triazin-2-yl)amino)carbonyl)amino)sulfonyl)-2-thiophenecarboxylic acid.

Symptoms of Exposure: May irritate eyes, nose, throat, and skin.

Formulation Types: Water-dispersible granules (75%).

Toxicity: LC_{50} (96 hr) for both bluegill sunfish and rainbow trout >100 mg/L; LC_{50} (48 hr) for *Daphnia magna* >1,000 mg/L (Humburg et al., 1989); acute oral LD_{50} for rats >5,000 mg/kg (Hartley and Kidd, 1987).

Uses: Postemergence herbicide used to control wild garlic and many broadleaf weeds in barley and spring wheat.

THIDIAZURON

Synonyms: Defolit; Dropp; 1-Phenyl-3-(1,2,3-thiadiazol-5-yl)urea; *N*-Phenyl-*N'*-(1,2,3-thiadiazol-5-yl)urea; SN 49537.

Structure:

Designations: CAS: 51707-55-2; mf: $C_9H_8N_4OS$; fw: 220.20; RTECS: YU1395000.

Properties: Colorless crystals. Mp: 213 °C (decomposes); pK_a: 8.86; K_d: 2.2-21; K_H: 3.2 x 10^{-13} atm·m³/mol at 25 °C (approximate - calculated from water solubility and vapor pressure); log K_{oc}: 2.86 (calculated); log K_{ow}: 1.77 at pH 7.3; S_o (g/L at 23 °C): acetone (8), benzene (0.035), chloroform (0.013), cyclohexanone (21.5), *N,N*-dimethylformamide (>500), dimethyl sulfoxide (>500), ethyl acetate (0.8), *n*-hexane (0.006), methanol (4.5); S_w: 20-31 mg/L at 23-25 °C; vp: 2.2 x 10^{-11} mmHg at 25 °C.

Transformation Products

Soil: The reported half-life in soil is approximately 26-144 d (Hartley and Kidd, 1987).

Photolytic: Rapidly converted to the photoisomer, 1-phenyl-3-(1,2,5-thiadiazol-3-yl)urea (Worthing and Hance, 1991). When thidiazuron adsorbed by soil was exposed to UV light (λ <290 nm), 1-phenyl-3-(1,2,5-thiadiazol-3-yl)urea formed as the major product (P-t$_{1/2}$ = 0.5 hr). In addition, several polar products formed but were not identified (Klehr et al., 1983).

Formulation Types: Wettable powder.

Toxicity: LC_{50} (96 hr) for bluegill sunfish, channel catfish and rainbow trout >1,000 mg/L (Hartley and Kidd, 1987); acute oral LD_{50} for rats >5,000 mg/kg (Hartley and Kidd, 1987), 5,350 mg/kg (RTECS, 1985).

Uses: Plant growth regulator used to defoliate cotton to facilitate harvesting.

THIODICARB

Synonyms: Bismethomyl thioether; CGA 45156; Dicarbosulf; **Dimethyl** N,N'-**(thiobis((methylimino)carbonyloxy))bis(ethanimidothioate);** Larvin; Lepicron; N,N'-(Thiobis(methylimino)carbonyloxy)bisethanimidothioic acid dimethyl ester; UC 51762.

Structure:

$$CH_3C = NOCN - S - NCON = CCH_3$$

with O (double bond) above each carbonyl carbon, SCH_3 below the first and last carbons, and CH_3 below the two nitrogen atoms.

Designations: CAS: 59669-26-0; mf: $C_{10}H_{18}N_4O_4S_3$; fw: 354.47; RTECS: KJ4301050.

Properties: White to light tan crystals. Mp: 168-172 °C; ρ: 1.4 at 20/4 °C; H-t$_{1/2}$: $\approx$ 9 d at pH 3; K_H: 4.3 x 10^{-7} atm·m^3/mol at 20-25 °C (approximate - calculated from water solubility and vapor pressure); log K_{oc}: 1.81-3.07; log K_{ow}: 1.2-1.6; S_o (g/kg at 25 °C): acetone (8), methanol (5), methylene chloride (150), xylene (3); S_w: 35 mg/L at 25 °C; vp: 3.23 x 10^{-5} mmHg at 20 °C.

Soil properties and adsorption data

Soil	K_d (mL/g)	f_{oc} (%)	K_{oc} (mL/g)	pH
Clay	14.00	1.21	1,160	--
Loam	1.34	0.75	178	8.1
Sand	0.58	0.46	125	5.8
Sand	0.16	0.25	489	--
Sandy loam	1.22	0.58	210	7.8
Sandy loam	1.34	0.40	335	--
Silty loam	4.47	1.21	371	--

Source: U.S. Department of Agriculture, 1990.

Transformation Products

Soil: Under aerobic and anaerobic soil conditions, thiodicarb degrades to methomyl and methomyl oxime (Hartley and Kidd, 1987). The reported half-life in various soils is 2-8 d (Hartley and Kidd, 1987).

Formulation Types: Water-soluble granules; wettable powder; suspension concentrate.

Toxicity: LC_{50} (96 hr) for rainbow trout 2.55 mg/L and bluegill sunfish 1.21 mg/L (Hartley and Kidd, 1987); acute oral LD_{50} for rats 66 mg/kg (in water) (Hartley and Kidd, 1987), 160 mg/kg (RTECS, 1985).

Use: Insecticide.

THIRAM

Synonyms: Aatack; Accelerator thiuram; Aceto TETD; Arasan; Arasan 70; Arasan 75; Arasan-M; Arasan 42-S; Arasan-SF; Arasan-SF-X; Aules; Bis(dimethylamino)carbonothioyl disulfide; Bis(dimethylthiocarbamoyl) disulfide; Bis(dimethylthiocarbamyl) disulfide; Chipco thiram 75; Cyuram DS; α,α'-Dithiobis(dimethylthio) formamide; N,N'-(Dithiodicarbonothioyl)bis(N-methylmethanamine); Ekagom TB; ENT 987; Falitram; Fermide; Fernacol; Fernasan; Fernasan A; Fernide; Flo pro T seed protectant; Hermal; Hermat TMT; Heryl; Hexathir; Kregasan; Mercuram; Methyl thiram; Methyl thiuram-disulfide; Methyl tuads; NA 2771; Nobecutan; Nomersan; Normersan; NSC 1771; Panoram 75; Polyram ultra; Pomarsol; Pomersol forte; Pomasol; Puralin; RCRA waste number U244; Rezifilm; Royal TMTD; Sadoplon; Spotrete; Spotrete-F; SQ 1489; Tersan; Tersan 75; Tetramethyldiurane sulphite; Tetramethylthiuram bisulfide; Tetramethylthiuram bisulphide; Tetramethylenethiuram disulfide; **Tetramethylthioperoxydicarbonic diamide**; Tetramethylthiocarbamoyl disulfide; Tetramethylthioperoxydicarbonic diamide; Tetramethylthiuram disulfide; Tetramethylthiuram disulphide; N,N-Tetramethylthiuram disulfide; N,N,N',N'-Tetramethylthiuram disulfide; Tetramethylthiurane disulphide; Tetramethylthiurum disulfide; Tetramethylthiurum disulphide; Tetrapom; Tetrasipton; Tetrathiuram disulfide; Tetrathiuram disulphide; Thillate; Thimer; Thiosan; Thiotex; Thiotox; Thiram 75; Thiramad; Thiram B; Thirasan; Thiulix; Thiurad; Thiuram; Thiuram D; Thiuramin; Thiuram M; Thiuram M rubber accelerator; Thiuramyl; Thylate; Tirampa; Tiuramyl; TMTD; TMTDS; Trametan; Tridipam; Tripomol; TTD; Tuads; Tuex; Tulisan; USAF B-30; USAF EK-2089; USAF P-5; Vancida TM-95; Vancide TM; Vulcafor TMTD; Vulkacit MTIC.

Structure:

$$
\underset{(CH_3)_2NC}{\overset{\overset{\displaystyle S}{\|}}{}} - SS - \underset{CN(CH_3)_2}{\overset{\overset{\displaystyle S}{\|}}{}}
$$

Designations: CAS: 137-26-8; DOT: 2771; mf: $C_6H_{12}N_2S_4$; fw: 269.35; RTECS: JO1400000.

Properties: Colorless to white to cream-colored crystals. Mp: 155.6 °C; bp: 310-315 °C at 15 mmHg; ρ: 1.29 at 20/4 °C; fl p: 88.9 °C; H-t$\frac{1}{2}$: 5.3 d at 25 °C and pH 7; S_o: soluble in carbon tetrachloride, chloroform, methylene chloride, acetone (1.2 wt %), alcohol (<0.2 wt %), benzene (2.5 wt %), ethyl ether (<0.2 wt %), methyl ethyl ketone, toluene, xylene but only slightly soluble in carbon disulfide; S_w: 30 ppm.

Soil properties and adsorption data

Soil	K_d (mL/g)	f_{oc} (%)	K_{oc} (mL/g)	pH
Alluvial-Rio Nacimiento	11.96	0.91	1,314	8.0
Brown Clay-Almanzora Alto	12.00	1.17	1,026	8.5
Brown Lime-Almanzora Bajo	12.93	1.48	874	8.9
Brown Lime-Los Velez	13.73	2.06	667	8.1
Desert-Campo de Tabernas	8.08	0.33	2,448	7.9
Rendzine-Andarax	9.15	0.65	1,408	8.1
Saline-Campo de Dalías	11.17	1.65	677	8.2
Volcanic-Campo de Nijar	4.81	0.37	1,300	8.7

Source: Valverde-García et al., 1988.

Transformation Products
Biological: In both soils and water, chemical and biological mediated reactions can transform thiram to compounds containing the mercaptan group (Alexander, 1981). Odeyemi and Alexander (1977) isolated three strains of *Rhizobium* sp. that degraded thiram. One of these strains, *Rhizobium meliloti*, metabolized thiram to yield dimethylamine (DMA) and carbon disulfide which formed spontaneously from dimethyldithiocarbamate (DMDT). The conversion of DMDT to DMA and carbon disulfide occurred via enzymatic and nonenzymatic mechanisms (Odeyemi and Alexander, 1977).

Soil: Decomposes in soils to carbon disulfide and dimethylamine (Sisler and Cox, 1954; Kaars Sijpesteijn et al., 1977). When a spodosol (pH 3.8) pretreated with thiram was incubated for 24 d at 30 °C and relative humidity of 60-90%, dimethylamine formed as the major product. Minor degradative products included nitrite ions (nitration reduction) and dimethylnitrosamine (Ayanaba et al., 1973).

Plant: Major plant metabolites are ethylene thiourea, thiram monosulfide, ethylene thiram disulfide, and sulfur (Hartley and Kidd, 1987).

Chemical/Physical: Emits toxic sulfur oxides when heated to decomposition (Lewis, 1990). Though no products were reported, the calculated hydrolysis half-life at 25 °C and pH 7 is 5.3 d (Ellington et al., 1988).

Exposure Limits: NIOSH REL: IDLH 1,500 mg/m³; ACGIH TLV: TWA 5 mg/m³.

Symptoms of Exposure: Irritates mucous membrane, dermatitis; with ethanol consumption: flush, erythema, pruritus, urticaria, headache, nausea, vomiting, diarrhea, weakness, dizziness, difficulty in breathing. May cause allergic reaction in contact with skin.

Formulation Types: Suspension concentrate; wettable powder; dry seed treatment; dustable powder; water dispersible granules; water suspension.

Toxicity: LC_{50} (96 hr) for rainbow trout 0.13 mg/L, bluegill sunfish 0.23 mg/L, and carp 4.0 mg/L (Hartley and Kidd, 1987); acute oral LD_{50} for rats 865 mg/kg (Hartley and Kidd, 1987), 560 mg/kg (RTECS, 1985).

Uses: Seed disinfectant; fungicide.

TOXAPHENE

Synonyms: Agricide maggot killer (F); Alltex; Alltox; Attac 4-2; Attac 4-4; Attac 6; Attac 6-3; Attac 8; Camphechlor; Camphochlor; Camphoclor; Chem-phene M5055; Chlorinated camphene; Chloro-camphene; Clor chem T-590; Compound 3956; Crestoxo; Crestoxo 90; ENT 9735; Estonox; Fasco terpene; Geniphene; Gy-phene; Hercules 3956; Hercules toxaphene; Huilex; Kamfochlor; M 5,055; Melipax; Motox; NA 2761; NCI-C00259; Octachlorocamphene; PCC; Penphene; Phenacide; Phenatox; Phenphane; Polychlorcamphene; Polychlorinated camphenes; Polychlorocamphene; RCRA waste number P123; Strobane-T; Strobane T-90; Synthetic 3956; Texadust; Toxakil; Toxon 63; Toxyphen; Vertac 90%; Vertac toxaphene 90.

Structure:

Designations: CAS: 8001-35-2; DOT: 2761; mf: $C_{10}H_{10}Cl_8$; fw: 413.82; RTECS: XW5250000.

Properties: Yellow, waxy solid with a chlorine/terpene-like odor. Mp: 65-90 °C; bp: >120 °C (decomposes); ρ: 1.65 at 25/4 °C; fl p: 28.9 °C (10% xylene solution); lel: 1.1% (in solvent); uel: 6.4% (in solvent); H-$t_{1/2}$: 10 yr at 25 °C and pH 7; K_H: 6.3 x 10^{-2} atm·m^3/mol; log K_{oc}: 3.18 (calculated); log K_{ow}: 3.23-5.50; S_o: soluble chloroform, 95% ethanol (5-10 mg/mL at 19 °C), n-hexane, mineral oil, petroleum oils, toluene, xylene, and most organic solvents; S_w: 550 µg/L at 20 °C; vp: 0.2-0.4 mmHg at 25 °C.

Transformation Products

Soil: Under reduced soil conditions, about 50% of the C-Cl bonds were cleaved (dechlorinated) by Fe^{2+} porphyrins forming two major toxicants having the molecular formulas $C_{10}H_{10}Cl_8$ (Toxicant A) and $C_{10}H_{11}Cl_7$ (Toxicant B). Toxicant A reacted with reduced hematin yielding two reductive dechlorination products ($C_{10}H_{11}Cl_7$), two dehydrodechlorination products ($C_{10}H_9Cl_7$) and two other products ($C_{10}H_{10}Cl_6$). Similarly, products formed from the reaction of Toxicant B with reduced hematin included two reductive dechlorination products ($C_{10}H_{12}Cl_6$), one dehydrochlorination product ($C_{10}H_{10}Cl_6$) and two products having the molecular formula $C_{10}H_{11}Cl_5$ (Khalifa et al., 1976).

Photolytic: Dehydrochlorination will occur after prolonged exposure to sunlight

405

releasing hydrochloric acid (U.S. Department of Health and Human Services, 1989). Two compounds isolated from toxaphene, 2-*exo*, 3-*exo*, 5,5,6-*endo*, 8,9,10,10-nonachloroborane, and 2-*exo*, 3-*exo*, 5,5,6-*endo*, 8,10,10-octachloroborane were irradiated with UV light (λ >290 nm) in a neutral aqueous solution and on a silica gel surface. Both compounds underwent reductive dechlorination, dehydrochlorination and/or oxidation to yield numerous products including bicyclo[2.1.1]hexane derivatives (Parlar, 1988).

Chemical/Physical: Emits toxic fumes of chlorides when heated to decomposition (Lewis, 1990). Though no products were reported, the calculated hydrolysis half-life at 25 °C and pH 7 is 10 yr (Ellington et al., 1988).

Exposure Limits: NIOSH REL: IDLH 200 mg/m^3; OSHA PEL: TWA 0.5 mg/m^3, STEL 1 mg/m^3; ACGIH TLV: TWA 0.5 mg/m^3, STEL 1 mg/m^3.

Symptoms of Exposure: Nausea, confusion, agitation, tremors, convulsions, unconscious, dry red skin.

Formulation Types: Dustable powder; emulsifiable concentrate; wettable powder.

Toxicity: LC$_{50}$ for young rainbow trout 0.2 mg/L and young pike 0.1 mg/L (Hartley and Kidd, 1987); acute oral LD$_{50}$ for rats 40-90 mg/kg (Hartley and Kidd, 1987), 55 mg/kg (RTECS, 1985).

Uses: Pesticide used primarily on cotton, lettuce, tomatoes, corn, peanuts, wheat, and soybean.

TRIADIMEFON

Synonyms: Amiral; Bay 6681 F; Bayleton; May-meb-6647; **1-(4-Chlorophen-oxy)-3,3-dimethyl-1-(1H-1,2,4-triazol-1-yl)-2-butanone**; 1-(4-Chlorophen-oxy)-3,3-(1,2,4-triazol-1-yl)butan-2-one; MEB 6447.

Structure:

Designations: CAS: 43121-43-3; mf: $C_{14}H_{16}ClN_3O_2$; fw: 293.75; RTECS: EL7100000.

Properties: Colorless crystals. Mp: 82.3 °C; ρ: 1.22 at 20/4 °C; K_H: 1.1 x 10^{-9} atm·m^3/mol at 20 °C (approximate - calculated from water solubility and vapor pressure); log K_{oc}: 2.28-2.73; log K_{ow}: 3.18; S_o at 20 °C: cyclohexanone (0.6-1.2 g/kg), n-hexane (10-20 g/L), methylene chloride (>200 g/L), 2-propanol (200-400 kg/kg), toluene (>200 g/L); S_w: 260 mg/L at 20 °C; vp: 7.5 x 10^{-7} mmHg at 20 °C.

Soil properties and adsorption data

Soil	K_d (mL/g)	f_{oc} (%)	K_{oc} (mL/g)	pH
Sandy loam	9.3	1.74	534	5.5
Sand	5.9	2.15	275	6.9
Sand	4.1	2.15	191	6.9
Sandy loam	5.3	1.74	305	5.5
Silty loam	3.5	1.22	287	6.7
Silty loam	2.4	1.22	197	6.7

Source: U.S. Department of Agriculture, 1990.

Transformation Products

Plant: In soils and plants, triadimefon degrades to triadimenol (Clark et al., 1978; Hartley and Kidd, 1987; Rouchaud et al., 1981). In barley plants, triadimefon was metabolized to triadimenol and p-chlorophenol (Rouchaud et al., 1981; Rouchaud, 1982). In the grains and straw of ripe barley, the concentrations (μg/kg) of triadimefon, triadimenol, and p-chlorophenol were 0.12, 0.26, 0.13 and 0.70, 1.71, 0.49, respectively (Rouchaud et al., 1981).

Photolytic: When triadimefon was subjected to UV light for 1 wk, *p*-chlorophenol, 4-chlorophenyl methyl carbamate and a 1,2,4-triazole formed as products (Clark et al., 1978).

Formulation Types: Paste; emulsifiable concentrate; wettable powder.

Toxicity: LC_{50} (96 hr) for rainbow trout 14 mg/L, bluegill sunfish 11 mg/L, and golden orfe 13.8 mg/L; LC_{50} (48 hr) for carp 7.6 mg/L (Hartley and Kidd, 1987); acute oral LD_{50} for male and female rats 568 and 313 mg/kg, respectively (Hartley and Kidd, 1987), 400 mg/kg (RTECS, 1985).

Uses: Systemic fungicide used to control mildews and rusts that attack coffee, cereals, stone fruit, grapes, and ornamentals.

TRIALLATE

Synonyms: Avadex BW; Buckle; CP 23426; *N*-Diisopropylthiocarbamic acid *S*-2,3,3-trichloro-2-propenyl ester; Diisopropyltrichloroallylthiocarbamate; Dipthal; Far-Go; *S*-2,3,3-Trichloroallyl diisopropylthiocarbamate; *S*-2,3,3-Trichloroallyl *N,N*-diisopropylthiocarbamate; *S*-2,3,3-(Trichloro-2-propenyl) bis(1-methylethyl)carbamothioate.

Structure:

$$[(CH_3)_2CH]_2NCOSCH_2CCl=CCl_2$$

Designations: CAS: 2303-17-5; mf: $C_{10}H_{16}Cl_3NOS$; fw: 304.70; RTECS: EZ8575000.

Properties: Colorless crystals or amber oil. Mp: 29-30 °C; bp: 117 °C at 0.3 mmHg (decomposes >200 °C); ρ: 1.273 at 25/15.6 °C; fl p: 90 °C, 95 °C (open cup); K_d: 5-35; K_H: 1.0 x 10^{-5} atm·m³/mol at 20-25 °C (approximate - calculated from water solubility and vapor pressure); log K_{oc}: 3.31 (calculated); log K_{ow}: 4.29; S_o: soluble in acetone, benzene, ethanol, ethyl acetate, ethyl ether, *n*-heptane, *n*-hexane, methanol, and many other organic solvents; S_w: 4 mg/L at 25 °C; vp: 1.2 x 10^{-4} mmHg at 20 °C.

Soil properties and adsorption data

Soil	K_d (mL/g)	f_{oc} (%)	K_{oc} (mL/g)	pH
Drummer silty clay loam	28.40	--	--	--
Dupo silty loam	18.50	--	--	--
Lintonia sandy loam	5.98	--	--	--
Spinks sandy loam	22.50	--	--	--

Source: U.S. Department of Agriculture, 1990.

Transformation Products

Soil: In an agricultural soil, $^{14}CO_2$ was the only biodegradation identified, however, bound residue and traces of benzene and water-soluble radioactivity were also detected in large amounts (Anderson and Domsch, 1980). In soil, triallate degrades via hydrolytic cleavage with the formation of dialkylamine, carbon dioxide, and mercaptan moieties. The mercaptan compounds are further degraded to the corresponding alcohol (Hartley and Kidd, 1987). The reported half-life in soil is 100 d (Jury et al., 1987).

Formulation Types: Emulsifiable concentrate (4 lb/gal); granules (10%).

Toxicity: LC_{50} (96 hr) for rainbow trout 1.2 mg/L and bluegill sunfish 1.3 mg/L (Hartley and Kidd, 1987); acute oral LD_{50} of technical triallate for rats is 1,100 mg/kg (Ashton and Monaco, 1991), 1,471 mg/kg (RTECS, 1985).

Uses: Herbicide used to control wild oats in lentils, barley, peas, and winter wheat.

TRICHLORFON

Synonyms: Aerol 1; Agroforotox; Anthion; Bay 15922; Bayer 15922; Bayer L 13/59; Bilarcil; Bovinox; Britten; Briton; Cekufon; Chlorak; Chlorfos; Chlorofos; Chlorophos; Chlorophose; Chlorophthalm; Chloroxyphos; Ciclosom; Combot; Combot equine; Danex; DEP; Depthon; DETF; Dimethoxy-2,2,2-trichloro-1-hydroxyethylphosphine oxide; *O,O*-Dimethyl-(1-hydroxy-2,2,2-tri-chloro)ethyl phosphate; Dimethyl-1-hydroxy-2,2,2-trichloroethyl phosphonate; *O,O*-Dimethyl-(1-hydroxy-2,2,2-trichloro)ethyl phosphonate; *O,O*-Dimethyl-1-oxy-2,2,2-trichloroethyl phosphonate; Dimethyltrichlorohydroxyethyl phosphonate; Dimethyl-2,2,2-trichloro-1-hydroxyethyl phosphonate; *O,O*-Dimethyl-2,2,2-trichloro-1-hydroxyethyl phosphonate; Dimetox; Dipterax; Dipterex; Dipterex 50; Diptevur; Ditrifon; Dylox; Dylox-metasystox-R; Dyrex; Dyvon; ENT 19763; Equino-acid; Equino-aid; Flibol E; Forotox; Foschlor; Foschlor 25; Foschlor R 50; 1-Hydroxy-2,2,2-trichloroethylphosphonic acid dimethyl ester; Hypodermacid; Leivasom; Loisol; Masoten; Mazoten; Methyl chlorophos; Metifonate; Metrifonate; Metriphonate; NA 2783; NCI-C54831; Neguvon; Neguvon A; Phoschlor; Phoschlor R50; Polfoschlor; Proxol; Ricifon; Ritsifon; Satox 20WSC; Soldep; Sotipox; Trichlorofon; 2,2,2-Trichloro-1-hydroxyethylphosphonate dimethyl ester; **2,2,2-Trichloro-1-hydroxyethylphosphonic acid dimethyl ester;** Trichlorophon; Trichlorphene; Trichlorphon; Trichlorphon FN; Trinex; Tugon; Tugon fly bait; Tugon stable spray; Vermicide Bayer 2349; Volfartol; Votexit; WEC 50; Wotexit.

Structure:

$$CH_3O \diagdown \underset{CH_3O \diagup}{\overset{O}{\underset{\|}{P}}} - \overset{OH}{\underset{|}{C}}HCCl_3$$

Designations: CAS: 52-68-6; mf: $C_4H_8ClO_4P$; fw: 257.45; RTECS: TA0700000.

Properties: Colorless to pale yellow crystals with an ethereal-like odor. Mp: 83-84 °C; bp: 100 °C at 0.1 mmHg; ρ: 1.73 at 20/4 °C; K_H: 1.7 x 10^{-11} atm·m^3/mol at 25 °C; log K_{oc}: 0.99-1.58; log K_{ow}: 0.43, 0.43-0.76; S_o (g/kg at 25 °C): benzene (152), chloroform (750), ethyl ether (170), *n*-hexane (0.80), *n*-pentane and *n*-heptane ($\approx$ 0.80); S_w: 120 g/L at 20 °C, 154 g/L at 25 °C; vp: 7.8 x 10^{-6} mmHg at 20 °C.

Soil properties and adsorption data

Soil	K_d (mL/g)	f_{oc} (%)	K_{oc} (mL/g)	pH
Sandy loam	9.25	0.81	31	6.4

continued

Soil	K_d (mL/g)	f_{oc} (%)	K_{oc} (mL/g)	pH
Sandy loam	0.08	0.81	10	6.4
Silty loam	0.40	1.04	38	5.5
Silty loam	0.35	1.04	34	5.5
Silty loam	0.51	2.67	19	5.4
Silty loam	0.58	2.67	22	5.4

Source: U.S. Department of Agriculture, 1990.

Transformation Products
Soil: Trichlorfon degraded in soil to dichlorvos (alkaline conditions) and desmethyl dichlorvos (Mattson et al., 1955).

Plant: In cotton leaves, the metabolites identified included dichlorvos, phosphoric acid, *O*-demethyl dichlorvos, *O*-demethyl trichlorfon, methyl phosphate, and dimethyl phosphate (Bull and Ridgway, 1969). Chloral hydrate and trichloro-ethanol were reported as possible breakdown products of trichlorfon in plants (Anderson et al., 1966).

Chemical/Physical: Emits toxic fumes of phosphorus oxides and chlorine when heated to decomposition (Sax and Lewis, 1987). Decomposed by hot water at pH <5 forming dichlorvos (Worthing and Hance, 1991). At 100 °C, trichlorfon begins to decompose to form chloral.

Symptoms of Exposure: Irritates eyes.

Formulation Types: Water-soluble powder; suspension concentrate; wettable powder; granules; granular bait; coating agent.

Toxicity: LC_{50} (96 hr) for rainbow trout 1.4 mg/l bluegill sunfish 0.26 mg/L; LC_{50} (48 hr) for carp 6.2 mg/L and goldfish >10 mg/L (Hartley and Kidd, 1987); acute oral LD_{50} for rats 560 mg/kg (Hartley and Kidd, 1987).

Uses: Insecticide used to control flies and roaches.

2,3,6-TRICHLOROBENZOIC ACID

Synonyms: Benzabar; Benzac; Benzac-1281; Fen-All; HC 1281; NCI-C60242; T-2; TBA; 2,3,6-TBA; TCB; 2,3,5-TCB; TCBA; 2,3,6-TCBA; Tribac; Trichlorobenzoic acid; Tryben; Trysben; Trysben 200; Zobar.

Structure:

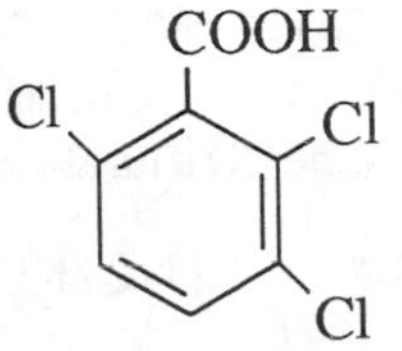

Designations: CAS: 50-31-7; mf: $C_7H_3Cl_3O_2$; fw: 225.47; RTECS: DH7700000.

Properties: Colorless to buff crystals. Mp: 125-126 °C (pure), 80-100 °C (technical); bp: decomposes; pK_a: <7; K_H: 9.3 x 10^{-7} atm·m^3/mol at 20-22 °C (approximate - calculated from water solubility and vapor pressure); log K_{oc}: 1.50 (calculated); log K_{ow}: 0.31 (calculated); S_o (g/L): acetone (607), benzene (238), chloroform (237), ethanol (637), methanol (717), xylene (210); S_w: 7.7 g/L at 22 °C; vp: 2.4 x 10^{-2} mmHg at 20 °C.

Transformation Products
Photolytic: Plimmer (1970) postulated that the irradiation of 2,3,6-trichlorobenzoic acid in methanol will undergo dechlorination forming 2,3-, 2,6-, and 3,6-dichlorobenzoic acids. Further irradiation may yield 2- and 3-chlorobenzoic acids which may ultimately form benzoic acid (Plimmer 1970).

Chemical/Physical: Reacts with alkalies and amines forming water-soluble salts.

Formulation Types: Solution.

Toxicity: Acute oral LD$_{50}$ for rats 1,500 mg/kg (Hartley and Kidd, 1987), 650 mg/kg (RTECS, 1985).

Uses: Systemic growth-regulator herbicide used for postemergence control of broadleaf perennial and annual weeds in cereals and grass seed crops.

413

TRIFLURALIN

Synonyms: Agreflan; Agriflan 24; Crisalin; Digermin; 2,6-Dinitro-*N,N*-dipropyl-4-trifluoromethylaniline; **2,6-Dinitro-*N,N*-dipropyl-4-(trifluoromethyl)-benzenamine;** 2,6-Dinitro-*N,N*-di-*n*-propyl-α,α,α-trifluoro-*p*-toluidine; 4-(Di-*n*-propylamino)-3,5-dinitro-1-trifluoromethylbenzene; *N,N*-Di-*n*-propyl-2,6-dinitro-4-trifluoromethylaniline; *N,N*-Dipropyl-4-trifluoromethyl-2,6-dinitro-aniline; Elancolan; L 36352; Lilly 36352; NCI-C00442; Nitran; Olitref; Trefanocide; Treficon; Treflam; Treflan; Treflanocide elancolan; Trifluoralin; α,α,α-Trifluoro-2,6-dinitro-*N,N*-dipropyl-*p*-toluidine; Trifluraline; Triflurex; Trikepin; Trim.

Structure:

$N(CH_2CH_2CH_3)_2$

O_2N NO_2

CF_3

Designations: CAS: 1582-09-8; DOT: 1609; mf: $C_{13}H_{16}F_3N_3O_4$; fw: 335.29; RTECS: XU9275000.

Properties: Yellow to orange crystals. Mp: 46-47 °C; bp: 139-140 °C at 4.2 mmHg; ρ: 1.294 at 25/4 °C; fl p: nonflammable; K_H: 4.84 x 10^{-5} atm·m^3/mol at 23 °C; log K_{oc}: 2.94-4.49; log K_{ow}: 5.07, 5.28, 5.34; S_o (g/L at 25 °C): acetone (>500), acetonitrile (>500), chloroform (>500), *N,N*-dimethylformamide (820), 1,4-dioxane (830), *n*-hexane (>500), methanol (20), methyl cellosolve (440), methyl ethyl ketone (880), xylene (810); S_w: 4 ppm at 27 °C; vp: 1.1 x 10^{-4} mmHg at 25 °C.

Soil properties and adsorption data

Soil	K_d (mL/g)	f_{oc} (%)	K_{oc} (mL/g)	pH
Cecil sandy loam	0.46	0.90	51	5.6
Catlin	58.40	2.01	3,000	6.2
Commerce	36.50	0.68	5,360	6.7
Eustis fine sand	0.24	0.56	43	5.6
Glendale sandy clay loam	1.60	0.50	178	7.4
Hickory Hill silt	999.00	3.27	30,550	--
Webster silty clay loam	2.93	3.87	76	7.3
Kanuma high clay	5.30	1.35	397	5.7
Tracy	52.00	1.12	4,650	6.2
Tsukuba clay loam	56.60	4.24	1,335	6.5

Source: Brown and Flagg, 1981; Kanazawa, 1989; McCall et al., 1981.

Transformation Products

Biological: Laanio et al. (1973) incubated $^{14}CF_3$-trifluralin with *Paecilomyces*, *Fusarium oxysporum*, or *Aspergillus fumigatus* and reported that <1% was converted to $^{14}CO_2$.

Soil: Anaerobic degradation in a Crowley silt loam yielded α,α,α-trifluoro-N^4,N^4-dipropyl-5-nitrotoluene-3,4-diamine, and α,α,α-trifluoro-N^4,N^4-dipropyltoluene-3,4,5-triamine (Parr and Smith, 1973). Probst and Tepe (1969) reported that trifluralin degradation in many soils was probably by chemical reduction of the nitro groups into amino groups. Reported degradation products in aerobic soils include α,α,α-trifluoro-2,6-dinitro-N-propyl-p-toluidine, α,α,α-trifluoro-2,6-dinitro-p-toluidine, α,α,α-trifluoro-5-nitrotoluene-3,4-diamine and α,α,α-trifluoro-N,N-dipropyl-5-nitrotoluene-3,4-diamine. Anaerobic degradation products identified include α,α,α-trifluoro-N,N-dipropyl-5-nitrotoluene-3,4-diamine, α,α,α-trifluoro-N,N-dipropyltoluene-3,4,5-triamine, α,α,α-trifluorotoluene-3,4,5-triamine, and α,α,α-trifluoro-N-propyltoluene-3,4,5-triamine. α,α,α-Trifluoro-5-nitro-N-propyl-toluene-3,4-diamine was identified in both aerobic and anaerobic soils (Probst et al., 1967). The following compounds were reported as major soil metabolites: α,α,α-trifluoro-2,6-dinitro-N-propyl-p-toluidine, α,α,α-trifluoro-2,6-dinitro-p-toluidine, α,α,α-trifluoro-5-nitrotoluene-3,4-diamine, α,α,α-trifluorotoluene-3,4,5-triamine, 2-ethyl-7-nitro-1-propyl-5-(trifluoromethyl)benzimidazole, 2-ethyl-7-nitro-5-(trifluoromethyl)benzimidazole, 7-nitro-1-propyl-5-(trifluoromethyl)benzimidazole, 4-(dipropylamino)-3,5-dinitrobenzoic acid, 2,2'-azoxybis(α,α,α-trifluoro-6-nitro-N-propyl-p-toluidine), 2,2'-azobis(α,α,α-trifluoro-6-nitro-N-propyl-p-toluidine), 2,6-dinitro-N,N-dipropyl-4-(trifluoromethyl)-m-anisidine, and α,α,α-trifluoro-2',6'-dinitro-N-propyl-p-propionotoluidine (Koskinen et al., 1984, 1985).

Golab et al. (1979) studied the degradation of trifluralin in soil over a 3-yr period. They found that the herbicide undergoes N-dealkylation, reduction of nitro substituents, followed by the formation cyclized products. Of the 28 transformations products identified, none exceeded 3% of the applied amount. These compounds were: α,α,α-trifluoro-2,6-dinitro-N-propyl-p-toluidine, α,α,α-trifluoro-2,6-dinitro-p-toluidine, α,α,α-trifluoro-5-nitro-N^4,N^4-dipropyltoluene-3,4-diamine, α,α,α-trifluoro-5-nitro-N^4-propyltoluene-3,4-diamine, α,α,α-trifluoro-5-nitrotoluene-3,4-diamine, α,α,α-trifluoro-N^4,N^4-dipropyltoluene-3,4,5-triamine, α,α,α-trifluoro-N^4-propyltoluene-3,4,5-triamine, α,α,α-trifluorotoluene-3,4,5-triamine, α,α,α-trifluoro-2'-hydroxyamino-6'-nitro-N-propyl-p-propionotoluidide, 2-ethyl-7-nitro-1-propyl-5-(trifluoromethyl)benzimidazole 3-oxide, 2-ethyl-7-nitro-5-(trifluoromethyl)benzimidazole 3-oxide, 2-ethyl-7-nitro-1-propyl-5-(trifluoromethyl)benzimidazole, 7-amino-2-ethyl-1-propyl-5-(trifluoromethyl)benzimidazole, 2-ethyl-7-nitro-5-(trifluoromethyl)benzimidazole, 7-amino-2-ethyl-5-(trifluoromethyl)benzimidazole, 7-nitro-1-propyl-

5-(trifluoromethyl)benzimidazole,   7-nitro-5-(trifluoromethyl)benzimidazole,   7-amino-5-(trifluoromethyl)benzimidazole,   α,α,α-trifluoro-2,6-dinitro-*p*-cresol,   4-(dipropylamino)-3,5-dinitrobenzoic   acid,   2,2′-azoxybis(α,α,α-trifluoro-6-nitro-*N*-propyl-*p*-toluidine),   *N*-propyl-2,2′-azoxybis(α,α,α-trifluoro-6-nitro-*p*-toluidine),   2,2′-azoxybis(α,α,α-trifluoro-6-nitro-*p*-toluidine),   2,2′-azobis(α,α,α-trifluoro-6-nitro-*N*-propyl-*p*-toluidine),   α,α,α-trifluoro-4,6-dinitro-5-(dipropylamino)-*o*-cresol,   α,α,α-trifluoro-2-hydroxyamino-6-nitro-*N*,*N*-dipropyl-*p*-toluidine,   α,α,α-trifluoro-2′,6′-dinitro-*N*-propyl-*p*-propionotoluidide,   and   α,α,α-trifluoro-2,6-dinitro-*N*-(propan-2-ol)-*N*-propyl-*p*-toluidine (Golab, et al., 1979). The reported half-life in soil is 132 d (Jury et al., 1987).

Plant: Trifluralin was absorbed by carrot roots in greenhouse soils pretreated with the herbicide (0.75 lb/acre). The major metabolite formed was α,α,α-trifluoro-2,6-dinitro-*N*-(*n*-propyltoluene)-*p*-toluidine (Golab et al., 1967). Two metabolites of trifluralin that were reported in goosegrass (*Eleucine indica*) were 3-methoxy-2,6-dinitro-*N*,*N*′-dipropyl-4-(trifluoromethyl)benzenamine   and   *N*-(2,6-dinitro-4-(trifluoromethyl)phenyl)-*N*-propylpropanamide (Duke et al., 1991).

Photolytic: Irradiation of trifluralin in *n*-hexane by laboratory light produced α,α,α-trifluoro-2,6-dinitro-*N*-propyl-*p*-toluidine and α,α,α-trifluoro-2,6-dinitro-*p*-toluidine. The sunlight irradiation of trifluralin in water yielded α,α,α-trifluoro-N^4,N^4-dipropyl-5-nitrotoluene-3,4-diamine,   α,α,α-trifluoro-N^4,N^4-dipropyl-toluene-3,4,5-triamine,   2-ethyl-7-nitro-5-trifluoromethylbenzimidazole,   2,3-dihydroxy-2-ethyl-7-nitro-1-propyl-5-trifluoromethylbenzimidazoline,   and   2-ethyl-7-nitro-5-trifluoromethylbenzimidazole.   2-Amino-6-nitro-α,α,α-trifluoro-*p*-toluidine   and   2-ethyl-5-nitro-7-trifluoromethylbenzimidazole   also   were reported as major products under acidic and basic conditions, respectively (Crosby and Leitis, 1973). In a later study, Leitis and Crosby (1974) reported that trifluralin was in aqueous solutions were very unstable to sunlight, especially in the presence of methanol. The photodecomposition of trifluralin involved oxidative *N*-dealkylation,   nitro   reduction,   and   reductive   cyclization.   The   principal photodecomposition products of trifluralin were 2-amino-6-nitro-α,α,α-trifluoro-*p*-toluidine,   2-ethyl-7-nitro-5-trifluoromethylbenzimidazole   3-oxide,   2,3-dihydroxy-2-ethyl-7-nitro-1-propyl-5-trifluoromethylbenzimidazole,   two azoxybenzenes. Under alkaline conditions, the principal photodecomposition product was 2-ethyl-7-nitro-5-trifluoromethyl-benzimidazole (Leitis and Crosby, 1974). When trifluralin was released in the atmosphere on a sunny day, it was rapidly converted to the photochemical 2,6-dinitro-*N*-propyl-α,α,α-trifluoro-*p*-toluidine (estimated $t_{1/2}$ = 20 min) (Woodrow et al., 1978). The vapor-phase photolysis of trifluralin was studied in the laboratory using a photoreactor which simulated sunlight conditions (Soderquist et al., 1975). Vapor-phase photoproducts of trifluralin were identified as 2,6-dinitro-*N*-propyl-α,α,α-trifluoro-*p*-toluidine, 2,6-dinitro-α,α,α-trifluoro-*p*-toluidine, 2-ethyl-7-nitro-1-propyl-5-trifluoromethylbenzimidazole, 2-ethyl-7-nitro-5-trifluoromethylbenzimidazole, and four benzimidazole precursors reported by Leitis and Crosby (1974). Similar photoproducts were also identified in air above both bare surface treated soil and soil incorporated fields (Soderquist et al., 1975). Sullivan et al. (1980) studied the UV photolysis of trifluralin in anaerobic benzene solutions. Products identified included the three azoxybenzene derivatives N-propyl-2,2'-azoxybis(α,α,α-trifluoro-6-nitro-p-toluidine), 2,2'-azoxybis(α,α,α-trifluoro-6-nitro-N-propyl-p-toluidine), and 2,2'-azoxybis(α,α,α-trifluoro-6-nitro-p-toluidine), and two azobenzene derivatives N-propyl-2,2'-azobis(α,α,α-trifluoro-6-nitro-p-toluidine) and 2,2'-azobis(α,α,α-trifluoro-6-nitro-N-propyl-p-toluidine) (Sullivan et al., 1980). ^{14}C-Labeled trifluralin on a silica plate was exposed to summer sunlight for 7.5 hr. Although 52% of trifluralin was recovered, no photodegradation products were identified. In soil, no significant photodegradation of trifluralin was observed after 9 hr of irradiation (Plimmer, 1978). When trifluralin on glass plates was irradiated for 4-6 hr, it photodegraded to 2,3-dihydroxy-2-ethyl-7-nitro-1-propyl-5-trifluoromethylbenzimidazoline and 2-ethyl-7-nitro-5-trifluoromethylbenzimidazole-3-oxide (Wright and Warren, 1965).

Chemical/Physical: Releases carbon monoxide, carbon dioxide, and ammonia when heated to 900 °C (Kennedy, 1972, 1972a). Incineration may also release hydrofluoric acid in the off-gases (Sittig, 1985).

Symptoms of Exposure: Irritates eyes and may cause skin sensitization in some individuals.

Formulation Types: Emulsifiable concentrate (4 lb/gal); granules (5 and 10%).

Toxicity: LC_{50} (96 hr) for young rainbow trout 10-40 μg/L and young bluegill sunfish 20-90 μg/L (Hartley and Kidd, 1987); LC_{50} (48 hr) for bluegill sunfish 19 ppb and rainbow trout 11 ppb (Verschueren, 1983); acute oral LD_{50} for mice is 5,000 mg/kg (RTECS, 1985).

Uses: Preemergence herbicide for controlling many grass and broadleaf weeds.

WARFARIN

Synonyms: 3-(Acetonylbenzyl)-4-hydroxycoumarin; 3-(α-Acetonylbenzyl)-4-hydroxycoumarin; Athrombine-K; Athrombin-K; Brumolin; Compound 42; Corax; Coumadin; Coumafen; Coumafene; Cov-r-tox; D-con; Dethmor; Dethnel; Eastern states duocide; Fasco fascrat powder; 1-(4'-Hydroxy-3'-coumarinyl)-1-phenyl-3-butanone; **4-Hydroxy-3-(3-oxo-1-phenylbutyl)-2H-1-benzopyran-2-one**; 4-Hydroxy-3-(1-phenyl-3-oxobutyl)coumarin; Kumader; Kumadu; Kypfarin; Liqua-tox; Mar-frin; Martin's mar-frin; Maveran; Mouse-pak; 3-(1'-Phenyl-2'-acetylethyl)-4-hydroxycoumarin; 3-α-Phenyl-β-acetylethyl-4-hydroxycoumarin; Prothromadin; Rat-a-way; Rat-b-gon; Rat-gard; Rat-kill; Rat & mice bait; Rat-mix; Rat-o-cide #2; Rat-ola; Ratorex; Ratox; Ratoxin; Ratron; Ratron G; Rats-no-more; Rat-trol; Rattunal; Rax; RCRA waste number P001; Rodafarin; Ro-deth; Rodex; Rodex blox; Rosex; Rough & ready mouse mix; Solfarin; Spray-trol brand roden-trol; Temus W; Tox-hid; Twin light rat away; Vampirinip II; Vampirin III; Waran; W.A.R.F. 42; Warfarat; Warfarin plus; Warfarin Q; Warf compound 42; Warficide.

Structure:

Designations: CAS: 81-81-2; DOT: 3027; mf: $C_{19}H_{16}O_4$; fw: 308.33; RTECS: GN4550000.

Properties: Colorless to white, odorless, tasteless crystals. Mp: 161 °C; bp: decomposes; H-t$_{1/2}$: 16 yr at 25 °C and pH 7; log K_{oc}: 2.96 (calculated); log K_{ow}: 3.20 (calculated); S_o: soluble in acetone, benzene, 1,4-dioxane, ethanol, ethylbenzene, ethyl ether, n-hexane, methyl ethyl ketone, toluene, xylene and other common organic solvents but is moderately soluble in methanol, isopropanol, and some oils; S_w: 17 mg/L at 20 °C.

Exposure Limits: NIOSH REL: IDLH 200 mg/m^3; ACGIH TLV: TWA 0.1 mg/m^3.

Symptoms of Exposure: Hematuria, back pain, hematoma arms, legs; epistaxis, bleeding lips, mucous membrane hemorrhage, abdominal pain, vomiting, fecal blood; petechial rash; abnormal hematology.

Formulation Types: Granular bait; tracking powder; gel; bait concentrate.

Toxicity: Acute oral LD_{50} for rats 186 mg/kg (Hartley and Kidd, 1987), 3 mg/kg (RTECS, 1985).

Uses: Rodenticide.

References

Abdel-Wahab, A.M., R.J. Kuhr, and J.E. Casida. "Fate of C^{14}-Carbonyl-Labeled Aryl Methylcarbamate Insecticide Chemicals in and on Bean Plants," *J. Agric. Food Chem.*, 14(3):290-298 (1966).

Abou-Assaf, N., J.R. Coats, M.E. Gray, and J.J. Tollefson. "Degradation of Isofenphos in Cornfields with Conservation Tillage Practices," *J. Environ. Sci. Health*, B21(6):425-446 (1986).

Acher, A.J. and S. Saltzman. "Dye-Sensitized Photooxidation of Bromacil in Water," *J. Environ. Qual.*, 9(2):190-194 (1980).

Acher, A.J., S. Saltzman, N. Brates, and E. Dunkelblum. "Photosensitized Decomposition of Terbacil in Aqueous Solutions," *J. Agric. Food Chem.*, 29(4):707-711 (1981).

Adams, C.D. and S.J. Randtke. "Ozonation Byproducts of Atrazine in Synthetic and Natural Waters," *Environ. Sci. Technol.*, 26(11):2218-2227 (1992).

Adewuyi, Y.G. and G.R. Carmichael. "Kinetics of Hydrolysis and Oxidation of Carbon Disulfide by Hydrogen Peroxide in Alkaline Medium and Application to Carbonyl Sulfide," *Environ. Sci. Technol.*, 21(2):170-177 (1987).

Adhya, T.K., Sudhakar-Barik, and N. Sethunathan. "Fate of Fenitrothion, Methyl Parathion, and Parathion in Anoxic Sulfur-Containing Soil Systems," *Pestic. Biochem. Physiol.*, 16(1):14-20 (1981).

Adhya, T.K., Sudhakar-Barik, and N. Sethunathan. "Stability of Commercial Formulation of Fenitrothion, Methyl Parathion, and Parathion in Anaerobic Soils," *J. Agric. Food Chem.*, 29(1):90-93 (1981a).

Ahmad, A., D.D. Walgenbach, and G.R. Sutter. "Degradation Rates of Technical Carbofuran and a Granular Formulation in Four Soils with Known Insecticide Use History," *Bull. Environ. Contam. Toxicol.*, 23(4/5):572-574 (1979).

Ahmed, M.K. and J.E. Casida. "Metabolism of Some Organophosphorus Insecticides by Microorganisms," *J. Econ. Entomol.*, 51(1):59-63 (1958).

Alexander, M. "Biodegradation of Chemicals of Environmental Concern," *Science (Washington, DC)*, 211(4478):132-138 (1981).

Alley, E.G., B.R. Layton, and J.P. Minyard, Jr. "Identification of the Photoproducts of the Insecticides Mirex and Kepone," *J. Agric. Food Chem.*, 22(3):442-445 (1974).

Altshuller, A.P. "Measurements of the Products of Atmospheric Photochemical Reactions in Laboratory Studies and in Ambient Air-Relationships Between Ozone and Other Products," *Atmos. Environ.*, 17(12):2383-2427 (1983).

Alva, A.K. and M. Singh. "Sorption-Desorption of Herbicides in Soils as Influenced by Electrolyte Cations and Ionic Strength," *J. Environ. Sci. Health*, B26(2):147-163 (1991).

Aly, O.M. and M.A. El-Dib. "Studies on the Persistence of Some Carbamate Insecticides in the Aquatic Environment-I. Hydrolysis of Sevin, Baygon, Pyrolam and Dimetilan in Waters," *Water Res.*, 5(12):1191-1205 (1971).

Ambrosi, D., P.C. Kearney, and J.A. Macchia. "Persistence and Metabolism of Oxadiazon in Soils," *J. Agric. Food Chem.*, 25(4):868-872 (1977).

Ambrosi, D., P.C. Kearney, and J.A. Macchia. "Persistence and Metabolism of Phosalone in Soil," *J. Agric. Food Chem.*, 25(2):342-347 (1977a).

Anderson, J.F., G.R. Stephenson, and C.T. Corke. "Atrazine and Cyanazine Activity in Ontario and Manitoba Soils," *Can. J. Plant Sci.*, 60:773-781 (1980a).

Anderson, J.J. and J.J. Dulka. "Environmental Fate of Sulfometuron Methyl in Aerobic Soils," *J. Agric. Food Chem.*, 33(4):596-602 (1985).

Anderson, J.P.E. and K.H. Domsch. "Microbial Degradation of the Thiolcarbamate Herbicide, Diallate, in Soils and by Pure Cultures by Soil Microorganisms," *Arch. Environ. Contam. Toxicol.*, 4(1):1-7 (1976).

Anderson, J.P.E. and K.H. Domsch. "Relationship Between Herbicide Concentration and the Rates of Enzymatic Degradation of ^{14}C-Diallate and ^{14}C-Triallate in Soil," *Arch. Environ. Contam. Toxicol.*, 9(3):259-268 (1980).

Anderson, R.J., C.A. Anderson, and T.J. Olson. "A Gas-Liquid Chromatographic Method for the Determination of Trichlorfon in Plant and Animal Tissues," *J. Agric. Food Chem.*, 14(5):508-512 (1966).

Andrawes, N.R., W.P. Bagley, and R.A. Herrett. "Fate and Carryover Properties of Temik Aldicarb Pesticide [2-Methyl-2-(methylthio)propionaldehyde O-(methylcarbamoyl)oxime] in Soil," *J. Agric. Food Chem.*, 19(4):727-730 (1971).

Andrawes, N.R., R.R. Romine, and W.P. Bagley. "Metabolism and Residues of Temik Aldicarb Pesticide in Cotton Foliage and under Field Conditions," *J. Agric. Food Chem.*, 21(3):379-386 (1973).

Angemar, Y., M. Rebhun, and M. Horowitz. "Adsorption, Phytotoxicity, and Leaching of Bromacil in Some Israeli Soils," *J. Environ. Qual.*, 13(2):321-326 (1984).

Archer, T.E. "Malathion Residues on Ladino Clover Seed Screenings Exposed to Ultraviolet Irradiation," *Bull. Environ. Contam. Toxicol.*, 6(2):142-143 (1971).

Archer, T.E., I.K. Nazer, and D.G. Crosby. "Photodecomposition of Endosulfan and Related Products in Thin Films by Ultraviolet Light Irradiation," *J. Agric. Food Chem.*, 20(5):954-956 (1972).

Archer, T.E., J.D. Stokes, and R.S. Bringhurst. "Fate of Carbofuran and Its Metabolites on Strawberries in the Environment," *J. Agric. Food Chem.*, 25(3):536-541 (1977).

Armstrong, D.E., G. Chesters, and R.R. Harris. "Atrazine Hydrolysis in Soil," *Soil Sci. Soc. Am. Proc.*, 31:61-66 (1967).

Arunachalam, K.D. and M. Lakshmanan. "Microbial Uptake and Accumulation of (^{14}C Carbofuran) 1,3-Dihydro-2,2-Dimethyl-7 Benzofuranylmethyl Carbamate in Twenty Fungal Strains Isolated by Miniecosystem Studies," *Bull. Environ. Contam. Toxicol.*, 41(1):127-134 (1988).

Ashton, F.M. and T.J. Monaco. *Weed Science: Principals and Practices* (New York: John Wiley & Sons, Inc., 1991), 466 p.

Attaway, H.H., N.D. Camper, and M.J.B. Paynter. "Anaerobic Microbial Degradation of Diuron by Pond Sediment," *Pestic. Biochem. Physiol.*, 17:96-101 (1982).

Attaway, H.H., M.J.B. Paynter, and N.D. Camper. "Degradation of Selected Phenylurea Herbicides by Anaerobic Pond Sediment," *J. Environ. Sci. Health,* B17:683–700 (1982a).

Ayanaba, A., W. Verstraete, and M. Alexander. "Formation of Dimethylnitrosamine, a Carcinogen and Mutagen, in Soils Treated with Nitrogen Compounds," *Soil Sci. Soc. Am. Proc.,* 37:565–568 (1973).

Bachmann, A., W.P. Wijnen, W. de Bruin, J.L.M. Huntjens, W. Roelofsen, and A.J.B. Zehnder. "Biodegradation of *Alpha-* and *Beta-*Hexachlorocyclohexane in a Soil Slurry under Different Redox Conditions," *Appl. Environ. Microbiol.,* 54(1):143–149 (1988).

Bank, S. and R.J. Tyrell. "Copper(II)-Promoted Aqueous Decomposition of Aldicarb," *J. Org. Chem.,* 50(24):4938–4943 (1985).

Barbash, J.E. and M. Reinhard. "Abiotic Dehalogenation of 1,2-Dichloroethane and 1,2-Dibromoethane in Aqueous Solution Containing Hydrogen Sulfide," *Environ. Sci. Technol.,* 23(11):1349–1358 (1989).

Barbeni, M., E. Pramauro, and E. Pelizzetti. "Photodegradation of Pentachlorophenol Catalyzed by Semiconductor Particles," *Chemosphere,* 14(2):195–208 (1985).

Barbeni, M., E. Pramauro, E. Pelizzetti, M. Vincenti, E. Borgarello, and N. Serpone. "Sunlight Photodegradation of 2,4,5-Trichlorophenoxyacetic acid and 2,4,5-Trichlorophenol on TiO_2. Identification of Intermediates and Degradation Pathway," *Chemosphere,* 16(6):1165–1179 (1987).

Bartha, R. "Biochemical Transformations of Anilide Herbicides in Soil," *J. Agric. Food Chem.,* 16(4):602–604 (1968).

Bartha, R., H.A.B. Linke, and D. Pramer. "Pesticide Transformations: Production of Chloroazobenzenes from Chloroanilines," *Science (Washington, DC),* 161(3841):582–583 (1968).

Bartha, R. and D. Pramer. "Pesticide Transformation to Aniline and Azo Compounds in Soil," *Science (Washington, DC),* 156(3782):1617–1618 (1967).

Bartl, P. and F. Korte. "Photochemisches und Thermisches Verhalten des Herbizids Sencor (4-Amino-6-*tert*-butyl-3-(methylthio)-1,2,4-triazin-5(4*H*)-on) als Festkörper und orf Oberflächen," *Chemosphere,* 4(3):173–176 (1975).

Bartley, W.J., N.R. Andrawes, E.L. Chancey, W.P. Bagley, and H.W. Spurr. "The Metabolism of Temik Aldicarb Pesticide [2-Methyl-2-methylthio)propionaldehyde *O*-(methylcarbamoyl)oxime] in the Cotton Plant," *J. Agric. Food Chem.,* 18(3):446–453 (1970).

Bartsch, E. "Diazinon. II. Residues in Plants, Soil and Water," *Residue Rev.,* 51:37–68 (1974).

Baude, F.J., H.L. Pease, and R.F. Holt. "Fate of Benomyl on Field Soil and Turf," *J. Agric. Food Chem.,* 22(3):413–418 (1974).

Baur, J.R. and R.W. Bovey. "Ultraviolet and Volatility Loss of Herbicides," *Arch. Environ. Contam.,* 2(3):275–288 (1974).

Baxter, J.N. and F.J. Johnston. "Photolysis Studies of the Chloroacetic Acids Using

Light of 2537 A," *Radiation Res.*, 33(2):303-310 (1968).

Baxter, R.M. "Reductive Dechlorination of Certain Chlorinated Organic Compounds by Reduced Hematin Compared with Their Behaviour in the Environment," *Chemosphere*, 21(4/5):451-458 (1990).

Beard, J.E. and G.W. Ware. "Fate of Endosulfan on Plants and Glass," *J. Agric. Food Chem.*, 17(2):216-220 (1969).

Beeman, R.W. and F. Matsumura. "Metabolism of *cis*- and *trans*-Chlordane by a Soil Microorganism," *J. Agric. Food Chem.*, 29(1):84-89 (1981).

Belafdal, O., M. Bergon, and J.P. Calmon. "Mechanism of Hydantoin Ring Opening in Prodione in Aqueous Media," *Pestic. Sci.*, 17:335-342 (1986).

Belasco, I.J. and H.L. Pease. "Investigation of Diuron- and Linuron-Treated Soils for 3,3',4,4'-Tetrachloroazobenzene," *J. Agric. Food Chem.*, 17(6):1414-1417 (1969).

Belasco, I.J. and W.P. Langsdorf. "Synthesis of C^{14}-Labeled Siduron and Its Fate in Soil," *J. Agric. Food Chem.*, 17(5):1004-1007 (1969).

Belay, N. and L. Daniels. "Production of Ethane, Ethylene, and Acetylene from Halogenated Hydrocarbons by Methanogenic Bacteria," *Appl. Environ. Microbiol.*, 53(7):1604-1610 (1987).

Bell, G.R. "Photochemical Degradation of 2,4-Dichlorophenoxyacetic Acid and Structurally Related Compounds," *Botan. Gaz.*, 118:133-136 (1956).

Bender, M.E. "The Toxicity of the Hydrolysis and Breakdown Products of Malathion to the Fathead Minnow (*Pimephales Promelas*, Rafinesque)," *Water Res.*, 3(8):571-582 (1969).

Benezet, H.I. and F. Matsumura. "Isomerization of γ-BHC to α-BHC in the Environment," *Nature (London)*, 243(5408):480-481 (1973).

Benoit-Guyod, J.L., D.G. Crosby, and J.B. Bowers. "Degradation of MCPA by Ozone and Light," *Water Res.*, 20(1):67-72 (1986).

Benson, W.R. "Photolysis of Solid and Dissolved Dieldrin," *J. Agric. Food Chem.*, 19(1):66-72 (1971).

Best, J.A. and J.B. Weber. "Disappearance of *s*-Triazines as Affected by Soil pH Using a Balance-Sheet Approach," *Weed Sci.*, 22(4):364-373 (1974).

Beynon, K.I., G. Stoydin, and A.N. Wright. "Comparison of the Breakdown of the Triazine Herbicides Cyanazine, Atrazine, and Simazine in Soil and in Maize," *Pestic. Biochem. Physiol.*, 2:153-161 (1972).

Biggar, J.W., D.R. Nielson, and W.R. Tillotson. "Movement of DBCP in Laboratory Soil Columns and Field Soils to Groundwater," *Environ. Geol.*, 5(3):127-131 (1984).

Bilkert, J.N. and P.S.C. Rao. "Sorption and Leaching of Three Nonfumigant Nematocides in Soils," *J. Environ. Sci. Health*, B29(1):1-26 (1985).

Boerner, H. "Untersuchungen über den Abbau von Afalon [*N*-(3,4-Dichloro-phenyl)-*N*'-methoxy-*N*'-methylharnstoff] und Aresin [*N*-(4-Chlorophenyl)-*N*'-methoxy-*N*'-methylharnstoff] im Boden," *Z. Pflanzenkr. Pflanzenschutz.*, 72:516-531 (1965).

Boerner, H. "Der Abbau von Harnstoffherbiziden im Bodeno," *Z. Pflanzenkr. Pflanzenschutz.*, 74:135-143 (1967).

Boesten, J.J.T.I., L.J.T. van der Pas, M. Leistra, J.H. Smelt, and N.W.H. Houx. "Transformation of ^{14}C-Labelled 1,2-Dichloropropane in Water-Saturated Subsoil Materials," *Chemosphere*, 24(8):993-1011 (1992).

Boethling, R.S. and M. Alexander. "Effect of Concentration of Organic Chemicals on Their Biodegradation by Natural Microbial Communities," *Appl. Environ. Microbiol.*, 37(6):1211-1216 (1979).

Bollag, J.-M. and S.-Y. Liu. "Hydroxylations of Carbaryl by Soil Fungi," *Nature (London)*, 236(5343):177-178 (1972).

Bollag, J.-M. and S.-Y. Liu. "Biological Transformation Processes of Pesticides" in *Pesticides in the Soil Environment: Processes, Impacts, and Modeling*, SSSA Book Series: 2, Cheng, H.H., Ed., (Madison, WI: Soil Science of America, Inc., 1990), pp. 169-211.

Borello, R., C. Minero, E. Pramauro, E. Pelizzetti, N. Serpone, and H. Hidaka. "Photocatalytic Degradation of DDT Mediated in Aqueous Semiconductor Slurries by Simulated Sunlight," *Environ. Toxicol. Chem.*, 8(11):997-1002 (1989).

Borsetti, A.P. and J.A. Roach. "Identification of Kepone Alteration Products in Soil and Mullet," *Bull. Environ. Contam. Toxicol.*, 20(2):241-247 (1978).

Bove, J.L. and P. Dalven. "Pyrolysis of Phthalic Acid Esters: Their Fate," *Sci. Tot. Environ.*, 36:313-318 (1984).

Boyle, T.P., E.F. Robinson-Wilson, J.D. Petty, and W. Weber. "Degradation of Pentachlorophenol in Simulated Lenthic Environment," *Bull. Environ. Contam. Toxicol.*, 24(2):177-184 (1980).

Breaux, E.J., J.E. Patanella, and E.F. Sanders. "Chloroacetanilide Herbicide Selectivity: Analysis of Glutathione and Homoglutathione in Tolerant, Susceptible and Safened Seedlings," *J. Agric. Food Chem.*, 35(4):474-478 (1987).

Briggs, G.G. "Theoretical and Experimental Relationships Between Soil Adsorption, Octanol-Water Partition Coefficients, Water Solubilities, Bioconcentration Factors, and the Parachor," *J. Agric. Food Chem.*, 29(5):1050-1059 (1981).

Briggs, G.G. and J.E. Dawson. "Hydrolysis of 2,6-Dichlorobenzonitrile in Soils," *J. Agric. Food Chem.*, 18(1):97-99 (1970).

Broadhurst, N.A., M.L. Montgomery, and V.H. Freed. "Metabolism of 2-Methoxy-3,6-dichlorobenzoic acid (Dicamba) by Wheat and Bluegrass Plants," *J. Agric. Food Chem.*, 14(6):585-588 (1966).

Bromilow, R.H., R.J. Baker, M.A.H. Freeman, and K. Görög. "The Degradation of Aldicarb and Oxamyl in Soil," *Pestic. Sci.*, 11(4):371-378 (1980).

Bromilow, R.H. and M. Leistra. "Measured and Simulated Behaviour of Aldicarb and its Oxidation Products in Fallow Soils," *Pestic. Sci.*, 11(4):389-395 (1980).

Brown, D.S. and E.W. Flagg. "Empirical Prediction of Organic Pollutant Sorption in Natural Sediments," *J. Environ. Qual.*, 10(3):382-386 (1981).

Bull, D.L. "Metabolism of UC-21149 [2-Methyl-2-(methylthio)propionaldehyde

O-(methylcarbamoyl)oxime] in Cotton Plants and Soil in the Field," *J. Econ. Entomol.*, 61:1598-1602 (1968).

Bull, D.L. "Fate and Efficacy of Acephate after Application to Plants and Insects," *J. Agric. Food Chem.*, 27(2):268-272 (1979).

Bull, D.L. and R.L. Ridgway. "Metabolism of Trichlorfon in Animals and Plants," *J. Agric. Food Chem.*, 17(4):837-841 (1969).

Bumpus, J.A. and S.D. Aust. "Biodegradation of DDT [1,1,1-Trichloro-2,2-bis(4-chlorophenyl)ethane] by the White Rot Fungus *Phanerochaete chrysosporium*," *Appl. Environ. Microbiol.*, 53(9):2001-2008 (1987).

Burczyk, L., K. Walczyk, and R. Burczyk. "Kinetics of Reaction of Water Addition to the Acrolein Double-Bond in Dilute Aqueous Solution," *Przem. Chem.*, 47(10):625-627 (1968).

Burge, W.D. "Anaerobic Decomposition of DDT in Soil. Acceleration by Volatile Components of Alfalfa," *J. Agric. Food Chem.*, 19(2):375-378 (1971).

Burkhard, N. and J.A. Guth. "Photodegradation of Atrazine, Atraton, and Ametryne in Aqueous Solution with Acetone as a Photosensitizer," *Pestic. Sci.*, 7:65 (1976).

Burkhard, N. and J.A. Guth. "Photolysis of Organophosphorus Insecticides on Soil Surfaces," *Pestic. Sci.*, 10(4):313-319 (1979).

Burkhard, N. and J.A. Guth. "Chemical Hydrolysis of 2-Chloro-4,6-bis(alkyl-amino)-1,3,5-triazine Herbicides and Their Breakdown in Soil Under the Influence of Adsorption," *Pestic. Sci.*, 12(1):45-52 (1981).

Burlinson, N.E., L.A. Lee, and D.H. Rosenblatt. "Kinetics and Products of Hydrolysis of 1,2-Dibromo-3-chloropropane," *Environ. Sci. Technol.*, 16(9):627-632 (1982).

Burton, W.B. and G.E. Pollard. "Rate of Photochemical Isomerization of Endrin in Sunlight," *Bull. Environ. Contam. Toxicol.*, 12(1):113-116 (1974).

Businelli, M., M. Patumi, and C. Marucchini. "Identification and Determination of Some Metalaxyl Degradation Products in Lettuce and Sunflower," *J. Agric. Food Chem.*, 32(3):644-647 (1984).

Bussacchini, V., B. Pouyet, and P. Meallier. "Photodegradation des Molecules Phytosanitaires. VI -Photodegradation du Phenmediphame," *Chemosphere*, 14(2):155-161 (1985).

Byast, T.H. and R.J. Hance. "Degradation of 2,4,5-T by South Vietnamese Soils Incubated in the Laboratory," *Bull. Environ. Contam. Toxicol.*, 14(1):71-76 (1975).

Calvert, J.G. and J.N. Pitts, Jr. *Photochemistry* (New York: John Wiley & Sons, Inc., 1966), 899 p.

Camper, N.D. "Effects of Pesticide Degradation Products on Soil Microflora" in *Pesticide Transformation Products. Fate and Significance in the Environment*, ACS Symposium Series 429, Somasundaram, L. and J.R. Coats, Eds., (New York: American Chemical Society, 1991), pp. 205-216.

Capriel, P., A. Haisch, and S.U. Khan. "Distribution and Nature of Bound

(Nonextractable) Residues of Atrazine in a Mineral Soil Nine Years after the Herbicide Application," *J. Agric. Food Chem.*, 33(4):567-569 (1985).

Casida, J.E., P.E. Gatterdam, L.W. Getzin, Jr., and R.K. Chapman. "Residual Properties of the Systemic Insecticide *O,O*-Dimethyl 1-Carbomethoxy-1-propen-2-yl Phosphate," *J. Agric. Food Chem.*, 4(3):236-243 (1956).

Casida, J.E., R.A. Gray, and H. Tilles. "Thiocarbamate Sulfoxides: Potent, Selective, and Biodegradable Herbicides," *Science (Washington, DC)*, 184(4136):573-574 (1974).

Castelfranco, P., C.L. Foy, and D.B. Deutsch. "Non-Enzymatic Detoxification of 2-Chloro-4,6-bis(ethylamino)-s-triazine by Extracts of *Zea mays*," *Weeds*, 9:580-591 (1961).

Castro, C.E. "The Rapid Oxidation of Iron(II) Porphyrins by Alkyl Halides. A Possible Mode of Intoxication of Organisms by Alkyl Halides," *J. Am. Chem. Soc.*, 86(2):2310-2311 (1964).

Castro, C.E. and N.O. Belser. "Hydrolysis of *cis*- and *trans*-1,3-Dichloropropene in Wet Soil," *J. Agric. Food Chem.*, 14(1):69-70 (1966).

Castro, C.E. and N.O. Belser. "Biodehalogenation. Reductive Dehalogenation of the Biocides Ethylene Dibromide, 1,2-Dibromo-3-chloropropane, and 2,3-Dibromobutane in Soil," *Environ. Sci. Technol.*, 2(10):779-783 (1968).

Castro, C.E. and N.O. Belser. "Photohydrolysis of Methyl Bromide and Chloropicrin," *J. Agric. Food Chem.*, 29(5):1005-1009 (1981).

Castro, C.E., R.S. Wade, and N.O. Belser. "Biodehalogenation. The Metabolism of Chloropicrin by *Pseudomonas* sp.," *J. Agric. Food Chem.*, 31(6):1184-1187 (1983).

Castro, T.F. and T. Yoshida. "Degradation of Organochlorine Insecticides in Flooded Soils in the Philippines," *J. Agric. Food Chem.*, 19(6):1168-1170 (1971).

Chacko, C.I., J.L. Lockwood, and M. Zabik. "Chlorinated Hydrocarbon Pesticides: Degradation by Microbes," *Science (Washington, DC)*, 154(3751):893-895 (1966).

Cessna, A.J. and D.C.G. Muir. "Photochemical Transformations" in *Environmental Chemistry of Herbicides. Volume II*, Grover, R. and A.J. Cessna, Eds. (Boca Raton, FL: CRC Press, Inc., 1991), pp. 199-263.

Chaigneau, M., G. LeMoan, and L. Giry. "Décomposition Pyrogénée du Bromurede Méthyle en L'absence D'oxygene et à 550 °C," *Comp. Rend. Ser. C.*, 263:259-261 (1966).

Chapman, R.A. and M. Cole. "Observations on the Influence of Water and Soil pH on the Persistence of Insecticides," *J. Environ. Sci. Health*, B17:487-504 (1982).

Chen, P.R., W.P. Tucker, and W.C. Dauterman. "Structure of Biologically Produced Malathion Monoacid," *J. Agric. Food Chem.*, 17(1):86-90 (1969).

Chesters, G., G.V. Simsiman, J. Levy, B.J. Alhajjar, R.N. Fathulla, and J.M. Harkin. "Environmental Fate of Alachlor and Metolachlor" in *Reviews of Environmental Contamination and Toxicology*, (New York: Springer-Verlag, Inc., 1989), pp. 1-74.

Chiba, M. and D.F. Veres. "Fate of Benomyl and Its Degradation Compound

Methyl 2-Benzimidazolecarbamate on Apple Foliage," *J. Agric. Food Chem.*, 29(3):588-590 (1981).

Chiou, C.T. and M. Manes. "Application of the Flory-Higgins Theory to the Solubility of Solids in Glyceryl Trioleate," *J. Chem. Soc., Faraday Trans. 1*, 82(1):243-246 (1986).

Chopra, N.M. and A.M. Mahfouz. "Metabolism of Endosulfan I, Endosulfan II, and Endosulfan Sulfate in Tobacco Leaf," *J. Agric. Food Chem.*, 25(1):32-36 (1977).

Chou, S.H. "Fate of Acylanilides in Soils and Polybrominated Biphenyls (PBBs) in Soils and Plants," Dissertation, Michigan State University, East Lansing, MI (1977).

Choudhry, G.G. and O. Hutzinger. "Acetone-Sensitized and Nonsensitized Photolyses of Tetra-, Penta-, and Hexachlorobenzenes in Acetonitrile-Water Mixtures: Photoisomerization and Formation of Several Products Including Polychlorobiphenyls," *Environ. Sci. Technol.*, 18(4):235-241 (1984).

Clapés, P., J. Soley, M. Vicente, J. Rivera, J. Caixach, and F. Ventura. "Degradation of MCPA by Photochemical Methods," *Chemosphere*, 15(4):395-401 (1986).

Clapp, D.W., D.V. Naylor, and G.C. Lewis. "The Fate of Disulfoton in Portneuf Silt Loam Soil," *J. Environ. Qual.*, 5(2):207-210 (1976).

Clark, T., D.R. Clifford, A.H.B. Deas, P. Gendle, and D.A.M. Watkins. "Photolysis, Metabolism, and Other Factors Influencing the Performance of Triadimefon as a Powdery Mildew Fungicide," *Pestic. Sci.*, 9:497-506 (1978).

Clemons, G.P. and H.D. Sisler. "Formation of a Fungitoxic Derivative from Benlate," *Phytopathology*, 59(5):705-706 (1969).

Coats, J.R. and N.L. O'Donnell-Jeffrey. "Toxicity of Four Synthetic Pyrethroid Insecticides to Rainbow Trout," *Bull. Environ. Contam. Toxicol.*, 23(1/2):250-255 (1979).

Colbert, F.O., V.V. Volk, and A.P. Appleby. "Sorption of Atrazine, Terbutryn, and GS-14254 on Natural and Lime-Amended Soils," *Weed Sci.*, 23(5):390-394 (1975).

Connors, T.F., J.D. Stuart, and J.B. Cope. "Chromatographic and Mutagenic Analyses of 1,2-Dichloropropane and 1,3-Dichloropropylene and Their Degradation Products," *Bull. Environ. Contam. Toxicol.*, 44(2):288-293 (1990).

Cook, A.M. and R. Hütter. "Ametryne and Prometryne as Sulfur Sources for Bacteria," *Appl. Environ. Microbiol.*, 43(4):781-786 (1982).

Coppedge, J.R., D.A. Lindquist, D.L. Bull, and H.W. Dorough. "Fate of 2-Methyl-2-(methylthio)propionaldehyde *O*-(Methylcarbamoyl) oxime (Temik) in Cotton Plants and Soil," *J. Agric. Food Chem.*, 15(5):902-910 (1967).

Corwin, D.L. and W.J. Farmer. "Nonsingle-Valued Adsorption-Desorption of Bromacil and Diquat by Freshwater Sediments," *Environ. Sci. Technol.*, 18(7):507-514 (1984).

Cowart, R.P., F.L. Bonner, and E.A. Epps, Jr. "Rate of Hydrolysis of Seven

Organophosphorus Pesticides," *Bull. Environ. Contam. Toxicol.*, 6(3):231-234 (1971).

Cremlyn, R.J. *Agrochemicals - Preparation and Mode of Action* (New York: John Wiley & Sons, Inc., 1991), 396 p.

Crosby, D.G. and N. Hamadmad. "The Photoreduction of Pentachlorobenzenes," *J. Agric. Food Chem.*, 19(6):1171-1174 (1971).

Crosby, D.G. and E. Leitis. "Photodecomposition of Chlorobenzoic Acids," *J. Agric. Food Chem.*, 17(5):1033-1035 (1969).

Crosby, D.G. and E. Leitis. "The Photodecomposition of Trifluralin in Water," *Bull. Environ. Contam. Toxicol.*, 10(4):237-241 (1973).

Crosby, D.G. and K.W. Moilanen. "Vapor-Phase Photodecomposition of Aldrin and Dieldrin," *Arch. Environ. Contam.*, 2(1):62-74 (1974).

Crosby, D.G. and C.S. Tang. "Photodecomposition of 3-(*p*-Chlorophenyl)-1-dimethylurea (Monuron)," *J. Agric. Food Chem.*, 17(5):1041-1044 (1969).

Crosby, D.G. and H.O. Tutass. "Photodecomposition of 2,4-Dichlorophenoxyacetic Acid," *J. Agric. Food Chem.*, 14(6):596-599 (1966).

Crosby, D.G. and A.S. Wong. "Photodecomposition of 2,4,5-Trichlorophenoxyacetic Acid (2,4,5-T) in Water," *J. Agric. Food Chem.*, 21(6):1052-1054 (1973).

Cserjsei, A.J. and E.L. Johnson. "Methylation of Pentachlorophenol by *Trichoderma virgatum*," *Can. J. Microbiol.*, 18:45-49 (1972).

Cupitt, L.T. "Fate of Toxic and Hazardous Materials in the Air Environment," Office of Research and Development, U.S. EPA Report-600/3-80-084 (1980), 28 p.

Dao, T.H. and T.L. Lavy. "Atrazine Adsorption on Soil as Influenced by Temperature, Moisture Content, and Electrolyte Concentration," *Weed Sci.*, 26(3):303-308 (1978).

Day, B.E., L.S. Jordan, and V.L. Jolliffe. "The Influence of Soil Characteristics on the Adsorption and Phytotoxicity of Simazine," *Weed Sci.*, 16(2):209-213 (1968).

Day, K.E. "Pesticide Transformation Products in Surface Waters" in *Pesticide Transformation Products, Fate and Significance in the Environment*, ACS Symposium Series 459, Somasundaram, L. and J.R. Coats, Eds. (Washington, DC: American Chemical Society, 1991), pp. 217-241.

Davis, J.T. and W.S. Hardcastle. "Biological Assay of Herbicides for Fish Toxicity," *Weeds*, 7:397-404 (1959).

Deeley, G.M., M. Reinhard, and S.M. Stearns. "Transformation and Sorption of 1,2-Dibromo-3-Chloropropane in Subsurface Samples Collected at Fresno, California," *J. Environ. Qual.*, 20(4):547-556 (1991).

Dick, W.A., R.O. Ankumah, G. McClung, and N. Abou-Assaf. "Enhanced Degradation of S-Ethyl *N,N*-Dipropylcarbamothioate in Soil and by an Isolated Soil Microorganism" in *Enhanced Biodegradation of Pesticides in the Environment*, ACS Symposium Series 426, Racke, K.D. and J.R. Coats, Eds. (Washington, DC: American Chemical Society, 1990), pp. 98-112.

DiGeronimo, M.J. and A.D. Antoine. "Metabolism of Acetonitrile and Propionitrile by *Nocardia rhodochrous* LL 100-21," *Appl. Environ. Microbiol.*, 31(6):900-906 (1976).

Dilling, W.L., L.C. Lickly, T.D. Lickly, P.G. Murphy, and R.L. McKellar. "Organic Photochemistry. 19. Quantum Yields for *O,O*-Diethyl *O*-(3,5,6-Trichloro-2-pyridinyl) Phosphorothioate (Chlorpyrifos) and 3,5,6-Trichloro-2-pyridinol in Dilute Aqueous Solutions and Their Environmental Phototransformation Rates," *Environ. Sci. Technol.*, 18(7):540-543 (1984).

Dobson, R.C., G.O. Throneberry, and T.E. Belling. "Residues of Established Alfalfa Treated with Granulated Phorate (Thimet) and Their Effect on Cattle Fed the Hay," *J. Econ. Entomol.*, 53:306-310 (1960).

Dorough, H.W., N.M. Randolph, and G.H. Wimbish. "Residual Nature of Certain Organophosphorus Insecticides in Grain Sorghum and Coastal Bermuda Grass," *Bull. Environ. Contam. Toxicol.*, 1(1):46-58 (1966).

Draper, W.M. and D.G. Crosby. "Solar Photooxidation of Pesticides in Dilute Hydrogen Peroxide," *J. Agric. Food Chem.*, 32(2):231-237 (1984).

Dreher, R.M. and B. Podratzki. "Development of an Enzyme Immunoassay for Endosulfan and Its Degradation Products," *J. Agric. Food Chem.*, 36(5):1072-1075 (1988).

Drinking Water Health Advisory. Pesticides. (Chelsea, MI: Lewis Publishers, Inc., 1989), 819 p.

Duff, W.G. and R.E. Menzer. "Persistence, Mobility, and Degradation of ^{14}C-Dimethoate in Soils," *Environ. Entomol.*, 2(3):309-318 (1973).

Duke, S.O., T.B. Moorman, and C.T. Bryson. "Phytotoxicity of Pesticide Degradation Products" in *Pesticide Transformation Products, Fate and Significance in the Environment*, ACS Symposium Series 459, Somasundaram, L. and J.R. Coats, Eds. (Washington, DC: American Chemical Society, 1991), pp. 188-204.

Dureja, P.S.W. and S.K. Mukerjee. "Photodecomposition of Isofenphos [*O*-Ethyl-*O*-(2-isopropoxycarbonyl)phenyl]isopropylphosphoramidothioate," *Toxicol. Environ. Chem.*, 19(3-4):187-192 (1989).

Dureja, P.S.W. and S. Walia. "Photodecomposition of Pendimethalin," *Pestic. Sci.*, 25(2):105-114 (1989).

Eichelberger, J.W. and J.J. Lichtenberg. "Persistence of Pesticides in River Water," *Environ. Sci. Technol.*, 5(6):541-544 (1971).

El-Dib, M.A. and O.S. Aly. "Persistence of Some Phenylamide Pesticides in the Aquatic Environment-I. Hydrolysis," *Water Res.*, 10(12):1047-1050 (1976).

El-Refai, A. and T.L. Hopkins. "Parathion Absorption, Translocation, and Conversion to Paraoxon in Bean Plants," *J. Agric. Food Chem.*, 14(6):588-592 (1966).

Ellington, J.J., F.E. Stancil, W.D. Payne, and C.D. Trusty. "Measurement of Hydrolysis Rate Constants for Evaluation of Hazardous Waste Land Disposal: Volume 3. Data on 70 Chemicals," Office of Research and Development, U.S. EPA Report-600/3-88-028 (1988), 233 p.

Elliott, S. "Effect of Hydrogen Peroxide on the Alkaline Hydrolysis of Carbon Disulfide," *Environ. Sci. Technol.*, 24(2):264-267 (1990).

Ellis, P.A. and N.D. Camper. "Aerobic Degradation of Diuron by Aquatic Microorganisms," *J. Environ. Sci. Health*, B17:277-290 (1982).

Elsner, E., D. Bieniek, W. Klein, and F. Korte. "Verteilung und Umwandlumg von Aldrin, Heptachlor, und Lindan in der Grunalge *Chlorella pyrenoidosa*," *Chemosphere*, 1:247-250 (1972).

Engelhardt, G., L. Oehlmann, K. Wagner, P.R. Wallnöfer, and M. Wiedermann. "Degradation of the Insecticide Azinphos-methyl in Soil and by Isolated Soil Bacteria," *J. Agric. Food Chem.*, 32(1):102-108 (1984).

Engelhardt, G. and P.R. Wallnöfer. "Microbial Transformation of Benzazimide, a Microbial Degradation Product of the Insecticide Azinphos-Methyl," *Chemosphere*, 12(7/8):955-960 (1983).

Engelhardt, G., P.R. Wallnöfer, and O. Hutzinger. "The Microbial Metabolism of Di-*n*-Butyl Phthalate and Related Dialkyl Phthalates," *Bull. Environ. Contam. Toxicol.*, 13(3):342-347 (1975).

Engelhardt, G., P.R. Wallnöfer, and R. Plapp. "Identification of *N,O*-Dimethylhydroxylamine as a Microbial Degradation Product of the Herbicide, Linuron," *Appl. Microbiol.*, 23:664-666 (1972).

Esser, H.O., G. Dupuis, E. Ebert, G. Marco, and C. Vogel. "*s*-Triazines" in *Herbicides: Chemistry, Degradation, and Mode of Action. Volume 1*, Kearney, P.C. and D.D. Kaufman, Eds., (New York: Marcel Dekker, Inc. 1975), pp. 129-208.

Farmer, W.J. and Y. Aochi. "Picloram Sorption by Soils," *Soil Sci. Soc. Am. Proc.*, 38:418-423 (1974).

Fathepure, B.Z., J.M. Tiedje, and S.A. Boyd. "Reductive Dechlorination of Hexachlorobenzene to Tri- and Dichlorobenzenes in Anaerobic Sewage Sludge," *Appl. Environ. Microbiol.*, 54(2):327-330 (1988).

Felsot, A. and P.A. Dahm. "Sorption of Organophosphorus and Carbamate Insecticides by Soil," *J. Agric. Food Chem.*, 27(3):557-563 (1979).

Felsot, A.S. and W.L. Pedersen. "Pesticidal Activity of Degradation Products Insecticides by Soil" in *Pesticide Transformation Products, Fate and Significance in the Environment*, ACS Symposium Series 459, Somasundaram, L. and J.R. Coats, Eds. (Washington, DC: American Chemical Society, 1991), pp. 172-187.

Felsot, A. and J. Wilson. "Adsorption of Carbofuran and Movement on Soil Thin Layers," *Bull. Environ. Contam. Toxicol.*, 24(5):778-782 (1980).

Ferris, I.G. and E.P. Lichtenstein. "Interactions between Agricultural Chemicals and Soil Microflora and Their Effects on the Degradation of [^{14}C]Parathion in a Cranberry Soil," *J. Agric. Food Chem.*, 28(5):1011-1019 (1980).

Feung, C.-S., R.H. Hamilton, and R.O. Mumma. "Metabolism of 2,4-Dichloro-phenoxyacetic Acid. V. Identification of Metabolites in Soybean Callus Tissue Cultures," *J. Agric. Food Chem.*, 21(4):637-640 (1973).

Feung, C.S, R.H. Hamilton, and F.H. Witham. "Metabolism of 2,4-Dichlorophenoxyacetic Acid by Soybean Cotyledon Callus Tissue Cultures," *J. Agric. Food Chem.*, 19(3):475-479 (1973).

Feung, C.S., R.H. Hamilton, F.H. Witham, and R.O. Mumma. "The Relative Amounts and Identification of Some 2,4-Dichlorophenoxyacetic Acid Metabolites Isolated from Soybean Cotyledon Callus Cultures," *Plant Physiol.*, 50:80-86 (1972).

Fleeker, J. and R. Steen. "Hydroxylation of 2,4-D in Several Weed Species," *Weed Sci.*, 19:507 (1971).

Frank, R., H.E. Braun, M. Van Hove Holdrimet, G.J. Sirons, and B.D. Ripley. "Agriculture and Water Quality in the Canadian Great Lakes Basin: V. Pesticide Use in 11 Agricultural Watersheds and Presence in Stream Water, 1975-1977," *J. Environ. Qual.*, 11(3):497-505 (1982).

Frank, P.A. and R.J. Demint. "Gas Chromatographic Analysis of Dalapon in Water," *Environ. Sci. Technol.*, 3(1):69-71 (1969).

Frankel, L.S., K.S. McCallum, and L. Collier. "Formation of Bis(chloromethyl) Ether from Formaldehyde and Hydrogen Chloride," *Environ. Sci. Technol.*, 8(4):356-359 (1974).

Freed, V.H., C.T. Chiou, and R. Haque. "Chemodynamics: Transport and Behavior of Chemicals in the Environment - A Problem in Environmental Health," *Environ. Health Perspect.*, 20:55-70 (1977).

Freed, V.H., C.T. Chiou, and D.W. Schmedding. "Degradation of Selected Organophosphate Pesticides in Water and Soil," *J. Agric. Food Chem.*, 27(4):706-708 (1979).

Freitag, D., L. Ballhorn, H. Geyer, and F. Korte. "Environmental Hazard Profile of Organic Chemicals," *Chemosphere*, 14(10):1589-1616 (1985).

Fries, G.F. "Degradation of Chlorinated Hydrocarbons under Anaerobic Conditions," in *Fate of Organic Pesticides in the Aquatic Environment, Advances in Chemistry Series*, R.F. Gould, Ed. (Washington, DC: American Chemical Society, 1972), pp. 256-270.

Fries, G.R., G.S. Marrow, and C.H. Gordon. "Metabolism of *o,p'*- and *p,p'*-DDT by Rumen Microorganisms," *J. Agric. Food Chem.*, 17(4):860-862 (1969).

Fuchs, A. and F.W. de Vries. "Bacterial Breakdown of Benomyl. II. Mixed Cultures," *Antonie van Leeuwenhoek*, 44(3/4):293-309 (1978).

Fuhremann, T.W. and E.P. Lichtenstein. "A Comparative Study of the Persistence, Movement, and Metabolism of Six Carbon-14 Insecticides in Soils and Plants," *J. Agric. Food Chem.*, 28(2):446-452 (1980).

Fukui, S., T. Hirayama, H. Shindo, and M. Nohara. "Photochemical Reaction of Biphenyl (BP) and *o*-Phenylphenol (OPP) with Nitrogen Monoxide (1)," *Chemosphere*, 9(12):771-775 (1980).

Funderburk, H.H., Jr. and G.A. Bozarth. "Review of the Metabolism and Decomposition of Diquat and Paraquat," *J. Agric. Food Chem.*, 15(4):563-567 (1967).

Gäb, S., H. Parlar, S. Nitz, K. Hustert, and F. Korte. "Beitrage zur Okologischen

Chemie. LXXXI. Photochemischer Abbau von Aldrin, Dieldrin and Photodieldrin als Festkorper im Sauerstoffstrom," *Chemosphere*, 3(5):183-186 (1974).

Gälli, R. and P.L. McCarty. "Biotransformation of 1,1,1-Trichloroethane, Trichloromethane, and Tetrachloromethane by a *Clostridium* sp.," *Appl. Environ. Microbiol.*, 55(4):837-844 (1989).

Gamar, Y. and J.K. Gaunt. "Bacterial Metabolism of 4-Chloro-2-methylphenoxy-acetate, Formation of Glyphosate by Side-Chain Cleavage," *Biochem. J.*, 122:527-531 (1971).

Gamerdinger, A.P., A.T. Lemley, and R.J. Wagenet. "Nonequilibrium Sorption and Degradation of Three 2-Chloro-s-Triazine Herbicides in Soil-Water Systems," *J. Environ. Qual.*, 20(4):815-822 (1991).

Gardiner, J.A., R.C. Rhodes, J.B. Adams, Jr., and E.J. Soboczenski. "Synthesis and Studies with 2-^{14}C-Labeled Bromacil and Terbacil," *J. Agric. Food Chem.*, 17(5):980-986 (1969).

Garner, W.Y. and R.E. Menzer. "Metabolism of N-Hydroxymethyl Dimethoate and N-Desmethyl Dimethoate in Bean Plants," *Pestic. Biochem. Physiol.*, 25:218-232 (1986).

Gaughan, L.C. and J.E. Casida. "Degradation of *trans*- and *cis*-Permethrin on Cotton and Bean Plants," *J. Agric. Food Chem.*, 26(3):525-528 (1978).

Gaynor, J.D. and V.V. Volk. "Surfactant Effects on Picloram Adsorption by Soils," *Weed Sci.*, 24(6):549-552 (1976).

Gear, J.R., J.G. Michel, and R. Grover. "Photochemical Degradation of Picloram," *Pestic. Sci.*, 13(2):189-194 (1982).

Geissbühler, H., C. Haselbach, and H. Aebi. "The Fate of N'-(4-Chloro-phenoxy)phenyl-N,N-dimethylurea. I. Adsorption and Leaching in Different Soils," *Weed Res.*, 3(2):140-153 (1963).

Geissbühler, H., C. Haselbach, H. Aebi, and L. Ebner. "The Fate of N'-(4-Chlorophenoxy)phenyl-N,N-dimethylurea. III. Breakdown in Soils and Plants," *Weed Res.*, 3(4):277-297 (1963a).

Geller, A. "Studies on the Degradation of Atrazine by Bacterial Communities Enriched from Various Biotypes," *Arch. Environ. Contam. Toxicol.*, 9:289-305 (1980).

Georgacakis, E. and M.A.Q. Khan. "Toxicity of the Photoisomers of Cyclodiene Insecticides to Freshwater Animals," *Nature (London)*, 233(5315):120-121 (1971).

Gerstl, Z. and L. Kliger. "Fractionation of the Organic Matter in Soils and Sediments and Their Contribution to the Sorption of Pesticides," *J. Environ. Sci. Health*, B25(6):729-741 (1990).

Gerstl, Z. and B. Yaron. "Behavior of Bromacil and Napropamide in Soils: I. Adsorption and Degradation," *Soil Sci. Soc. Am. J.*, 47(3):474-478 (1983).

Getzin, L.W. "Metabolism of Diazinon and Zinophos in Soils," *J. Econ. Entomol.*, 60(2):505-508 (1967).

Getzin, L.W. "Degradation of Chlorpyrifos in Soil: Influence of Autoclaving, Soil

Moisture, and Temperature," *J. Econ. Entomol.*, 74(2):158-162 (1981).

Getzin, L.W. "Dissipation of Chlorpyrifos from Dry Soil Surfaces," *J. Econ. Entomol.*, 74(6):707-713 (1981a).

Getzin, L.W. and J.C.H. Shanks. "Persistence, Degradation of Bioactivity of Phorate and Its Oxidative Analog in Soil," *J. Econ. Entomol.*, 63(1):52-58 (1970).

Gibson, W.P. and R.G. Burns. "The Breakdown of Malathion in Soil and Soil Components," *Microbial Ecol.*, 3:219-230 (1977).

Gilbert, M. "Fate of Chlorothalonil in Apple Foliage and Fruit," *J. Agric. Food Chem.*, 24(5):1004-1007 (1976).

Gilderhus, P.A. "Effects of Diquat on Bluegills and Their Food Organisms," *Progr. Fish-Cult.*, 29(2):67-74 (1967).

Glass, R.L. Adsorption of Glyphosate by Soils and Clay Minerals," *J. Agric. Food Chem.*, 35(4):497-500 (1987).

Glotfelty, D.E., M.S. Majewski, and J.N. Seiber. "Distribution of Several Organophosphorus Insecticides and Their Oxygen Analogues in a Foggy Atmosphere," *Environ. Sci. Technol.*, 24(3):353-357 (1990).

Gohre, K. and G.C. Miller. "Photooxidation of Thioether Pesticides on Soil Surfaces," *J. Agric. Food Chem.*, 34(4):709-713 (1986).

Golab, T., W.A. Althaus, and H.L. Wooten. "Fate of [^{14}]Trifluralin in Soil," *J. Agric. Food Chem.*, 27(1):163-179 (1979).

Golab, T., R.J. Herberg, S.J. Parka, and J.B. Tepe. "The Metabolism of Carbon-14 Diphenamid in Strawberry Plants," *J. Agric. Food Chem.*, 14(6):593-596 (1966).

Golab, T., R.J. Herberg, S.J. Parka, and J.B. Tepe. "Metabolism of Carbon-14 Trifluralin in Carrots," *J. Agric. Food Chem.*, 15(4):638-641 (1967).

Gomaa, H.M., I.H. Suffet, and S.D. Faust. "Kinetics of Hydrolysis of Diazinon and Diazoxon," *Residue Rev.*, 29:171-190 (1969).

Gomez, J., C. Bruneau, N. Soyer, and A. Brault. "Identification of Thermal Degradation Products from Diuron and Iprodione," *J. Agric. Food Chem.*, 30(1):180-182 (1982).

González, V., J.H. Ayala, and A.M. Afonso. "Degradation of Carbaryl in Natural Waters: Enhanced Hydrolysis Rate in Micellar Solution," *Bull. Environ. Contam. Toxicol.*, 42(2):171-178 (1992).

Goring, C.A.I. "Control of Nitrification by 2-Chloro-6-(trichloromethyl) pyridine," *Soil Sci.*, 93:211-218 (1962).

Goring, C.A.I. and J.W. Hamaker. "Degradation and Movement of Picloram in Soil and Water," *Down to Earth*, 20(4):3-5 (1971).

Gormley, O.R. and R.F. Spalding. "Sources and Concentrations of Nitrate-Nitrogen in the Groundwater of the Central Platte Region, Nebraska," *Groundwater*, 17(3):291-301 (1979).

Goswami, K.P. and R.E. Green. "Microbial Degradation of the Herbicide Atrazine and Its 2-Hydroxy Analog in Submerged Soils," *Environ. Sci. Technol.*, 5(5):426-429 (1971).

Gowda, T.K.S. and N. Sethunathan. "Persistence of Endrin in Indian Rice Soils under Flooded Conditions," *J. Agric. Food Chem.*, 24(4):750-753 (1976).

Graham-Bryce, I.J. "Adsorption of Disulfoton by Soil," *J. Sci. Food Agric.*, 18:72-77 (1967).

Grayson, B.T. "Hydrolysis of Cyanazine and Related Diaminochloro-1,3,5-triazines; Hydrolysis in Sulfuric Acid Solutions," *Pestic. Sci.*, 11(5):493-505 (1980).

Green, R.E. and S.R. Obien. "Herbicide Equilibrium in Soils in Relation to Soil Water Content," *Weed Sci.*, 17(4):514-519 (1969).

Grover, R. "Adsorption of Picloram by Soil Colloids and Various Other Adsorbants," *Weed Sci.*, 19(4):417-418 (1971).

Grover, R. "Adsorption and Desorption of Urea Herbicides in Soils," *Can. J. Soil Sci.*, 55:127-135 (1975).

Grover, R. "Mobility of Dicamba, Picloram, and 2,4-D in Soil Columns," *Weed Sci.*, 25(2):159-162 (1977).

Grover, R. and R.J. Hance. "Adsorption of Some Herbicides by Soil and Roots," *Can. J. Plant Sci.*, 49:378-380 (1969).

Grover, R. and A.E. Smith. "Adsorption Studies with the Acid and Dimethylamine Forms of 2,4-D and Dicamba," *Can. J. Soil Sci.*, 54:179-186 (1974).

Guenzi, W.D. and W.E. Beard. "Anaerobic Biodegradation of DDT to DDD in Soil," *Science (Washington, DC)*, 156(3778):1116-1117 (1967).

Gysin, H. and A. Margot. "Chemistry and Toxicological Properties of O,O-Diethyl-O-(2-isopropyl-4-methyl-6-pyrimidinyl) Phosphorothioate (Diazinon)," *J. Agric. Food Chem.*, 6(12):900-903 (1958).

Hagin, R.D., D.L. Linscott, and J.E. Dawson. "2,4-D Metabolism in Resistant Grasses," *J. Agric. Food Chem.*, 18(5):848-850 (1970).

Hall, R.C., C.S. Giam, and M.G. Merkle. "The Photolytic Degradation of Picloram," *Weed Res.*, 8(4):292-297 (1968).

Hamilton, R.H., J. Hurter, J.K. Hall, and C.D. Ercegovich. "Metabolism of 2,4-Dichlorophenoxyacetic Acid and 2,4,5-Trichlorophenoxyacetic Acid on Bean Plants," *J. Agric. Food Chem.*, 19(3):480-483 (1971).

Hance, R.J. "The Adsorption of Urea and Some of Its Derivatives by a Variety of Soils," *Weed Res.*, 5:98-107 (1965).

Hance, R.J. "The Speed of Attainments of Sorption Equilibria in Some Systems Involving Herbicides," *Weed Res.*, 7:29-36 (1967).

Hansen, J.L. and M.H. Spiegel. "Hydrolysis Studies of Aldicarb, Aldicarb Sulfoxide and Aldicarb Sulfone," *Environ. Toxicol. Chem.*, 2(2):147-153 (1983).

Hanst, P.L. and B.W. Gay, Jr. "Photochemical Reactions Among Formaldehyde, Chlorine, and Nitrogen Dioxide in Air," *Environ. Sci. Technol.*, 11(12):1105-1109 (1977).

Hapeman-Somich, C.J. "Mineralization of Pesticide Degradation Products" in *Pesticide Transformation Products, Fate and Significance in the Environment*, ACS Symposium Series 459, Somasundaram, L. and J.R. Coats, Eds.

(Washington, DC: American Chemical Society, 1991), pp. 133-147.

Hargrove, R.S. and M.G. Merkle. "The Loss of Alachlor from Soil," *Weed Sci.*, 19(6):652-654 (1971).

Harris, C.I. "Adsorption, Movement, and Phytotoxicity of Monuron and s-Triazine Herbicides in Soils," *Weeds*, 14:6-10 (1966).

Harris, C.I. "Fate of 2-Chloro-s-triazine Herbicides in Soil," *J. Agric. Food Chem.*, 15(1):157-162 (1967).

Harris, C.I. and T.J. Sheets. "Influence of Soil Properties on Adsorption and Phytotoxicity of CIPC, Diuron, and Simazine," *Weeds*, 13:215-219 (1965).

Harris, C.I., E.A. Woolson, and B.E. Hummer. "Dissipation of Herbicides at Three Soil Depths," *Weed Sci.*, 17(1):27-31 (1969).

Hartley, D. and H. Kidd, Eds. *The Agrochemicals Handbook*, 2nd ed. (England: Royal Society of Chemistry, 1987).

Harvey, J., Jr., J.J. Dulka, and J.J. Anderson. "Properties of Sulfometuron Methyl Affecting Its Environmental Fate: Aqueous Hydrolysis and Photolysis, Mobility and Adsorption on Soils, and Bioaccumulation Potential," *J. Agric. Food Chem.*, 33(4):590-596 (1985).

Hatzios, K.K. "Biotransformations of Herbicides in Higher Plants" in *Environmental Chemistry of Herbicides. Volume II*, Grover, R. and A.J. Cessna, Eds. (Boca Raton, FL: CRC Press, Inc., 1991), pp. 141-185.

Hazardous Substances Data Bank. National Library of Medicine, Toxicology Information Program (1989).

Helling, C.S. "Pesticide Mobility in Soils II. Applications of Soil Thin-Layer Chromatography," *Soil Sci. Soc. Am. Proc.*, 35:737-743 (1971).

Helweg, A. "Microbial Breakdown of the Fungicide Benomyl," *Soil Biol. Biochem.*, 4:377-378 (1972).

Henderson, C., Q.H. Pickering, and C.M. Tarzwell. "Relative Toxicity of Ten Chlorinated Hydrocarbon Insecticides to Four Species of Fish," *Trans. Am. Fish Soc.*, 88(1):23-32 (1959).

Heritage, A.D. and I.C. MacRae. "Degradation of Lindane by Cell-free Preparations of *Clostridium sphenoides*," *Appl. Environ. Microbiol.*, 34(2):222-224 (1977).

Heritage, A.D. and I.C. MacRae. "Identification of Intermediates Formed During the Degradation of Hexachlorocyclohexanes by *Clostridium sphenoides*," *Appl. Environ. Microbiol.*, 33(6):1295-1297 (1977a).

Herrett, R.A. and W.P. Bagley. "The Metabolism and Translocation of 3-Amino-1,2,4-triazole by Canada Thistle," *J. Agric. Food Chem.*, 12(1):17-20 (1964).

Herrmann, M., D. Kotzias, and F. Korte. "Photochemical Behavior of Chlorsulfuron in Water and In Adsorbed Phase," *Chemosphere*, 14(1):3-8 (1985).

Hiltbold, A.E. and G.A. Buchanan. "Influence of Soil pH on Persistence of Atrazine in the Field," *Weed Sci.*, 25(6):515-520 (1977).

Hirsch, M. and O. Hutzinger. "Naturally Occurring Proteins from Pond Water Sensitize Hexachlorobenzene Photolysis," *Environ. Sci. Technol.*, 23(10):1306-

1307 (1989).

Hoagland, R.E. "Effects of Glyphosate on Metabolism of Phenolic Compounds. VI. Effects of Glyphosine and Glyphosate Metabolites on Phenylalanine Ammonia-Lyase Activity, Growth, and Protein, Chlorophyll, and Anthocuanin Levels in Soybean (Glycine mas) Seedlings Phytotoxicity," *Weed Sci.*, 28:393-400 (1980).

Holmstead, R.L, J.E. Casida, L.O. Ruzo, and D.G. Fullmer. "Pyrethroid Photodecomposition: Permethrin," *J. Agric. Food Chem.*, 26(3):590-595 (1978).

Holt, R.F. "Determination of Hexazinone and Metabolite Residues Using Nitrogen-Selective Gas Chromatography," *J. Agric. Food Chem.*, 29(1):165-172 (1981).

Humburg, N.E., S.R. Colby, R.G. Lym, E.R. Hill, W.J. McAvoy, L.M. Kitchen, and R. Prasad, Eds. *Herbicide Handbook of the Weed Science Society of America*, 6th ed. (Champaign, IL: Weed Science Society of America, 1989), 301 p.

Hustert, K. and P.N. Moza. "Photokatalytischer Abbau von Phthalaten an Titandioxid in Wässriger Phase," *Chemosphere*, 17(9):1751-1754 (1988).

Hwang, H.-M., R.E. Hodson, and R.F. Lee. "Degradation of Phenol and Chlorophenols by Sunlight and Microbes in Estuarine Water," *Environ. Sci. Technol.*, 20(10):1002-1007 (1986).

Ibrahim, F.B., J.M. Gilbert, R.T. Evans, and J.C. Cavagnol. "Decomposition of Di-Syston (*O,O*-Diethyl *S*-[2-(ethylthio)ethyl] Phosphorodithioate) on Fertilizers by Infrared, Gas-Liquid Chromatography, and Thin-Layer Chromatography," *J. Agric. Food Chem.*, 17(2):300-305 (1969).

Ide, A., Y. Ueno, K. Takaichi, and H. Watanabe. "Photochemical Reaction of 2,3-Dichloro-1,4-naphthoquinone (Dichlone) in the Presence of Oxygen," *Agric. Biol. Chem.*, 43(7):1387-1394 (1979).

Iizuka, H. and T. Masuda. "Residual Fate of Chlorphenamidine in Rice Plant and Paddy Soil," *Bull. Environ. Contam. Toxicol.*, 22(6):745-749 (1979).

Ivie, G.W. and D.L. Bull. "Photodegradation of *O*-Ethyl *O*-[4-(Methylthio)phenyl] *S*-Propyl Phosphorodithioate (BAY NTN 9306)," *J. Agric. Food Chem.*, 24(5):1053-1057 (1976).

Ivie, G.W. and J.E. Casida. "Enhancement of Photoalteration of Cyclodiene Insecticide Chemical Residues by Rotenone," *Science (Washington, DC)*, 167(3925):1620-1622 (1970).

Ivie, G.W. and J.E. Casida. "Photosensitizers for the Accelerated Degradation of Chlorinated Cyclodienes and Other Insecticide Chemicals Exposed to Sunlight on Bean Leaves," *J. Agric. Food Chem.*, 19(3):410-416 (1971).

Ivie, G.W. and J.E. Casida. "Sensitized Photodecomposition and Photosensitizer Activity of Pesticide Chemicals Exposed to Sunlight on Silica Gel Chromatoplates," *J. Agric. Food Chem.*, 19(3):405-409 (1971a).

Ivie, G.W., J.R. Knox, S. Khalifa, I. Yamamoto, and J.E. Casida. "Novel Photoproducts of Heptachlor Epoxide, *trans*-Chlordane, and *trans*-Nonachlor," *Bull. Environ. Contam. Toxicol.*, 7(6):376-383 (1972).

Jamet, P. and M.A. Piedallu. "Mouvement du Carbofuran dans Different Types de Sols," *Phytiatrie-Phytopharmacie*, 24:279–296 (1975).

Janko, Z., H. Stefaniak, and E. Czerwinska. "Microbiological decomposition of Diuron [*N,N*-Dimethyl-*N'*-(3,4-dichlorophenyl)urea]," [Chemical Abstracts 78:25271n] *Pr. Inst. Przem. Org.*, 2:241–257 (1970).

Janssen, D.B., D. Jager, and B. Wilholt. "Degradation of *n*-Haloalkanes and α,ω-Dihaloalkanes by Wild Type and Mutants of *Acinetobacter* sp. Strain GJ70," *Appl. Environ. Microbiol.*, 53(3):561–566 (1987).

Jensen, S., R. Göthe, and M.-O. Kindstedt. "Bis(*p*-chlorophenyl)acetonitrile (DDN), a New DDT Derivative formed in Anaerobic Digested Sewage Sludge and Lake Sediment," *Nature (London)*, 240(5381):421–422 (1972).

Johnsen, R.E. "DDT Metabolism in Microbial Systems," *Residue Rev.*, 61:1–28 (1976).

Johnson-Logan, L.R., R.E. Broshears, and S.J. Klaine. "Partitioning Behavior and the Mobility of Chlordane in Groundwater," *Environ. Sci. Technol.*, 26(11):2234–2239 (1992).

Joiner, R.L. and K.P. Baetcke. "Parathion: Persistence on Cotton and Identification of Its Photoalteration Products," *J. Agric. Food Chem.*, 21(3):391–396 (1973).

Joiner, R.L., H.W. Chambers, /and K.P. Baetcke. "Toxicity of Parathion and Several of its Photoalteration Products to Boll Weevils," *Bull. Environ. Contam. Toxicol.*, 6(3):220–224 (1971).

Jones, A.S. "Metabolism of Aldicarb by Five Soil Fungi," *J. Agric. Food Chem.*, 24(1):115–117 (1976).

Jones, T.W., W.M. Kemp, J.C. Stevenson, and J.C. Means. "Degradation of Atrazine in Estuarine Water/Sediment Systems and Soils," *J. Environ. Qual.*, 11(4):632–638 (1982).

Junk, G.A., R.F. Spalding, and J.J. Richard. "Areal, Vertical, and Temporal Differences in Ground Water Chemistry: II. Organic Constituents," *J. Environ. Qual.*, 9(3):479–483 (1980).

Jury, W.A., H. Elabd, and M. Resketo. "Field Study of Napropamide Movement Through Unsaturated Soil," *Water Resour. Res.*, 22(5):749–755 (1986).

Jury, W.A., D.D. Focht, and W.J. Farmer. "Evaluation of Pesticide Pollution Potential from Standard Indices of Soil-Chemical Adsorption and Biodegradation," *J. Environ. Qual.*, 16(4):422–428 (1987).

Kaars Sijpesteijn, A., H.M. Dekhuijzen, and J.W. Vonk. "Biological Conversion of Fungicides in Plants and Microorganisms" in *Antifungal Compounds*, (New York: Marcel Dekker, Inc., 1977), pp. 91–147.

Kale, S.P., N.B.K. Murthy, and K. Raghu. "Effect of Carbofuran, Carbaryl, and Their Metabolites on the Growth of *Rhizobium* sp. and *Azotobacter chroococcum*," *Bull. Environ. Contam. Toxicol.*, 42(5):769–772 (1989).

Kanazawa, J. "Relationship Between the Soil Sorption Constants for Pesticides and Their Physiochemical Properties," *Environ. Toxicol. Chem.*, 8(6):477–484 (1989).

Kaneda, Y., K. Nakamura, H. Nakahara, and M. Iwaida. "Degradation of

Organochlorine Pesticides with Calcium Hypochlorite," [Chemical Abstracts 82:94190e]: *I. Eisei Kagaku*, 20:296-299 (1974).

Kansouh, A.S.H. and T.L. Hopkins. "Diazinon Absorption, Translocation, and Metabolism in Bean Plants," *J. Agric. Food Chem.*, 16(3):446-450 (1968).

Kapoor, I.P., R.L. Metcalf, R.F. Nystrom, and G.K. Sangha. "Comparative Metabolism of Methoxychlor, Methiochlor, and DDT in Mouse, Insects, and in a Model Ecosystem," *J. Agric. Food Chem.*, 18(6):1145-1152 (1970).

Karickhoff, S.W., D.S. Brown, and T.A. Scott. "Sorption of Hydrophobic Pollutants on Natural Sediments," *Water Res.*, 13(3):241-248 (1979).

Karinen, J.F., J.G. Lamberton, N.E. Stewart, and L.C. Terriere. "Persistence of Carbaryl in the Marine Estuarine Environment. Chemical and Biological Stability in Aquarium Systems," *J. Agric. Food Chem.*, 15(1):148-156 (1967).

Katz, M. "Acute Toxicity of Some Organic Insecticides to Three Species of Salmonids and to Threespine Stickleback," *Trans. Am. Fish Soc.*, 90(3):264-268 (1961).

Kaufman, D.D. and P.C. Kearney. "Microbial Degradation of *s*-Triazine Herbicides," *Residue Rev.*, 32:235-265 (1970).

Kaufman, D.D., J.R. Plimmer, P.C. Kearney, J. Blake, and F.S. Guardia. "Chemical Versus Microbial Decomposition of Amitrole in Soil," *Weed Sci.*, 16:266-272 (1968).

Kaufman, D.D., B.A. Russell, C.S. Helling, and A.J. Kayser. "Movement of Cypermethrin, Decamethrin, Permethrin, and Their Degradation Products in Soil," *J. Agric. Food Chem.*, 29(2):239-245 (1981).

Kay, B.D. and D.E. Elrick. "Adsorption and Movement of Lindane in Soils," *Soil Sci.*, 104(5):314-322 (1967).

Kazano, H., P.C. Kearney, and D.D. Kaufman. "Metabolism of Methylcarbamate Insecticides in Soils," *J. Agric. Food Chem.*, 20(5):975-979 (1972).

Kearney, P.C. and D.D. Kaufman. *Herbicides: Chemistry, Degradation and Mode of Action* (New York: Marcel Dekker, Inc., 1976), 1036 p.

Kearney, P.C., E.A. Woolson, J.R. Plimmer, and A.R. Isensee. "Decontamination of Pesticide Residues in Soils," *Residue Rev.*, 29:137-149 (1969).

Keith, L.H. and D.B. Walters. *The National Toxicology Program's Chemical Data Compendium - Volume II. Chemical and Physical Properties* (Chelsea, MI: Lewis Publishers, Inc., 1992), 1642 p.

Kenaga, E.E. "Toxicological and Residue Data Useful in the Environmental Safety Evaluation of Dalapon," *Residue Rev.*, 53:109-151 (1974).

Kennedy, M.V., B.J. Stojanovic, and F.L. Shuman, Jr. "Chemical and Thermal Aspects of Pesticide Disposal," *J. Environ. Qual.*, 1(1):63-65 (1972).

Kennedy, M.V., B.J. Stojanovic, and F.L. Shuman, Jr. "Analysis of Decomposition Products of Pesticides," *J. Agric. Food Chem.*, 20(2):341-343 (1972a).

Khalifa, S., R.L. Holmstead, and J.E. Casida. "Toxaphene Degradation by Iron(II) Protoporphyrin Systems," *J. Agric. Food Chem.*, 24(2):277-282 (1976).

Khalil, M.A.K. and R.A. Rasmussen. "Global Sources, Lifetimes and Mass

Balances of Carbonyl Sulfide (OCS) and Carbon Disulfide (CS$_2$) in the Earth's Atmosphere," *Atmos. Environ.*, 18(9):1805-1813 (1984).

Khan, S. and V. Bansal. "Thermodynamics of Adsorption of Methyl-2-(dimethylamino)-N-[{(methylamino)carbonyl}oxy]-2-oxoethanimidothioate on Saturated Cu- and Zn-Montmorillonites," *J. Coll. Interface Sci.*, 78(2):554-558 (1980).

Khan, S.U. "Kinetics of Hydrolysis of Atrazine in Aqueous Fulvic Acid Solution," *Pestic. Sci.*, 9(1):39-43 (1978).

Khan, S.U. "Distribution and Characteristics of Bound Residues of Prometryn in an Organic Soil," *J. Agric. Food Chem.*, 30(1):175-179 (1982).

Khan, S.U. and D.S. Gamble. "Ultraviolet Irradiation on an Aqueous Solution of Prometryn in the Presence of Humic Materials," *J. Agric. Food Chem.*, 31(5):1099-1104 (1983).

Khan, S.U. and H.A. Hamilton. "Extractable and Bound (Nonextractable) Residues of Prometryn and Its Metabolites in an Organic Soil," *J. Agric. Food Chem.*, 28(1):126-132 (1980).

Khan, S.U. and P.B. Marriage. "Residues of Atrazine and Its Metabolites in an Orchard Soil and Their Uptake by Oat Plants," *J. Agric. Food Chem.*, 25(6):1408-1413 (1977).

Khan, S.U. and M. Schnitzer. "UV Irradiation of Atrazine in Aqueous Fulvic Acid Solution," *J. Environ. Sci. Health*, B13:299-310 (1978).

King, P.H. and P.L. McCarty. "A Chromatographic Model for Predicting Pesticide Migration in Soils," *Soil Sci.*, 106(4):248-261 (1968).

Kishi, H., N. Kogure, and Y. Hashimoto. "Contribution of Soil Constituents in Adsorption Coefficient of Aromatic Compounds, Halogenated Alicyclic and Aromatic Compounds to Soil," *Chemosphere*, 21(7):867-876 (1990).

Kjeldsen, P., J. Kjølholt, B. Schultz, T.H. Christensen, and J.C. Tjell. "Sorption and Degradation of Chlorophenols, Nitrophenols and Organophosphorus Pesticides in the Subsoil under Landfills," *J. Contam. Hydrol.*, 6(2):165-184 (1990).

Klehr, M., J. Iwan, and J. Riemann. "An Experimental Approach to the Photolysis of Pesticides Adsorbed on Soil: Thidiazuron," *Pestic. Sci.*, 14(4):359-366 (1983).

Klein, W., J. Kohli, I. Weisgerber, and F. Korte. "Fate of Aldrin-^{14}C in Potatoes and Soil under Outdoor Conditions," *J. Agric. Food Chem.*, 21(2):152-156 (1973).

Knoevenagel, K. and R. Himmelreich. "Degradation of Compounds Containing Carbon Atoms by Photo-Oxidation in the Presence of Water," *Arch. Environ. Contam. Toxicol.*, 4:324-333 (1976).

Knuutinen, J., H. Palm, H. Hakala, J. Haimi, V. Huhta, and J. Salminen. "Polychlorinated Phenols and Their Metabolites in Soil and Earthworms of Sawmill Environment," *Chemosphere*, 20(6):609-623 (1990).

Kobayashi, H. and B.E. Rittman. "Microbial Removal of Hazardous Organic Compounds," *Environ. Sci. Technol.*, 16(3):170A-183A (1982).

Kobrinsky, P.C. and M.R. Martin. "High-Energy Methyl Radicals; the Photolysis of Methyl Bromide at 1850 Å," *J. Chem. Phys.*, 48(12):5728-5729 (1968).

Kolbe, A., A. Bernasch, M. Stock, H.R. Schütte, and W. Dedek. "Persistence of the Insecticide Dimethoate in Three Different Soils under Laboratory Conditions," *Bull. Environ. Contam. Toxicol.*, 46(4):492-498 (1991).

Konrad, J.G., G. Chesters, and D.E. Armstrong. "Soil Degradation of Malathion, a Phosphorothioate Insecticide," *Soil Sci. Soc. Am. Proc.*, 33:259-262 (1969).

Kopczynski, S.L., A.P. Altshuller, and F.D. Sutterfield. "Photochemical Reactivities of Aldehyde-Nitrogen Oxide Systems," *Environ. Sci. Technol.*, 8(10):909-918 (1974).

Korte, F., G. Ludwig, and J. Vogel. "Umwandlung von Aldrin-[^{14}C] and Dieldrin-[^{14}C] durch Mikroorganismen, Leberhomogenate, und Moskito-Larven," *Liebigs Ann. Chem.*, 656:135-140 (1962).

Koskinen, W.C., H.R. Leffler, J.E. Oliver, P.C. Kearney, and C.G. McWhorter. "Effect of Trifluralin Soil Metabolites on Cotton Boll Components and Fiber and Seed Properties," *J. Agric. Food Chem.*, 33(5):958-961 (1985).

Koskinen, W.C., J.E. Oliver, P.C. Kearney, and C.G. McWhorter. "Effect of Trifluralin Soil Metabolites on Cotton Growth and Yield," *J. Agric. Food Chem.*, 32(6):1246-1248 (1984).

Kotronarou, A., G. Mills, and M.R. Hoffmann. "Decomposition of Parathion in Aqueous Solution by Ultrasonic Irradiation," *Environ. Sci. Technol.*, 26(7):1460-1462 (1992).

Kozak, J., J.B. Weber, and T.J. Sheets. "Adsorption of Prometryn and Metolachlor by Selected Soil Organic Matter Fractions," *Soil Sci.*, 136(2):94-101 (1983).

Krause, A., W.G. Hancock, R.D. Minard, A.J. Freyer, R.C. Honeycutt, H.M.L. Baron, D.L. Paulson, S.-Y. Liu, and J.-M. Bollag. "Microbial Transformation of the Herbicide Metolachlor by a Soil Actinomycete," *J. Agric. Food Chem.*, 33(4):584-589 (1985).

Krause, A.A. and H.D. Niemczyk. "Gas-Liquid Chromatographic Analysis of Chlorthal-Dimethyl Herbicide and Its Degradates in Turfgrass Thatch and Soil using a Solid-Phase Extraction Technique," *J. Environ. Sci. Health*, B25(5):587-606 (1990).

Kriegman-King, M.R. and M. Reinhard. "Transformation of Carbon Tetrachloride in the Presence of Sulfide, Biotite, and Vermiculite," *Environ. Sci. Technol.*, 26(11):2198-2206 (1992).

Krumzdorov, A.M. "Transformation of 2-Methoxy-3,6-dichlorobenzoic Acid (Dicamba) in Corn Plants," *Agrokhimya*, 7:128-133 (1974).

Kuhr, R.J. "Metabolism of Methylcarbamate Insecticide Chemicals in Plants," *J. Agric. Food Chem. Suppl.*, 44-49 (1968).

Kulshrestha, G. and S.B. Singh. "Influence of Soil Moisture and Microbial Activity on Pendimethalin Degradation," *Bull. Environ. Contam. Toxicol.*, 48(2):269-274 (1992).

Laanio, T.L., P.C. Kearney, and D.D. Kaufman. "Microbial Metabolism of Dinitramine," *Pestic. Biochem. Physiol.*, 3:271-277 (1973).

Lamoureux, G.L., R.H. Shimabukuro, H.R. Swanson, and D.S. Frear. "Metabolism of 2-Chloro-4-ethylamino-6-isopropylamino-s-triazine (Atrazine) in Excised Sorghum Leaf Sections," *J. Agric. Food Chem.*, 18(1):81-86 (1970).

Lamoureux, G.L., L.E. Stafford, and F.S. Tanaka. "Metabolism of 2-Chloro-N-isopropylacetanilide (Propachlor) in the Leaves of Corn, Sorghum, Sugarcane, and Barley," *J. Agric. Food Chem.*, 19(2):346-350 (1971).

Lamparski, L.L., R.H. Stehl, and R.L. Johnson. "Photolysis of Pentachlorophenol-Treated Wood. Chlorinated Dibenzo-p-dioxin Formation," *Environ. Sci. Technol.*, 14(2):196-200 (1980).

Langlois, B.E. "Reductive Dechlorination of DDT by *Escherichia coli*," *J. Dairy Sci.*, 50:1168-1170 (1967).

Laplanche, A., G. Martin, and F. Tonnard. "Ozonation Schemes of Organophosphorus Pesticides. Application in Drinking Water Treatment," *Ozone: Sci. Engrg.*, 6:207-219 (1984).

Leafe, E.L. "Metabolism and Selectivity of Plant-Growth Regulator Herbicides," *Nature (London)*, 193(4814):485-486 (1962).

Leavitt, D.D. and M.A. Abraham. "Acid-Catalyzed Oxidation of 2,4-Dichlorophenoxyacetic Acid by Ammonium Nitrate in Aqueous Solution," *Environ. Sci. Technol.*, 24(4):566-571 (1990).

LeBaron, H.M., J.E. McFarland, B.J. Simoneaux, and E. Ebert. "Metolachlor" in *Herbicides: Chemistry, Degradation, and Mode of Action. Volume 3*, Kearney, P.C. and D.D. Kaufman, Eds., (New York: Marcel Dekker, Inc., 1988), pp. 336-383.

Lee, C.-C., R.E. Green, and W.J. Apt. "Transformation and Adsorption of Fenamiphos, F. Sulfoxide and F. Sulfone in Molokai Soil and Simulated Movement with Irrigations," *J. Contam. Hydrol.*, 1:211-225 (1986).

Lee, J.K. "Degradation of the Herbicide, Alachlor, by Soil Microorganisms. III. Degradation under an Upland Soil Condition," *J. Korean Agric. Chem. Soc.*, 29:182-189 (1986).

Lee, P.W., S.M. Stearns, H. Hernandez, W.R. Powell, and M.V. Naidu. "Fate of Dicrotophos in the Soil Environment," *J. Agric. Food Chem.*, 37(4):1169-1174 (1989).

Leger, D.A. and V.N. Mallet. "New Degradation Products and a Pathway for the Degradation of Aminocarb [4-(Dimethylamino)-3-methylphenyl N-Methyl-carbamate] in Purified Water," *J. Agric. Food Chem.*, 36(1):185-189 (1988).

Leinster, P., R. Perry, and R.J. Young. "Ethylenebromide in Urban Air," *Atmos. Environ.*, 12:2382-2398 (1978).

Leistra, M. "Distribution of 1,3-Dichloropropene in Soil," *J. Agric. Food Chem.*, 18(6):1124-1126 (1970).

Leitis, E. and D.G. Crosby. "Photodecomposition of Trifluralin," *J. Agric. Food Chem.*, 22(5):842-848 (1974).

Leland, H.V., W.N. Bruce, and N.F. Shimp. "Chlorinated Hydrocarbon Insecticides in Sediments in Southern Lake Michigan," *Environ. Sci. Technol.*, 7(9):833-838

(1973).

Leuck, D.G. and M.C. Bowman. "Residues of Phorate and Five of Its Metabolites: Their Persistence in Forage Corn and Grass," *J. Econ. Entomol.*, 63:1838-1842 (1970).

Lewis, R.J., Sr. *Rapid Guide to Hazardous Chemicals in the Workplace* (New York: Van Nostrand Reinhold, 1990), 286 p.

Lemley, A.T., R.J. Wagenet, and W.Z. Zhong. "Sorption and Degradation of Aldicarb and Its Oxidation Products in a Soil-water Flow System as a Function of pH and Temperature," *J. Environ. Qual.*, 17(3):408-414 (1988).

Leoni, V., C.B. Hollick, D.E. D'Alessandro, R.J. Collinson, and S. Merolli. "The Soil Degradation of Chlorpyrifos and the Significance of Its Presence in the Superficial Water in Italy," *Agrochimica*, 25:414-426 (1981).

Li, C.F. and R.L. Bradley. "Degradation of Chlorinated Hydrocarbon Pesticides in Milk and Butter Oil by Ultraviolet Energy," *J. Dairy Sci.*, 52:27-30 (1969).

Li, C.Y. and E.E. Nelson. "Persistence of Benomyl and Captan and Their Effects on Microbial Activity in Field Soils," *Bull. Environ. Contam. Toxicol.*, 34(4):533-540 (1985).

Li, G.-C. and G.T. Felbeck, Jr. "Atrazine Hydrolysis as Catalyzed by Humic Acids," *Soil Sci*, 114(3):201-208 (1972).

Liang, T.T. and E.P. Lichtenstein. "Effect of Light, Temperature, and pH on the Degradation of Azinphosmethyl," *J. Econ. Entomol.*, 65:315-321 (1972).

Lichtenstein, E.P. "'Bound' Residues in Soils and Transfer of Soil Residues in Crops," *Residue Rev.*, 76:147-153 (1980).

Lichtenstein, E.P., T.W. Fuhremann, and K.R. Schulz. "Effect of Sterilizing Agents on Persistence of Parathion and Diazinon in Soils and Water," *J. Agric. Food Chem.*, 16(5):870-873 (1968).

Lichtenstein, E.P., T.W. Fuhremann, and K.R. Schulz. "Persistence and Vertical Distribution of DDT, Lindane, and Aldrin Residues, 10 and 15 years after a Single Soil Application," *J. Agric. Food Chem.*, 19(4):718-721 (1971).

Lichtenstein, E.P., T.W. Fuhremann, and K.R. Schulz. "Translocation and Metabolism of [^{14}C]Phorate as Affected by Percolating Water in a Model Soil-Plant Ecosystem," *J. Agric. Food Chem.*, 22(6):991-996 (1974).

Lichtenstein, E.P., T.T. Liang, and M.K. Koeppe. "Effects of Fertilizers, Captafol, and Atrazine on the Fate and Translocation of [^{14}C]Fonofos and [^{14}C]Parathion in a Soil-Plant Microcosm," *J. Agric. Food Chem.*, 30(5):871-878 (1982).

Lieberman, M.T. and M. Alexander. "Microbial and Nonenzymatic Steps in the Decomposition of Dichlorvos (2,2-Dichlorovinyl *O,O*-Dimethyl Phosphate)," *J. Agric. Food Chem.*, 31(2):265-267 (1983).

Lin, S., M.T. Lukasewycz, R.J. Liukkonen, and R.M. Carlson. "Facile Incorporation of Bromine into Aromatic Systems under Conditions of Water Chlorination," *Environ. Sci. Technol.*, 18(12):985-986 (1984).

Linke, H.A.B. and R. Bartha. "Transformation Products of the Herbicide Propanil in Soil: A Balance Study," in *Agric. Indust. Microbiol. Abstr., Bacteriol. Proc. 70,*

(Washington, DC: American Society of Microbiology, 1970), p. 9.

Liu, D., W.M.J. Strachan, K. Thomson, and K. Kwasniewska. "Determination of the Biodegradability of Organic Compounds," *Environ. Sci. Technol.*, 15(7):788–793 (1981).

Liu, L.C., H. Cibes-Viadé, and F.K.S. Koo. "Adsorption of Ametryne and Diuron by Soils," *Weed Sci.*, 18(4):470–474 (1970).

Liu, S.-Y. and J.M. Bollag. "Metabolism of Carbaryl by a Soil Fungus," *J. Agric. Food Chem.*, 19(3):487–490 (1971).

Liu, S.-Y. and J.M. Bollag. "Carbaryl Decomposition to 1-Naphthyl Carbamate by *Aspergillus terreus*," *Pestic. Biochem. Physiol.*, 1:366–372 (1971a).

Loekke, H. "Residues in Carrots Treated with Linuron," *Pestic. Sci.*, 5:749–757 (1974).

Løkke, H. "Sorption of Selected Organic Pollutants in Danish Soils," *Ecotoxicol. Environ. Saf.*, 8(5):395–409 (1984).

Lopez, C.E. and J.I. Kirkwood. "Isolation of Microorganisms from a Texas Soil Capable of Degrading Urea Derivative Herbicides," *Soil Sci. Soc. Am. Proc.*, 38:309–312 (1974).

Lopez-Gonzales, J.De D. and C. Valenzuela-Calahorro. "Associated Decomposition of DDT to DDE in the Diffusion of DDT on Homoionic Clays," *J. Agric. Food Chem.*, 18(3):520–523 (1970).

Lord, K.A., G.G. Briggs, M.C. Neale, and R. Manlove. "Uptake of Pesticides from Water and Soil by Earthworms," *Pestic. Sci.*, 11(4):401–408 (1980).

Lowder, S.W. and J.B. Weber. "Atrazine Efficacy and Longevity as Affected by Tillage, Liming, and Fertilizer Type," *Weed Sci.*, 30:273–280 (1982).

Lu, P.-Y., R.L. Metcalf, A.S. Hirwe, and J.W. Williams. "Evaluation of Environmental Distribution and Fate of Hexachlorocyclopentadiene, Chlordene, Heptachlor, and Heptachlor Epoxide in a Model Ecosystem," *J. Agric. Food Chem.*, 23(5):967–973 (1975).

Lucier, G.W. and R.E. Menzer. "Metabolism of Dimethoate in Bean Plants in Relation to Its Mode of Application," *J. Agric. Food Chem.*, 16(6):936–945 (1968).

Lucier, G.W. and R.E. Menzer. "Nature of Oxidative Metabolites of Dimethoate Formed in Rats, Liver Microsomes, and Bean Plants," *J. Agric. Food Chem.*, 18(4):698–704 (1970).

Lund-Høie, K. and H.O. Friestad. "Photodegradation of the Herbicide Glyphosate in Water," *Bull. Environ. Contam. Toxicol.*, 36:723–729 (1986).

Lyman, W.J., W.F. Reehl, and D.H. Rosenblatt. *Handbook of Chemical Property Estimation Methods: Environmental Behavior of Organic Compounds* (New York: McGraw-Hill, Inc., 1982).

Mabey, W. and T. Mill. "Critical Review of Hydrolysis of Organic Compounds in Water under Environmental Conditions," *J. Phys. Chem. Ref. Data*, 7(2):383–415 (1978).

Macalady, D.L. and N.L. Wolfe. "New Perspectives on the Hydrolytic Degradation of the Organophosphorothioate Insecticide Chlorpyrifos," *J. Agric. Food Chem.*, 31(6):1139-1147 (1983).

Macalady, D.L. and N.L. Wolfe. "Effects of Sediment Sorption and Abiotic Hydrolyses. 1. Organophosphorothioate Esters," *J. Agric. Food Chem.*, 33(2):167-173 (1985).

Macalady, D.L., P.G. Tratnyek, and T.J. Grundy. "Abiotic Reduction Reactions of Anthropogenic Organic Chemicals in Anaerobic Systems: A Critical Review," *J. Contam. Hydrol.*, 1(1):1-28 (1986).

Macek, K.J. and W.A. McAllister. "Insecticide Susceptibility of Some Common Fish Family Representatives," *Trans. Am. Fish Soc.*, 99(1):20-27 (1970).

MacNamara, G. and S.J. Toth. "Adsorption of Linuron and Malathion by Soils and Clay Minerals," *Soil Sci.*, 109(4):234-240 (1970).

MacRae, I.C. "Microbial Metabolism of Pesticides and Structurally Related Compounds," *Rev. Environ. Contam. Toxicol.*, 109:2-87 (1989).

MacRae, I.C. and M. Alexander. "Microbial Degradation of Selected Herbicides in Soil," *J. Agric. Food Chem.*, 13(1):72-75 (1965).

MacRae, I.C., K. Raghu, and E.M. Bautista. "Anaerobic Degradation of the Insecticide Lindane by *Clostridium* sp," *Nature (London)*, 221(5183):859-860 (1969).

MacRae, I.C., K. Raghu, and T.F. Castro. "Persistence and Biodegradation of Four Common Isomers of Benzene Hexachloride in Submerged Soils," *J. Agric. Food Chem.*, 15(5):911-914 (1967).

Madhun, Y.A. and V.H. Freed. "Degradation of the Herbicides Bromacil, Diuron and Chlortoluron in Soil," *Chemosphere*, 16(5):1003-1011 (1987).

Madhun, Y.A., V.H. Freed, J.L. Young, and S.C. Fang. "Sorption of Bromacil, Chlortoluron, and Diuron by Soils," *Soil Sci. Soc. Am. J.*, 50:1467-1471 (1986).

Majka, J.T. and T.L. Lavy. "Adsorption, Mobility, and Degradation of Cyanazine and Diuron in Soils," *Weed Sci.*, 25(5):401-406 (1977).

Mansour, M., E. Feicht, and P. Méallier. "Improvement of the Photostability of Selected Substances in Aqueous Medium," *Toxicol. Environ. Contam.*, 20-21:139-147 (1989).

Mansour, M., S. Thaller, and F. Korte. "Action of Sunlight on Parathion," *Bull. Environ. Contam. Toxicol.*, 30(3):358-364 (1983).

Martens, R. "Degradation of Endosulfan-8,9-^{14}C in Soil under Different Conditions," *Bull. Environ. Contam. Toxicol.*, 17(4):438-446 (1977).

Martin, G., A. Laplanche, J. Morvan, Y. Wei, and C. LeCloirec. "Action of Ozone on Organo-Nitrogen Products," in *Proceedings Symposium on Ozonization: Environmental Impact and Benefit* (Paris, France: International Ozone Association, 1983), pp. 379-393.

Massini, P. "Movement of 2,6-Dichlorobenzonitrile in Soils and in Plants in Relation to Its Physical Properties," *Weed Res.*, 1:142-146 (1961).

Mathur, S.P. and J.G. Saha. "Microbial Degradation of Lindane-C^{14} in a Flooded Sandy Loam Soil," *Soil Sci.*, 120(4):301-307 (1975).

Mathur, S.P. and J.G. Saha. "Degradation of Lindane-^{14}C in a Mineral Soil and in an Organic Soil," *Bull. Environ. Contam. Toxicol.*, 17(4):424-430 (1977).

Matsumura, F. and G.M. Boush. "Degradation of Insecticides by a Soil Fungus, *Trichoderma viride*," *J. Econ. Entomol.*, 61:610-612 (1968).

Matsumura, F., G.M. Boush, and A. Tai. "Breakdown of Dieldrin in the Soil by a Microorganism," *Nature (London)*, 219(5157):965-967 (1968).

Matsumura, F., K.C. Patil, and G.M. Boush. "DDT Metabolized by Microorganisms from Lake Michigan," *Nature (London)*, 230(5292):324-325 (1971).

Matsunaka, S. "Activation and Inactivation of Herbicides by Higher Plants," *Residue Rev.*, 25:45-58 (1969).

Matsuo, H. and J.E. Casida. "Photodegradation of two Dinitrophenolic Pesticide Chemicals, Dinobuton and Dinoseb, Applied to Bean Leaves," *Bull. Environ. Contam. Toxicol.*, 5(1):72-78 (1970).

Mattson, A.M., J.T. Spillane, and G.W. Pearce. "Dimethyl 2,2-Dichlorovinyl Phosphate (DDVP), an Organophosphorus Compound Highly Toxic to Insects," *J. Agric. Food Chem.*, 3(4):319-321 (1955).

Maule, A., S. Plyte, and A.V. Quirk. "Dehalogenation of Organochlorine Insecticides by Mixed Anaerobic Microbial Populations," *Pestic. Biochem. Physiol.*, 27:229-236 (1987).

Mazzochi, P.H. and M.P. Rao. "Photolysis of 3-(*p*-Chlorophenyl)-1,1-dimethylurea (Monuron) and 3-Phenyl-1,1-dimethylurea (Fenuron)," *J. Agric. Food Chem.*, 20(5):957-959 (1972).

McBride, K.E., J.W. Kenny, and D.M. Stalker. "Metabolism of the Herbicide Bromoxynil by *Klebsiella pneumoniae* supus. ozaenae," *Appl. Environ. Microbiol.*, 52(2):325-330 (1986).

McCall, P.J., R.L. Swann, D.A. Laskowski, S.A. Vrona, S.M. Unger, and H.J. Dishburger. "Prediction of Chemical Mobility in Soil from Sorption Coefficients," in *Aquatic Toxicology and Hazard Assessment, Fourth Conference, ASTM STP 737*, Branson, D.R. and K.L. Dickson, Eds. (Philadelphia, PA: American Society for Testing and Materials, 1981), pp. 49-58.

McCall, P.J., S.A. Vrona, and S.S. Kelley. "Fate of Uniformly Carbon-14 Ring Labeled 2,4,5-Trichlorophenoxyacetic Acid and 2,4-Dichlorophenoxy Acid," *J. Agric. Food Chem.*, 29(1):100-107 (1981a).

McGahen, L.L. and J.M. Tiedje. "Metabolism of Two New Acylanilide Herbicides, Antor Herbicide (H-22234) and Dual (Metolachlor) by the Soil Fungus *Chaetomium globosum*," *J. Agric. Food Chem.*, 26(2):414-419 (1978).

McGlamery, M.D. and F.W. Slife. "The Adsorption and Desorption of Atrazine as Affected by pH, Temperature, and Concentration," *Weeds*, 14:237-239 (1966).

McGuire, R.R., M.J. Zabik, R.D. Schuetz, and R.D. Flotard. "Photochemistry of Bioactive Compounds. Photochemical Reactions of Heptachlor: Kinetics and Mechanisms," *J. Agric. Food Chem.*, 20(4):856-861 (1972).

McMahon, P.B., F.H. Chapelle, and M.J. Jagucki. "Atrazine Mineralization Potential of Alluvial-Aquifer Sediments under Aerobic Conditions," *Environ.*

Sci. Technol., 26(8):1556–1559 (1992).

Medley, D.R. and E.L. Stover. "Effects of Ozone on the Biodegradability of Biorefractory Pollutants," *J. Water Pollut. Control Fed.*, 55(5):489–494 (1983).

Meikle, R.W., N.H. Kurihara, and D.H. DeVries. "The Photocomposition Rates in Dilute Aqueous Solution and on a Surface, and the Volatilization Rate from a Surface," *Arch. Environ. Contam. Toxicol.*, 12(2):189–193 (1983).

Meikle, R.W., E.A. Williams, and C.T. Redemann. "Metabolism of Tordon Herbicide (4-Amino-3,5,6-trichloropicolinic Acid) in Cotton and Decomposition in Soil," *J. Agric. Food Chem.*, 14(4):384–387 (1966).

Meikle, R.W. and C.R. Youngson. "The Hydrolysis Rate of Chlorpyrifos, *O,O*-Diethyl *O*-(3,5,6-trichloro-2-pyridyl) Phosphorothioate, and Its Dimethyl Analog, Chloropyrifos-Methyl in Dilute Aqueous Solution," *Arch. Environ. Contam. Toxicol.*, 7(1):13–22 (1978).

Meikle, R.W., C.R. Youngson, R.T. Hedlund, C.A.I. Goring, and W.W. Addington. "Decomposition of Picloram by Soil Microorganisms: A Proposed Reaction Sequence," *Weed Sci.*, 22:263–268 (1974).

Menn, J.J. and G.G. Still. "Metabolism of Insecticides and Herbicides in Higher Plants," *CRC Crit. Rev. Toxicol.*, 5:1–21 (1977).

Menzer, R.E. and L.P. Dittman. "Residues in Spinach Grown in Disulfoton and Phorate-Treated Soil," *J. Econ. Entomol.*, 61(1):225–229 (1968).

Menzer, R.E., E.L. Fontanilla, and L.P. Dittman. "Degradation of Disulfoton and Phorate in Soil Influenced by Environmental Factors and Soil Type," *Bull. Environ. Contam. Toxicol.*, 5(1):1–5 (1970).

Metcalf, R.L. "A Century of DDT," *J. Agric. Food Chem.*, 21(4):511–519 (1973).

Metcalf, R.L., T.R. Fukuto, C. Collins, K. Borck, S. Abd El-Aziz, R. Munoz, and C.C. Cassil. "Metabolism of 2,2-Dimethyl-2,3-dihydrobenzofuranyl-7 *N*-Methylcarbamate (Furadan) in Plants, Insects, and Mammals," *J. Agric. Food Chem.*, 16(2):300–311 (1968).

Metcalf, R.L., T.R. Fukuto, C. Collins, K. Borck, J. Burk, H.T. Reynolds, and M.F. Osman. "Metabolism of 2-Methyl-(2-methylthio)propionaldehyde *O*-(Methylcarbamoyl)oxime in Plant and Insect," *J. Agric. Food Chem.*, 14(6):579–584 (1966).

Metcalf, R.L., T.R. Fukuto, and R.B. March. "Plant Metabolism of Dithiosystox and Thimet," *J. Econ. Entomol.*, 50:338–345 (1957).

Metcalf, R.L., G.K. Sangha, and I.P. Kapoor. "Model Ecosystem for the Evaluation of Pesticide Biodegradability and Ecological Magnification," *Environ. Sci. Technol.*, 5(8):709–713 (1971).

Mick, D.L. and P.A. Dahm. "Metabolism of Parathion by Two Species of *Rhizobium*," *J. Econ. Entomol.*, 63:1155–1159 (1970).

Mikesell, M.D. and S.A. Boyd. "Reductive Dechlorination of the Pesticides 2,4-D, 2,4,5-T, and Pentachlorophenol in Anaerobic Sludges," *J. Environ. Qual.*, 14(3):337–340 (1985).

Milano, J.C., A. Guibourg, and J.L. Vernet. "Non Biological Evolution, in Water, of Some Three- and Four-Carbon Atoms Organohalogenated Compounds: Hydrolysis and Photolysis," *Water Res.*, 22(12):1553-1562 (1988).

Miles, C.J. "Degradation Products of Sulfur-Containing Pesticides in Soil and Water" in *Pesticide Transformation Products. Fate and Significance in the Environment*, ACS Symposium Series 459, Somasundaram, L. and J.R. Coats, Eds. (Washington, DC: American Chemical Society, 1991), pp. 61-74.

Miles, C.J. "Degradation of Aldicarb, Aldicarb Sulfoxide, and Aldicarb Sulfone in Chlorinated Water," *Environ. Sci. Technol.*, 25(10):1774-1779 (1991a).

Miles, C.J. and J.J. Delfino. "Fate of Aldicarb, Aldicarb Sulfoxide, and Aldicarb Sulfone in Floridan Groundwater," *J. Agric. Food Chem.*, 33(3):455-460 (1985).

Miles, C.J. and S. Takashima. "Fate of Malathion and O,O,S-Trimethyl Phosphorothioate By-Product in Hawaiian Soil and Water," *Arch. Environ. Contam. Toxicol.*, 20(3):325-329 (1991).

Miles, C.J., M.L. Trehy, and R.A. Yost. "Degradation of N-Methylcarbamate and Carbamoyl Oxime Pesticides in Chlorinated Water," *Bull. Environ. Contam. Toxicol.*, 41(6):838-843 (1988).

Miles, J.R.W. and P. Moy. "Degradation of Endosulfan and Its Metabolites by a Mixed Culture of Soil Microorganisms," *Bull. Environ. Contam. Toxicol.*, 23(1/2):13-19 (1979).

Miles, J.R.W., C.M. Tu, and C.R. Harris. "Metabolism of Heptachlor and Its Degradation Products by Soil Microorganisms," *J. Econ. Entomol.*, 62:1334 (1969).

Miles, J.R.W., C.M. Tu, and C.R. Harris. "Persistence of Eight Organophosphorus Insecticides in Sterile and Nonsterile Mineral and Organic Soils," *Bull. Environ. Contam. Toxicol.*, 23(3):312-318 (1979).

Miller, G.C. and R.G. Zepp. "Photoreactivity of Aquatic Pollutants Sorbed on Suspended Sediments," *Environ. Sci. Technol.*, 13(7):860-863 (1979).

Minero, C., E. Pramauro, E. Pelizzetti, M. Dolci, and A. Marchesini. "Photosensitized Transformations of Atrazine under Simulated Sunlight in Aqueous Humic Acid Solution," *Chemosphere*, 24(11):1597-1606 (1992).

Miyazaki, S., G.M. Boush, and F. Matsumura. "Metabolism of ^{14}C-Chlorobenzilate and ^{14}C-Chloropropylate by *Rhodotorula gracilis*," *Appl. Microbiol.*, 18(6):972-976 (1969).

Miyazaki, S., G.M. Boush, and F. Matsumura. "Microbial Degradation of Chlorobenzilate (Ethyl 4,4'-Dichlorobenzilate) and Chloropropylate (Isopropyl 4,4'-Dichlorobenzilate)," *J. Agric. Food Chem.*, 18(1):87-91 (1970).

Miyazaki, S., H.C. Sikka, and R.S. Lynch. "Metabolism of Dichlobenil by Microorganisms in the Aquatic Environment," *J. Agric. Food Chem.*, 23(3):365-368 (1975).

Moilanen, K.W. and D.G. Crosby. "Photodecomposition of 3'4'-Dichloropropionanilide (Propanil)," *J. Agric. Food Chem.*, 20(5):950-953 (1972).

Moilanen, K.W. and D.G. Crosby. "The Photodecomposition of Bromacil," *Arch. Environ. Contam. Toxicol.*, 2(1):3-8 (1974).

Moilanen, K.W., D.G. Crosby, J.R. Humphrey, and J.W. Giles. "Vapor-Phase Photodecomposition of Chloropicrin (Trichloronitromethane)," *Tetrahedron*, 34(22):3345-3349 (1978).

Montgomery, J.H. *Groundwater Chemicals Desk Reference - Volume 2* (Chelsea, MI: Lewis Publishers, Inc., 1991), 944 p.

Montgomery, J.H. and L.M. Welkom. *Groundwater Chemicals Desk Reference* (Chelsea, MI: Lewis Publishers, Inc., 1990), 640 p.

Montgomery, M.L. and V.H. Freed. "Metabolism of Triazine Herbicides by Plants," *J. Agric. Food Chem.*, 12(1):11-14 (1964).

Moore, J.W. and S. Ramamoorthy. *Organic Chemicals in Natural Waters - Applied Monitoring and Impact Assessment* (New York: Springer-Verlag, Inc., 1984), 289 p.

Morrison, R.T. and R.N. Boyd. *Organic Chemistry* (Boston, MA: Allyn & Bacon, Inc., 1971), 1258 p.

Mosher, D.R. and A.M. Kadoum. "Effects of Four Lights on Malathion Residues on Glass Beads, Sorghum Grain, and Wheat Grain," *J. Econ. Entomol.*, 65:847-850 (1972).

Mosier, A.R. and W.D. Guenzi. "Picloram Photolytic Decomposition," *J. Agric. Food Chem.*, 21(5):835-837 (1973).

Mosier, A.R., W.D. Guenzi, and L.L. Miller. "Photochemical Decomposition of DDT by a Free Radical Mechanism," *Science (Washington, DC)*, 164(3883):1083-1085 (1969).

Mostafa, I.Y., M.R.E. Bahig, I.M.I. Fakhr, and Y. Adam. "Metabolism of Organophosphorus Pesticides. XIV. Malathion Breakdown by Soil Fungi," *Z. Naturforsch. B.*, 27:1115-1116 (1972).

Muir, D.C.G. "Dissipation and Transformations in Water and Sediment" in *Environmental Chemistry of Herbicides. Volume II*, Grover, R. and A.J. Cessna, Eds. (Boca Raton, FL: CRC Press, Inc., 1991), pp. 1-87.

Muir, D.C.G. and B.E. Baker. "Detection of Triazine Herbicides and Their Degradation Products in Tile-Drain Water from Fields under Corn (Maize) Production," *J. Agric. Food Chem.*, 21(1):122-125 (1976).

Muir, D.C.G. and B.E. Baker. "The Disappearance and Movement of Three Triazine Herbicides and Several of Their Degradation Products in Soil under Field Conditions," *Weed Res.*, 18:111-120 (1978).

Mulla, M.S., L.S. Mian, and J.A. Kawecki. "Distribution, Transport and Fate of the Insecticides Malathion and Parathion in the Environment," *Residue Rev.*, 81:116-125 (1981).

Munnecke, D.M. and D.P.H. Hsieh. "Pathways of Microbial Metabolism of Parathion," *Appl. Environ. Microbiol.*, 31(1):63-69 (1976).

Munshi, H.B., K.V.S. Rama Rao, and R. M. Iyer. "Characterization of Products of Ozonolysis of Acrylonitrile in Liquid Phase," *Atmos. Environ.*, 23(9):1945-1948 (1989).

Munshi, H.B., K.V.S. Rama Rao, and R.M. Iyer. "Rate Constants of the Reactions of Ozone with Nitriles, Acrylates and Terpenes in Gas Phase," *Atmos. Environ.*, 23(9):1971-1976 (1989a).

Murray, D.S., P.W. Santelman, and J.M. Davidson. "Comparative Adsorption, Desorption, and Mobility of Dipropetryn and Prometryn in Soil," *J. Agric. Food Chem.*, 23(3):578-582 (1975).

Murthy, N.B.K., D.D. Kaufman, and G.F. Fries. "Degradation of Pentachlorophenol (PCP) in Aerobic and Anaerobic Soil," *J. Environ. Sci. Health*, B14(1):1-14 (1979).

Musoke, G.M.S., D.J. Roberts, and M. Cooke. "Heterogeneous Hydrodechlorination of Chlordan," *Bull. Environ. Contam. Toxicol.*, 28(4):467-472 (1982).

Nair, D.R. and J.L. Schnoor. "Effect of Two Electron Acceptors on Atrazine Mineralization Rates in Soil," *Environ. Sci. Technol.*, 26(11):2298-2300 (1992).

Nakajima, S., N. Nakato, and T. Tani. "Microbial Transformation of 2,4-D and Its Analog," *Chem. Pharm. Bull.*, 21:671-673 (1973).

Neudorf, S. and M.A.Q. Khan. "Pick Up and Metabolism of DDT, Dieldrin, and Photodieldrin by a Fresh Water Alga (*Ankistrodesmus amalloides*) and a Microcrustacean (*Daphnia pulex*)," *Bull. Environ. Contam. Toxicol.*, 13:443-450 (1975).

Newland, L.W., G. Chesters, and G.B. Lee. "Degradation of γ-BHC in Simulated Lake Impoundments as Affected by Aeration," *J. Water Pollut. Control Fed.*, 41(5):R174-R188 (1969).

Newton, M., K.M. Howard, B.R. Kelpsas, R. Danhaus, C.M. Lottman, and S. Dubelman. "Fate of Glyphosate in an Oregon Forest Ecosystem," *J. Agric. Food Chem.*, 32(5):1144-1151 (1984).

Nilles, G.P. and M.J. Zabik. "Photochemistry of Bioactive Compounds. Multiphase Photodegradation and Mass Spectral Analysis of Basagran," *J. Agric. Food Chem.*, 23(3):410-415 (1975).

Niimi, A.J. "Solubility of Organic Chemicals in Octanol, Triolein and Cod Liver Oil and Relationships Between Solubility and Partition Coefficients," *Water Res.*, 25(12):1515-1521 (1991).

Nkedi-Kizza, P., P.S.C. Rao, and J.W. Johnson. "Adsorption of Diuron and 2,4,5-T on Soil Particle-Size Separates," *J. Environ. Qual.*, 12(2):195-197 (1983).

Novick, N.J., R. Mukherjee, and M. Alexander. "Metabolism of Alachlor and Propachlor in Suspensions of Pretreated Soils and in Samples from Ground Water Aquifers," *J. Agric. Food Chem.*, 34(4):721-725 (1986).

Obien, S.R. and R.E. Green. "Degradation of Atrazine in Four Hawaiian Soils," *Weed Sci.*, 17(4):509-514 (1969).

O'Connell, K.M., E.J. Breaux, and R.T. Fraley. "Different Rates of Metabolism of Two Chloroacetanilide Herbicides in Pioneer 3220 Corn," *Plant Physiol.*, 86:359-363 (1988).

O'Connor, G.A. and J.U. Anderson. "Soil Factors Affecting the Adsorption of 2,4,5-T," *Soil Sci. Soc. Am. Proc.*, 38:433-436 (1974).

Odeyemi, O. and M. Alexander. "Resistance of *Rhizobium* Strains to Phygon,

Spergon, and Thiram," *Appl. Environ. Microbiol.*, 33(4):784-790 (1977).

Otto, S., P. Beutel, N. Decker, and R. Huber. "Investigations in the Degradation of Bentazon in Plant and Soil," *Adv. Pestic. Sci.*, 3:551-556 (1978).

Ou, L.-T., K.S.V. Edvarsson, and P.S.C. Rao. "Aerobic and Anaerobic Degradation of Aldicarb in Soils," *J. Agric. Food Chem.*, 33(1):72-78 (1985).

Ou, L.-T., D.H. Gancarz, W.B. Wheeler, P.S.C. Rao, and J.M. Davidson. "Influence of Soil Temperature and Soil Moisture on Degradation and Metabolism of Carbofuran in Soils," *J. Environ. Qual.*, 11(2):293-298 (1982).

Ou, L.-T. and P.S.C. Rao. "Degradation and Metabolism of Oxamyl and Phenamiphos in Soils," *J. Environ. Sci. Health*, B21(1):25-40 (1986).

Owen, W.J. and B. Donzel. "Oxidative Degradation of Chlortoluron, Propiconazole, and Metalaxyl in Suspension Cultures of Various Crop Plants," *Pestic. Biochem. Physiol.*, 26:75-89 (1986).

Pape, B.E. and M.J. Zabik. "Photochemistry of Bioactive Compounds. Photochemistry of Selected 2-Chloro- and 2-Methylthio-4,6-di(alkylamino)-s-Triazine Herbicides," *J. Agric. Food Chem.*, 18(2):202-207 (1970).

Pape, B.E. and M.J. Zabik. "Photochemistry of Bioactive Compounds. Solution-Phase Photochemistry of Symmetrical Triazines," *J. Agric. Food Chem.*, 20(2):316-320 (1972).

Pardue, J.R., E.A. Hansen, R.P. Barron, and J.-Y.T. Chen. "Diazinon Residues on Field-Sprayed Kale. Hydroxydiazinon-A New Alteration Product of Diazinon," *J. Agric. Food Chem.*, 18(3):405-408 (1970).

Paris, D.F. and D.L. Lewis. "Chemical and Microbial Degradation of Ten Selected Pesticides in Aquatic Systems," *Residue Rev.*, 45:95-124 (1973).

Paris, D.F., D.L. Lewis, and N.L. Wolfe. "Rates of Degradation of Malathion by Bacteria Isolated from Aquatic System," *Environ. Sci. Technol.*, 9(2):135-138 (1975).

Parlar, H. "Photoinduced Reactions of Two Toxaphene Compounds in Aqueous Medium and Adsorbed on Silica Gel," *Chemosphere*, 17(11):2141-2150 (1988).

Parr, J.F. and S. Smith. "Degradation of Trifluralin under Laboratory Conditions and Soil Anaerobiosis," *Soil Sci.*, 115(1):55-63 (1973).

Parr, J.F. and S. Smith. "Degradation of DDT in an Everglades Muck as Affected by Lime, Ferrous Ion, and Anaerobiosis," *Soil Sci.*, 118(1):45-52 (1974).

Patil, K.C. and F. Matsumura. "Degradation of Endrin, Aldrin, and DDT by Soil Microorganisms," *Appl. Microbiol.*, 19:879-881 (1970).

Patil, K.C., F. Matsumura, and G.M. Boush. "Metabolic Transformation of DDT, Dieldrin, Aldrin, and Endrin by Marine Microorganisms," *Environ. Sci. Technol.*, 6(7):629-632 (1972).

Patumi, M., C. Marucchini, M. Businelli, and F. Tafuri. "Effetto di Ripetuti Trattamenti con Atrazina Sulla sua Persistenza in Terreni Destinati alla Monosuccessione Di Mais," *Agrochimica*, 25(2):162-167 (1981).

Peek, D.C. and A.P. Appleby. "Phytotoxicity, Adsorption, and Mobility of Metribuzin and Its Ethylthio Analog as Influenced by Soil Properties," *Weed*

Sci., 37(3):419-423 (1989).

Pelizzetti, E., V. Maurino, C. Minero, V. Carlin, E. Pramauro, O. Zerbinati, and M.L. Tosato. "Photocatalytic Degradation of Atrazine and Other s-Triazine Herbicides," *Environ. Sci. Technol.*, 24(10):1559-1565 (1990).

Pelizzetti, E., C. Minero, V. Carlin, M. Vincenti, E. Pramauro, and M. Dolci. "Identification of Photocatalytic Degradation Pathways of 2-Cl-s-Triazine Herbicides and Detection of Their Decomposition Intermediates," *Chemosphere*, 24(7):891-910 (1992).

Pereira, W.E. and C.E. Rostad. "Occurrence, Distributions, and Transport of Herbicides and Their Degradation Products in the Lower Mississippi River and Its Tributaries," *Environ. Sci. Technol.*, 24(9):1400-1406 (1990).

Perscheid, M., H. Schlueter, and K. Ballschmiter. "Aerober Abbau von Endosulfan Durch Bodenmikroorganismen," *Zeitschrift Naturforsch*, 28:761-763 (1973).

Petrier, C., M. Micolle, G. Merlin, J.-L. Luche, and G. Reverdy. "Characteristics of Pentachlorophenate Degradation in Aqueous Solution by Means of Ultrasound," *Environ. Sci. Technol.*, 26(8):1639-1642 (1992).

Pettigrew, C.A., M.J.B. Paynter, and N.D. Camper. Anaerobic Microbial Degradation of the Herbicide Propanil," *Soil Biol. Biochem.*, 17:815-818 (1985).

Peyton, T.O., R.V. Steel, and W.R. Mabey. "Carbon Disulfide, Carbonyl Sulfide: Literature Review and Environmental Assessment," U.S. EPA Report PB-257-947/2 (1976), 64 p.

Pfaender, F.K. and M. Alexander. "Extensive Microbial Degradation of DDT *in Vitro* and DDT Metabolism by Natural Communities," *J. Agric. Food Chem.*, 20(4):842-846 (1972).

Pfaender, F.K. and M. Alexander. "Effect of Nutrient Additions on the Apparent Cometabolism of DDT," *J. Agric. Food Chem.*, 21(3):397-399 (1973).

Pickering, Q.H., C. Henderson, and A.E. Lemke. "The Toxicity of Organic Phosphorus Insecticides to Different Species of Warmwater Fishes," *Trans. Am. Fish Soc.*, 91(2):175-184 (1962).

Pignatello, J.J. "Microbial Degradation of 1,2-Dibromoethane in Shallow Aquifer Materials," *J. Environ. Qual.*, 16(4):307-312 (1987).

Pillai, C.G.P., J.D. Weete, and D.E. Davis. "Metabolism of Atrazine by *Spartina alterniflora*. 1. Chloroform-Soluble Metabolites," *J. Agric. Food Chem.*, 25(4):852-855 (1977).

Plimmer, J.R. "The Photochemistry of Halogenated Herbicides," *Residue Rev.*, 33:47-74 (1970).

Plimmer, J.R. "Photolysis of TCDD and Trifluralin on Silica and Soil," *Bull. Environ. Contam. Toxicol.*, 20(1):87-92 (1978).

Plimmer, J.R. and B.E. Hummer. "Photolysis of Amiben (3-Amino-2,5-dichlorobenzoic Acid) and Its Methyl Ester," *J. Agric. Food Chem.*, 17(1):83-85 (1969).

Plimmer, J.R., P.C. Kearney, H. Chisaka, J.B. Yount, and U.I. Klingebiel. "1,3-Bis(3,4-dichlorophenyl)triazene from Propanil in Soils," *J. Agric. Food Chem.*,

18(5):859-861 (1970).

Plimmer, J.R., P.C. Kearney, D.D. Kaufman, and F.S. Guardia. "Amitrole Decomposition by Free Radical-Generating Systems and by Soils," *J. Agric. Food Chem.*, 15(6):996-999 (1967).

Plimmer, J.R., P.C. Kearney, and D.W. Von Endt. "Mechanism of Conversion of DDT to DDD by *Aerobacter aerogenes*," *J. Agric. Food Chem.*, 16(4):594-597 (1968).

Plimmer, J.R. and U.J. Klingebiel. "Photolysis of Hexachlorobenzene," *J. Agric. Food Chem.*, 24(4):721-723 (1976).

Plimmer, J.R., U.J. Klingebiel, and B.E. Hummer. "Photooxidation of DDT and DDE," *Science (Washington, DC)*, 167(3914):67-69 (1970).

Pogány, E., P.R. Wallnöfer, W. Ziegler, and W. Mücke. "Metabolism of *o*-Nitroaniline and Di-*n*-butyl Phthalate in Cell Suspension Cultures of Tomatoes," *Chemosphere*, 21(4/5):557-562 (1990).

Pollero, R. and S.C. dePollero. "Degradation of DDT by a Soil Amoeba," *Bull. Environ. Contam. Toxicol.*, 19(3):345-350 (1978).

Probst, G.W., T. Golab, R.J. Herberg, F.J. Holzer, S.J. Parka, C. van der Schans, and J.B. Tepe. "Fate of Trifluralin in Soils and Plants," *J. Agric. Food Chem.*, 15(4):592-599 (1967).

Probst, G.W. and J.B. Tepe. "Trifluralin and Related Compounds" in *Degradation of Herbicides*, Kearney, P.C. and D.D. Kaufman, Eds., (New York: Marcel Dekker, Inc., 1969), pp. 255-282.

Putnam, T.B., D.D. Bills, and L.M. Libbey. "Identification of Endosulfan Based on the Products of Laboratory Photolysis," *Bull. Environ. Contam. Toxicol.*, 13(6):662-665 (1975).

Que Hee, S.S. and R.G. Sutherland. *The Phenoxyalkanoic Herbicides. Vol. 1. Chemistry, Analysis, and Environmental Pollution* (Boca Raton, FL: CRC Press, Inc, 1981), 321 p.

Quirke, J.M.E., A.S.M. Marei, and G. Eglinton. "The Degradation of DDT and Its Degradative Products by Reduced Iron (III) Porphyrins and Ammonia," *Chemosphere*, 8(3):151-155 (1979).

Racke, K.D. and J.R. Coats. "Enhanced Degradation of Isofenphos by Soil Microorganisms," *J. Agric. Food Chem.*, 35(1):94-99 (1987).

Racke, K.D. and J.R. Coats. "Enhanced Degradation and the Comparative Fate of Carbamate Insecticides in Soil," *J. Agric. Food Chem.*, 36(5):1067-1072 (1988).

Racke, K.D., J.R. Coats, and K.R. Titus. "Degradation of Chlorpyrifos and Its Hydrolysis Product, 3,5,6-Trichloro-2-pyridinol, in Soil," *J. Environ. Sci. Health*, B23(6):527-539 (1988).

Raghu, K. and I.C. MacRae. "Biodegradation of the Gamma Isomer of Benzene Hexachloride in Submerged Soils," *Science (Washington, DC)*, 154(3746):263-264 (1966).

Raha, P. and A.K. Das. "Photodegradation of Carbofuran," *Chemosphere*, 21(1/2):99-106 (1990).

Rajagopal, B.S., G.P. Brahmaprakash, B.R. Reddy, U.D. Singh, and N. Sethunathan. "Effect and Persistence of Selected Carbamate Pesticides in Soil," *Residue Rev.*, 93:1–199 (1984).

Rajagopal, B.S., K. Chendrayan, B.R. Reddy, and N. Sethunathan. "Persistence of Carbaryl in Flooded Soils and Its Degradation by Soil Enrichment Cultures," *Plant Soil*, 73(1):35–45 (1973).

Rajagopal, B.S., S. Panda, and N. Sethunathan. "Accelerated Degradation of Carbaryl and Carbofuran in a Flooded Soil Pretreated with Hydrolysis Products, 1-Naphthol and Carbofuran Phenol," *Bull. Environ. Contam. Toxicol.*, 36(6):827–832 (1986).

Rajagopal, B.S., V.R. Rao, G. Nagendrappa, and N. Sethunathan. "Metabolism of Carbaryl and Carbofuran by Soil-Enrichment and Bacterial Cultures," *Can. J. Microbiol.*, 30(12):1458–1466 (1984a).

Ralls, J.W., D.R. Gilmore, and A. Cortes. "Fate of Radioactive O,O-Diethyl O-(2-Isopropyl-4-methylpyrimidin-6-yl) Phosphorothioate on Field-Grown Experimental Crops," *J. Agric. Food Chem.*, 14(4):387–293 (1966).

Ramakrishna, C., T.K.S. Gowda, and N. Sethunathan. "Effect of Benomyl and Its Hydrolysis Products, MBC and AB, on Nitrification in a Flooded Soil," *Bull. Environ. Contam. Toxicol.*, 21(3):328–333 (1979).

Ramanand, K., S. Panda, M. Sharmila, T.K. Adhya, and N. Sethunathan. "Development and Acclimatization of Carbofuran-Degrading Soil Enrichment Cultures at Different Temperatures," *J. Agric. Food Chem.*, 36(1):200–205 (1988).

Ramanand, K., M. Sharmila, and N. Sethunathan. "Mineralization of Carbofuran by a Soil Bacterium," *Appl. Environ. Microbiol.*, 54(8):2129–2133 (1988a).

Randall, T.L. and P.V. Knopp. "Detoxification of Specific Organic Substances by Wet Oxidation," *J. Water Pollut. Control Fed.*, 52(8):2117–2130 (1980).

Rao, P.S.C. and J.M. Davidson. "Adsorption and Movement of Selected Pesticides at High Concentrations in Soils," *Water Res.*, 13(4):375–380 (1979).

Rao, P.S.C. and J.M. Davidson. "Retention and Transformation of Selected Pesticides and Phosphorus in Soil-Water Systems: A Critical Review," Office of Research and Development, U.S. EPA Report-600/3-82-060 (1982), 321 p.

Reddy, B.R. and N. Sethunathan. "Mineralization of Parathion in the Rice Rhizosphere," *Appl. Environ. Microbiol.*, 45(3):826–829 (1983).

Reddy, K.N., M. Singh, and A.K. Alva. "Sorption and Leaching of Bromacil and Simazine in Florida Flatwoods Soils," *Bull. Environ. Contam. Toxicol.*, 48(5):662–670 (1992).

Redemann, C.T., R.W. Meikle, P. Hamilton, V.S. Banks, and C.R. Youngson. "The Fate of 4-Amino-3,5,6-Trichloropicolinic Acid in Spring Wheat and Soil," *Bull. Environ. Contam. Toxicol.*, 3(2):80–96 (1968).

Reduker, S., C.G. Uchrin, and G. Winnett. "Characteristics of the Sorption of Chlorothalonil and Azinphos-Methyl to a Soil from a Commercial Cranberry Bog," *Bull. Environ. Contam. Toxicol.*, 41(5):633–641 (1988).

"Registry of Toxic Effects of Chemical Substances," U.S. Department of Health and Human Services, National Institute for Occupational Safety and Health (1985), 2050 p.

Reinert, K.H. M.L. Hinman, and J.H. Rodgers. "Fate of Endothall During the Pay Mayse Lake, Texas Aquatic Plant Management Program," *Arch. Environ. Contam. Toxicol.*, 17:195-199 (1988).

Reinert, K.H. and J.H. Rodgers. "Fate and Persistence of Aquatic Herbicides," *Rev. Environ. Contam. Toxicol.*, 98:61-98 (1987).

Rejtö, M., S. Saltzman, and A.J. Acher. "Photodecomposition of Propachlor," *J. Agric. Food Chem.*, 32(2):226-230 (1984).

Rejtö, M., S. Saltzman, A.J. Acher, and L. Muszkat. "Identification of Sensitized Photooxidation Products of s-Triazine Herbicides in Water," *J. Agric. Food Chem.*, 31(1):138-142 (1983).

Reuber, M.D. "Carcinogenicity of Lindane," *Environ. Res.*, 19:460-491 (1979).

Rhodes, R.C. "Metabolism of [2-^{14}C]Terbacil in Alfalfa," *J. Agric. Food Chem.*, 25(5):1066-1068 (1977).

Rhodes, R.C. "Studies with ^{14}C-Labeled Hexazinone in Water and Bluegill Sunfish," *J. Agric. Food Chem.*, 28(2):306-310 (1980).

Rhodes, R.C. "Soil Studies with ^{14}C-Labeled Hexazinone," *J. Agric. Food Chem.*, 28(2):311-315 (1980a).

Rhodes, R.C., I.J. Belasco, and H.L. Pease. "Determination and Mobility of Agrochemicals on Soils," *J. Agric. Food Chem.*, 18(3):524-528 (1970).

Rhodes, R.C. and J.D. Long. "Run-off and Mobility Studies on Benomyl in Soils and Turf," *Bull. Environ. Contam. Toxicol.*, 12(4):385-393 (1974).

Richard, Y. and L. Bréner. "Removal of Pesticides from Drinking Water by Ozone," in *Handbook of Ozone Technology and Applications, Volume II. Ozone for Drinking Water Treatment*, Rice, A.G. and A. Netzer, Eds. (Montvale, MA: Butterworth Publishers, 1984), pp. 77-97.

Rippen, G., M. Ilgenstein, and W. Klöpffer. "Screening of the Adsorption Behavior of New Chemicals: Natural Soils and Model Adsorbents," *Ecotoxicol. Environ. Saf.*, 6(3):236-245 (1982).

Roberts, H.A. and B.J. Wilson. "Adsorption of Chlorpropham by Different Soils," *Weed Res.*, 5(4):348-350 (1965).

Robinson, J., A. Richardson, B. Bush, and K.E. Elgar. "A Photo-Isomerization Product of Dieldrin," *Bull. Environ. Contam. Toxicol.*, 1(4):127-132 (1966).

Rosen, J.D. and W.F. Carey. "Preparation of the Photoisomers of Aldrin and Dieldrin," *J. Agric. Food Chem.*, 16(3):536-537 (1968).

Rosen, J.D., R.F. Strusz, and C.C. Still. "Photolysis of Phenylurea Herbicides," *J. Agric. Food Chem.*, 17(2):206-207 (1969).

Rosen, J.D. and D.J. Sutherland. "The Nature and Toxicity of the Photoconversion Products of Aldrin," *Bull. Environ. Contam. Toxicol.*, 2(1):1-9 (1967).

Rosen, J.D., J. Sutherland, and G.R. Lipton. "The Photochemical Isomerization of Dieldrin and Endrin and Effects on Toxicity," *Bull. Environ. Contam. Toxicol.*,

1(4):133–140 (1966).

Rosenfield, C. and W. Van Valkenburg. "Decomposition of (*O,O*-Dimethyl-*O*-2,4,5-trichlorophenyl) Phosphorothioate (Ronnel) Adsorbed on Bentonite and Other Clays," *J. Agric. Food Chem.*, 13(1):68–72 (1972).

Ross, R.D. and D.G. Crosby. "The Photooxidation of Aldrin in Water," *Chemosphere*, 4(5):277–282 (1975).

Ross R.D. and D.G. Crosby. "Photooxidant Activity in Natural Waters," *Environ. Toxicol. Chem.*, 4(6):773–778 (1985).

Rouchaud, J., C. Moons, and J.A. Meyer. "The Products of Metabolism of [^{14}-C]Triadimefon in the Grain and in the Straw of Ripe Barley," *Bull. Environ. Contam. Toxicol.*, 27(4):543–550 (1981).

Rouchaud, J., C. Moons, and J.A. Meyer. "Metabolism of ^{14}C Triadimefon in Barley Shoots," *Pestic. Sci.*, 13(2):169–176 (1982).

Rouchaud, J., P. Roucourt, and A. Vanachter. "Hydrolytic Biodegradation of Chlorothalonil in the Soil and Cabbage Crops," *Toxicol. Environ. Chem.*, 17:59–68 (1988).

Roy, D.N., S.K. Konar, D.A. Charles, J.C. Feng, R. Prasad, and R.A. Campbell. "Determination of Persistence, Movement, and Degradation of Hexazinone in Selected Canadian Boreal Forest Soils," *J. Agric. Food Chem.*, 37(2):443–447 (1989).

Rueppel, M.L., B.B. Brightwell, J. Schaefer, and J.T. Marvel. "Metabolism and Degradation of Glyphosate in Soil and Water," *J. Agric. Food Chem.*, 25(3):517–528 (1977).

Ruzo, L.O., J.K. Lee, and M.J. Zabik. "Solution-Phase Photodecomposition of Several Substituted Diphenyl Ether Herbicides," *J. Agric. Food Chem.*, 28(6):1289–1292 (1980).

Ruzo, L.O. and J.E. Casida. "Photochemistry of Thiocarbamate Herbicides: Oxidative and Free Radical Processes of Thiobencarb and Diallate," *J. Agric. Food Chem.*, 33(2):272–276 (1985).

Sahoo, A., S.K. Sahu, M. Sharmila, and N. Sethunathan. "Persistence of Carbamate Insecticides, Carbosulfan and Carbofuran in Soils as Influenced by Temperature and Microbial Activity," *Bull. Environ. Contam. Toxicol.*, 44(6):948–954 (1990).

Sahu, S.K., K.K. Patnaik, and N. Sethunathan. "Dehydrochlorination of δ-Isomer of Hexachlorocyclohexane by a Soil Bacterium, *Pseudomonas* sp.," *Bull. Environ. Contam. Toxicol.*, 48(2):265–268 (1992).

Sanborn, J.R., B.M. Francis, and R.L. Metcalf. "The Degradation of Selected Pesticides in Soil: A Review of the Published Literature," Office of Research and Development, U.S. EPA Report-600/9-77-022 (1977), 616 p.

Sanderman, W., H. Stockmann, and R. Casten. "Über die Pyrolyse des Pentachlorophenols," *Chem. Ber.*, 90:690–692 (1957).

Sanders, H.O. "Toxicities of Some Herbicides to Six Species of Freshwater Crustaceans," *J. Water Poll. Control Fed.*, 42(8):1544–1550 (1970).

Sanders, H.O. and O.B. Cope. "The Relative Toxicities of Several Pesticides to

Naiads of Three Species of Stoneflies," *Limnol. Oceanogr.*, 13(1):112-117 (1968).

Saravanja-Bozanic, V., S. Gäb, K. Hustert, and F. Korte. "Reacktionen von Aldrin, Chlorden, und 2,2-Dichlorobiphenyl mit $O(^3P)$)," *Chemosphere*, 6(1):21-26 (1977).

Sax, N.I. *Dangerous Properties of Industrial Materials* (New York: Van Nostrand Reinhold Co., 1984), 3124 p.

Sax, N.I. and R.J. Lewis, Sr., Eds. *Hazardous Chemicals Desk Reference* (New York: Van Nostrand Reinhold Co., 1987), 1084 p.

Schellenberg, K., C. Leuenberger, and R.P. Schwarzenbach. "Sorption of Chlorinated Phenols by Natural Sediments and Aquifer Materials," *Environ. Sci. Technol.*, 18(9):652-657 (1984).

Scheunert, I., D. Vockel, J. Schmitzer, and F. Korte. "Biomineralization Rates of ^{14}C-Labelled Organic Chemicals in Aerobic and Anaerobic Suspended Soil," *Chemosphere*, 16(5):1031-1041 (1987).

Schimmel, S.C., R.L. Garnas, J.M. Patrick, Jr., and J.C. Moore. "Acute Toxicity, Bioconcentration, and Persistence of AC 222,705, Benthiocarb, Chlorpyrifos, Fenvalerate, Methyl Parathion, and Permethrin in the Estuarine Environment," *J. Agric. Food Chem.*, 31(1):104-113 (1983).

Schliebe, K.A., O.C. Burnside, and T.L. Lavy. "Dissipation of Amiben," *Weeds*, 13:321-325 (1965).

Schumacher, H.G., H. Parlar, W. Klein, and F. Korte. "Beitrage zur Okologischen Chemie. LV. Photochemische Reaktion von Endosulfan," *Chemosphere*, 3(2):65-70 (1974).

Sethunathan, N. "Organic Matter and Parathion Degradation in Flooded Soil," *Soil Biol. Biochem.*, 5(5):641-644 (1973).

Sethunathan, N. and I.C. MacRae. "Persistence and Biodegradation of Diazinon in Submerged Soils," *J. Agric. Food Chem.*, 17(2):221-225 (1969).

Sethunathan, N. and M.D. Pathak. "Development of a Diazinon-Degrading Bacterium in Paddy Water after Repeated Applications of Diazinon," *Can. J. Microbiol.*, 17(5):699-702 (1971).

Sethunathan, N. and M.D. Pathak. "Increased Biological Hydrolysis of Diazinon after Repeated Application in Rice Paddies," *J. Agric. Food Chem.*, 20(3):586-589 (1972).

Sethunathan, N., R. Siddaramappa, K.P. Rajaram, S. Barik, and P.A. Wahid. "Parathion: Residues in Soil and Water," *Residue Rev.*, 68:91-122 (1977).

Sethunathan, N. and T. Yoshida. "Fate of Diazinon in Submerged Soil. Accumulation of Hydrolysis Product," *J. Agric. Food Chem.*, 17(6):1192-1195 (1969).

Sethunathan, N. and T. Yoshida. "A *Flavobacterium* sp. that Degrades Diazinon and Parathion," *Can. J. Microbiol.*, 19(5):873-875 (1973).

Sethunathan, N. and T. Yoshida. "Degradation of Chlorinated Hydrocarbons by *Clostridium* sp. Isolated from Lindane-Amended, Flooded Soil," *Plant Soil*, 38:663-666 (1973a).

Sharmila, M., K. Ramanand, and N. Sethunathan. "Effect of Yeast Extract on the Degradation of Organophosphorus Insecticides by Soil Enrichment and Bacterial Cultures," *Can. J. Microbiol.*, 35(12):1105-1110 (1989).

Sharom, M.S. and L.V. Edgington. "Mobility and Dissipation of Metalaxyl in Tobacco Soils," *Can. J. Plant Sci.*, 66:761-771 (1986).

Sharom, M.S., J.R.W. Miles, J.W. Harris, and F.L. McEwen. "Persistence of 12 Insecticides in Water," *Water Res.*, 14(8):1089-1093 (1980).

Sharom, M.S. and K.R. Solomon. "Adsorption-Desorption, Degradation, and Distribution of Permethrin in Aqueous Systems," *J. Agric. Food Chem.*, 29(6):1122-1125 (1981).

Shelton, D.R., S.A. Boyd, and J.M. Tiedje. "Anaerobic Biodegradation of Phthalic Acid Esters in Sludge," *Environ. Sci. Technol.*, 18(2):93-97 (1984).

Shevchenko, M.A., P.N. Taran, and P.V. Marchenko. "Technology of Water Treatment and Demineralization," *Soviet J. Water Chem. Technol.*, 4(4):53-71 (1982).

Shimabukuro, R.H. "Significance of Atrazine Dealkylation in Root and Shoot of Pea Plants," *J. Agric. Food Chem.*, 15(4):557-562 (1967).

Shin, Y.-O., J.J. Chodan, and A.R. Wolcott. "Adsorption of DDT by Soils, Soil Fractions, and Biological Materials," *J. Agric. Food Chem.*, 18(6):1129-1133 (1970).

Siddaramappa, R., K.P. Rajaram, and N. Sethunathan. "Degradation of Parathion by Bacteria Isolated from Flooded Soil," *Appl. Microbiol.*, 26(6):846-849 (1973).

Sikka, H.C. and J. Saxena. "Metabolism of Endothall by Aquatic Microorganisms," *J. Agric. Food Chem.*, 21(3):402-406 (1973).

Simsiman, G.V. and G. Chesters. "Persistence of Diquat in the Aquatic Environment," *Water Res.*, 10(2):105-112 (1976).

Singh, A.K. and P.K. Seth. "Degradation of Malathion by Microorganisms Isolated from Industrial Effluents," *Bull. Environ. Contam. Toxicol.*, 43(1):28-35 (1989).

Singh, N., P.A. Wahid, M.V.R. Murty, and N. Sethunathan. "Sorption-Desorption of Methyl Parathion, Fenitrothion, and Carbofuran in Soils," *J. Environ. Sci. Health*, B25(6):713-728 (1990).

Singh, R.P. and M. Chiba. "Solubility of Benomyl in Water at Different pHs and Its Conversion to Methyl 2-Benzimidazolecarbamate, 3-Butyl-2,4-dioxo[1,2-*a*]-*s*-triazinobenzimidazole, and 1-(2-Benzimidazolyl)-3-*n*-butylurea," *J. Agric. Food Chem.*, 33(1):63-67 (1985).

Singh, S.B. and G. Kulshrestha. "Microbial Degradation of Pendimethalin," *J. Environ. Sci. Health*, B26(3):309-321 (1991).

Singmaster, J.A., III. "Environmental Behavior of Hydrophobic Pollutants in Aqueous Solutions," PhD Thesis, University of California, Davis CA (1975).

Sirons, G.J., R. Frank, and T. Sawyer. "Residues of Atrazine, Cyanazine, and Their Phytotoxic Metabolites in a Clay Loam Soil," *J. Agric. Food Chem.*, 21(6):1016-1019 (1973).

Sisler, H.D. and C.E. Cox. "Effects of Tetramethyl Thiuram Disulfide on

Metabolism of *Fusarium roseum*," *Am. J. Bot.*, 41:338 (1954).

Sittig, M. *Handbook of Toxic and Hazardous Chemicals and Carcinogens* (Park Ridge, NJ: Noyes Publications, 1985), 950 p.

Skipper, H.D., C.M. Gilmour, and W.R. Furtick. "Microbial Versus Chemical Degradation of Atrazine in Soils," *Soil Sci. Soc. Am. Proc.*, 31:653-656 (1967).

Skipper, H.D. and V.V. Volk. "Biological and Chemical Degradation of Atrazine in Three Oregon Soils," *Weed Sci.*, 20(4):344-347 (1972).

Skurlatov, Y.I., R.G. Zepp, and G.L. Baughman. "Photolysis Rates of (2,4,5-Trichlorophenoxy)acetic Acid and 4-Amino-3,5,6-trichloropicolinic Acid in Natural Waters," *J. Agric. Food Chem.*, 31(5):1065-1071 (1983).

Slade, P. and A.E. Smith. "Photochemical Degradation of Diquat," *Nature (London)*, 213(5069):919-920 (1967).

Smelt, J.H., M. Leistra, N.W.H. Houx, and A. Dekker. "Conversion Rates of Aldicarb and its Oxidation Products in Soils. III. Aldicarb," *Pestic. Sci.*, 9(4):293-300 (1978).

Smith, A.E. "Breakdown of the Herbicide Dicamba and Its Degradation Product 3,6-Dichlorosalicylic Acid in Prairie Soils," *J. Agric. Food Chem.*, 22(4):601-605 (1974).

Smith, A.E. "Degradation of the Herbicide Diclofop-Methyl in Prairie Soils," *J. Agric. Food Chem.*, 25(4):893-898 (1977).

Smith, A.E. "Transformation of [^{14}C]Diclofop-Methyl in Small Field Plots," *J. Agric. Food Chem.*, 27(6):1145-1148 (1979).

Smith, A.E. "Identification of 2,4-Dichloroanisole and 2,4-Dichlorophenol as Soil Degradation Products of Ring-Labelled [^{14}C]2,4-D," *Bull. Environ. Contam. Toxicol.*, 34(2):150-157 (1985).

Smith, A.E. and J. Grove. "Photochemical Degradation of Diquat in Dilute Aqueous Solution and on Silica Gel," *J. Agric. Food Chem.*, 17(3):609-613 (1969).

Smith, J.G., S.-F. Lee, and A. Netzer. "Model Studies in Aqueous Chlorination: The Chlorination of Phenols in Dilute Aqueous Solutions," *Water Res.*, 10(11):985-990 (1976).

Smith, L.R. and J. Dragun. "Degradation of Volatile Chlorinated Aliphatic Priority Pollutants in Groundwater," *Environ. Int.*, 19(4):291-298 (1984).

Smith, R.V. and J.P. Rosazza. "Microbial Models of Mammalian Metabolism. Aromatic Hydroxylation," *Arch. Biochem. Biophys.*, 161(2):551-558 (1974).

Snider, E.H. and F.C. Alley. "Kinetics of the Chlorination of Biphenyl under Conditions of Waste Treatment Processes," *Environ. Sci. Technol.*, 13(10):1244-1248 (1979).

Soderquist, C.J., J.B. Bowers, and D.G. Crosby. "Dissipation of Molinate in a Rice Field," *J. Agric. Food Chem.*, 25(4):940-945 (1977).

Soderquist, C.J. and D.G. Crosby. "Dissipation of 4-Chloro-2-methylphenoxyacetic Acid (MCPA) in a Rice Field," *Pestic. Sci.*, 6(1):17-33 (1975).

Soderquist, C.J., D.G. Crosby, K.W. Moilanen, J.N. Seiber, and J.E. Woodrow. "Occurrence of Trifluralin and Its Photoproducts in Air," *J. Agric. Food Chem.*, 23(2):304-309 (1975).

Solon, J.M. and J.H. Nair. "The Effect of a Sublethal Concentration of LAS on the Acute Toxicity of Various Phosphate Pesticides to the Fathead Minnow *Pimephales promelas* Rafinesque," *Bull. Environ. Contam. Toxicol.*, 5(5):408-413 (1970).

Somasundaram, L. and J.R. Coats. "Interactions Between Pesticides and Their Major Degradation Products," in *Pesticide Transformation Products. Fate and Significance in the Environment*, ACS Symposium Series 459, Somasundaram, L. and J.R. Coats, Eds., (Washington, DC: American Chemical Society, 1991), pp. 162-171.

Somasundaram, L., J.R. Coats, and K.D. Racke. "Degradation of Pesticides in Soil as Influenced by the Presence of Hydrolysis Metabolites," *J. Environ. Sci. Health*, B24(5):457-478 (1989).

Somasundaram, L., J.R. Coats, K.D. Racke, and V.M. Shanbhag. "Mobility of Pesticides and Their Metabolites in Soil," *Environ. Toxicol. Chem.*, 10(2):185-194 (1991).

Somich, C.J., P.C. Kearney, M.T. Muldoon, and S. Elsasser. "Enhanced Soil Degradation of Alachlor by Treatment with Ultraviolet Light and Ozone," *J. Agric. Food Chem.*, 36(6):1322-1326 (1988).

Sonobe, H., L.R. Kamps, E.P. Mazzola, and J.A.G. Roach. "Isolation and Identification of a New Conjugated Carbofuran Metabolite in Carrots: Angelic Acid Ester of 3-Hydroxycarbofuran," *J. Agric. Food Chem.*, 29(6):1125-1129 (1981).

Spencer, W.F., J.D. Adams, T.D. Shoup, and R.C. Spear. "Conversion of Parathion to Paraoxon on Soil Dusts and Clay Minerals as Affected by Ozone and UV Light," *J. Agric. Food Chem.*, 28(2):366-371 (1980).

Stalker, D.M., K.E. McBride, and L.D. Malyj. "Herbicide Resistance in Transgenic Plants Expressing a Bacterial Detoxification Gene," *Science (Washington, DC)*, 242(4877):419-423 (1988).

Steenson, T.I. and N. Walker. "The Pathway of Breakdown of 2,4-Dichloro- and 4-Chloro-2-methylphenoxyacetic Acid by Bacteria," *J. Gen. Microbiol.*, 16:146-155 (1957).

Steinwandter, H. "Experiments on Lindane Metabolism in Plants III. Formation of β-HCH," *Bull. Environ. Contam. Toxicol.*, 20(4):535-536 (1978).

Steller, W.A. and W.W. Brand. "Analysis of Dimethoate-Treated Grapes for the *N*-Hydroxymethyl and De-*N*-Methyl Metabolites and for Their Sugar Adducts," *J. Agric. Food Chem.*, 22(3):445-449 (1974).

Stepp, T.D., N.D. Camper, and M.J.B. Paynter. "Anaerobic Microbial Degradation of Selected 2,4-Dihalogenated Aromatic Compounds," *Pestic. Biochem. Physiol.*, 23:256-260 (1985).

Stewart, D.K.R. and K.G. Cairns. "Endosulfan Persistence in Soil and Uptake by

Potato Tubers," *J. Agric. Food Chem.*, 22(6):984-986 (1974).

Still, G.G. "Metabolism of 3,4-Dichloropropionanilide in Plants: The Metabolic Fate of the 3,4-Dichloroaniline Moiety," *Science (Washington, DC)*, 159(3818):992-993 (1968).

Stojanovic, B.J., M.V. Kennedy, and F.L. Shuman, Jr. "Mild Thermal Degradation of Pesticides," *J. Environ. Qual.*, 1:397-401 (1972).

Struif, B., L. Weil, and K.-E. Quentin. "Verhalten Herbizider Phenoxyalkan-carbonäuren bei der Wasseraufbereitung mit Ozon," *Zeit. fur Wasser und Abwasser-Forschung*, 3/4:118-127 (1978).

Su, F., J.G. Calvert, and J.H. Shaw. "Mechanism of the Photooxidation of Gaseous Formaldehyde," *J. Phys. Chem.*, 83(25):3185-3191 (1979).

Subba-Rao, R.V. and M. Alexander. "Effect of DDT Metabolites on Soil Respiration and on an Aquatic Alga," *Bull. Environ. Contam. Toxicol.*, 25(2):215-220 (1980).

Sud, R.K., K.A. Sud, and K.G. Gupta. "Degradation of Sevin (1-Naphthyl *N*-methylcarbamate) by *Achromobacter* sp.," *Arch. Microbiol.*, 87:353-358 (1972).

Sudhakar-Barik and N. Sethunathan. "Biological Hydrolysis of Parathion in Natural Ecosystems," *J. Environ. Qual.*, 7(3):346-348 (1978).

Sudhakar-Barik, R. Siddaramappa, and N. Sethunathan. "Metabolism of Nitrophenols by Bacteria Isolated from Parathion-Amended Flooded Soil," *Antonie van Leeuwenhoek*, 42(4):461-470 (1976).

Sudhakar-Barik, R. Siddaramappa, P.A. Wahid, and N. Sethunathan. "Conversion of *p*-Nitrophenol to 4-Nitrocatechol by a *Pseudomonas* sp.," *Antonie van Leeuwenhoek*, 44(2):171-176 (1978a).

Sudhakar-Barik, P.A. Wahid, C. Ramakrishna, and N. Sethunathan. "A Change in the Degradation Pathway of Parathion after Repeated Applications to Flooded Soil," *J. Agric. Food Chem.*, 27(6):1391-1392 (1979).

Suffet, I.H., S.D. Faust, and W.F. Carey. "Gas-Liquid Chromatographic Separation of Some Organophosphate Pesticides, Their Hydrolysis Products, and Oxons," *Environ. Sci. Technol.*, 1(8):639-643 (1967).

Suflita, J.M., J. Stout, and J.M. Tiedje. "Dechlorination of (2,4,5-Trichloro-phenoxy)acetic acid by Anaerobic Microorganisms," *J. Agric. Food Chem.*, 32(2):218-221 (1984).

Sukop, M. and C.G. Cogger. "Adsorption of Carbofuran. Metalaxyl, and Simazine: K_{oc} Evaluation and Relation to Soil Transport," *J. Environ. Sci. Health*, B27(5):565-590 (1992).

Sullivan, R.G., H.W. Knoche, and J.C. Markle. "Photolysis of Trifluralin: Characterization of Azobenzene and Azoxybenzene Photodegradation Products," *J. Agric. Food Chem.*, 28(4):746-755 (1980).

Surber, E.W. and Q.H. Pickering. "Acute Toxicity of Endothal, Diquat, Hyamine, Dalapon, and Silvex to Fish," *Progr. Fish-Cult.*, 24(4):164-171 (1962).

Swoboda, A.R. and G.W. Thomas. "Movement of Parathion in Soil Columns," *J. Agric. Food Chem.*, 16(6):923-927 (1968).

Szalkowski, M.B. and D.E. Stallard. "Effect of pH on the Hydrolysis of Chlorothalonil," *J. Agric. Food Chem.*, 25(1):208-210 (1977).

Szeto, S.Y. and M.J. Brown. "Gas-Liquid Chromatographic Methods for the Determination of Disulfoton, Phorate, Oxydemeton-methyl and Their Toxic Metabolites in Asparagus Tissue and Soil," *J. Agric. Food Chem.*, 30(6):1082-1086 (1982).

Szeto, S.Y., R.S. Vernon, and M.J. Brown. "Degradation of Disulfoton in Soil and Its Translocation into Asparagus," *J. Agric. Food Chem.*, 31(2):217-220 (1983).

Talbert, R.E. and O.H. Fletchall. "The Adsorption of Some *s*-Triazines in Soils," *Weeds*, 13:46-52 (1965).

Talbert, R.E., R.L. Runyan, and H.R. Baker. "Behavior of Amiben and Dinoben Derivatives in Arkansas Soils," *Weed Sci.*, 18(1):10-15 (1970).

Tanaka, F.S., B.L. Hoffer, and R.G. Wien. "Detection of Halogenated Biphenyls from Sunlight Photolysis of Chlorinated Herbicides in Aqueous Solution," *Pestic. Sci.*, 16:265-270 (1985).

Tanaka, F.S., R.G. Wien, and B.L. Hoffer. "Biphenyl Formation in the Photolysis of 3-(4-Chlorophenyl)-1,1-dimethylurea in Aqueous Solution," *J. Agric. Food Chem.*, 29(6):1153-1158 (1981).

Tanaka, F.S., R.G. Wien, and B.L. Hoffer. "Investigation of the Mechanism and Pathway of Biphenyl Formation in the Photolysis of Monuron," *J. Agric. Food Chem.*, 30(5):957-963 (1982).

Tanaka, F.S., R.G. Wien, and B.L. Hoffer. "Biphenyl Formation in the Photolysis of Aqueous Herbicide Solutions," *Ind. Eng. Chem. Prod. Res. Dev.*, 23(1):1-5 (1984).

Tanaka, F.S., R.G. Wien, and E.R. Mansager. "Photolytic Demethylation of Monuron and Demethylmonuron in Aqueous Solution," *Pestic. Sci.*, 13(3):287-294 (1982a).

Tanaka, F.S., R.G. Wien, and E.R. Mansager. "Effect of Nonionic Surfactants on the Photochemistry of 3-(4-Chlorophenyl)-1,1-dimethylurea in Aqueous Solution," *J. Agric. Food Chem.*, 27(4):774-779 (1979).

Tanaka, F.S., R.G. Wien, and E.R. Mansager. "Survey for Surfactant Effects on the Photodegradation of Herbicides in Aqueous Media," *J. Agric. Food Chem.*, 29(2):227-230 (1981).

Tanaka, F.S., R.G. Wien, and R.G. Zaylskie. "Photolysis of 3-(4-Chlorophenyl)-1,1-dimethylurea in Dilute Aqueous Solution," *J. Agric. Food Chem.*, 25(5):1068-1072 (1977).

Tiedje, J.M. and M.L. Hagedorn. "Degradation of Alachlor by a Soil Fungus, *Chaetomium globosum*," *J. Agric. Food Chem.*, 23(1):77-81 (1975).

Timms, P. and I.C. MacRae. "Conversion of Fensulfothion by *Klebsiella pneumoniae* to Fensulfothion Sulfide by Selected Microbes," *Aust. J. Biol. Sci.*, 35:661-668 (1982).

Timms, P. and I.C. MacRae. "Reduction of Fensulfothion and Accumulation of the Product, Fensulfothion Sulfide, by Selected Microbes," *Bull. Environ. Contam.*

Toxicol., 31(1):112-115 (1983).

Tomkiewicz, M.A., A. Groen, and M. Cocivera. "Electron Paramagnetic Resonance Spectra of Semiquinone Intermediates Observed during the Photooxidation of Phenol in Water," *J. Am. Chem. Soc.*, 93(25):7102-7103 (1971).

Trehy, M.L., R.A. Yost, and J.J. McCreary. "Determination of Aldicarb, Aldicarb Oxime, and Aldicarb Nitrile in Water by Gas Chromatography/Mass Spectrometry," *Anal. Chem.*, 56(8):1281-1285 (1984).

Tsao, R. and M. Eto. "Photoreactions of the Herbicide Naproanilide and the Effect of Some Photosensitizers," *J. Environ. Sci. Health*, B25(5):569-585 (1990).

Tu, C.M. "Influence of Pesticides and Some of the Oxidized Analogues on Microbial Populations, Nitrification and Respiration Activities in Soil," *Bull. Environ. Contam. Toxicol.*, 24(1):13-19 (1980).

Tu, C.M. "Utilization and Degradation of Lindane by Soil Microorganisms," *Arch. Microbiol.*, 108(3):259-263 (1976).

Tuazon, E.C., R. Atkinson, A.M. Winer, and J.N. Pitts, Jr. "A Study of the Atmospheric Reactions of 1,3-Dichloropropene and Other Selected Organochlorine Compounds," *Arch. Environ. Contam. Toxicol.*, 13(6):691-700 (1984).

U.S. Department of Agriculture. Agricultural Research Service Pesticide Properties Database. Systems Research Laboratory, Beltsville, MD (1990).

U.S. Department of Health and Human Services. *Hazardous Substances Data Bank.* National Library of Medicine, Toxnet File (Bethedsa, MD: National Institute of Health, 1989).

Uyeta, M., S. Taue, K. Chikasawa, and M. Mazaki. "Photoformation of Polychlorinated Biphenyls from Chlorinated Benzenes," *Nature (London)*, 264(5586):583-584 (1976).

Valverde-García, A., E. González-Pradas, M. Villafranca-Sánchez, F. del Rey-Bueno, and A. García-Rodriguez. "Adsorption of Thiram and Dimethoate on Almeria Soils," *Soil Sci. Soc. Am. J.*, 52(6):1571-1574 (1988).

Venkateswarlu, K., T.K.S. Gowda, and N. Sethunathan. "Persistence and Biodegradation of Carbofuran in Flooded Soils," *J. Agric. Food Chem.*, 25(3):533-536 (1977).

Venkateswarlu, K. and N. Sethunathan. "Metabolism of Carbofuran in Rice Straw-amended and Unamended Rice Soils," *J. Environ. Qual.*, 8(3):365-368 (1979).

Venkateswarlu, K. and N. Sethunathan. "Degradation of Carbofuran by *Azospirillium lipoferum* and *Streptomyces* spp. Isolated from Flooded Alluvial Soil," *Bull. Environ. Contam. Toxicol.*, 33(5):556-560 (1984).

Verschueren, K. *Handbook of Environmental Data on Organic Chemicals* (New York: Van Nostrand Reinhold Co., 1983), 1310 p.

Vogel, T.M. and M. Reinhard. "Reaction Products and Rates of Disappearance of Simple Bromoalkanes, 1,2-Dibromopropane, and 1,2-Dibromoethane in Water," *Environ. Sci. Technol.*, 20(10):992-997 (1986).

Vontor, T., J. Socha, and M. Vecera. "Kinetics and Mechanism of Hydrolysis of 1-Naphthyl, *N*-Methyl carbamate and *N,N*-Dimethyl Carbamates," *Collect. Czech.*

Chem. Commun., 37:2183-2196 (1972).

Wahid, P.A., C. Ramakrishna, and N. Sethunathan. "Instantaneous Degradation of Parathion in Anaerobic Soils," *J. Environ. Qual.*, 9(1):127-130 (1980).

Wahid, P.A. and N. Sethunathan. "Sorption-Desorption of Parathion in Soils," *J. Agric. Food Chem.*, 26(1):101-105 (1978).

Wahid, P.A. and N. Sethunathan. "Involvement of Hydrogen Sulfide in the Degradation of Parathion in Flooded Acid Sulphate Soil," *Nature (London)*, 282(5737):401-402 (1979).

Walker, A. and P.A. Brown. "The Relative Persistence in Soil of Five Acetanilide Herbicides," *Bull. Environ. Contam. Toxicol.*, 34:143-149 (1985).

Walker, A. and D.V. Crawford. "Diffusion Coefficients for Two Triazine Herbicides in Six Soils," *Weed Res.*, 10(2):126-132 (1970).

Walker, C.R. "Toxicological Effects of Herbicides on the Fish Environment," *Water Sewer. Works*, 111(3):113-116 (1964).

Walker, W.W. "Chemical and Microbiological Degradation of Malathion and Parathion in an Estuarine Environment," *J. Environ. Qual.*, 5(2):210-216 (1976).

Walker, W.W. and B.J. Stojjanovic. "Malathion Degradation by an *Arthrobacter* sp.," *J. Environ. Qual.*, 3(1):4-10 (1974).

Wallnöefer, P.R., S. Safe, and O. Hutzinger. "Microbial Demethylation and Debutynylation of Four Phenylurea Herbicides," *Pestic. Biochem. Physiol.*, 3:253-258 (1973).

Wang, Y.-S., E.L. Madsen, and M. Alexander. "Microbial Degradation by Mineralization or Cometabolism Determined by Chemical Concentration and Environment," *J. Agric. Food Chem.*, 33(3):495-499 (1985).

Wanner, O., T. Egli, T. Fleischmann, K. Lanz, P. Reichert, and R.P. Schwarzenbach. "Behavior of the Insecticides Disulfoton and Thiometon in the Rhine River: A Chemodynamic Study," *Environ. Sci. Technol.*, 23(10):1232-1242 (1989).

Wauchope, R.D. and R. Haque. "Effects of pH, Light and Temperature on Carbaryl in Aqueous Media," *Bull. Environ. Contam. Toxicol.*, 9(5):257-261 (1973).

Way, M.J. and N.E.A. Scopes. "Studies on the Persistence and Effects on Soil Fauna and Some Soil-Applied Systemic Insecticides," *Ann. Appl. Biol.*, 62:199-214 (1968).

Weber, J.B. and H.D. Coble. "Microbial Decomposition of Diquat Adsorbed on Montmorillonite and Kaolinite Clays," *J. Agric. Food Chem.*, 16(3):475-478 (1968).

Wedemeyer, G.A. "Dechlorination of DDT by *Aerobacter aerogenes*" *Science (Washington, DC)*, 152(3722):647 (1966).

Wei, L.Y. "Gas Chromatographic and Mass Spectrometric Characterization of Terbufos sulfoxide and Its Hydration Product," *J. Environ. Sci. Health*, B25(5):607-613 (1990).

Weiss, U.M., I. Scheunert, W. Klein, and F. Korte. "Fate of Pentachlorophenol-^{14}C in Soil under Controlled Conditions," *J. Agric. Food Chem.*, 30(6):1191-1194 (1982).

Werner, R.A. "Distribution and Toxicity of Root Absorbed ^{14}C-Orthene and Its Metabolites in Loblolly Pine Seedlings," *J. Econ. Entomol.*, 67(5):588-591 (1974).

Wildung, R.E., G. Chesters, and D.E. Armstrong. "Chloramben (Amiben) Degradation in Soil," *Weed Res.*, 8(3):213-225 (1968).

Williams, J.D.H. "Adsorption and Desorption of Simazine by Some Rothamsted Soils," *Weed Res.*, 8(4):327-335 (1968).

Windholz, M., S. Budavari, R.F. Blumetti, and E.S. Otterbein, Eds., *The Merck Index*, 10th ed. (Rahway, NJ: Merck & Co., 1983), 1463 p.

Winkelmann, D.A. and S.J. Klaine. "Atrazine Metabolite Behavior in Soil-Core Microcosms" in *Pesticide Transformation Products. Fate and Significance in the Environment*, ACS Symposium Series 459, Somasundaram, L. and J.R. Coats, Eds., (Washington, DC: American Chemical Society, 1991), pp. 75-92.

Witkonton, S. and C.D. Ercegovich. "Degradation of N'-(4-Chloro-*o*-tolyl)-*N,N*-dimethylformamidine in Six Different Fruit," *J. Agric. Food Chem.*, 20(3):569-573 (1972).

Wolf, D.C. and J.P. Martin. "Microbial Degradation of Bromacil-2-^{14}C and Terbacil-2-^{14}C," *Soil Sci. Soc. Am. Proc.*, 38:921-925 (1974).

Wolfe, N.L. "Abiotic Transformations of Pesticides in Natural Waters and Sediments," in *Fate of Pesticides and Chemicals in the Environment*, Schnoor, J.L., Ed., (New York: John Wiley & Sons, Inc., 1992), pp. 93-104.

Wolfe, N.L., U. Mingelgrin, and G.C. Miller. "Abiotic Transformations in Water, Sediments, and Soil" in *Pesticides in the Soil Environment: Processes, Impacts, and Modeling*, SSSA Book Series: 2, Cheng, H.H., Ed., (Madison, WI: Soil Science Society of America, Inc., 1990), pp. 103-168.

Wolfe, N.L., W.C. Steen, and L.A. Burns. "Phthalate Ester Hydrolysis: Linear Free Energy Relationships," *Chemosphere*, 9(7/8):403-408 (1980).

Wolfe, N.L., R.G. Zepp, G.L. Baughman, R.C. Fincher, and J.A. Gordon. "Chemical and Photochemical Transformations of Selected Pesticides in Aquatic Systems," U.S. EPA Report 600/3-76-067 (1976), 141 p.

Wolfe, N.L., R.G. Zepp, J.A. Gordon, G.L. Baughman, and D.M. Cline. "Kinetics of Chemical Degradation of Malathion in Water," *Environ. Sci. Technol.*, 11(1):88-93 (1977).

Wolfe, N.L., R.G. Zepp, and D.F. Paris. "Use of Structure-Reactivity Relationships to Estimate Hydrolytic Persistence of Carbamate Pesticides," *Water Res.*, 12(8):561-563 (1978).

Wolfe, N.L., R.G. Zepp, and D.F. Paris. "Carbaryl, Propham, and Chlorpropham: A Comparison of the Rates of Hydrolysis and Photolysis with the Rate of Biolysis," *Water Res.*, 12(8):565-571 (1978a).

Wolfe, N.L., R.G. Zepp, D.F. Paris, G.L. Baughman, and R.C. Hollis. "Methoxychlor and DDT Degradation in Water: Rates and Products," *Environ. Sci. Technol.*, 11(12):1077-1081 (1977a).

Wong, A.S. and D.G. Crosby. "Photodecomposition of Pentachlorophenol in Water," *J. Agric. Food Chem.*, 29(1):125-130 (1981).

Woodrow, J.E., D.G. Crosby, T. Mast, K.W. Moilanen, and J.N. Seiber. "Rates of Transformation of Trifluralin and Parathion Vapors in Air," *J. Agric. Food Chem.*, 26(6):1312-1316 (1978).

Worthing, C.R. and R.J. Hance, Eds. *The Pesticide Manual - A World Compendium*, 9th ed. (Great Britain: British Crop Protection Council, 1991), 1141 p.

Wright, W.L. and G.F. Warren. "Photochemical Decomposition of Trifluralin," *Weeds*, 13:329-331 (1965).

Yaron, B., B. Heuer, and Y. Birk. "Kinetics of Azinphosmethyl Losses in the Soil Environment," *J. Agric. Food Chem.*, 22(3):439-441 (1974).

Yih, R.Y., D.H. McRae, and H.F. Wilson. "Metabolism of 3',4'-Dichloropropion-anilide: 3,4-Dichloroaniline-Lignin Complex in Rice Plants," *Science (Washington, DC)*, 161(3839):376-377 (1968).

Young, J.C. and S.U. Khan. "Kinetics of Nitrosation of the Herbicide Glyphosate," *J. Environ. Sci. Health*, B13:59-72 (1978).

Yu, C.-C., D.J. Hansen, and G.M. Booth. "Fate of Dicamba in a Model Ecosystem," *Bull. Environ. Contam. Toxicol.*, 13(3):280-283 (1975).

Yu, C.-C. and J.R. Sanborn. "The Fate of Parathion in a Model Ecosystem," *Bull. Environ. Contam. Toxicol.*, 13(5):543-550 (1975).

Yule, W.N., M. Chiba, and H.V. Morley. "Fate of Insecticide Residues. Decomposition of Lindane in Soil," *J. Agric. Food Chem.*, 15(6):1000-1004 (1967).

Zabik, M.J., R.D. Schuetz, W.L. Burton, and B.E. Pape. "Photochemistry of Bioreactive Compounds. Studies of a Major Product of Endrin," *J. Agric. Food Chem.*, 19(2):308-313 (1971).

Zbozinek, J.V. "Environmental Transformations of DPA, SOPP, Benomyl, and TBZ," *Residue Rev.*, 92:113-155 (1984).

Zepp, R.G., G.L. Baughman, and P.F. Schlotzhauer. "Comparison of Photo-chemical Behavior of Various Humic Substances in Water. I. Sunlight Induced Reactions of Aquatic Pollutants Photosensitized by Humic Substances," *Chemosphere*, 10:109-117 (1981).

Zepp, R.G., N.L. Wolfe, J.A. Gordon, and R.C. Fincher. "Light-Induced Transformations of Methoxychlor in Aquatic Systems," *J. Agric. Food Chem.*, 24(4):727-733 (1976).

Zhong, W.Z., A.T. Lemley, and R.J. Wagenet. *Evaluation of Pesticides in Ground Water*, ACS Symposium Series 315, Garner, W.J., R.C. Honeycutt, and H.N. Nigg, Eds., (Washington, DC: American Chemical Society, 1986), pp. 61-67.

Zimdahl, R.L. and S.K. Clark. "Degradation of Three Acetanilide Herbicides in Soil," *Weed Sci.*, 30:545-548 (1982).

Conversion Factors

To convert	Into	Multiply by
acre–feet	feet^3	4.356×10^4
	gallons (U.S.)	3.529×10^5
	inches^3	7.527×10^7
	liters	1.233×10^6
	meters^3	1,233
	yards^3	1,613
acre–feet/day	feet^3/second	0.5042
	gallons (U.S.)/minute	226.3
	liters^3/second	14.28
	meters^3/day	1,234
	meters^3/second	0.01428
acres	feet^2	43,560
	hectares	0.4047
	inches^2	6.273×10^6
	kilometers^2	4.047×10^{-3}
	meters^2	4,047
	miles^2	1.563×10^{-3}
atmospheres	bars	1.01325
	inches of Hg	29.92126
	millibars	1013.25
	millimeters of Hg	760
	millimeters of water	1.033227×10^4
	pascals	1.01325×10^5
centimeters	feet	0.03281
	millimeters	10
	inches	0.3937
	yards	0.01094
centimeters^2	feet^2	0.001076
	inches^2	0.155
	meters^2	1×10^{-4}
	yards^2	1.196×10^{-4}
centimeters^3	fluid ounces	0.03381
	feet^3	3.5314×10^{-5}
	inches^3	0.06102
	liters	0.001
	ounces (U.S., fluid)	0.03381

To convert	Into	Multiply by
centimeters/second	feet/day	2,835
	feet/minute	1.1969
	feet/second	0.03281
	kilometers/hour	0.036
	liters/meter/second	9.985
	meters/minute	0.6
	miles/minute	3.728×10^{-4}
Darcy	centimeters/second	9.66×10^{-4}
	feet/second	3.173×10^{-5}
	liters/meter/second	8.58×10^{-3}
feet	centimeters	30.48
	inches	12
	kilometers	3.048×10^{-4}
	meters	0.3048
	miles	1.894×10^{-4}
	millimeters	304.8
	yards	0.333
$feet^2$	acres	2.296×10^{-5}
	hectares	9.29×10^{-9}
	$inches^2$	144
	$kilometers^2$	9.29×10^{-8}
	$meters^2$	9.29×10^{-2}
	$miles^2$	3.587×10^{-8}
	$yards^2$	0.1111
$feet^3$	acre-feet	2.296×10^{-5}
	gallons (U.S.)	7.481
	$inches^3$	1,728
	liters	28.32
	$meters^3$	2.832×10^{-3}
	$yards^3$	3.704×10^{-2}
feet/day	feet/second	1.157×10^{-5}
	meters/second	3.528×10^{-6}
$feet^3$/foot/day	gallons/foot/day	7.48052
	liters/meter/day	92.903
	$meters^3$/meter/day	0.0929

To convert	Into	Multiply by
feet^3/foot^2/day	feet^3/foot^2/minute	6.944 x 10^{-4}
	gallons/foot^2/day	7.4805
	inches^3/inch^2/hour	0.5
	liters/meter^2/day	304.8
	meters^3/meter^2/day	0.3048
	millimeters^3/inch^2/hour	0.5
	millimeters^3/millimeter^2/hour	25.4
feet/second	centimeters/second	0.5080
	feet/day	86,400
	feet/hour	3,600
	gallons (U.S.)/foot^2/day	5.737 x 10^5
	kilometers/hour	1.097
	meters/second	0.3048
	miles/hour	0.6818
feet^3/second	acre-feet/day	1.983
	feet^3/minute	60.0
	gallons (U.S.)/minute	448.8
	liters/second	28.32
	meters^3/second	0.02832
	meters^3/day	2,447
ounces (avoirdupois)	grams	28.35
	kilograms	0.0284
	pounds	0.0625
gallons (U.S.)	acre-feet	3.068 x 10^{-6}
	feet^3	0.1337
	fluid ounces	128.0
	liters	3.785
	meters^3	3.785 x 10^{-3}
	yards^3	4.951 x 10^{-3}

To convert	Into	Multiply by
gallons/foot/day	feet^3/foot/day	0.13368
	liters/meter/day	12.42
	meters^3/meter/day	0.01242
gallons (U.S.)/foot^2/day	centimeters/second	4.717×10^{-5}
	Darcy	0.05494
	feet/day	0.13368
	feet/second	1.547×10^{-6}
	gallons/foot^2/ minute	6.944×10^{-4}
	liters/meter^2/ day	40.7458
	meters/day	0.0407458
	meters/minute	2.83×10^{-5}
	meters/second	4.716×10^{-7}
gallons (U.S.)/foot^2/minute	meters/day	58.67
	meters/second	0.06791
gallons (U.S.)/minute	acre-feet/day	4.419×10^{-3}
	feet^3/second	2.228×10^{-3}
	feet^3/hour	8.0208
	liters/second	6.309×10^{-2}
	meters^3/day	5.45
	meters^3/second	6.309×10^{-5}
grams/centimeter^3	kilograms/meter^3	1000
	pounds/feet^3	62.428
	pounds/gallon (U.S.)	8.345
grams	kilograms	0.001
	ounces (avoirdupois)	0.03527
	pounds	0.022046
grams/liter	grains/gallon (U.S.)	58.4178
	grams/centimeter^3	0.001
	kilograms/meter^3	1
	pounds/feet^3	0.0624
	pounds/inch^3	3.61×10^{-5}
	pounds/gallon (U.S.)	8.35×10^{-3}

To convert	Into	Multiply by
grams/meter^3	grains/feet^3	0.4370
	milligrams/liter	1.0
	pounds/gallon (U.S.)	8.345×10^{-5}
	pounds/inch^3	7.433×10^{-3}
hectares	acres	2.471
	feet^2	1.076×10^5
	inches^2	1.55×10^7
	kilometers^2	0.01
	meters^2	10,000
	miles^2	3.861×10^{-3}
	yards^2	11,959.90
inches	centimeters	2.54
	feet	0.8333
	kilometers	2.54×10^{-5}
	meters	2.54×10^{-2}
	miles	1.578×10^{-5}
	millimeters	25.4
	yards	2.778×10^{-2}
inches^2	acres	1.594×10^{-8}
	centimeters^2	6.4516
	feet^2	6.944×10^{-3}
	hectares	6.452×10^{-8}
	kilometers^2	6.452×10^{-10}
	meters^2	6.452×10^{-4}
	millimeters^2	645.16
inches^3	acre-feet	1.329×10^{-8}
	feet^3	5.787×10^{-4}
	gallons (U.S.)	4.329×10^{-3}
	liters	1.639×10^{-2}
	meters^3	1.639×10^{-5}
	milliliters	16.387
	yards^3	2.143×10^{-5}
kilograms	grams	1,000
	ounces (avoirdupois)	35.28
	pounds	2.205
	tons (metric)	0.001

To convert	Into	Multiply by
kilograms/meter^3	grams/centimeter^3	0.001
	grams/liter	1.0
	pounds/inch^3	3.613 x 10^{-5}
	pounds/feet^3	0.0624
kilometers	feet	3,281
	inches	39,370
	meters	1,000
	miles	0.6214
	millimeters	1 x 10^6
kilometers^2	acres	247.1
	hectares	100
	inches^2	1.55 x 10^9
	meters^2	1 x 10^6
	miles^2	0.3861
kilometers/hour	feet/day	78,740
	feet/second	0.9113
	meters/second	0.2778
	miles/hour	0.6214
liters	acre-feet	8.106 x 10^{-7}
	feet^3	3.531 x 10^{-2}
	fluid ounces	33.814
	gallons (U.S.)	0.2642
	inches^3	61.02
	meters^3	0.001
	yards^3	1.308 x 10^{-3}
liters/second	acre-feet/day	7.005 x 10^{-2}
	feet^3/second	3.531 x 10^{-2}
	gallons (U.S.)/minute	15.85
	meters^3/day	86.4
meters	feet	3.28084
	inches	39.3701
	kilometers	0.001
	miles	6.214 x 10^{-4}
	millimeters	1,000
	yards	1.0936

To convert	Into	Multiply by
meters^2	acres	2.471×10^{-4}
	feet^2	10.76
	hectares	1×10^{-4}
	inches^2	1.550
	kilometers	1×10^{-6}
	miles^2	3.861×10^{-7}
	yards^2	1.196
meters^3	acre-feet	8.106×10^{-4}
	feet^3	35.31
	gallons (U.S.)	264.2
	inches^3	6.102×10^4
	liters	1,000
	yards^3	1.308
meters/second	feet/minute	196.8
	feet/day	283,447
	kilometers/hour	3.6
	feet/second	3.281
	miles/hour	2.237
meters^3/day	acre-feet/day	6.051×10^6
	feet^3/second	3.051×10^6
	gallons (U.S.)/minute	1.369×10^9
	liters/second	8.64×10^7
miles	feet	5,280
	kilometers	1.609
	meters	1,609
	millimeters	1.609×10^6
miles^2	acres	640
	feet^2	2.778×10^7
	hectares	259
	inches^2	4.014×10^9
	kilometers^2	2.59
	meters^2	2.59×10^6
miles/hour	feet/day	1.267×10^5
	kilometers/hour	1.609
	meters/second	0.447

To convert	Into	Multiply by
milligrams/liter	grams/meter^3	1.0
	parts/million	1.0
	pounds/feet^3	6.2428 x 10^{-5}
milliliters	centimeters^3	1.0
	liters	0.001
millimeters	centimeters	0.1
	feet	3.281 x 10^{-3}
	inches	0.03937
	kilometers	1.0 x 10^{-6}
	meters	1.0 x 10^{-3}
	miles	6.214 x 10^{-7}
millimeters^2	centimeters^2	0.01
	inches^2	0.00155
millimeters^3	centimeters^3	0.001
	inches^3	6.102 x 10^{-5}
	liters	1 x 10^{-6}
millimeters of Hg	atmospheres	1.316 x 10^{-3}
	pascals	133.3224
	torrs	1.0
ounces (avoirdupois)	grams	28.35
	kilograms	0.02835
	pounds	0.0625
pascals	atmospheres	9.869 x 10^{-6}
	millimeters of Hg	7.501 x 10^{-3}
pounds (avoirdupois)	kilograms	0.4535
	ounces (avoirdupois)	16
pounds/centimeter^3	pounds/inch^3	0.0361
pounds/feet^3 (con't)	grams/centimeter^3	0.016
	grams/liter	27,680
	pounds/inch^3	5.787 x 10^{-4}
	pounds/gallon (U.S.)	0.1337

To convert	Into	Multiply by
pounds/inch^3	pounds/centimeter^3	27.68
	pounds/feet^3	1,728
	pounds/gallon (U.S.)	231
pounds/gallon (U.S.)	grams/centimeter^3	0.1198
	grams/liter	119.8
	pounds/feet^3	7.481
	pounds/inch^3	4.329×10^{-3}
torrs	atmospheres	1.316×10^{-3}
	millimeters of Hg	1.0
	pascals	133.322
yards	centimeters	91.44
	fathoms	0.5
	feet	3.0
	inches	36.0
	meters	0.9144
	miles	5.682×10^{-4}
yards^3	acre-feet	6.198×10^{-4}
	feet^3	27
	gallons (U.S.)	202
	inches^3	4.666×10^4
	liters	764.6
	meters^3	.7646
years (normal calendar)	hours	8,760
	minutes	5.256×10^5
	seconds	3.1536×10^7
	weeks	52.1428

U.S. EPA Approved Test Methods[a]

The numbers that follow each compound name below are the U.S. EPA approved test methods based on information obtained from the "Guide to Environmental Analytical Methods" (Schenectady, NY: Genium Publishing Corp., 1992) and the U.S. EPA's Sampling and Analysis Methods Database. The database information is available in hard copy from Keith, L.H. *Compilation of EPA's Sampling and Analysis Methods* (Chelsea, MI: Lewis Publishers, Inc., 1991), 803 p.

Chemical Name	Test Methods
Acrolein	8030, 8240
Acrylonitrile	8030, 8240
Alachlor	525
Aldrin	508, 525, 608, 625, SM-6410, SM-6630, 8080, 8270
Anilazine	8270
Atrazine	525
Azinphos-methyl	8140, 8270
α-BHC	508, 608, 625, SM-6410, SM-6630, 8080, 8270
β-BHC	508, 608, 625, SM-6410, SM-6630, 8080, 8270
δ-BHC	508, 608, 625, SM-6410, SM-6630, 8080, 8270
Bis(2-chloroethyl)ether	625, SM-6410, 8270
Bis(2-chloroisopropyl)ether	625, SM-6410, 8270
Bromoxynil	8270
Carbaryl	8270
Carbofuran	8270
Carbon disulfide	8240
Carbon tetrachloride	502.1, 502.2, 601, 524.1, 524.2, 624, SM-6210 SM-6230, 8010, 8240
Carbophenothion	8270
Chloramben	515.1
Chlordane (technical)	508, 608, SM-6630, 8080, 8270
Chlorobenzilate	508, 8270
Chlorothalonil	508
Chlorpyrifos	508, 8140
Crotoxyphos	8270
2,4-D	515.1, 8150
Dalapon	8150
p,p'-DDD	508, 608, SM-6630, 8080, 8270
p,p'-DDE	508, 608, SM-6630, 8080, 8270
p,p'-DDT	508, 608, SM-6630, 8080, 8270
Diallate	8270
Diazinon	8140
1,2-Dibromo-3-chloropropane	502.2, 504, 524.1, 524.2, SM-6210, 8240
Di-*n*-butyl phthalate	525, 625, SM-6410, 8060, 8270

Chemical Name	Test Methods
Dicamba	515.1, 8150
Dichlone	8270
1,2-Dichloropropane	502.1, 502.2, 524.1, 524.2, 601, 624, SM-6210
SM-6230, 8010, 8240	
cis-1,3-Dichloropropylene	502.1, 502.2, 524.1, 601, 624, SM-6210
SM-6230, 8010, 8240	
trans-1,3-Dichloropropylene	502.1, 502.2, 524.1, 601, 624, SM-6210
SM-6230, 8010, 8240	
Dichlorvos	8140, 8270
Dicrotophos	8270
Dieldrin	508, 608, SM-6630, 8080, 8270
Dimethoate	8270
Dimethyl phthalate	525, 625, SM-6410, 8060, 8270
4,6-Dinitro-o-cresol	8040
Dinoseb	515.1, 8150, 8270
Disulfoton	8140, 8270
α-Endosulfan	508, 608, 625, SM-6410, SM-6630, 8080, 8270
β-Endosulfan	508, 608, 625, SM-6410, SM-6630, 8080, 8270
Endosulfan sulfate	508, 608, 625, SM-6410, SM-6630, 8080, 8270
Endrin	508, 525, 608, 625, SM-6410, SM-6630, 8080, 8270
Endrin aldehyde	508, 608, 625, SM-6410, SM-6630, 8080, 8270
EPN	8270
Ethion	8270
Ethoprop	8140
Ethylene dibromide	502.1, 502.2, 504, 524.1, 524.2, SM-6210, 8240
Fensulfothion	8140, 8270
Fenthion	8140, 8270
Heptachlor	508, 525, 608, 625, SM-6410, SM-6630, 8080, 8270
Heptachlor epoxide	508, 525, 608, 625, SM-6410, SM-6630, 8080, 8270
Hexachlorobenzene	508, 525, 625, SM-6410, 8120, 8270
Kepone	8270
Lindane	508, 525, 608, SM-6630, 8080, 8270
Malathion	8270
MCPA	8150
Methoxychlor	508, 525, SM-6630, 8080, 8270
Methyl bromide	502.1, 502.2, 524.1, 524.2, 601, 624, SM-6210, SM-6230
8010, 8240	
Mevinphos	8140, 8270
Monocrotophos	8270
Naled	8140, 8270
Parathion	8270

Chemical Name	Test Methods
Pentachlorobenzene	8270
Pentachlorophenol	525, 625, SM-6410, 8040, 8270
Phorate	8140, 8270
Phosalone	8270
Phosmet	8270
Phosphamidon	8270
Picloram	515.1
Propachlor	508
Propyzamide	8270
Ronnel	8140
Simazine	525
Strychnine	525
Sulprofos	8140
2,4,5-T	515.1, 8150, SM-6640
Terbufos	8270
Tetraethyl pyrophosphate	8270
Toxaphene	508, 525, 608, 625, SM-6630, 8080, 8270
Trifluralin	508, 8270

a) Methods beginning with the prefix "SM-" refer to standard methods found in 17th edition of the "Standard Methods for the Examination of Water and Waste Water", published by the American Water Works Association in 1989.

CAS Index

RTECS Number Index

NT2600000   Chlorothalonil
NU8050000   Biphenyl
PA2975000   Hexachlorobenzene
PA4900000   Methyl bromide
PA7625000   Methyl isothiocyanate
PB6300000   Chloropicrin
PB9450000   Heptachlor epoxide
PB9705000   *trans*-Chlordane
PB9800000   Chlordane
PC0175000   *cis*-Chlordane
PC0700000   Heptachlor
PC8575000   Kepone
QL7525000   Dichlone
RB9275000   α-Endosulfan, β-Endosulfan
RN7875000   Endothall
RO0874000   Oxadiazon
RP2300000   Oxamyl
RP4550000   Carboxin
SJ9800000   Dinoseb
SM6300000   Pentachlorophenol
SZ7100000   Ethephon
SZ9640000   Fosetyl-aluminum
TA0700000   Trichlorfon
TA5950000   Fonofos
TB1925000   EPN
TB3675000   Fenamiphos
TB4970000   Methamidophos
TB4760000   Acephate
TB9450000   Naled
TC0350000   Dichlorvos
TC2800000   Phosphamidon
TC3850000   Dicrotophos
TC4375000   Monocrotophos
TD5165000   Dialifos
TD5175000   Phosalone
TD5250000   Carbophenothion
TD7200000   Terbufos
TD9275000   Disulfoton
TD9450000   Phorate
TE1750000   Dimethoate
TE1925000   Azinphos-methyl
TE2100000   Methidathion
TE2275000   Phosmet

Empirical Formula Index

CCl_3NO_2	Chloropicrin
CCl_4	Carbon tetrachloride
CH_2O	Formaldehyde
CH_3Br	Methyl bromide
CS_2	Carbon disulfide
C_2H_3NS	Methyl isothiocyanate
$C_2H_4Br_2$	Ethylene dibromide
$C_2H_4N_4$	Amitrole
$C_2H_6ClO_3P$	Ethephon
$C_2H_8NO_2PS$	Methamidophos
$C_3H_3Cl_2NaO_2$	Dalapon-sodium
C_3H_3N	Acrylonitrile
$C_3H_4Cl_2$	cis-1,2-Dichloropropylene
	trans-1,2-Dichloropropylene
C_3H_4O	Acrolein
$C_3H_5Br_2Cl$	1,2-Dibromo-3-chloropropane
$C_3H_6Cl_2$	1,2-Dichloropropane
$C_3H_8NO_5P$	Glyphosate
$C_3H_{11}N_2O_4P$	Fosamine-ammonium
$C_4H_4N_2O_2$	Maleic hydrazide
$[C_4H_6N_2S_2Mn]_xZn_y$	Mancozeb
$C_4H_7Br_2Cl_2O_4P$	Naled
$C_4H_7Cl_2O_4P$	Dichlorvos
$C_4H_8ClO_4P$	Trichlorfon
$C_4H_8Cl_2O$	Bis(2-chloroethyl)ether
$C_4H_{10}NO_3PS$	Acephate
$C_5H_{10}N_2O_2S$	Methomyl
$C_5H_{10}N_2S_2$	Dazomet
$C_5H_{12}NO_3PS_2$	Dimethoate
C_6Cl_6	Hexachlorobenzene
$C_6H_3Cl_4N$	Nitrapyrin
C_6HCl_5	Pentachlorobenzene
C_6HCl_5O	Pentachlorophenol
$C_6H_3Cl_3N_2O_2$	Picloram
$C_6H_6Cl_6$	α-BHC
	β-BHC
	δ-BHC
	Lindane
$C_6H_{10}N_6$	Cyromazine
$C_6H_{11}N_2O_4PS_3$	Methidathion
$C_6H_{12}Cl_2O$	Bis(2-chloroisopropyl)ether
$C_6H_{12}N_2S_4$	Thiram
$C_6H_{15}O_4PS_2$	Oxydemeton-methyl

$C_6H_{18}AlO_9P_3$	Fosetyl-aluminum
$C_7H_3Br_2NO$	Bromoxynil
$C_7H_3Cl_2N$	Dichlobenil
$C_7H_3Cl_3O_2$	2,3,6-Trichlorobenzoic acid
$C_7H_5Cl_2NO_2$	Chloramben
$C_7H_6N_2O_5$	4,6-Dinitro-o-cresol
$C_7H_{12}ClN_5$	Simazine
$C_7H_{13}N_3O_3S$	Oxamyl
$C_7H_{13}O_6P$	Mevinphos
$C_7H_{14}NO_5P$	Monocrotophos
$C_7H_{14}N_2O_2S$	Aldicarb
$C_7H_{17}O_2PS_3$	Phorate
$C_8Cl_4N_2$	Chlorothalonil
$C_8H_5Cl_3O_3$	2,4,5-T
$C_8H_6Cl_2O_3$	2,4-D
	Dicamba
$C_8H_8Cl_3O_3PS$	Ronnel
$C_8H_{10}N_2O_4S$	Asulam
$C_8H_{10}O_5$	Endothall
$C_8H_{10}O_7P_2$	Tetraethyl pyrophosphate
$C_8H_{12}ClNO$	Allidochlor
$C_8H_{14}ClN_5$	Atrazine
$C_8H_{14}N_4OS$	Metribuzin
$C_8H_{16}NO_5P$	Dicrotophos
$C_8H_{19}O_2PS_3$	Disulfoton
$C_8H_{19}O_2PS_2$	Ethoprop
$C_8H_{20}O_5P_2S_2$	Sulfotepp
$C_9H_5Cl_3N_4$	Anilazine
$C_9H_6Cl_6O_3S$	α-Endosulfan
	β-Endosulfan
$C_9H_6Cl_6O_4S$	Endosulfan sulfate
$C_9H_8N_4OS$	Thidiazuron
$C_9H_9ClO_3$	MCPA
$C_9H_9Cl_2NO$	Propanil
$C_9H_{10}Cl_2N_2O$	Diuron
$C_9H_{10}Cl_2N_2O_2$	Linuron
$C_9H_{11}ClN_2O$	Monuron
$C_9H_{11}Cl_3NO_3PS$	Chlorpyrifos
$C_9H_{13}BrN_2O_2$	Bromacil
$C_9H_{13}ClN_2O_2$	Terbacil
$C_9H_{13}ClN_6$	Cyanazine
$C_9H_{16}ClN_5$	Propazine
$C_9H_{16}N_4OS_2$	Tebuthiuron

$C_9H_{17}NOS$	Molinate
$C_9H_{17}N_2S$	Ametryn
$C_9H_{18}FeN_3S_6$	Ferbam
$C_9H_{19}NOS$	EPTC
$C_9H_{21}O_2PS_3$	Terbufos
$C_9H_{22}O_4P_2S_4$	Ethion
$C_{10}Cl_{10}O$	Kepone
$C_{10}H_4Cl_2O_2$	Dichlone
$C_{10}H_5Cl_7$	Heptachlor
$C_{10}H_5Cl_7O$	Heptachlor epoxide
$C_{10}H_6Cl_4O_4$	Chlorthal-dimethyl
$C_{10}H_6Cl_8$	Chlordane
	cis-Chlordane
	trans-Chlordane
$C_{10}H_6N_2OS_2$	Quinomethionate
$C_{10}H_7N_3S$	Thiabendazole
$C_{10}H_{10}Cl_8$	Toxaphene
$C_{10}H_{10}O_4$	Dimethyl phthalate
$C_{10}H_{11}F_3N_2O$	Fluometuron
$C_{10}H_{12}ClNO_2$	Chlorpropham
$C_{10}H_{12}N_2O_3S$	Bentazone
$C_{10}H_{12}N_2O_5$	Dinoseb
$C_{10}H_{12}N_3O_3PS_2$	Azinphos-methyl
$C_{10}H_{13}ClN_2$	Chlordimeform
$C_{10}H_{13}NO_2$	Propham
$C_{10}H_{14}NO_5PS$	Parathion
$C_{10}H_{15}OPS_2$	Fonofos
$C_{10}H_{15}O_3PS_2$	Fenthion
$C_{10}H_{16}Cl_3NOS$	Triallate
$C_{10}H_{17}Cl_2NOS$	Diallate
$C_{10}H_{18}N_4O_4S_3$	Thiodicarb
$C_{10}H_{19}ClNO_5P$	Phosphamidon
$C_{10}H_{19}N_5O$	Prometon
$C_{10}H_{19}N_5S$	Prometryn
	Terbutryn
$C_{10}H_{19}O_6PS_2$	Malathion
$C_{10}H_{21}NOS$	Pebulate
$C_{11}H_{10}N_2S$	ANTU
$C_{11}H_{12}NO_4PS_2$	Phosmet
$C_{11}H_{13}F_3N_2O_3$	Mefluidide
$C_{11}H_{13}NO_4$	Bendiocarb
$C_{11}H_{14}ClNO$	Propachlor
$C_{11}H_{15}BrClO_3PS$	Profenofos

Synonym Index

A 42 *see* Aspon
A 361 *see* Atrazine
A 363 *see* Aminocarb
A 1068 *see* Chlordane
A 1093 *see* Ametryn
A 2079 *see* Simazine
Aafertis *see* Ferbam
Aahepta *see* Heptachlor
Aalindan *see* Lindane
AAT *see* Parathion
Aatack *see* Thiram
AATP *see* Parathion
Aatrex *see* Atrazine
Aatrex 4L *see* Atrazine
Aatrex 4LC *see* Atrazine
Aatrex nine-o *see* Atrazine
Aatrex 80W *see* Atrazine
AC 3422 *see* Ethion, Parathion
AC 3911 *see* Phorate
AC 12880 *see* Dimcthoate
AC 18682 *see* Dimethoate
AC 84777 *see* Difenzoquat methyl sulfate
AC 92100 *see* Terbufos
AC 92553 *see* Pendimethalin
AC 222705 *see* Flucythrinate
Acar *see* Chlorobenzilate
Acaraben *see* Chlorobenzilate
Acaraben 4E *see* Chlorobenzilate
Acarithion *see* Carbophenothion
Acaron *see* Chlordimeform
Accelerate *see* Endothall
Accelerator thiuram *see* Thiram
Acephate
Acephate-met *see* Methamidophos
3-(Acetonylbenzyl)-4-hydroxycoumarin *see* Warfarin
3-(α-Acetonylbenzyl)-4-hydroxycoumarin *see* Warfarin
Aceto TETD *see* Thiram
Acetylene dibromide *see* Ethylene dibromide
Acetylphosphoramidothioic acid O,S-dimethyl ester *see* Acephate
N-Acetylphosphoramidothioic acid O,S-dimethyl ester *see* Acephate
ACP 322 *see* Naptalam
ACP-M-728 *see* Chloramben
Acquinite *see* Chloropicrin

Acraldehyde *see* Acrolein
Acritet *see* Acrylonitrile
Acrolein
Acrylaldehyde *see* Acrolein
Acrylic aldehyde *see* Acrolein
Acrylon *see* Acrylonitrile
Acrylonitrile
Acrylonitrile monomer *see* Acrylonitrile
Acutox *see* Pentachlorophenol
Aerol 1 *see* Trichlorfon
AF 101 *see* Diuron
Afalon *see* Linuron
Afalon inuron *see* Linuron
Aficide *see* Lindane
Agreflan *see* Trifluralin
Agricide maggot killer (F) *see* Toxaphene
Agriflan 24 *see* Trifluralin
Agrisol G-20 *see* Lindane
Agritan *see* p,p'-DDT
Agritox *see* MCPA
Agroceres *see* Heptachlor
Agrocide *see* Lindane
Agrocide 2 *see* Lindane
Agrocide 6G *see* Lindane
Agrocide 7 *see* Lindane
Agrocide III *see* Lindane
Agrocide WP *see* Lindane
Agroforotox *see* Trichlorfon
Agronexit *see* Lindane
Agrotect *see* 2,4-D
Agroxon *see* MCPA
Agroxone *see* 2,4-D, MCPA
AI3-29158 *see* Permethrin
Akar *see* Chlorobenzilate
Akar 50 *see* Chlorobenzilate
Akar 338 *see* Chlorobenzilate
Akarithion *see* Carbophenothion
Aktikon *see* Atrazine
Aktikon PK *see* Atrazine
Aktinit A *see* Atrazine
Aktinit PK *see* Atrazine
Aktinit S *see* Simazine
Alachlor

Alanap *see* Naptalam
Alanap 10G AT *see* Naptalam
Alanap L *see* Naptalam
Alanape *see* Naptalam
Alanex *see* Alachlor
Albrass *see* Propachlor
Aldecarb *see* Aldicarb
Aldrec *see* Aldrin
Aldrex *see* Aldrin
Aldrex 30 *see* Aldrin
Aldrin
Aldrite *see* Aldrin
Aldrosol *see* Aldrin
Alfatox *see* Diazinon
Algistat *see* Dichlone
Alidochlor *see* Allidochlor
Aliette *see* Fosetyl-aluminum
Alkron *see* Parathion
Alleron *see* Parathion
Allidochlor
Alltex *see* Toxaphene
Alltox *see* Toxaphene
Ally *see* Metsulfuron-methyl
Ally 20DF *see* Metsulfuron-methyl
Allyl aldehyde *see* Acrolein
Alochlor *see* Alachlor
Altox *see* Aldrin
Aluminum tris(*O*-ethyl phosphonate) *see* Fosetyl-aluminum
Alvit *see* Dieldrin
Amatin *see* Hexachlorobenzene
Amaze *see* Isofenphos
Amazol *see* Amitrole
Ambiben *see* Chloramben
Ambush *see* Aldicarb
Ambush *see* Permethrin
Amchem 68-250 *see* Ethephon
Amdon grazon *see* Picloram
Ameisenatod *see* Lindane
Ameisenmittel merck *see* Lindane
American Cyanamid 3422 *see* Parathion
American Cyanamid 3911 *see* Phorate
American Cyanamid 4049 *see* Malathion
American Cyanamid 12880 *see* Dimethoate

Anicon M *see* MCPA
Anilazin *see* Anilazine
Anilazine
Anofex *see* p,p'-DDT
Anthion *see* Trichlorfon
Anticarie *see* Hexachlorobenzene
Antinonin *see* 4,6-Dinitro-*o*-cresol
Antinonnon *see* 4,6-Dinitro-*o*-cresol
ANTU
Anturat *see* ANTU
Apadrin *see* Monocrotophos
Apamidon *see* Phosphamidon
Aparasin *see* Lindane
Apavap *see* Dichlorvos
Apavinphos *see* Mevinphos
Aphalon *see* Linuron
Aphamite *see* Parathion
Aphtiria *see* Lindane
Aplidal *see* Lindane
Apl-Luster *see* Thiabendazole
Appa *see* Phosmet
Aprocarb *see* Propoxur
Apron 2E *see* Metalaxyl
Aquacide *see* Diquat
Aqua-kleen *see* 2,4-D
Aqualin *see* Acrolein
Aqualine *see* Acrolein
Aquathol *see* Endothall
Aquazine *see* Simazine
Aquazine 80W *see* Simazine
Arasan *see* Thiram
Arasan 42-S *see* Thiram
Arasan 70 *see* Thiram
Arasan 75 *see* Thiram
Arasan-M *see* Thiram
Arasan-SF *see* Thiram
Arasan-SF-X *see* Thiram
Arbitex *see* Lindane
Arborol *see* 4,6-Dinitro-*o*-cresol
Arbotect *see* Thiabendazole
Ardap *see* Cypermethrin
Aretit *see* Dinoseb
Argezin *see* Atrazine

Arilate *see* Benomyl
Arkotine *see* p,p'-DDT
Arprocarb *see* Propoxur
Arthodibrom *see* Naled
Arylam *see* Carbaryl
Asilan *see* Asulam
ASP 47 *see* Sulfotepp
ASP 51 *see* Aspon
Aspon
Aspon-chlordane *see* Chlordane
Assure *see* Quizalofop-methyl
Astrobot *see* Dichlorvos
Asulam
Asulox *see* Asulam
Asulox 40 *see* Asulam
Asulfox F *see* Asulam
AT *see* Amitrole
AT-90 *see* Amitrole
ATA *see* Amitrole
Atazinax *see* Atrazine
ATCP *see* Picloram
Atgard *see* Dichlorvos
Atgard C *see* Dichlorvos
Atgard V *see* Dichlorvos
Athrombin-K *see* Warfarin
Athrombine-K *see* Warfarin
AT liquid *see* Amitrole
Atranex *see* Atrazine
Atrasine *see* Atrazine
Atratol A *see* Atrazine
Atrazin *see* Atrazine
Atrazine
Atred *see* Atrazine
Atrex *see* Atrazine
Attac 4-2 *see* Toxaphene
Attac 4-4 *see* Toxaphene
Attac 6 *see* Toxaphene
Attac 6-3 *see* Toxaphene
Attac 8 *see* Toxaphene
Aules *see* Thiram
Avadex *see* Diallate
Avadex BW *see* Triallate
Avenge *see* Difenzoquat methyl sulfate

Avicade *see* Cypermethrin
Avolin *see* Dimethyl phthalate
Azaplant *see* Amitrole
Azinphos-methyl
Azodrin *see* Monocrotophos
Azodrin insecticide *see* Monocrotophos
Azofene *see* Phosalone
Azolan *see* Amitrole
Azole *see* Amitrole
Azotox *see* p,p'-DDT
B 404 *see* Parathion
B 622 *see* Anilazine
B 1776 *see* Butifos
B 29493 *see* Fenthion
B 37344 *see* Methiocarb
B-Selektonon *see* 2,4-D
B-Selektonon M *see* MCPA
Banex *see* Dicamba
Ban-Hoe *see* Propham
Bantu *see* ANTU
Banvel *see* Dicamba
Banvel CST *see* Dicamba
Banvel D *see* Dicamba
Banvel herbicide *see* Dicamba
Banvel II herbicide *see* Dicamba
Banvel 4S *see* Dicamba
Banvel 4WS *see* Dicamba
Barricade *see* Cypermethrin
Barrier 2G *see* Dichlobenil
Barrier 50W *see* Dichlobenil
Basagran *see* Bentazone
Basaklor *see* Heptachlor
Basamid *see* Dazomet
Basamid G *see* Dazomet
Basamid-granular *see* Dazomet
Basamid P *see* Dazomet
Basamid-puder *see* Dazomet
Basanite *see* Dinoseb
Basfapon F *see* Dalapon-sodium
BAS 351-H *see* Bentazone
Basudin *see* Diazinon
Basudin 10 G *see* Diazinon
Batazina *see* Simazine

Bay 6681 F *see* Triadimefon
Bay 9010 *see* Propoxur
Bay 9026 *see* Methiocarb
Bay 9027 *see* Azinphos-methyl
Bay 15922 *see* Trichlorfon
Bay 17147 *see* Azinphos-methyl
Bay 19149 *see* Dichlorvos
Bay 19639 *see* Disulfoton
Bay 21097 *see* Oxydemeton-methyl
Bay 25141 *see* Fensulfothion
Bay 29493 *see* Fenthion
Bay 30130 *see* Propanil
Bay 36205 *see* Quinomethionate
Bay 37344 *see* Methiocarb
Bay 39007 *see* Propoxur
Bay 44646 *see* Aminocarb
Bay 61597 *see* Metribuzin
Bay 68138 *see* Fenamiphos
Bay 70143 *see* Carbofuran
Bay 71628 *see* Methamidophos
Bay 92114 *see* Isofenphos
Baycid *see* Fenthion
Bay dic 1468 *see* Metribuzin
Bay E-393 *see* Sulfotepp
Bay E-605 *see* Parathion
Bayer 4964 *see* Quinomethionate
Bayer 5080 *see* Aminocarb
Bayer 6159H *see* Metribuzin
Bayer 6443H *see* Metribuzin
Bayer 9007 *see* Fenthion
Bayer 9027 *see* Azinphos-methyl
Bayer 15922 *see* Trichlorfon
Bayer 17147 *see* Azinphos-methyl
Bayer 19639 *see* Disulfoton
Bayer 21097 *see* Oxydemeton-methyl
Bayer 24493 *see* Fenthion
Bayer 25141 *see* Fensulfothion
Bayer 36205 *see* Quinomethionate
Bayer 37344 *see* Methiocarb
Bayer 39007 *see* Propoxur
Bayer 44646 *see* Aminocarb
Bayer 71268 *see* Methamidophos
Bayer 94337 *see* Metribuzin

Bayer E 393 *see* Sulfotepp
Bayer L 13/59 *see* Trichlorfon
Bayer S 767 *see* Fensulfothion
Bayer S 1752 *see* Fenthion
Bay FCR 1272 *see* Cyfluthrin
Baygon *see* Propoxur
Bay-Hox-1901 *see* Ethiofencarb
Bayleton *see* Triadimefon
Bay-ntn-9306 *see* Sulprofos
Bay-sra-12869 *see* Isofenphos
Baytex *see* Fenthion
Baythroid *see* Cyfluthrin
Baythroid H *see* Cyfluthrin
Bazinon *see* Diazinon
Bazuden *see* Diazinon
BBC *see* Benomyl
BBC 12 *see* 1,2-Dibromo-3-chloropropane
BBX *see* Lindane
BCF-bushkiller *see* 2,4,5-T
BCIE *see* Bis(2-chloroisopropyl)ether
BCMEE *see* Bis(2-chloroisopropyl)ether
Beet-Kleen *see* Chlorpropham, Propham
Beetomax *see* Phenmedipham
Beetup *see* Phenmedipham
Belmark *see* Fenvalerate
Belt *see* Chlordane
Bencarbate *see* Bendiocarb
Bendex *see* Fenbutatin oxide
Bendiocarb
Bendioxide *see* Bentazone
Benfos *see* Dichlorvos
Ben-hex *see* Lindane
Benlat *see* Benomyl
Benlate *see* Benomyl
Benlate 50 *see* Benomyl
Benlate 50W *see* Benomyl
Benomyl
Benomyl 50W *see* Benomyl
Bentanex *see* Desmedipham
Bentazon *see* Bentazone
Bentazone
Bentox 10 *see* Lindane
Benzabar *see* 2,3,5-Trichlorobenzoic acid

Benzac *see* 2,3,5-Trichlorobenzoic acid

Benzac-1281 *see* 2,3,5-Trichlorobenzoic acid

1,2-Benzenedicarboxylate *see* Di-*n*-butyl phthalate

1,2-Benzenedicarboxylic acid dibutyl ester *see* Di-*n*-butyl phthalate

o-Benzenedicarboxylic acid dibutyl ester *see* Di-*n*-butyl phthalate

Benzene-*o*-dicarboxylic acid di-*n*-butyl ester *see* Di-*n*-butyl phthalate

1,2-Benzenedicarboxylic acid dimethyl ester *see* Dimethyl phthalate

Benzene hexachloride *see* Lindane

Benzene-*cis*-hexachloride *see* β-BHC

trans-α-Benzene hexachloride *see* β-BHC

Benzene-γ-hexachloride *see* Lindane

α-Benzene hexachloride *see* α-BHC

β-Benzene hexachloride *see* β-BHC

γ-Benzene hexachloride *see* Lindane

δ-Benzene hexachloride *see* δ-BHC

Benzene hexachloride-α-isomer *see* α-BHC

Benzilan *see* Chlorobenzilate

Benzinoform *see* Carbon tetrachloride

Benz-o-chlor *see* Chlorobenzilate

Benzoepin *see* α-Endosulfan, *see* β-Endosulfan

Benzoic acid *o*-((3-(4,6-dimethyl-2-pyrimidinyl)ureido)sulfonyl) methyl
    ester *see* Sulfometuron-methyl

Benzophosphate *see* Phosalone

Benzotriazinedithiophosphoric acid dimethoxy ester *see* Azinphos-methyl

S-((3-Benzoxazolinyl-6-chloro-2-oxo)methyl) *O,O*-diethyl phosphorodi-
    thioate *see* Phosalone

Benzphos *see* Phosalone

Beosit *see* α-Endosulfan, β-Endosulfan

Bercema Fertam 50 *see* Ferbam

Bermat *see* Chlordimeform

Betanal *see* Phenmedipham

Betanal AM *see* Desmedipham

Betanal E *see* Phenmedipham

Betanex *see* Desmedipham

Betosip *see* Phenmedipham

Bexol *see* Lindane

Bexton *see* Propachlor

Bexton 4L *see* Propachlor

BFV *see* Formaldehyde

BHC *see* Lindane

α-BHC

β-BHC

δ-BHC

γ-BHC *see* Lindane
BH 2,4-D *see* 2,4-D
BH MCPA *see* MCPA
BI 58 *see* Dimethoate
BI 58 EC *see* Dimethoate
Bibenzene *see* Biphenyl
Bibesol *see* Dichlorvos
Bicep *see* Metolachlor
Bidirl *see* Dicrotophos
Bidrin *see* Dicrotophos
Bifenox
Bifex *see* Propoxur
Bilarcil *see* Trichlorfon
Biloborb *see* Monocrotophos
Bilobran *see* Monocrotophos
Bio 5462 *see* α-Endosulfan, β-Endosulfan
Biocide *see* Acrolein
Biphenyl
1,1'-Biphenyl *see* Biphenyl
2,2-Bis(*p*-anisyl)-1,1,1-trichloroethane *see* Methoxychlor
S-1,2-Bis(carbethoxy)ethyl-*O*,*O*-dimethyl dithiophosphate *see* Malathion
Bis(2-chloroethyl)ether
Bis(β-chloroethyl)ether *see* Bis(2-chloroethyl)ether
Bis(2-chloroisopropyl)ether
Bis(β-chloroisopropyl)ether *see* Bis(2-chloroisopropyl)ether
Bis(2-chloro-1-methylethyl)ether *see* Bis(2-chloroisopropyl)ether
1,1-Bis(4-chlorophenyl)-2,2-dichloroethane *see* *p*,*p'*-DDD
1,1-Bis(*p*-chlorophenyl)-2,2-dichloroethane *see* *p*,*p'*-DDD
2,2-Bis(4-chlorophenyl)-1,1-dichloroethane *see* *p*,*p'*-DDD
2,2-Bis(*p*-chlorophenyl)-1,1-dichloroethane *see* *p*,*p'*-DDD
2,2-Bis(4-chlorophenyl)-1,1-dichloroethene *see* *p*,*p'*-DDE
2,2-Bis(*p*-chlorophenyl)-1,1-dichloroethene *see* *p*,*p'*-DDE
1,1-Bis(4-chlorophenyl)-2,2-dichloroethylene *see* *p*,*p'*-DDE
1,1-Bis(*p*-chlorophenyl)-2,2-dichloroethylene *see* *p*,*p'*-DDE
1,1-Bis(*p*-chlorophenyl)-2,2,2-trichloroethane *see* *p*,*p'*-DDT
2,2-Bis(4-chlorophenyl)-1,1,1-trichloroethane *see* *p*,*p'*-DDT
2,2-Bis(*p*-chlorophenyl)-1,1,1-trichloroethane *see* *p*,*p'*-DDT
α,α-Bis(*p*-chlorophenyl)-β,β,β-trichloroethane *see* *p*,*p'*-DDT
Bis-*O*,*O*-diethylphosphoric anhydride *see* Tetraethyl pyrophosphate
Bis-*O*,*O*-diethylphosphorothionic anhydride *see* Sulfotepp
Bis(*S*-(dimethoxyphosphinothioyl)mercapto)methane *see* Ethion
Bis(dimethylamino)carbonothioyl disulfide *see* Thiram
Bis(dimethylthiocarbamoyl) disulfide *see* Thiram

Bovinox *see* Trichlorfon
Bovizole *see* Thiabendazole
Boygon *see* Propoxur
Bravo *see* Chlorothalonil
Bravo 6F *see* Chlorothalonil
Bravo-W-75 *see* Chlorothalonil
Brevinyl *see* Dichlorvos
Brevinyl E50 *see* Dichlorvos
Briton *see* Trichlorfon
Britten *see* Trichlorfon
Brittox *see* Bromoxynil
Brodan *see* Chlorpyrifos
Bromacil
Bromax *see* Bromacil
Bromax 4G *see* Bromacil
Bromax 4L *see* Bromacil
Bromazil *see* Bromacil
Bromchlophos *see* Naled
Bromex *see* Naled
Brominal *see* Bromoxynil
Brominal M & Plus *see* MCPA
Brominex *see* Bromoxynil
Brominil *see* Bromoxynil
5-Bromo-3-*sec*-butyl-6-methyluracil *see* Bromacil
O-(4-Bromo-2-chlorophenyl) *O*-ethyl *S*-propyl phosphorothioate *see*
   Profenofos
Bromoflor *see* Ethephon
Bromofume *see* Ethylene dibromide
Brom-o-gas *see* Methyl bromide
Brom-o-gaz *see* Methyl bromide
Bromomethane *see* Methyl bromide
5-Bromo-6-methyl-3-(1-methylpropyl)-2,4(1*H*3*H*)-pyrimidinedione *see*
   Bromacil
5-Bromo-6-methyl-3-(1-methylpropyl)uracil *see* Bromacil
Bromoxynil
Bromoxynil octanoate
Bronate *see* Bromoxynil octanoate
Bronco *see* Alachlor
Broxynil *see* Bromoxynil
Brulan *see* Tebuthiuron
Brumolin *see* Warfarin
Brush buster *see* Dicamba
Brush-off 445 low volatile brush killer *see* 2,4,5-T

Brush-rhap *see* 2,4-D, 2,4,5-T
Brushtox *see* 2,4,5-T
Brygou *see* Propoxur
Buckle *see* Triallate
Buctril *see* Bromoxynil, Bromoxynil octanoate
Buctril 20 *see* Bromoxynil
Buctril 21 *see* Bromoxynil
Buctril industrial *see* Bromoxynil
Bud-nip *see* Chlorpropham
Bullet *see* Alachlor
Bunt-cure *see* Hexachlorobenzene
Bunt-no-more *see* Hexachlorobenzene
Burtolin *see* Maleic hydrazide
Bushwacker *see* Tebuthiuron
Butaphene *see* Dinoseb
2-Butenoic acid 3-((dimethoxyphosphinyl)oxy)methyl ester *see* Mevinphos
Butifos
Butilate *see* Butylate
Butilchlorofos *see* Bromoxynil
Butiphos *see* Butifos
(1-((Butylamino)carbonyl)-1*H*-benzimidazol-2-yl)carbamic acid methyl ester
    see Benomyl
2-*tert*-Butylamino-4-ethylamino-6-methylmercapto-*s*-triazine *see* Terbutryn
2-*tert*-Butylamino-4-ethylamino-6-methylthio-*s*-triazine *see* Terbutryn
Butylate
1-(Butylcarbamoyl)-2-benzimidazolecarbamic acid methyl ester *see* Benomyl
3-*tert*-Butyl-5-chloro-6-methyluracil *see* Terbacil
2-*tert*-Butyl-4-(2-2,4-dichloro-5-isopropyloxyphenyl)-1,3,4-oxadiazolin-5-
    one *see* Oxadiazon
1-Butyl-3-(3,4-dichlorophenyl)-1-methylurea *see* Neburon
N-Butyl-*N'*-(3,4-dichlorophenyl)-*N*-methylurea *see* Neburon
2-*sec*-Butyl-2,4-dinitrophenol *see* Dinoseb
Butylethylthiocarbamic acid *S*-propyl ester *see* Pebulate
Butyl phosphorotrithioate *see* Butifos
Butyl phthalate *see* Di-*n*-butyl phthalate
n-Butyl phthalate *see* Di-*n*-butyl phthalate
1-(5-*t*-Butyl-1,3,4-thiadiazol-2-yl)-1,3-dimethylurea *see* Tebuthiuron
S-((*tert*-Butylthio)methyl) *O,O*-diethyl phosphorodithioate *see* Terbufos
BW-21-Z *see* Permethrin
C 570 *see* Phosphamidon
C 709 *see* Dicrotophos
C 1414 *see* Monocrotophos
C 1983 *see* Chloroxuron

C 2059 *see* Fluometuron
C 8514 *see* Chlordimeform
Caldon *see* Dinoseb
Caliper *see* Simazine
Caliper 90 *see* Simazine
Calmathion *see* Malathion
Campaprim A 1544 *see* Amitrole
Camphechlor *see* Toxaphene
Camphochlor *see* Toxaphene
Camphoclor *see* Toxaphene
Camposan *see* Ethephon
Candex *see* Atrazine
Cannon *see* Alachlor
Canogard *see* Dichlorvos
Caparol *see* Prometryn
Caparol 80W *see* Prometryn
Capsine *see* 4,6-Dinitro-*o*-cresol
Carbacryl *see* Acrylonitrile
Carbamate *see* Ferbam
Carbamic acid ethylenebis(dithio-, manganese zinc complex) *see* Mancozeb
Carbamine *see* Carbaryl
Carbanilic acid isopropyl ester *see* Propham
Carbanolate *see* Aldicarb
Carbaryl
Carbatox *see* Carbaryl
Carbatox 60 *see* Carbaryl
Carbatox 75 *see* Carbaryl
Carbethoxy malathion *see* Malathion
Carbetovur *see* Malathion
Carbetox *see* Malathion
Carbicron *see* Dicrotophos
Carbofos *see* Malathion
Carbofuran
2-Carbomethoxy-1-methylvinyl dimethyl phosphate *see* Mevinphos
α-Carbomethoxy-1-methylvinyl dimethyl phosphate *see* Mevinphos
2-Carbomethoxy-1-propen-2-yl dimethyl phosphate *see* Mevinphos
Carbon bisulfide *see* Carbon disulfide
Carbon bisulphide *see* Carbon disulfide
Carbon chloride *see* Carbon tetrachloride
Carbon disulfide
Carbon disulphide *see* Carbon disulfide
Carbon sulfide *see* Carbon disulfide
Carbon sulphide *see* Carbon disulfide

Carbona *see* Carbon tetrachloride
Carbon tet *see* Carbon tetrachloride
Carbon tetrachloride
Carbophenothion
Carbophos *see* Malathion
Carbothialdin *see* Dazomet
Carbothialdine *see* Dazomet
5-Carboxanilido-2,3-dihydro-6-methyl-1,4-oxathiin *see* Carboxin
Carboxin
Carboxine *see* Carboxin
Carfene *see* Azinphos-methyl
Carmazine *see* Mancozeb
Carpolin *see* Carbaryl
Carsoron *see* Dichlobenil
Carylderm *see* Carbaryl
Carzol *see* Chlordimeform
Casaron *see* Dichlobenil
Casoron *see* Dichlobenil
Casoron 133 *see* Dichlobenil
Casoron G *see* Dichlobenil
Casoron G-4 *see* Dichlobenil
Casoron G-10 *see* Dichlobenil
Casoron W-50 *see* Dichlobenil
CAT *see* Simazine
CCN52 *see* Cypermethrin
CD-68 *see* Chlordane
CDAA *see* Allidochlor
CDAAT *see* Allidochlor
CDM *see* Chlordimeform
CDT *see* Simazine
Cekiuron *see* Diuron
Cekubaryl *see* Carbaryl
Cekufon *see* Trichlorfon
Cekusan *see* Dichlorvos
Cekusan *see* Simazine
Cekuthoate *see* Dimethoate
Cekuzina-S *see* Simazine
Cekuzina-T *see* Atrazine
Celanex *see* Lindane
Celfume *see* Methyl bromide
Celluflex DPB *see* Di-*n*-butyl phthalate
Celmide *see* Ethylene dibromide
Celthion *see* Malathion

Chiptox *see* MCPA

Chlorak *see* Trichlorfon

Chloramben

Chloran *see* Lindane

Chlorbenzilat *see* Chlorobenzilate

Chlorbenzilate *see* Chlorobenzilate

Chlordan *see* Chlordane

α-Chlordan *see* trans-Chlordane

γ-Chlordan *see* Chlordane

cis-Chlordan *see* trans-Chlordane

Chlordane

α-Chlordane *see* cis-Chlordane, trans-Chlordane

β-Chlordane *see* cis-Chlordane

γ-Chlordane *see* trans-Chlordane

***cis*-Chlordane**

α(cis)-Chlordane *see* trans-Chlordane

***trans*-Chlordane**

Chlordecone *see* Kepone

Chlordimeform

Chloresene *see* Lindane

Chlorethephon *see* Ethephon

Chlorex *see* Bis(2-chloroethyl)ether

Chlorfenamidine *see* Chlordimeform

Chlorfenidim *see* Monuron

Chlorfos *see* Trichlorfon

Chloridan *see* Chlordane

Chlor-IFC *see* Chlorpropham

Chlorinated camphene *see* Toxaphene

Chlorindan *see* Chlordane

Chlor-IPC *see* Chlorpropham

Chlor kil *see* Chlordane

Chloroalonil *see* Chlorothalonil

(o-Chloroanilo)dichlorotriazine *see* Anilazine

Chlorobenzilate

Chlorobenzylate *see* Chlorobenzilate

2-Chloro-4,6-bis(ethylamino)-1,3,5-triazine *see* Simazine

2-Chloro-4,6-bis(ethylamino)-s-triazine *see* Simazine

2-Chloro-4,6-bis(isopropylamino)-s-triazine *see* Propazine

6-Chloro-N,N'-bis(1-methylethyl)-1,3,5-triazine-2,4-diamine *see* Propazine

5-Chloro-3-*tert*-butyl-6-methyluracil *see* Terbacil

Chloro-camphene *see* Toxaphene

3-Chlorochlordene *see* Heptachlor

1-Chloro-2-(β-chloroethoxy)ethane *see* Bis(2-chloroethyl)ether

1-Chloro-2-(β-chloroisopropoxy)propane *see* Bis(2-chloroisopropyl)ether

4-Chloro-*o*-cresolxyacetic acid *see* MCPA

2-Chloro-4-(1-cyano-1-methylethylamino)-6-ethylamino-1,3,5-triazine *see* Cyanazine

Chlorodane *see* Chlordane

2-Chloro-*N,N*-diallylacetamide *see* Allidochlor

α-Chloro-*N,N*-diallylacetamide *see* Allidochlor

1-Chloro-2,3-dibromopropane *see* 1,2-Dibromo-3-chloropropane

3-Chloro-1,2-dibromopropane *see* 1,2-Dibromo-3-chloropropane

2-Chloro-*N*-(4,6-dichloro-1,3,5-triazin-2-yl)aniline *see* Anilazine

2-Chloro-3-(diethylamino)-1-methyl-3-oxo-1-propenyl dimethyl phosphate *see* Phosphamidon

2-Chloro-2-diethylcarbamoyl-1-methylvinyl dimethylphosphate *see* Phosphamidon

2-Chloro-2′,6′-diethyl-*N*-(methoxymethyl)acetanilide *see* Alachlor,

2-Chloro-*N*-(2,6-diethylphenyl)-*N*-(methoxymethyl)acetamide *see* Alachlor

6-Chloro-*N′,N′*-diethyl-1,3,5-triazine-2,4-diamine *see* Simazine

6-Chloro-N^2,N^4-diethyl-1,3,5-triazine-2,4-diyldiamine *see* Simazine

S-(2-Chloro-1-(1,3-dihydro-1,3-dioxo-2*H*-isoindol-2-yl)ethyl) *O,O*-diethyl phosphorodithioate, *see* Dialifos

6-Chloro-*N,N′*-diisopropyl-1,3,5-triazine-2,4-diyldiamine *see* Propazine

5-Chloro-3-(1,1-dimethylethyl)-6-methyl-2,4(1*H*,3*H*)-pyrimidinedione *see* Terbacil

4′-Chloro-2,2-dimethylvaleranilide *see* Monalide

4′-Chloro-α,α-dimethylvaleranilide *see* Monalide

2-Chloro-*N,N*-di-2-propenylacetamide *see* Allidochlor

2-Chloroethanephosphonic acid *see* Ethephon

2-Chloro-4-ethylamineisopropylamine-*s*-triazine *see* Atrazine

1-Chloro-3-ethylamino-5-isopropylamino-2,4,6-triazine *see* Atrazine

1-Chloro-3-ethylamino-5-isopropylamino-*s*-triazine *see* Atrazine

2-Chloro-4-ethylamino-6-isopropylamino-1,3,5-triazine *see* Atrazine

2-Chloro-4-ethylamino-6-isopropylamino-*s*-triazine *see* Atrazine

2-(4-Chloro-6-ethylamino-*s*-triazine-2-ylamino)-2-methylpropionitrile *see* Cyanazine

2-((Chloro-6-(ethylamino)-1,3,5-triazin-2-yl)amino)-2-methylpropanenitrile *see* Cyanazine

2-((4-Chloro-6-(ethylamino)-*s*-triazin-2-yl)amino)-2-methylpropionitrile *see* Cyanazine

Chloroethyl ether *see* Bis(2-chloroethyl)ether

2-Chloroethyl ether *see* Bis(2-chloroethyl)ether

(β-Chloroethyl)ether *see* Bis(2-chloroethyl)ether

6-Chloro-N^2-ethyl-N^4-isopropyl-1,3,5-triazine-2,4-diamine *see* Atrazine

2-Chloro-6′-ethyl-*N*-(2-methoxy-1-methylethyl)acet-*o*-toluidide *see*

Metolachlor

α-Chloro-2′-ethyl-6′-methyl-*N*-(1-methyl-2-methoxyethyl)acetanilide *see* Metolachlor

2-Chloro-*N*-(2-ethyl-6-methylphenyl)-*N*-(2-methoxy-1-methylethyl)-acetamide *see* Metolachlor

(2-Chloroethyl)phosphonic acid *see* Ethephon

Chlorofos *see* Trichlorfon

Chloro-IFK *see* Chlorpropham

Chloro-IPC *see* Chlorpropham

2-Chloro-*N*-isopropylacetanilide *see* Propachlor

α-Chloro-*N*-isopropylacetanilide *see* Propachlor

2-Chloroisopropyl ether *see* Bis(2-chloroisopropyl)ether

β-Chloroisopropyl ether *see* Bis(2-chloroisopropyl)ether

S-(6-Chloro-3-(mercaptomethyl)-2-benzoxazolinone) *O,O*-diethyl phosphorothiolothionate *see* Phosalone

2-Chloro-*N*-(((4-methoxy-6-methyl-1,3,5-triazin-2-yl)amino)carbonyl)-benzenesulfonamide *see* Chlorsulfuron

(2-Chloro-1-methylethyl)ether *see* Bis(2-chloroisopropyl)ether

4-Chloro-2-methylphenoxyacetic acid *see* MCPA

N′-(4-Chloro-2-methylphenyl)-*N,N*-dimethylmethanimidamide *see* Chlordimeform

S-((6-Chloro-2-oxo-3(2*H*)-benzoxazolyl)methyl) *O,O*-diethyl phosphorodithioate *see* Phosalone

Chlorophen *see* Pentachlorophenol

Chlorophenamidin *see* Chlordimeform

Chlorophenamidine *see* Chlordimeform

Chlorophenothan *see* *p,p*′-DDT

Chlorophenothane *see* *p,p*′-DDT

Chlorophenotoxum *see* *p,p*′-DDT

1-(4-Chlorophenoxy)-3,3-dimethyl-1-(1*H*-1,2,4-triazol-1-yl)-2-butanone *see* Triadimefon

3-(*p*-(*p*-Chlorophenoxy)phenyl)-1,1-dimethylurea *see* Chloroxuron

N′-(4-(4-Chlorophenoxy)phenyl)-*N,N*-dimethylurea *see* Chloroxuron

1-(4-Chlorophenoxy)-3,3-(1,2,4-triazol-1-yl)butan-2-one *see* Triadimefon

N-(((4-Chlorophenyl)amino)carbonyl)-2,6-difluorobenzamide *see* Diflubenzuron

(3-Chlorophenyl)carbamic acid 1-methylethyl ester *see* Chlorpropham

trans-5-(4-Chlorophenyl)-*N*-cyclohexyl-4-methyl-2-oxo-3-thiazolidine-carboxamide *see* Hexythiazox

1-(4-Chlorophenyl)-3-(2,6-difluorobenzoyl)urea *see* Diflubenzuron

N-(4-Chlorophenyl)-2,2-dimethylpentanamide *see* Monalide

1-(4-Chlorophenyl)-3,3-dimethylurea *see* Monuron

1-(*p*-Chlorophenyl)-3,3-dimethylurea *see* Monuron

3-(4-Chlorophenyl)-1,1-dimethylurea *see* Monuron
3-(*p*-Chlorophenyl)-1,1-dimethylurea *see* Monuron
N-(4-Chlorophenyl)-*N'*,*N'*-dimethylurea *see* Monuron
N'-(4-Chlorophenyl)-*N*,*N*-dimethylurea *see* Monuron
N'-(*p*-Chlorophenyl)-*N*,*N*-dimethylurea *see* Monuron
N-(4-Chlorophenyl)-2,2-dimethylvaleramide *see* Monalide
1-((*o*-Chlorophenyl)sulfonyl)-3-(4-methoxy-6-methyl-*s*-triazin-2-yl) urea
    see Chlorsulfuron
S-((4-Chlorophenyl)thio)methyl *O*,*O*-diethyl phosphorodithioate *see*
    Carbophenothion
S-((*p*-Chlorophenyl)thio)methyl *O*,*O*-diethyl phosphorodithioate *see*
    Carbophenothion
S-(4-Chlorophenylthiomethyl)diethyl phosphorothiolothionate *see*
    Carbophenothion
Chlorophos *see* Trichlorfon
Chlorophose *see* Trichlorfon
S-(2-Chloro-1-phthalimidoethyl) *O*,*O*-diethyl phosphorodithioate *see*
    Dialifos
Chlorophthalm *see* Trichlorfon
Chlor-o-pic *see* Chloropicrin
Chloropicrin
Chloropropham *see* Chlorpropham
2-Chloro-4-(2-propylamino)-6-ethylamino-*s*-triazine *see* Atrazine
Chlorothal *see* Chlorthal-dimethyl
4-Chloro-*o*-toloxyacetic acid *see* MCPA
N'-(4-Chloro-*o*-tolyl)-*N*,*N*-dimethylformamidine *see* Chlordimeform
2-Chloro-6-(trichloromethyl)pyridine *see* Nitrapyrin
Chloroxifenidim *see* Chloroxuron
Chloroxone *see* 2,4-D
Chloroxuron
Chloroxyphos *see* Trichlorfon
Chlorphenamidine *see* Chlordimeform
Chlorpropham
Chlorpyrifos
Chlorpyrifos-ethyl *see* Chlorpyrifos
Chlorsulfuron
Chlorthal-dimethyl
Chlorthiepin *see* α-Endosulfan, β-Endosulfan
Chlortox *see* Chlordane
Chlorvinphos *see* Dichlorvos
Chwastox *see* MCPA
CIBA 570 *see* Phosphamidon
CIBA 709 *see* Dicrotophos

Cornox-M *see* MCPA
Corodane *see* Chlordane
Corothion *see* Parathion
Corthion *see* Parathion
Corthione *see* Parathion
Cortilan-neu *see* Chlordane
Cotneon *see* Azinphos-methyl
Cotnion methyl *see* Azinphos-methyl
Cotofor *see* Dipropetryn
Cotoran *see* Fluometuron
Cotoran multi *see* Metolachlor
Cotoran multi 50WP *see* Fluometuron
Cottonex *see* Fluometuron
Cotton-Pro *see* Prometryn
Coumadin *see* Warfarin
Coumafen *see* Warfarin
Coumafene *see* Warfarin
Counter *see* Terbufos
Counter 15G soil insecticide *see* Terbufos
Counter 15G soil insecticide-nematocide *see* Terbufos
Cov-r-tox *see* Warfarin
CP 6343 *see* Allidochlor
CP 15336 *see* Diallate
CP 23426 *see* Triallate
CP 31393 *see* Propachlor
CP 50144 *see* Alachlor
Crag 974 *see* Dazomet
Crag fungicide 974 *see* Dazomet
Crag nematocide *see* Dazomet
Crag sevin *see* Carbaryl
Crag 85W *see* Dazomet
Crestoxo *see* Toxaphene
Crestoxo 90 *see* Toxaphene
Crisalin *see* Trifluralin
Crisatine *see* Ametryn
Crisatrina *see* Atrazine
Crisazine *see* Atrazine
Crisodrin *see* Monocrotophos
Crisulfan *see* α-Endosulfan, β-Endosulfan
Crisuron *see* Diuron
Crolean *see* Acrolein
Croneton *see* Ethiofencarb
Crop rider *see* 2,4-D

Crotilin *see* 2,4-D

Crotoxyphos

Cryptogil OL *see* Pentachlorophenol

Crystal Propanil-4 *see* Propanil

Crysthion 2L *see* Azinphos-methyl

Crysthyon *see* Azinphos-methyl

Curacron *see* Profenofos

Cyanazine

Cyanoethylene *see* Acrylonitrile

Cyano(4-fluoro-3-phenoxyphenyl)methyl 3-(2,2-dichloroethenyl)-2,2-di-
methylcyclopropanecarboxylate *see* Cyfluthrin

α-Cyano-3-phenoxybenzyl-2-(4-chlorophenyl)-3-methylbutyrate *see*
Fenvalerate

(+)-α-Cyano-3-phenoxybenzyl 2,2-dimethyl-3-(2,2-dichlorovinyl)cyclo-
propane carboxylate, *see* Cypermethrin

(*S*-(*R**,*R**))-Cyano(3-phenoxyphenyl)methyl 4-chloro-α-(1-methylethyl)-
benzeneacetate *see* Esfenvalerate

Cyano(3-phenoxyphenyl)methyl 3-(2,2-dichloroethenyl)-2,2-dimethyl-
cyclopropanecarboxylate, *see* Cypermethrin

(*RS*)-Cyano-(3-phenoxyphenyl)methyl (*S*)-4-(difluoromethoxy)phenyl)-α-
(1-methylethyl)benzeneacetate *see* Flucythrinate

Cyanophos *see* Dichlorvos

Cyazin *see* Atrazine

Cybolt *see* Flucythrinate

Cyclic *S*,*S*-(6-methyl-2,3-quinoxalinediyl) dithiocarbonate *see*
Quinomethionate

Cycloate

Cyclodan *see* α-Endosulfan, β-Endosulfan

3-Cyclohexyl-6-(dimethylamino)-1-methyl-1,3,5-triazine-2,4(1*H*,3*H*)dione
see Hexazinone

3-Cyclohexyl-6-(dimethylamino)-1-methyl-*s*-triazine-2,4(1*H*,3*H*)dione *see*
Hexazinone

2-Cyclopropylamino-4,6-diamino-*s*-triazine *see* Cyromazine

N-Cyclopropyl-1,3,5-triazine-2,4,6-triamine *see* Cyromazine

Cyfluthrin

Cygon *see* Dimethoate

Cygon 4E *see* Dimethoate

Cygon insecticide *see* Dimethoate

Cymbush *see* Cypermethrin

Cynogan *see* Bromacil

Cyodrin *see* Crotoxyphos

Cyperkill *see* Cypermethrin

Cypermethrin

Cypona *see* Dichlorvos
Cypona E.C. *see* Crotoxyphos
Cyromazine
Cythion *see* Malathion
Cytrol *see* Amitrole
Cytrol Amitrole-T *see* Amitrole
Cytrole *see* Amitrole
Cyuram DS *see* Thiram
2,4-D
D 50 *see* 2,4-D
D 735 *see* Carboxin
D 1410 *see* Oxamyl
D 1991 *see* Benomyl
D-90-A *see* Monalide
DAC 893 *see* Chlorthal-dimethyl
DAC 2787 *see* Chlorothalonil
Dacamine *see* 2,4-D, 2,4,5-T
2,4-D acid *see* 2,4-D
Daconil *see* Chlorothalonil
Daconil 2787 *see* Chlorothalonil
Daconil 2787 flowable fungicide *see* Chlorothalonil
Dacosoil *see* Chlorothalonil
Dacthal *see* Chlorthal-dimethyl
Dacthalor *see* Chlorthal-dimethyl
Dagadip *see* Carbophenothion
Dailon *see* Diuron
Dalapon *see* Dalapon-sodium
Dalapon-sodium
Dalapon sodium salt *see* Dalapon-sodium
Danex *see* Trichlorfon
Danthion *see* Parathion
Daphene *see* Dimethoate
Dasanit *see* Fensulfothion
DATC *see* Diallate
Dawson 100 *see* Methyl bromide
Dazomet
Dazomet-powder BASF *see* Dazomet
Dazzel *see* Diazinon
DBCP *see* 1,2-Dibromo-3-chloropropane
DBD *see* Azinphos-methyl
DBE *see* Ethylene dibromide
DBH *see* Lindane
2,6-DBN *see* Dichlobenil

DBP *see* Di-*n*-butyl phthalate

DCB *see* Dichlobenil

2,3-DCDT *see* Diallate

DCEE *see* Bis(2-chloroethyl)ether

DCMO *see* Carboxin

DCMU *see* Diuron

D-con *see* Warfarin

DCPA *see* Chlorthal-dimethyl, Propanil

DDD *see* *p,p'*-DDD

4,4'-DDD *see* *p,p'*-DDD

***p,p'*-DDD**

DDE *see* *p,p'*-DDE

4,4'-DDE *see* *p,p'*-DDE

***p,p'*-DDE**

DDP *see* Parathion

DDT *see* *p,p'*-DDT

4,4'-DDT *see* *p,p'*-DDT

***p,p'*-DDT**

DDT dehydrochloride *see* *p,p'*-DDE

DDVF *see* Dichlorvos

DDVP *see* Dichlorvos

Debroussaillant 600 *see* 2,4-D

Debroussaillant concentre *see* 2,4,5-T

Debroussaillant super concentre *see* 2,4,5-T

Decabane *see* Dichlobenil

1,2,3,5,6,7,8,9,10,10-Decachloro[$5.2.1.0^{2,6}.0^{3,9}.0^{5,8}$]decano-4-one *see* Kepone

Decachloroketone *see* Kepone

Decachloro-1,3,4-metheno-2*H*-cyclobuta[*cd*]pentalen-2-one *see* Kepone

Decachlorooctahydrokepone-2-one *see* Kepone

Decachlorooctahydro-1,3,4-metheno-2*H*-cyclobuta[*cd*]pentalen-2-one *see* Kepone

1,1a,3,3a,4,5,5a,5b,6-Decachlorooctahydro-1,3,4-metheno-2*H*-cyclo-buta[*cd*]pentalen-2-one, *see* Kepone

Decachloropentacyclo[$5.2.1.0^{2,6}.0^{3,9}.0^{5,8}$]decan-3-one *see* Kepone

Decachloropentacyclo[$5.2.1.0^{2,6}.0^{4,10}.0^{5,9}$]decan-3-one *see* Kepone

Decachlorotetracyclodecanone *see* Kepone

Decachlorotetrahydro-4,7-methanoindeneone *see* Kepone

Decamine *see* 2,4-D

Decamine 4T *see* 2,4,5-T

Decrotox *see* Crotoxyphos

De-cut *see* Maleic hydrazide

Dedelo *see* *p,p'*-DDT

Dedevap *see* Dichlorvos

Ded-weed *see* 2,4-D, MCPA
Ded-weed brush killer *see* 2,4,5-T
Ded-weed LV-6 brush-kil and T-5 brush-kil *see* 2,4,5-T
Ded-weed LV-69 *see* 2,4-D
Def *see* Butifos
Def defoliant *see* Butifos
De-Fend *see* Dimethoate
Defolit *see* Thidiazuron
Degrassan *see* 4,6-Dinitro-*o*-cresol
De-green *see* Butifos
Deiquat *see* Diquat
Dekrysil *see* 4,6-Dinitro-*o*-cresol
Demeton-methyl sulfoxide *see* Oxydemeton-methyl
Demeton-*O*-methyl sulfoxide *see* Oxydemeton-methyl
Demeton-*S*-methyl sulfoxide *see* Oxydemeton-methyl
Demos-L40 *see* Dimethoate
Denapon *see* Carbaryl
Deoval *see* *p,p'*-DDT
DEP *see* Trichlorfon
Depthon *see* Trichlorfon
Deriban *see* Dichlorvos
Dermaphos *see* Ronnel
Derribante *see* Dichlorvos
Desapon *see* Diazinon
Des-i-cate *see* Endothall
Desmedipham
Desormone *see* 2,4-D
Detal *see* 4,6-Dinitro-*o*-cresol
DETF *see* Trichlorfon
Dethmor *see* Warfarin
Dethnel *see* Warfarin
Detmol-extrakt *see* Lindane
Detmol MA *see* Malathion
Detmol MA 96% *see* Malathion
Detmol U.A. *see* Chlorpyrifos
Detox *see* *p,p'*-DDT
Detox 25 *see* Lindane
Detoxan *see* *p,p'*-DDT
Devicarb *see* Carbaryl
Devigon *see* Dimethoate
Devikol *see* Dichlorvos
Devoran *see* Lindane
Devrinol *see* Napropamide

Devrinol 2EC *see* Napropamide
Devrinol 10G *see* Napropamide
Devrinol 50WP *see* Napropamide
Dextrone *see* Diquat
Dfluron *see* Diflubenzuron
Dialifor *see* Dialifos
Dialifos
Diallate
Di-allate *see* Diallate
Diallylchloroacetamide *see* Allidochlor
N,N-Diallylchloroacetamide *see* Allidochlor
N,N-Diallyl-2-chloroacetamide *see* Allidochlor
N,N-Diallyl-α-chloroacetamide *see* Allidochlor
Diamide *see* Diphenamid
Dianate *see* Dicamba
2,2-Di-*p*-anisyl-1,1,1-trichloroethane *see* Methoxychlor
Dianon *see* Diazinon
Diater *see* Diuron
Diaterr-fos *see* Diazinon
Diazatol *see* Diazinon
Diazide *see* Diazinon
Diazinon
Dibovan *see* *p,p'*-DDT
Dibrom *see* Naled
Dibromochloropropane *see* 1,2-Dibromo-3-chloropropane
1,2-Dibromo-3-chloropropane
2,6-Dibromo-4-cyanophenol *see* Bromoxynil
2,6-Dibromo-4-cyanophenyl octanoate *see* Bromoxynil octanoate
1,2-Dibromo-2,2-dichloroethyldimethyl phosphate *see* Naled
Dibromoethane *see* Ethylene dibromide
1,2-Dibromoethane *see* Ethylene dibromide
α,β-Dibromoethane *see* Ethylene dibromide
sym-Dibromoethane *see* Ethylene dibromide
3,5-Dibromo-4-hydroxybenzonitrile *see* Bromoxynil
3,5-Dibromo-4-hydroxyphenylcyanide *see* Bromoxynil
3,5-Dibromo-4-octanoyloxybenzonitrile *see* Bromoxynil octanoate
Dibutyl-1,2-benzenedicarboxylate *see* Di-*n*-butyl phthalate
Dibutyl phthalate *see* Di-*n*-butyl phthalate
Di-*n*-butyl phthalate
DIC 1468 *see* Metribuzin
Dicamba
Dicambe *see* Dicamba
Dicarbam *see* Carbaryl

S-1,2-Dicarbethoxyethyl-*O,O*-dimethyl dithiophosphate *see* Malathion
Dicarboethoxyethyl-*O,O*-dimethyl phosphorodithioate *see* Malathion
Dicarbosulf *see* Thiodicarb
Dichlobenil
Dichlone
Dichlor-fenidim *see* Diuron
Dichlorman *see* Dichlorvos
Dichloroallyl diisopropylthiocarbamate *see* Diallate
S-Dichloroallyl diisopropylthiocarbamate *see* Diallate
Dichloroallyl *N,N*-diisopropylthiolcarbamate *see* Diallate
3,6-Dichloro-*o*-anisic acid *see* Dicamba
4,4′-Dichlorobenzilate *see* Chlorobenzilate
4,4′-Dichlorobenzilic acid ethyl ester *see* Chlorobenzilate
2,6-Dichlorobenzonitrile *see* Dichlobenil
1,1-Dichloro-2,2-bis(*p*-chlorophenyl)ethane *see* *p,p*′-DDD
1,1-Dichloro-2,2-bis(*p*-chlorophenyl)ethylene *see* *p,p*′-DDE
Dichlorochlordene *see* Chlordane
2,4-Dichloro-6-(2-chloroanilo)-1,3,5-triazine *see* Anilazine
2,4-Dichloro-6-*o*-chloroanilo-*s*-triazine *see* Anilazine
4,6-Dichloro-*N*-(2-chlorophenyl)-1,3,5-triazin-2-amine *see* Anilazine
1,1-Dichloro-2,2-di(4-chlorophenyl)ethane *see* *p,p*′-DDD
1,1-Dichloro-2,2-di(*p*-chlorophenyl)ethane *see* *p,p*′-DDD
Dichlorodiethyl ether *see* Bis(2-chloroethyl)ether
2,2′-Dichlorodiethyl ether *see* Bis(2-chloroethyl)ether
β,β′-Dichlorodiethyl ether *see* Bis(2-chloroethyl)ether
Dichlorodiisopropyl ether *see* Bis(2-chloroisopropyl)ether
S-2,3-Dichlorodiisopropylthiocarbamate *see* Diallate
3,5-Dichloro-*N*-(1,1-dimethyl-2-propynyl)benzamide *see* Propyzamide
3,5-Dichloro-*N*-(1,1-dimethylpropynyl)benzamide *see* Propyzamide
Dichlorodiphenyldichloroethane *see* *p,p*′-DDD
4,4′-Dichlorodiphenyldichloroethane *see* *p,p*′-DDD
p,p′-Dichlorodiphenyldichloroethane *see* *p,p*′-DDD
Dichlorodiphenyldichloroethylene *see* *p,p*′-DDE
p,p′-Dichlorodiphenyldichloroethylene *see* *p,p*′-DDE
Dichlorodiphenyltrichloroethane *see* *p,p*′-DDT
4,4′-Dichlorodiphenyltrichloroethane *see* *p,p*′-DDT
p,p′-Dichlorodiphenyltrichloroethane *see* *p,p*′-DDT
3-(2,2-Dichloroethenyl)-2,2-dimethylcyclopropanecarboxylic acid (3-
    phenoxyphenyl)methyl ester *see* Permethrin
2,2-Dichloroethenyl dimethyl phosphate *see* Dichlorvos
1,1′-(Dichloroethenylidene)bis(4-chlorobenzene) *see* *p,p*′-DDE
2,2-Dichloroethenyl phosphoric acid dimethyl ester *see* Dichlorvos
Dichloroether *see* Bis(2-chloroethyl)ether

(Z)-1,3-Dichloropropene *see cis*-1,3-Dichloropropylene

1,3-Dichloroprop-1-ene *see cis*-1,3-Dichloropropylene, *trans*-1,3-Dichloropropylene

cis-1,3-Dichloro-1-propene *see cis*-1,3-Dichloropropylene

(E)-1,3-Dichloro-1-propene *see trans*-1,3-Dichloropropylene

trans-1,3-Dichloro-1-propene *see trans*-1,3-Dichloropropylene

(Z)-1,3-Dichloro-1-propene *see cis*-1,3-Dichloropropylene

Dichloropropionanilide *see* Propanil

3′,4′-Dichloropropionanilide *see* Propanil

2,2-Dichloropropionic acid sodium salt *see* Dalapon-sodium

α,α-Dichloropropionic acid sodium salt *see* Dalapon-sodium

cis-1,2-Dichloropropylene

cis-1,3-Dichloro-1-propylene *see cis*-1,3-Dichloropropylene

trans-1,2-Dichloropropylene

trans-1,3-Dichloro-1-propylene *see trans*-1,3-Dichloropropylene

2,2-Dichlorovinyl dimethyl phosphate *see* Dichlorvos

2,2-Dichlorovinyl dimethyl phosphoric acid ester *see* Dichlorvos

Dichlorovos *see* Dichlorvos

Dichlorvos

Diclofop-methyl

Dicophane *see p,p′*-DDT

Dicopur *see* 2,4-D

Dicopur-M *see* MCPA

Dicotex *see* MCPA

Dicotox *see* 2,4-D

Dicrotophos

1,3-Dicyanotetrachlorobenzene *see* Chlorothalonil

Didigam *see p,p′*-DDT

Didimac *see p,p′*-DDT

Dieldrin

Dieldrite *see* Dieldrin

Dieldrix *see* Dieldrin

Diethion *see* Ethion

1,2-Di(ethoxycarbonyl)ethyl-*O,O*-dimethyl phosphorodithioate *see* Malathion

S-1,2-Di(ethoxycarbonyl)ethyl dimethyl phosphorothiolothionate *see* Malathion

(E)-3-(Diethylamino)-1-methyl-3-oxo-1-propenyl dimethyl phosphate *see* Dicrotophos

O,O-Diethyl *S*-(6-chlorobenzoxazolinyl-3-methyl) dithiophosphate *see* Phosalone

O,O-Diethyl *S*-((6-chloro-2-oxobenzoxazolin-3-yl)methyl) phosphorodithioate *see* Phosalone

O,O-Diethyl *S*-(6-chloro-2-oxobenzoxazolin-3-yl)methyl phosphoro-

O,O-Diethyl-*O*-4-nitrophenyl thionophosphate *see* Parathion

O,O-Diethyl-*O*-*p*-nitrophenyl thionophosphate *see* Parathion

Diethyl-*p*-nitrophenyl thiophosphate *see* Parathion

O,O-Diethyl-*O*-*p*-nitrophenyl thiophosphate *see* Parathion

Diethylparathion *see* Parathion

O,O-Diethyl phosphorodithioate, *S*-ester with 6-chloro-3-(mercaptomethyl)-2-benzoxazolinone, *see* Phosalone

O,O-Diethyl-*O*-3,5,6-trichloro-2-pyridyl phosphorothioate *see* Chlorpyrifos

Dif 4 *see* Diphenamid

Difenzoquat methyl sulfate

Diflubenzuron

4-(Difluoromethoxy)-α-(1-methylethyl)benzeneacetic acid cyano(3-phenoxyphenyl)methyl ester *see* Flucythrinate

Difonate *see* Fonofos

Digermin *see* Trifluralin

2,3-Dihydro-5-carboxanilido-6-methyl-1,4-oxatiin *see* Carboxin

9,10-Dihydro-8a,10-diazoniaphenanthrene dibromide *see* Diquat

Dihydro-8a,10a-diazoniaphenanthrene-(1,1'-ethylene-2,2'-bipyridylium)dibromide *see* Diquat

2,3-Dihydro-2,2-dimethyl-7-benzofuranol methylcarbamate *see* Carbofuran

5,6-Dihydrodipyrido[1,2*a*2,1*c*]pyrazinium dibromide *see* Diquat

S-(2-Dihydro-5-methoxy-2-oxo-1,3,4-thiadiazol-3-methyl)dimethyl phosphorothiolothionate *see* Methidathion

S-2-Dihydro-5-methoxy-2-oxo-1,3,4-thiadiazol-3-ylmethyl *O,O*-dimethyl phosphorodithionate, *see* Methidathion

2,3-Dihydro-6-methyl-1,4-oxathiin-5-carboxanilide *see* Carboxin

5,6-Dihydro-2-methyl-1,4-oxathiin-3-carboxanilide *see* Carboxin

5,6-Dihydro-2-methyl-*N*-phenyl-1,4-oxathiin-3-carboxamide *see* Carboxin

S-(3,4-Dihydro-4-oxobenzo(α)(1,2,3)triazin-3-ylmethyl)-*O,O*-dimethyl phosphorodithioate *see* Azinphos-methyl

S-(3,4-Dihydro-4-oxo-1,2,3-benzotriazin-3-ylmethyl)-*O,O*-dimethyl phosphorodithioate *see* Azinphos-methyl

1,2-Dihydro-3,6-pyradizinedione *see* Maleic hydrazide

1,2-Dihydroxypyridazine-3,6-dione *see* Maleic hydrazide

1,2-Dihydro-3,6-pyridizinedione *see* Maleic hydrazide

6,7-Dihydropyrido(1,2-α;2',1'-*c*)pyrazinedium dibromide *see* Diquat

Diisobutylthiocarbamic acid *S*-ethyl ester *see* Butylate

Diisocarb *see* Butylate

2,6-Diisopropylamino-4-methoxytriazine *see* Prometon

N,N'-Diisopropyl-6-methoxy-1,3,5-triazine-2,4-diyldiamine *see* Prometon

N,N'-Diisopropyl-6-methylthio-1,3,5-triazine-2,4-diyldiamine *see* Prometryn

N-Diisopropylthiocarbamic acid *S*-2,3,3-trichloro-2-propenyl ester *see*

Triallate
Diisopropyltrichloroallylthiocarbamate *see* Triallate
Dikotes *see* MCPA
Dikotex *see* MCPA
Dilene *see* *p,p'*-DDD
Dimate 267 *see* Dimethoate
Dimaz *see* Disulfoton
Dimecron *see* Phosphamidon
Dimecron 100 *see* Phosphamidon
Dimetate *see* Dimethoate
Dimethoate
Dimethoate-267 *see* Dimethoate
Dimethoat tecvhnisch 95% *see* Dimethoate
Dimethogen *see* Dimethoate
Dimethoxy-DDT *see* Methoxychlor
Dimethoxy-DT *see* Methoxychlor
p,p'-Dimethoxydiphenyltrichloroethane *see* Methoxychlor
2,2-Di-(*p*-methoxyphenyl)-1,1,1-trichloroethane *see* Methoxychlor
Di(*p*-methoxyphenyl)trichloromethyl methane *see* Methoxychlor
((Dimethoxyphosphinothioyl)thio)butanedioic acid diethyl ester *see* Malathion
N-Dimethoxyphosphinothioylthiomethyl)phthalimide *see* Phosmet
3-((Dimethoxyphosphinyl)oxy)-2-butenoic acid methyl ester *see* Mevinphos
3-((Dimethoxyphosphinyl)oxy)-2-butenoic acid 1-phenylethyl ester *see*
    Crotoxyphos
3-(Dimethoxyphosphinyloxy)-*N,N*-dimethyl-*cis*-crotonamide *see*
    Dicrotophos
3-(Dimethoxyphosphinyloxy)-*N*-methylisocrotonamide *see* Monocrotophos
3-(Dimethoxyphosphinyloxy)-*N,N*-dimethylisocrotonamide *see* Dicrotophos
Dimethoxy-2,2,2-trichloro-1-hydroxyethylphosphine oxide *see* Trichlorfon
O,S-Dimethylacetylphosphoroamidothioate *see* Acephate
4-Dimethylamine-*m*-cresyl methylcarbamate *see* Aminocarb
4-Dimethylamino-3-cresyl methylcarbamate *see* Aminocarb
2-(Dimethylamino)-*N*-(((methylamino)carbonyl)oxy)-2-oxoethanimidothioic
    acid methyl ester, *see* Oxamyl
3-(Dimethylamino)-1-methyl-3-oxo-1-propenyl dimethyl phosphate *see*
    Dicrotophos
4-(Dimethylamino)-3-methylphenol methylcarbamate *see* Aminocarb
2-Dimethylamino-1-(methylthio)glyoxal *O*-methylcarbamoylmonoxime *see*
    Oxamyl
4-(Dimethylamino)-*m*-tolyl methylcarbamate *see* Aminocarb
O,O-Dimethyl-*S*-(benzaziminomethyl) dithiophosphate *see* Azinphos-
    methyl
Dimethyl-1,2-benzenedicarboxylate *see* Dimethyl phthalate

Dimethylbenzeneorthodicarboxylate *see* Dimethyl phthalate

2,2-Dimethylbenzo-1,3-dioxol-4-ol methylcarbamate *see* Bendiocarb

2,2-Dimethyl-1,3-benzodioxol-4-ol methylcarbamate *see* Bendiocarb

2,2-Dimethyl-1,3-benzodioxol-4-ol *N*-methylcarbamate *see* Bendiocarb

O,O-Dimethyl-*S*-(1,2,3-benzotriazinyl-4-keto)methyl phosphorodithioate *see* Azinphos-methyl

O,O-Dimethyl-*S*-1,2-bis(ethoxycarbonyl)ethyldithiophosphate *see* Malathion

Dimethylcarbamodithioc acid iron complex *see* Ferbam

Dimethylcarbamodithioc acid iron(3+) salt *see* Ferbam

cis-2-Dimethylcarbamoyl-1-methylvinyl dimethylphosphate *see* Dicrotophos

O,O-Dimethyl-*O*-(2-carbomethoxy-1-methylvinyl)phosphate *see* Mevinphos

Dimethyl-1-carbomethoxy-1-propen-2-yl phosphate *see* Mevinphos

O,O-Dimethyl 1-carbomethoxy-1-propen-2-yl phosphate *see* Mevinphos

Dimethyl 2-chloro-2-diethylcarbamoyl-1-methylvinyl phosphate *see* Phosphamidon

O,O-Dimethyl *O*-(2-chloro-2-(*N,N*-diethylcarbamoyl)-1-methylvinyl) phosphate *see* Phosphamidon

N,N-Dimethyl-*N'*-(4-chlorophenyl)urea *see* Monuron

N,N-Dimethyl-*N'*-(*p*-chlorophenyl)urea *see* Monuron

2,2-Dimethyl-7-coumaranyl *N*-methylcarbamate *see* Carbofuran

Dimethyl 1,2-dibromo-2,2-dichloroethyl phosphate *see* Naled

O,O-Dimethyl-*O*-(1,2-dibromo-2,2-dichloroethyl)phosphate *see* Naled

O,O-Dimethyl-*S*-(1,2-dicarbethoxyethyl)dithiophosphate *see* Malathion

O,O-Dimethyl-*S*-(1,2-dicarbethoxyethyl)phosphorodithioate *see* Malathion

O,O-Dimethyl-*S*-(1,2-dicarbethoxyethyl)thiothionophosphate *see* Malathion

O,O-Dimethyl *O*-(2,2-dichloro-1,2-dibromoethyl)phosphate *see* Naled

Dimethyl 2,2-dichloroethenyl phosphate *see* Dichlorvos

1,1-Dimethyl-3-(3,4-dichlorophenyl)urea *see* Diuron

Dimethyl dichlorovinyl phosphate *see* Dichlorvos

Dimethyl 2,2-dichlorovinyl phosphate *see* Dichlorvos

O,O-Dimethyl *O*-(2,2-dichlorovinyl)phosphate *see* Dichlorvos

O,O-Dimethyl-*S*-1,2-di(ethoxycarbamyl)ethyl phosphorodithioate *see* Malathion

2,2-Dimethyl-2,3-dihydro-7-benzofuranyl-*N*-methylcarbamate *see* Carbofuran

O,O-Dimethyl-*S*-(3,4-dihydro-4-keto-1,2,3-benzotriazinyl-3-methyl) dithiophosphate *see* Azinphos-methyl

2,2-Dimethyl-*N,N*-dimethylacetamide *see* Diphenamid

O,O-Dimethyl *O*-(*N,N*-dimethylcarbamoyl-1-methylvinyl) phosphate *see* Dicrotophos

O,O-Dimethyl-*O*-(1,4-dimethyl-3-oxo-4-azapent-1-enyl)phosphate *see* Dicrotophos

N,N-Dimethyldiphenylacetamide *see* Diphenamid

N,N-Dimethyl-2,2-diphenylacetamide *see* Diphenamid
N,N-Dimethyl-α,α-diphenylacetamide *see* Diphenamid
1,2-Dimethyl-3,5-diphenyl-1*H*-pyrazolium methyl sulfate *see* Difenzoquat
   methyl sulfate
Dimethyldithiocarbamic acid iron salt *see* Ferbam
Dimethyldithiocarbamic acid iron(3+) salt *see* Ferbam
O,O-Dimethyldithiophosphate dimethylmercaptosuccinate *see* Malathion
Dimethyldithiophosphoric acid *N*-methylbenzazimide ester *see* Azinphos-
   methyl
O,O-Dimethyl dithiophosphorylacetic acid *N*-monomethylamide salt *see*
   Dimethoate
O,S-Dimethyl ester amide of aminothioate *see* Methamidophos
O,O-Dimethyl *S*-(2-eththionylethyl) phosphorothioate *see* Oxydemeton-
   methyl
Dimethyl *S*-(2-eththionylethyl) thiophosphate *see* Oxydemeton-methyl
N-(1,1-Dimethylethyl)-*N'*-ethyl-6-(methylthio)-1,3,5-triazine-2,4-diamine
   see Terbutryn
O,O-Dimethyl *S*-ethylsulphinylethyl phosphorothioate *see* Oxydemeton-
   methyl
O,O-Dimethyl *S*-2-(ethylsulfinyl)ethyl phosphorothioate *see* Oxydemeton-
   methyl
O,O-Dimethyl *S*-2-(ethylsulfinyl)ethyl thiophosphate *see* Oxydemeton-
   methyl
N-(5-(1,1-Dimethylethyl)-1,3,4-thiadiazol-2-yl)-*N,N'*-dimethylurea *see*
   Tebuthiuron
S-(((1,1-Dimethylethyl)thio)methyl) *O,O*-diethyl phosphorodithioate *see*
   Terbufos
Dimethylformocarbothialdine *see* Dazomet
Dimethyl-1-hydroxy-2,2,2-trichloroethyl phosphonate *see* Trichlorfon
O,O-Dimethyl-(1-hydroxy-2,2,2-trichloro)ethyl phosphate *see* Trichlorfon
O,O-Dimethyl-(1-hydroxy-2,2,2-trichloro)ethyl phosphonate *see* Trichlorfon
Dimethyl 2-methoxycarbonyl-1-methylvinyl phosphate *see* Mevinphos
Dimethyl methoxycarbonylpropenyl phosphate *see* Mevinphos
Dimethyl (1-methoxycarboxypropen-2-yl)phosphate *see* Mevinphos
O,O-Dimethyl *S*-(5-methoxy-1,3,4-thiadiazolinyl-3-methyl) dithiophosphate
   see Methidathion
O,O-Dimethyl *S*-(2-methoxy-1,3,4-thiadiazol-5(4*H*)-onyl-4-methyl)phos-
   phorodithioate *see* Methidathion
O,O-Dimethyl *S*-(2-(methylamino)-2-oxoethyl) phosphorodithioate *see*
   Dimethoate
O,O-Dimethyl *S*-(*N*-methylcarbamoylmethyl) dithiophosphate *see*
   Dimethoate
O,O-Dimethyl *S*-(*N*-methylcarbamoylmethyl) phosphorodithioate *see*

Dimethoate

O,O-Dimethyl-*O*-(2-*N*-methylcarbamoyl-1-methylvinyl) phosphate *see* Monocrotophos

N,N-Dimethyl-α-methylcarbamoyloxyimino-α-(methylthio)acetamide *see* Oxamyl

N',N'-Dimethyl-*N*-((methylcarbamoyl)oxy)-1-thiooxamimidic acid methyl ester *see* Oxamyl

O,O-Dimethyl *S*-(*N*-methylcarbamylmethyl) thiothionophosphate *see* Dimethoate

O,O-Dimethyl *O*-(1-methyl-2-carboxy-α-phenylethyl)vinyl phosphate *see* Crotoxyphos

O,O-Dimethyl *O*-(1-methyl-2-carboxyvinyl)phosphate *see* Mevinphos

N,N-Dimethyl-*N'*-(2-methyl-4-chlorophenyl)formamidine *see* Chlordimeform

O,O-Dimethyl *O*-4-(methylmercapto)-3-methylphenyl phosphorothioate *see* Fenthion

O,O-Dimethyl *O*-4-(methylmercapto)-3-methylphenyl thiophosphate *see* Fenthion

(*E*)-Dimethyl 1-methyl-3-(methylamino)-3-oxo-1-propenyl phosphate *see* Monocrotophos

Dimethyl-1-methyl-2-(methylcarbamoyl)vinyl phosphate *see* Monocrotophos

Dimethyl (*E*)-1-methyl-2-(methylcarbamoyl)vinyl phosphate *see* Monocrotophos

O,O-Dimethyl *O*-(3-methyl-4-methylmercaptophenyl) phosphorothioate *see* Fenthion

O,O-Dimethyl *O*-(3-methyl-4-methylthiophenyl) phosphorothioate *see* Fenthion

Dimethyl-*cis*-1-methyl-2-(1-phenylethoxycarbonyl)vinyl phosphate *see* Crotoxyphos

O,O-Dimethyl *O*-(4-methylthio-3-methylphenyl) phosphorothioate *see* Fenthion

3,5-Dimethyl-4-(methylthio)phenyl methylcarbamate *see* Methiocarb

3,5-Dimethyl-4-methylthiophenyl *N*-methylcarbamate *see* Methiocarb

O,O-Dimethyl *O*-(4-methylthio-*m*-tolyl) phosphorothioate *see* Fenthion

O,O-Dimethyl-*S*-(*N*-monomethyl)carbamyl methyldithiophosphate *see* Dimethoate

O,O-Dimethyl-*S*-(4-oxo-3*H*-1,2,3-benzotriazine-3-methyl) phosphorodithioate *see* Azinphos-methyl

O,O-Dimethyl-*S*-(4-oxobenzotriazino-3-methyl) phosphorodithioate *see* Azinphos-methyl

O,O-Dimethyl-*S*-(4-oxo-1,2,3-benzotriazino-3-methyl) thiothionophosphate *see* Azinphos-methyl

O,O-Dimethyl *S*-((4-oxo-1,2,3-benzotriazin-3-yl)methyl) phosphorodithioate *see* Azinphos-methyl

O,O-Dimethyl *S*-((4-oxo-1,2,3-benzotriazin-3(4*H*)-yl)methyl) phosphorodithioate *see* Azinphos-methyl

2-(2,2-Dimethyl-1-oxopropyl)-1*H*-indene-1,3(2*H*)-dione *see* Pindone

O,O-Dimethyl-1-oxy-2,2,2-trichloroethyl phosphonate *see* Trichlorfon

N,N-Dimethyl-α-phenylbenzeneacetamide *see* Diphenamid

N-(2,6-Dimethylphenyl)-*N*-methoxyacetyl)alanine methyl ester *see* Metalaxyl

N-(2,6-Dimethylphenyl)-*N*-methoxyacetyl)-DL-alanine methyl ester *see* Metalaxyl

Dimethyl phosphate ester with 3-hydroxy-*N,N*-dimethyl-*cis*-crotonamide *see* Dicrotophos

Dimethyl phosphate ester with (*E*)-3-hydroxy-*N,N*-dimethylcrotonamide *see* Dicrotophos

Dimethyl phosphate ester of 3-hydroxy-*N*-methyl-*cis*-crotonamide *see* Monocrotophos

Dimethyl phosphate of chloro-*N,N*-diethyl-3-hydroxycrotonamide *see* Phosphamidon

Dimethyl phosphate of 3-hydroxy-*N,N*-dimethyl-*cis*-crotonamide *see* Dicrotophos

Dimethyl phosphate of α-methylbenzyl 3-hydroxy-*cis*-crotonate *see* Crotoxyphos

Dimethyl phosphate of methyl-3-hydroxy-*cis*-crotonate *see* Mevinphos

O,S-Dimethyl phosphoramidothioate *see* Methamidophos

O,O-Dimethyl phosphorodithioate *S*-ester with *N*-(mercapto-methyl)phthalimide *see* Phosmet

Dimethyl phthalate

O,O-Dimethyl *S*-phthalimidomethyl phosphorodithioate *see* Phosmet

2-(((((4,6-Dimethyl-2-pyrimidinyl)amino)carbonyl)amino)sulfonyl)benzoic acid methyl ester, *see* Sulfometuron-methyl

Dimethyl 2,3,5,6-tetrachloro-1,4-benzenedicarboxylate *see* Chlorthal-dimethyl

Dimethyl tetrachloroterephthalate *see* Chlorthal-dimethyl

Dimethyl 2,3,5,6-tetrachloroterephthalate *see* Chlorthal-dimethyl

Dimethyl-2*H*-1,3,5-tetrahydrothiadiazine-2-thione *see* Dazomet

3,5-Dimethyl-1,2,3,5-tetrahydro-1,3,5-thiadiazinethione-2 *see* Dazomet

3,5-Dimethyl-1,3,5,2*H*-tetrahydrothiadiazine-2-thione *see* Dazomet

3,5-Dimethyltetrahydro-1,3,5-2*H*-thiadiazine-2-thione *see* Dazomet

3,5-Dimethyltetrahydro-1,3,5-thiadiazine-2-thione *see* Dazomet

3,5-Dimethyltetrahydro-2*H*-1,3,5-thiadiazine-2-thione *see* Dazomet

Dimethyl *N,N*'-(thiobis((methylimino)carbonyloxy))bis(ethanimidothioate) *see* Thiodicarb

3,5-Dimethyl-2-thionotetrahydro-1,3,5-thiadiazine *see* Dazomet

Dimethyltrichlorohydroxyethyl phosphonate *see* Trichlorfon
Dimethyl-2,2,2-trichloro-1-hydroxyethyl phosphonate *see* Trichlorfon
O,O-Dimethyl-2,2,2-trichloro-1-hydroxyethyl phosphonate *see* Trichlorfon
O,O-Dimethyl-*O*-2,4,5-trichlorophenyl phosphorothioate *see* Ronnel
Dimethyl trichlorophenyl thiophosphate *see* Ronnel
O,O-Dimethyl *O*-(2,4,5-trichlorophenyl)thiophosphate *see* Ronnel
1,1-Dimethyl-3-(3-trifluoromethylphenyl)urea *see* Fluometuron
N,N-Dimethyl-*N'*-(3-(trifluoromethyl)phenyl)urea *see* Fluometuron
N-(2,4-Dimethyl-5-((((trifluoromethyl)sulfonyl)amino)phenyl)acetamide *see*
    Mefluidide
1,1-Dimethyl-3-(α,α,α-trifluoro-*m*-tolyl)urea *see* Fluometuron
Dimeton *see* Dimethoate
Dimetox *see* Trichlorfon
Dimevur *see* Dimethoate
Dimid *see* Diphenamid
Dimilin *see* Diflubenzuron
Dimpylate *see* Diazinon
Dinitro *see* Dinoseb
Dinitro-3 *see* Dinoseb
Dinitrobutylphenol *see* Dinoseb
2-Dinitro-6-*sec*-butylphenol *see* Dinoseb
4,6-Dinitro-2-*sec*-butylphenol *see* Dinoseb
4,6-Dinitro-*o*-*sec*-butylphenol *see* Dinoseb
Dinitrocresol *see* 4,6-Dinitro-*o*-cresol
Dinitro-*o*-cresol *see* 4,6-Dinitro-*o*-cresol
2,4-Dinitro-*o*-cresol *see* 4,6-Dinitro-*o*-cresol
3,5-Dinitro-*o*-cresol *see* 4,6-Dinitro-*o*-cresol
Dinitrodendtroxal *see* 4,6-Dinitro-*o*-cresol
2,6-Dinitro-*N,N*-dipropyl-4-trifluoromethylaniline *see* Trifluralin
2,6-Dinitro-*N,N*-dipropyl-4-(trifluoromethyl)benzenamine *see* Trifluralin
2,6-Dinitro-*N,N*-di-*n*-propyl-α,α,α-trifluoro-*p*-toluidine *see* Trifluralin
3,5-Dinitro-2-hydroxytoluene *see* 4,6-Dinitro-*o*-cresol
Dinitrol *see* 4,6-Dinitro-*o*-cresol
Dinitromethyl cyclohexyltrienol *see* 4,6-Dinitro-*o*-cresol
2,4-Dinitro-2-methylphenol *see* 4,6-Dinitro-*o*-cresol
2,4-Dinitro-6-methylphenol *see* 4,6-Dinitro-*o*-cresol
4,6-Dinitro-2-methylphenol *see* 4,6-Dinitro-*o*-cresol
4,6-Dinitro-2-(1-methyl-*n*-propyl)phenol *see* Dinoseb
Dinitrosol *see* 4,6-Dinitro-*o*-cresol
Dinoc *see* 4,6-Dinitro-*o*-cresol
Dinoseb
S-((1,3-Dioxo-2*H*-isoindol-2-yl)methyl) *O,O*-dimethyl phosphorodithioate
    see Phosmet

Dinoxol *see* 2,4-D, 2,4,5-T
Dinurania *see* 4,6-Dinitro-*o*-cresol
Di-on *see* Diuron
Diphenamid
Diphenamide *see* Diphenamid
Diphenyl *see* Biphenyl
Diphenylamide *see* Diphenamid
Diphenyltrichloroethane *see* *p,p'*-DDT
Diphosphoric acid tetraethyl ester *see* Tetraethyl pyrophosphate
Dipofene *see* Diazinon
Dipram *see* Propanil
Dipropetryn
Dipropetryne *see* Dipropetryn
4-(Di-*n*-propylamino)-3,5-dinitro-1-trifluoromethylbenzene *see* Trifluralin
Dipropylcarbamothioic acid *S*-ethyl ester *see* EPTC
N,N-Di-*n*-propyl-2,6-dinitro-4-trifluoromethylaniline *see* Trifluralin
N,N-Dipropylthiocarbamic acid *S*-ethyl ester *see* EPTC
N,N-Dipropyl-4-trifluoromethyl-2,6-dinitroaniline *see* Trifluralin
Dipterax *see* Trichlorfon
Dipterex *see* Trichlorfon
Dipterex 50 *see* Trichlorfon
Diptevur *see* Trichlorfon
Dipthal *see* Triallate
Diquat
Direx *see* Diuron
Direx 4L *see* Diuron
Direz *see* Anilazine
Disulfaton *see* Disulfoton
Disulfoton
Di-syston *see* Disulfoton
Disystox *see* Disulfoton
Dithane M 45 *see* Mancozeb
Dithane S 60 *see* Mancozeb
Dithane SPC *see* Mancozeb
Dithane ultra *see* Mancozeb
Dithio *see* Sulfotepp
α,α'-Dithiobis(dimethylthio) formamide *see* Thiram
Dithiocarbonic anhydride *see* Carbon disulfide
Dithiodemeton *see* Disulfoton
N,N'-(Dithiodicarbonothioyl)bis(*N*-methylmethanamine) *see* Thiram
Dithione *see* Sulfotepp
Dithiophos *see* Sulfotepp
Dithiophosphoric acid tetraethyl ester *see* Sulfotepp

Dithiosystox *see* Disulfoton
Dithiotep *see* Sulfotepp
Di(tri-(2,2-dimethyl-2-phenylpropyl)tin)oxide *see* Fenbutatin oxide
Ditrifon *see* Trichlorfon
Diurex *see* Diuron
Diurol *see* Amitrole, Diuron
Diurol 5030 *see* Amitrole
Diuron
Diuron 4L *see* Diuron
Diuron 80W *see* Diuron
Divipan *see* Dichlorvos
Dixon *see* Phosphamidon
Dizinon *see* Diazinon
DMA-4 *see* 2,4-D
DMDT *see* Methoxychlor
4,4'-DMDT *see* Methoxychlor
p,p'-DMDT *see* Methoxychlor
DMP *see* Dimethyl phthalate
DMSP, Fensulfothion
DMTD *see* Methoxychlor
DMTP *see* Fenthion, Methidathion
DMTT *see* Dazomet
DMU *see* Diuron
DN *see* 4,6-Dinitro-*o*-cresol
DN 289 *see* Dinoseb
DNBP *see* Dinoseb
DNC *see* 4,6-Dinitro-*o*-cresol
DN-dry mix no. 2 *see* 4,6-Dinitro-*o*-cresol
DNOC *see* 4,6-Dinitro-*o*-cresol
Dnosbp *see* Dinoseb
DNSBP *see* Dinoseb
DNTP *see* Parathion
Dodat *see* *p,p'*-DDT
Dolco mouse cereal *see* Strychnine
Dol granule *see* Lindane
Dolochlor *see* Chloropicrin
Domatol *see* Amitrole
Domatol 88 *see* Amitrole
Dormone *see* 2,4-D
Dow ET 14 *see* Ronnel
Dow ET 57 *see* Ronnel
Dow general *see* Dinoseb
Dow general weed killer *see* Dinoseb

Dow MCP amine weed killer *see* MCPA
Dow pentachlorophenol DP-2 antimicrobial *see* Pentachlorophenol
Dow selective weed killer *see* Dinoseb
Dowcide 7 *see* Pentachlorophenol
Dowco-163 *see* Nitrapyrin
Dowco-179 *see* Chlorpyrifos
Dowfume *see* Methyl bromide
Dowfume 40 *see* Ethylene dibromide
Dowfume EDB *see* Ethylene dibromide
Dowfume MC-2 *see* Methyl bromide
Dowfume MC-2 soil fumigant *see* Methyl bromide
Dowfume MC-33 *see* Methyl bromide
Dowfume W-8 *see* Ethylene dibromide
Dowfume W-85 *see* Ethylene dibromide
Dowfume W-90 *see* Ethylene dibromide
Dowfume W-100 *see* Ethylene dibromide
Dowicide 7 *see* Pentachlorophenol
Dowicide EC-7 *see* Pentachlorophenol
Dowicide G *see* Pentachlorophenol
Dowklor *see* Chlordane
Dowpon *see* Dalapon-sodium
DPA *see* Propanil
2,2-DPA *see* Dalapon-sodium
DPP *see* Parathion
DPX 1108 *see* Fosamine-ammonium
DPX 1410 *see* Oxamyl
DPX 3674 *see* Hexazinone
DPX 4189 *see* Chlorsulfuron
DPX 5648 *see* Sulfometuron-methyl
DPX M6316 *see* Thiameturon-methyl
Draza *see* Methiocarb
Drexel parathion 8E *see* Parathion
Drexel-super P *see* Maleic hydrazide
Drinox *see* Aldrin, Heptachlor
Drinox H-34 *see* Heptachlor
Dropp *see* Thidiazuron
DU 112307 *see* Diflubenzuron
Du Pont 326 *see* Linuron
Du Pont 732 *see* Terbacil
Du Pont 1179 *see* Methomyl
Du Pont 1991 *see* Benomyl
Du Pont herbicide 326 *see* Linuron
Du Pont herbicide 732 *see* Terbacil

E-D-BEE *see* Ethylene dibromide
Edco *see* Methyl bromide
Eerex granular weed killer *see* Bromacil
Eerex water soluble concentrate weed killer *see* Bromacil
Effusan *see* 4,6-Dinitro-*o*-cresol
Effusan 3436 *see* 4,6-Dinitro-*o*-cresol
Efosite-AL *see* Fosetyl-aluminum
EI 103 *see* Tebuthiuron
EI 3911 *see* Phorate
EI 12880 *see* Dimethoate
Ekagom TB *see* Thiram
Ekatin WF & WF ULV *see* Parathion
Ekatox *see* Parathion
EL 103 *see* Tebuthiuron
EL 4049 *see* Malathion
Elancolan *see* Trifluralin
Elaol *see* Di-*n*-butyl phthalate
Elgetol *see* 4,6-Dinitro-*o*-cresol, Dinoseb
Elgetol 30 *see* 4,6-Dinitro-*o*-cresol
Elgetol 318 *see* Dinoseb
Elipol *see* 4,6-Dinitro-*o*-cresol
Elmasil *see* Amitrole
Embafume *see* Methyl bromide
Embark *see* Mefluidide
Embathion *see* Ethion
Emcepan *see* MCPA
Emisol *see* Amitrole
Emisol 50 *see* Amitrole
Emisol F *see* Amitrole
Emmatos *see* Malathion
Emmatos extra *see* Malathion
Empal *see* MCPA
Emulsamine BK *see* 2,4-D
Emulsamine E-3 *see* 2,4-D
Endocel *see* α-Endosulfan, β-Endosulfan
3,6-Endooxohexahydrophthalic acid *see* Endothall
Endosol *see* α-Endosulfan, β-Endosulfan
Endosulfan *see* α-Endosulfan, β-Endosulfan
Endosulfan I *see* α-Endosulfan
Endosulfan II *see* β-Endosulfan
α-Endosulfan
β-Endosulfan
Endosulfan sulfate

Endosulphan *see* α-Endosulfan, see β-Endosulfan
Endothal *see* Endothall
Endothall
Endrex *see* Endrin
Endrin
Endyl *see* Carbophenothion
Enide *see* Diphenamid
Enide 50 *see* Diphenamid
Enide 90 *see* Diphenamid
ENT 54 *see* Acrylonitrile
ENT 154 *see* 4,6-Dinitro-*o*-cresol
ENT 262 *see* Dimethyl phthalate
ENT 987 *see* Thiram
ENT 1122 *see* Dinoseb
ENT 1506 *see* *p,p'*-DDT
ENT 1716 *see* Methoxychlor
ENT 3776 *see* Dichlone
ENT 4225 *see* *p,p'*-DDD
ENT 4504 *see* Bis(2-chloroethyl)ether
ENT 4705 *see* Carbon tetrachloride
ENT 7796 *see* Lindane
ENT 8538 *see* 2,4-D
ENT 9232 *see* α-BHC
ENT 9233 *see* β-BHC
ENT 9234 *see* δ-BHC
ENT 9735 *see* Toxaphene
ENT 9932 *see* Chlordane
ENT 14689 *see* Ferbam
ENT 15108 *see* Parathion
ENT 15152 *see* Heptachlor
ENT 15349 *see* Ethylene dibromide
ENT 15406 *see* 1,2-Dichloropropane
ENT 15949 *see* Aldrin
ENT 16225 *see* Dieldrin
ENT 16273 *see* Sulfotepp
ENT 16391 *see* Kepone
ENT 16894 *see* Aspon
ENT 17034 *see* Malathion
ENT 17251 *see* Endrin
ENT 17798 *see* EPN
ENT 18060 *see* Chlorpropham
ENT 18596 *see* Chlorobenzilate
ENT 18771 *see* Tetraethyl pyrophosphate

ENT 18870 *see* Maleic hydrazide
ENT 19507 *see* Diazinon
ENT 19763 *see* Trichlorfon
ENT 20738 *see* Dichlorvos
ENT 20852 *see* Bromoxynil
ENT 22324 *see* Mevinphos
ENT 23233 *see* Azinphos-methyl
ENT 23284 *see* Ronnel
ENT 23437 *see* Disulfoton
ENT 23708 *see* Carbophenothion
ENT 23969 *see* Carbaryl
ENT 23979 *see* α-Endosulfan, β-Endosulfan
ENT 24042 *see* Phorate
ENT 24105 *see* Ethion
ENT 24482 *see* Dicrotophos
ENT 24650 *see* Dimethoate
ENT 24717 *see* Crotoxyphos
ENT 24945 *see* Fensulfothion
ENT 24964 *see* Oxydemeton-methyl
ENT 24988 *see* Naled
ENT 25455 *see* Amitrole
ENT 25515 *see* Phosphamidon
ENT 25540 *see* Fenthion
ENT 25552 *see* Chlordane
ENT 25584 *see* Heptachlor epoxide
ENT 25606 *see* Quinomethionate
ENT 25671 *see* Propoxur
ENT 25705 *see* Phosmet
ENT 25726 *see* Methiocarb
ENT 25784 *see* Aminocarb
ENT 25796 *see* Fonofos
ENT 26058 *see* Anilazine
ENT 27093 *see* Aldicarb
ENT 27129 *see* Monocrotophos
ENT 27163 *see* Phosalone
ENT 27164 *see* Carbofuran
ENT 27193 *see* Methidathion
ENT 27311 *see* Chlorpyrifos
ENT 27318 *see* Ethoprop
ENT 27320 *see* Dialifos
ENT 27335 *see* Chlordimeform
ENT 27341 *see* Methomyl
ENT 27396 *see* Methamidophos

ENT 27567 *see* Chlordimeform
ENT 27572 *see* Fenamiphos
ENT 27738 *see* Fenbutatin oxide
ENT 27822 *see* Acephate
ENT 29054 *see* Diflubenzuron
Entex *see* Fenthion
Entomoxan *see* Lindane
Envert 171 *see* 2,4-D
Envert DT *see* 2,4-D
Envert T *see* 2,4,5-T
EP 30 *see* Pentachlorophenol
EP 161E *see* Methylisothiocyanate
EP 333 *see* Chlordimeform
EP 452 *see* Phenmedipham
EP 475 *see* Desmedipham
Epal *see* Fosetyl-aluminum
EPN
EPN 300 *see* EPN
3,6-Epoxycyclohexane-1,2-dicarboxylic acid *see* Endothall
3,6-*endo*-Epoxy-1,2-cyclohexanedicarboxylic acid *see* Endothall
Epoxy heptachlor *see* Heptachlor epoxide
Eprofil *see* Thiabendazole
Eptam *see* EPTC
EPTC
Equigard *see* Dichlorvos
Equigel *see* Dichlorvos
Equino-acid *see* Trichlorfon
Equino-aid *see* Trichlorfon
Equizole *see* Thiabendazole
Erade *see* Quinomethionate
Eradex *see* Chlorpyrifos
Erazidon *see* Quinomethionate
Erban *see* Propanil
Erbanil *see* Propanil
Escort *see* Metsulfuron-methyl
Esfenvalerate
Estercide T-2 and T-245 *see* 2,4,5-T
Esteron *see* 2,4-D, 2,4,5-T
Esteron 245 BE *see* 2,4,5-T
Esteron 76 BE *see* 2,4-D
Esteron 99 *see* 2,4-D
Esteron 99 concentrate *see* 2,4-D
Esteron 44 weed killer *see* 2,4-D

Ethyl-*N,N*-diisobutylthiolcarbamate *see* Butylate

S-Ethyl dipropylcarbamothioate *see* EPTC

O-Ethyl *S,S*-dipropyl phosphorodithioate *see* Ethoprop

S-Ethyl dipropylthiocarbamate *see* EPTC

S-Ethyl di-*n*-propylthiocarbamate *see* EPTC

S-Ethyl-*N,N*-di-*n*-propylthiolcarbamate *see* EPTC

Ethylene aldehyde *see* Acrolein

1,1-Ethylene-2,2-bipyridylium dibromide *see* Diquat

1,1'-Ethylene-2,2'-bipyridylium dibromide *see* Diquat

Ethylene bromide *see* Ethylene dibromide

Ethylene bromide glycol dibromide *see* Ethylene dibromide

Ethylene dibromide

1,2-Ethylene dibromide *see* Ethylene dibromide

Ethyl-4,4-dichlorobenzilate *see* Chlorobenzilate

Ethyl-4,4'-dichlorodiphenyl glycollate *see* Chlorobenzilate

S-Ethyl *N*-ethyl *N*-cyclohexylthiolcarbamate *see* Cycloate

O,O-Ethyl *S*-2-((ethylthio)ethyl) phosphorodithioate *see* Disulfoton

S-Ethyl hexahydro-1*H*-azepine-1-carbothioate *see* Molinate

S-Ethyl *N,N*-hexamethylenethiocarbamate *see* Molinate

Ethyl-2-hydroxy-2,2-bis(4-chlorophenyl)acetate *see* Chlorobenzilate

O-Ethyl *O*-(2-isopropoxycarbonyl)phenyl isopropylphosphoramidothioate *see* Isofenphos

N-Ethyl-*N*'-isopropyl-6-methylthio-1,3,5-triazine-2,4-diyldiamine *see* Ametryn

2-Ethylmercapotomethylphenyl-*N*-methylcarbamate *see* Ethiofencarb

Ethyl methylene phosphorodithioate *see* Ethion

N-Ethyl-*N*'-(1-methylethyl)-6-(methylthio)-1,3,5-triazine-2,4-diamine *see* Ametryn

O-Ethyl *O*-(4-(methylmercapto)phenyl) *S*-*n*-propylphosphorothionothiolate *see* Sulprofos

2-Ethyl-6-methyl-1-*N*-(2-methoxy-1-methylethyl)chloroacetanilide *see* Metolachlor

Ethyl 3-methyl-4-(methylthio)phenyl (1-methylethyl)phosphoramidate *see* Fenamiphos

O-Ethyl *O*-(4-methylthio)phenyl) phosphorodithioic acid *S*-propyl ester *see* Sulprofos

O-Ethyl *O*-(4-methylthio)phenyl) *S*-propyl phosphorodithioate *see* Sulprofos

Ethyl-4-methylthio-*m*-tolyl isopropyl phosphoramidate *see* Fenamiphos

Ethyl *p*-nitrophenyl benzenethionophosphate *see* EPN

Ethyl *p*-nitrophenyl benzenethiophosphonate *see* EPN

Ethyl *p*-nitrophenyl ester *see* EPN

Ethyl *p*-nitrophenyl phenylphosphonothioate *see* EPN

O-Ethyl *O*-4-nitrophenyl phenylphosphonothioate *see* EPN

Extrar *see* 4,6-Dinitro-*o*-cresol
E-Z-Off *see* Butifos
F 735 *see* Carboxin
F 1991 *see* Benomyl
F 2966 *see* Mancozeb
FA *see* Formaldehyde
Fair-2 *see* Maleic hydrazide
Fair-30 *see* Maleic hydrazide
Fair-plus *see* Maleic hydrazide
Fair PS *see* Maleic hydrazide
Falitram *see* Thiram
Famfos *see* Phosphamidon
Fannoform *see* Formaldehyde
Far-Go *see* Triallate
Farmco *see* 2,4-D
Farmco atrazine *see* Atrazine
Farmco diuron *see* Diuron
Farmco fence rider *see* 2,4,5-T
Farmco propanil *see* Propanil
Fasciolin *see* Carbon tetrachloride
Fasco fascrat powder *see* Warfarin
Fasco terpene *see* Toxaphene
Fasco Wy-hoe *see* Chlorpropham
Fatal *see* Chlorthal-dimethyl
FB/2 *see* Diquat
FDA 1541 *see* EPTC
FDN *see* Diphenamid
Fecama *see* Dichlorvos
Felan *see* Molinate
Fen-All *see* 2,3,5-Trichlorobenzoic acid
Fenam *see* Diphenamid
Fenamin *see* Atrazine
Fenamine *see* Amitrole, Atrazine
Fenamiphos
Fenatrol *see* Atrazine
Fenavar *see* Amitrole
Fenbutatin oxide
Fence rider *see* 2,4,5-T
Fenchchlorphos *see* Ronnel
Fenchlorfos *see* Ronnel
Fenchlorophos *see* Ronnel
Fender *see* Phenmedipham
Fenmedifam *see* Phenmedipham

FMC 1240 *see* Ethion
FMC 5462 *see* α-Endosulfan, β-Endosulfan
FMC 30980 *see* Cypermethrin
FMC 33297 *see* Permethrin
FMC 41665 *see* Permethrin
FMC 45497 *see* Cypermethrin
FMC 45806 *see* Cypermethrin
Folbex *see* Chlorobenzilate
Folbex smoke-strips *see* Chlorobenzilate
Folidol *see* Parathion
Folidol E605 *see* Parathion
Folidol E & E 605 *see* Parathion
Fonofos
Fonophos *see* Fonofos
Fore *see* Mancozeb
Foredex 75 *see* 2,4-D
Forlin *see* Lindane
Formal *see* Malathion
Formaldehyde
Formalin *see* Formaldehyde
Formalin 40 *see* Formaldehyde
Formalith *see* Formaldehyde
Formic aldehyde *see* Formaldehyde
Formol *see* Formaldehyde
Formula 40 *see* 2,4-D
Forotox *see* Trichlorfon
Forron *see* 2,4,5-T
Forstan *see* Quinomethionate
Forst U 46 *see* 2,4,5-T
Fortex *see* 2,4,5-T
Forthion *see* Malathion
Fortion NM *see* Dimethoate
Fortrol *see* Cyanazine
Forturf *see* Chlorothalonil
Fosamine-ammonium
Foschlor *see* Trichlorfon
Foschlor 25 *see* Trichlorfon
Foschlor R 50 *see* Trichlorfon
Fosdrin *see* Mevinphos
Fosetyl AL *see* Fosetyl-aluminum
Fosetyl-aluminum
Fos-fall 'A' *see* Butifos
Fosfamid *see* Dimethoate

Fyde *see* Formaldehyde
Fyfanon *see* Malathion
G 25 *see* Chloropicrin
G 301 *see* Diazinon
G 338 *see* Chlorobenzilate
G 996 *see* Ethephon
G 23992 *see* Chlorobenzilate
G 24480 *see* Diazinon
G 27692 *see* Simazine
G 30027 *see* Atrazine
G 30028 *see* Propazine
G 31435 *see* Prometon
G 34161 *see* Prometryn
G 34162 *see* Ametryn
Galecron *see* Chlordimeform
Gallogama *see* Lindane
Gamacid *see* Lindane
Gamaphex *see* Lindane
Gamene *see* Lindane
Gamiso *see* Lindane
Gammahexa *see* Lindane
Gammalin *see* Lindane
Gammexene *see* Lindane
Gammopaz *see* Lindane
Gamonil *see* Carbaryl
Gardentox *see* Diazinon
Garnitan *see* Linuron
Garrathion *see* Carbophenothion
Garvox *see* Bendiocarb
GC-1189 *see* Kepone
Gearphos *see* Parathion
Gebutox *see* Dinoseb
Geigy 338 *see* Chlorobenzilate
Geigy 13005 *see* Methidathion
Geigy 24480 *see* Diazinon
Geigy 27692 *see* Simazine
Geigy 30027 *see* Atrazine
Geigy 30028 *see* Propazine
Geigy GS-13005 *see* Methidathion
Genate Plus 6.7EC *see* Butylate
General chemicals 1189 *see* Kepone
Geniphene *see* Toxaphene
Genithion *see* Parathion

Genitox *see* p,p′-DDT
Germain's *see* Carbaryl
Gesafid *see* p,p′-DDT
Gesafram *see* Prometon
Gesafram 50 *see* Prometon
Gesagard *see* Prometryn
Gesamil *see* Propazine
Gesapax *see* Ametryn
Gesapon *see* p,p′-DDT
Gesaprim *see* Atrazine
Gesaran *see* Simazine
Gesarex *see* p,p′-DDT
Gesarol *see* p,p′-DDT
Gesatop *see* Simazine
Gesatop 50 *see* Simazine
Gesfid *see* Mevinphos
Gesoprim *see* Atrazine
Gestid *see* Mevinphos
Gexane *see* Lindane
Glazd penta *see* Pentachlorophenol
Glean *see* Chlorsulfuron
Glean 20DF *see* Chlorsulfuron
Glycol bromide *see* Ethylene dibromide
Glycol dibromide *see* Ethylene dibromide
Glycophen *see* Iprodione
Glycophene *see* Iprodione
Glyphosate
Goliath *see* Phenmedipham
Gothnion *see* Azinphos-methyl
GPKh *see* Heptachlor
Gramevin *see* Dalapon-sodium
Granox NM *see* Hexachlorobenzene
Granurex *see* Neburon
Granutox *see* Phorate
Grascide *see* Propanil
Graslan *see* Tebuthiuron
Grazon *see* Picloram
Green-daisen M *see* Mancozeb
Griffex *see* Atrazine
Grisol *see* Tetraethyl pyrophosphate
Grundier arbezol *see* Pentachlorophenol
GS 13005 *see* Methidathion
GS 14260 *see* Terbutryn

Hexathir *see* Thiram
Hexatox *see* Lindane
Hexaverm *see* Lindane
Hexavin *see* Carbaryl
Hexazinone
Hexicide *see* Lindane
Hexyclan *see* Lindane
Hexythiazox
Hexylthiocarbam *see* Cycloate
HGI *see* Lindane
HHDN *see* Aldrin
Hibrom *see* Naled
Higalnate *see* Molinate
Hildan *see* α-Endosulfan, β-Endosulfan
Hilthion *see* Malathion
Hilthion 25WDP *see* Malathion
HOCH *see* Formaldehyde
Hoe 2671 *see* α-Endosulfan, β-Endosulfan
Hoe 2810 *see* Linuron
Hoe 23408 *see* Diclofop-methyl
Hoe-Grass *see* Diclofop-methyl
Hoelon *see* Diclofop-methyl
Hoelon 3EC *see* Diclofop-methyl
Hokmate *see* Ferbam
Horbadox *see* Pendimethalin
Hormotuho *see* MCPA
Hornotuho *see* MCPA
Hortex *see* Lindane
HOX 1901 *see* Ethiofencarb
HS 14260 *see* Terbutryn
Huilex *see* Toxaphene
Hungazin *see* Atrazine
Hungazin DT *see* Simazine
Hungazin PK *see* Atrazine
HW 920 *see* Diuron
Hydout *see* Endothall
Hydram *see* Molinate
Hydrothal *see* Endothall
Hydrothal-47 *see* Endothall
Hydrothal-191 *see* Endothall
m-Hydroxycarbanilic acid methyl ester *m*-methylcarbanilate *see*
     Phenmedipham
1-(4'-Hydroxy-3'-coumarinyl)-1-phenyl-3-butanone *see* Warfarin

Invisi-gard *see* Propoxur
Ipaner *see* 2,4-D
IPC *see* Propham
Iprodione
Iron flowable *see* Ferbam
Iron tris(dimethyldithiocarbamate) *see* Ferbam
Iscobrome *see* Methyl bromide
Iscobrome D *see* Ethylene dibromide
Isocarb *see* Propoxur
Isodrin epoxide *see* Endrin
Isofenphos
β-Isomer *see* β-BHC
γ-Isomer *see* Lindane
Isomethylsystox sulfoxide *see* Oxydemeton-methyl
Isophenphos *see* Isofenphos
IsoPPC *see* Propham
O-Isopropoxyphenyl methylcarbamate *see* Propoxur
O-Isopropoxyphenyl *N*-methylcarbamate *see* Propoxur
Isopropylamino-*o*-ethyl-(4-methylmercapto)-3-methylphenyl)phosphate *see*
      Fenamiphos
3-Isopropyl-1*H*-2,1,3-benzothiadiazin-4(3*H*)-one-2,2-dioxide *see* Bentazone
1-Isopropyl carbamoyl-3-(3,5-dichlorophenyl)hydantoin *see* Iprodione
Isopropyl carbanilate *see* Propham
Isopropyl carbanilic acid ester *see* Propham
N-Isopropyl-2-chloroacetanilide *see* Propachlor
N-Isopropyl-α-chloroacetanilide *see* Propachlor
Isopropyl 3-chlorocarbanilate *see* Chlorpropham
Isopropyl *m*-chlorocarbanilate *see* Chlorpropham
Isopropyl 3-chlorophenylcarbamate *see* Chlorpropham
Isopropyl-*N*-(3-chlorophenyl)carbamate *see* Chlorpropham
Isopropyl-*N*-*m*-chlorophenylcarbamate *see* Chlorpropham
O-Isopropyl *N*-(3-chlorophenyl)carbamate *see* Chlorpropham
2,3-Isopropylidenedioxyphenyl methylcarbamate *see* Bendiocarb
O-2-Isopropyl-4-methylpyrimidinyl-*O,O*-diethyl phosphorothioate *see*
      Diazinon
Isopropylmethylpyrimidinyl diethyl thiophosphate *see* Diazinon
Isopropyl phenylcarbamate *see* Propham
O-Isopropyl *N*-phenylcarbamate *see* Propham
Isopropyl *N*-phenylurethane *see* Propham
Isopropyl salicylate *O*-ester with *O*-ethylisopropylphosphoramidothioate *see*
      Isofenphos
Isothiocyanatomethane *see* Methylisothiocyanate
Isothiocyanic acid methyl ester *see* Methylisothiocyanate

Isotox *see* Lindane
Itopaz *see* Ethion
Ivalon *see* Formaldehyde
Ivoran *see* *p,p'*-DDT
Ixodex *see* *p,p'*-DDT
Jacutin *see* Lindane
Jalan *see* Molinate
JF 5705F *see* Cypermethrin
Jolt *see* Ethoprop
Jonnix *see* Asulam
Julin's carbon chloride *see* Hexachlorobenzene
K III *see* 4,6-Dinitro-*o*-cresol
K IV *see* 4,6-Dinitro-*o*-cresol
Kafil super *see* Cypermethrin
Kamfochlor *see* Toxaphene
Kamposan *see* Ethephon
Karamate *see* Mancozeb
Karbam black *see* Ferbam
Karbaspray *see* Carbaryl
Karbatox *see* Carbaryl
Karbofos *see* Malathion
Karbosep *see* Carbaryl
Karlan *see* Ronnel
Karmex *see* Diuron, Monuron
Karmex diuron herbicide *see* Diuron
Karmex DW *see* Diuron
Karsan *see* Formaldehyde
Kayafume *see* Methyl bromide
Kayazinon *see* Diazinon
Kayazol *see* Diazinon
Kemate *see* Anilazine
Kemifam *see* Phenmedipham
Kemolate *see* Phosmet
Kepone
Kerb *see* Propyzamide
Kerb 50W *see* Propyzamide
Kestrel *see* Permethrin
Killax *see* Tetraethyl pyrophosphate
Kilmite 40 *see* Tetraethyl pyrophosphate
Kiloseb *see* Dinoseb
Kilsem *see* MCPA
Kleer-lot *see* Amitrole
Kloben *see* Neburon

4K-2M *see* MCPA
KMH *see* Maleic hydrazide
Kokotine *see* Lindane
Kolphos *see* Parathion
Kopfume *see* Ethylene dibromide
Kop-mite *see* Chlorobenzilate
Kopsol *see* p,p'-DDT
Kop-thiodan *see* α-Endosulfan, β-Endosulfan
Kop-thion *see* Malathion
Korlan *see* Ronnel
Korlane *see* Ronnel
K-pin *see* Picloram
Krecalvin *see* Dichlorvos
Kregasan *see* Thiram
Krenite *see* Fosamine-ammonium
Krenite brush control agent *see* Fosamine-ammonium
Kresamone *see* 4,6-Dinitro-*o*-cresol
Krezone *see* MCPA
Krezotol 50 *see* 4,6-Dinitro-*o*-cresol
Krotiline *see* 2,4-D
Krovar I *see* Bromacil
Krovar II *see* Bromacil
Krysid *see* ANTU
Kumader *see* Warfarin
Kumadu *see* Warfarin
Kwell *see* Lindane
Kwik-kil *see* Strychnine
Kwit *see* Ethion
Kypchlor *see* Chlordane
Kypfarin *see* Warfarin
Kypfos *see* Malathion
Kypthion *see* Parathion
L 11/6 *see* Phorate
L 395 *see* Dimethoate
L 34314 *see* Diphenamid
L 36352 *see* Trifluralin
Lanex *see* Fluometuron
Lannate *see* Methomyl
Lannate L *see* Methomyl
Lariat *see* Alachlor
Larvacide 100 *see* Chloropicrin
Larvin *see* Thiodicarb
Lasso *see* Alachlor

Lasso II *see* Alachlor
Lasso EC *see* Alachlor
Lauxtol *see* Pentachlorophenol
Lauxtol A *see* Pentachlorophenol
Lawn-keep *see* 2,4-D
Lazo *see* Alachlor
Lebaycid *see* Fenthion
Legumex DB *see* MCPA
Leivasom *see* Trichlorfon
Lemonene *see* Biphenyl
Lendine *see* Lindane
Lentox *see* Lindane
Lepicron *see* Thiodicarb
Lethalaire G 52 *see* Tetraethyl pyrophosphate
Lethalaire G 54 *see* Parathion
Lethalaire G 57 *see* Sulfotepp
Lethox *see* Carbophenothion
Leuna M *see* MCPA
Lexone *see* Metribuzin
Lexone DF *see* Metribuzin
Lexone 4L *see* Metribuzin
Leyspray *see* MCPA
LFA 2043 *see* Iprodione
Lidenal *see* Lindane
Lilly 34314 *see* Diphenamid
Lilly 36352 *see* Trifluralin
Lindafor *see* Lindane
Lindagam *see* Lindane
Lindagrain *see* Lindane
Lindagranox *see* Lindane
Lindan *see* Dichlorvos
Lindane
α-Lindane *see* α-BHC
β-Lindane *see* β-BHC
δ-Lindane *see* δ-BHC
γ-Lindane *see* Lindane
Lindapoudre *see* Lindane
Lindatox *see* Lindane
Lindosep *see* Lindane
Line rider *see* 2,4,5-T
Linex 4L *see* Linuron
Linormone *see* MCPA
Linorox *see* Linuron

Lintox *see* Lindane
Linurex *see* Linuron
Linuron
Linuron 4L *see* Linuron
Lipan *see* 4,6-Dinitro-*o*-cresol
Liqua-tox *see* Warfarin
Lirobetarex *see* Monuron
Liro CIPC *see* Chlorpropham
Lirohex *see* Tetraethyl pyrophosphate
Liroprem *see* Pentachlorophenol
Lirothion *see* Parathion
Loisol *see* Trichlorfon
Lombristop *see* Thiabendazole
Lorex *see* Linuron
Lorexane *see* Lindane
Lorox *see* Linuron
Lorox DF *see* Linuron
Lorox L *see* Linuron
Lorox linuron weed killer *see* Linuron
Lorsban *see* Chlorpyrifos
LS 74783 *see* Fosetyl-aluminum
Lurgo *see* Dimethoate
Lysoform *see* Formaldehyde
M 40 *see* MCPA
M 74 *see* Disulfoton
M 140 *see* Chlordane
M 410 *see* Chlordane
M 5055 *see* Toxaphene
M&B 10064 *see* Bromoxynil
M&B 10731 *see* Bromoxynil octanoate
Maax *see* Propazine
Macrondray *see* 2,4-D
Mafu *see* Dichlorvos
Mafu strip *see* Dichlorvos
Magnacide *see* Acrolein
MAH *see* Maleic hydrazide
Maintain 3 *see* Maleic hydrazide
Malacide *see* Malathion
Malafor *see* Malathion
Malagran *see* Malathion
Malakill *see* Malathion
Malamar *see* Malathion
Malamar 50 *see* Malathion

Malaphele *see* Malathion
Malaphos *see* Malathion
Malasol *see* Malathion
Malaspray *see* Malathion
Malathion
Malathion E50 *see* Malathion
Malathion LV concentrate *see* Malathion
Malathion ULV concentrate *see* Malathion
Malathiozoo *see* Malathion
Malathon *see* Malathion
Malathyl LV concentrate & ULV concentrate *see* Malathion
Malatol *see* Malathion
Malatox *see* Malathion
Maldison *see* Malathion
Maleic acid hydrazide *see* Maleic hydrazide
Maleic hydrazide
Maleic hydrazide 30% *see* Maleic hydrazide
Malein 30 *see* Maleic hydrazide
N,N-Maleohydrazine *see* Maleic hydrazide
Malix *see* α-Endosulfan, β-Endosulfan
Malmed *see* Malathion
Malphos *see* Malathion
Maltox *see* Malathion
Maltox MLT *see* Malathion
Malzid *see* Maleic hydrazide
Mancofol *see* Mancozeb
Mancozeb
Maneb-zinc *see* Mancozeb
Manoseb *see* Mancozeb
Manzate 200 *see* Mancozeb
Manzeb *see* Mancozeb
Manzin *see* Mancozeb
Manzin 80 *see* Mancozeb
Maralate *see* Methoxychlor
Mar-frin *see* Warfarin
Marlate *see* Methoxychlor
Marlate 50 *see* Methoxychlor
Marmer *see* Diuron
Martin's mar-frin *see* Warfarin
Marvex *see* Dichlorvos
Masoten *see* Trichlorfon
Matacil *see* Aminocarb
Mataven *see* Difenzoquat methyl sulfate

Maveran *see* Warfarin
May–meb–6647 *see* Triadimefon
Mazide *see* Maleic hydrazide
Mazoten *see* Trichlorfon
MB *see* Methyl bromide
MB 9057 *see* Asulam
MB 10064 *see* Bromoxynil
MBC *see* Benomyl
M–B–C Fumigant *see* Methyl bromide
MBR 12325 *see* Mefluidide
MBX *see* Methyl bromide
MC–4379 *see* Bifenox
2M–4C *see* MCPA
2M–4CH *see* MCPA
MCP *see* MCPA
MCPA
2,4–MCPA *see* MCPA
MDBA *see* Dicamba
ME 1700 *see* p,p'-DDD
MEB 6447 *see* Triadimefon
MEBR *see* Methyl bromide
ME4 Brominal *see* Bromoxynil
Mediben *see* Dicamba
Mefluidide
Melipax *see* Toxaphene
Mendrin *see* Endrin
Meniphos *see* Mevinphos
Menite *see* Mevinphos
Mephanac *see* MCPA
Mercaptodimethur *see* Methiocarb
3-(Mercaptomethyl)-1,2,3-benzotriazin-4(3H)-one-O,O-dimethyl
    phosphorodithioate-S-ester *see* Azinphos-methyl
Mercaptophos *see* Fenthion
Mercaptosuccinic acid diethyl ester *see* Malathion
Mercaptothion *see* Malathion
Mercuram *see* Thiram
Merex *see* Kepone
Merkazin *see* Prometryn
Mertec *see* Thiabendazole
Mertect *see* Thiabendazole
Mertect 160 *see* Thiabendazole
Mesomile *see* Methomyl
Mesurol *see* Methiocarb

Metachlor *see* Alachlor

Metafume *see* Methyl bromide

Metaisosystox sulfoxide *see* Oxydemeton-methyl

Metalaxil *see* Metalaxyl

Metalaxyl

Metamidofos estrella *see* Methamidophos

Metasol TK-100 *see* Thiabendazole

Metasystemox *see* Oxydemeton-methyl

Metasystox-R *see* Oxydemeton-methyl

Metaxon *see* MCPA

Metelilachlor *see* Metolachlor

Methachlor *see* Alachlor

Methamidophos

Methanal *see* Formaldehyde

Methanedithiol-*S*,*S*-diester with *O*,*O*-diethyl phosphorodithioate *see* Ethion

Methane tetrachloride *see* Carbon tetrachloride

6,9-Methano-2,4,3-benzodioxathiepin *see* Endosulfan sulfate

Methidathion

Methidathion 50S *see* Methidathion

Methiocarb

Methogas *see* Methyl bromide

Methomyl

Methoxcide *see* Methoxychlor

Methoxo *see* Methoxychlor

Methoxone *see* MCPA

2-Methoxy-4,6-bis(isopropylamino)-1,3,5-triazine *see* Prometon

6-Methoxy-*N*,*N*'-bis(1-methylethyl)-1,3,5-triazine-2,4-diamine *see* Prometon

3-Methoxycarbonylaminophenyl 3-methylcarbanilate *see* Phenmedipham

3-((Methoxycarbonyl)amino)phenyl (3-methylphenyl)carbamate *see* Phenmedipham

2-Methoxycarbonyl-1-methylvinyl dimethyl phosphate *see* Mevinphos

cis-2-Methoxycarbonyl-1-methylvinyl dimethyl phosphate *see* Mevinphos

1-Methoxycarbonyl-1-propen-2-yl dimethyl phosphate *see* Mevinphos

Methoxychlor

4,4'-Methoxychlor *see* Methoxychlor

p,*p*'-Methoxychlor *see* Methoxychlor

Methoxy-DDT *see* Methoxychlor

2-Methoxy-3,6-dichlorobenzoic acid *see* Dicamba

Methoxydiuron *see* Linuron

1-Methoxy-1-methyl-3-(3,4-dichlorophenyl)urea *see* Linuron

N-(Methoxy(methylthio)phosphinoyl)acetamide *see* Acephate

2-(((((4-Methoxy-6-methyl-1,3,5-triazin-2-yl)amino)carbonyl)amino)-

sulfonyl)benzoic acid *see* Metsulfuron-methyl

3-(((((4-Methoxy-6-methyl-1,3,5-triazin-2-yl)amino)carbonyl)amino)-sulfonyl)-2-thiophenecarboxylic acid methyl ester *see* Thiameturon-methyl

S-((5-Methoxy-2-oxo-1,3,4-thiadiazol-3(2*H*)-yl)methyl) *O,O*-dimethyl phosphorodithioate *see* Methidathion

Methoxypropazine *see* Prometon

Methyl aldehyde *see* Formaldehyde

Methyl *N*-(4-aminobenzenesulfonyl)carbamate *see* Asulam

N-(((Methylamino)carbonyl)oxy)ethanimidothioate *see* Methomyl

N-(((Methylamino)carbonyl)oxy)ethanimidothioic acid methyl ester *see* Methomyl

Methyl((4-aminophenyl)sulfonyl)carbamate *see* Asulam

Methylazinphos *see* Azinphos-methyl

N-Methylbenzazimide, dimethyldithiophosphoric acid ester *see* Azinphos-methyl

1-Methylbenzyl-3-(dimethoxyphosphinyloxo)isocrotonate *see* Crotoxyphos

α-Methyl benzyl-3-(dimethoxyphosphinyloxy)-*cis*-crotonate *see* Crotoxyphos

α-Methylbenzyl 3-hydroxycrotonate dimethyl phosphate *see* Crotoxyphos

Methyl bromide

Methyl (1-((butylamino)carbonyl)-1*H*-benzimidazol-2-yl)carbamate *see* Benomyl

Methyl 1-(butylcarbamoyl)-2-benzimidazolylcarbamate *see* Benomyl

Methylcarbamate-1-naphthalenol *see* Carbaryl

Methylcarbamate-1-naphthol *see* Carbaryl

Methyl carbamic acid 2,3-dihydro-2,2-dimethyl-7-benzofuranyl ester *see* Carbofuran

Methyl carbamic acid 4-(methylthio)-3,5-xylyl ester *see* Methiocarb

Methyl carbamic acid 1-naphthyl ester *see* Carbaryl

S-Methylcarbamoylmethyl *O,O*-dimethyl phosphorodithioate *see* Dimethoate

N-((Methylcarbamoyl)oxy)thioacetimidic acid methyl ester *see* Methomyl

2-Methyl-4-chlorophenoxyacetic acid *see* MCPA

Methyl chlorophos *see* Trichlorfon

1-(2-Methylcychohexyl)-3-phenylurea *see* Siduron

N-(2-Methylcyclohexyl)-*N'*-phenylurea *see* Siduron

Methyl demeton-*O*-sulfoxide *see* Oxydemeton-methyl

Methyl 5-(2,4-dichlorophenoxy)-2-nitrobenzoate *see* Bifenox

Methyl 2-(4-(2,4-dichlorophenoxy)phenoxy)propanoate *see* Diclofop-methyl

Methyl 3-(dimethoxyphosphinyloxy)crotonate *see* Mevinphos

Methyl 2-(dimethylamino)-*N*-(((methylamino)carbonyl)oxy)-2-oxo-ethanimidothioate *see* Oxamyl

Methyl-1-(dimethylcarbamoyl)-*N*-((methylcarbamoyl)oxy)thioformimidate
  see Oxamyl

S-Methyl 1-(dimethylcarbamoyl)-*N*-((methylcarbamoyl)oxy)thioformimidate
  see Oxamyl

Methyl-*N'*,*N'*-dimethyl-*N*-((methylcarbamoyl)oxy)-1-thiooxamimidate *see*
  Oxamyl

Methyl *N*-(2,6-dimethylphenyl)-*N*-(methoxyacetyl)-DL-alaninate *see*
  Metalaxyl

2-Methyl-4,6-dinitrophenol *see* 4,6-Dinitro-*o*-cresol

6-Methyl-2,4-dinitrophenol *see* 4,6-Dinitro-*o*-cresol

6-Methyl-1,3-dithiolo[4,5-*b*]quinoxalin-2-one *see* Quinomethionate

S,*S'*-Methylene bis(*O*,*O*-diethyl phosphorodithioate) *see* Ethion

Methylene glycol *see* Formaldehyde

Methylene oxide *see* Formaldehyde

S,*S'*-Methylene *O*,*O*,*O'*,*O'*-tetraethyl phosphorodithioate *see* Ethion

2-(1-Methylethoxy)phenol methylcarbamate *see* Propoxur

3-(1-Methylethyl)-1*H*-2,1,3-benzothiadiazin-4(3*H*)-one-2,2-dioxide *see*
  Bentazone

1-Methylethyl (3-chlorophenyl)carbamate *see* Chlorpropham

1-Methylethyl-2-((ethoxy((1-methylethyl)amino)phosphino-
  thioyl)oxy)benzoate *see* Isofenphos

1-(Methylethyl)ethyl 3-methyl-4-(methylthio)phenyl phosphoramidate *see*
  Fenamiphos

N-(1-Methylethyl)-*N*-phenylacetamide *see* Propachlor

1-Methylethyl phenylcarbamate *see* Propham

Methyl guthion *see* Azinphos-methyl

Methyl *m*-hydroxycarbanilate *m*-methylcarbanilate *see* Phenmedipham

N-Methyl-2-isopropoxyphenylcarbamate *see* Propoxur

Methyl isothiocyanate

2-Methylmercapto-4,6-bis(isopropylamino)-*s*-triazine *see* Prometryn

4-Methylmercapto-3,5-dimethylphenyl *N*-methylcarbamate *see* Methiocarb

2-Methylmercapto-4-ethylamino-*s*-triazine *see* Ametryn

4-Methylmercapto-3-methylphenyl dimethyl thiophosphate *see* Fenthion

Methyl-*N*-(((methylamino)carbonyl)oxy)ethanimidothioate *see* Methomyl

Methyl-*N*-((methylcarbamoyl)oxy)thioacetimidate *see* Methomyl

S-Methyl *N*-((methylcarbamoyl)oxy)thioacetimidate *see* Methomyl

Methyl *O*-(methylcarbamoyl)thioacethohydroxamate *see* Methomyl

cis-1-Methyl-(2-methylcarbamoyl)vinyl phosphate *see* Monocrotophos

Methyl 3-(3-methylcarbaniloyloxy)carbanilate *see* Phenmedipham

2-Methyl-2-(methylthio)propanal *O*-((methylamino)carbonyl)oxime *see*
  Aldicarb

2-Methyl-2-(methylthio)propionaldehyde *O*-(methylcarbamoyl)oxime *see*
  Aldicarb

Methyl mustard oil *see* Methylisothiocyanate
N-Methyl-1-naphthylcarbamate *see* Carbaryl
N-Methyl-α-naphthylcarbamate *see* Carbaryl
N-Methyl-α-naphthylurethan *see* Carbaryl
6-Methyl-2-oxo-1,3-dithiolo[4,5-*b*]quinoxaline *see* Quinomethionate
(Methylphenyl)carbamic acid 3-((methoxycarbonyl)amino)phenyl ester *see*
    Phenmedipham
Methyl phthalate *see* Dimethyl phthalate
2-(1-Methylpropyl)-4,6-dinitrophenol *see* Dinoseb
6-Methyl-2,3-quinoxaline dithiocarbonate *see* Quinomethionate
6-Methyl-2,3-quinoxalinedithiol cyclic carbonate *see* Quinomethionate
6-Methyl-2,3-quinoxalinedithiol cyclic dithiocarbonate *see* Quinomethionate
6-Methyl-2,3-quinoxalinedithiol cyclic *S,S*-dithiocarbonate *see*
    Quinomethionate
6-Methylquinoxaline-2,3-dithiolcyclocarbonate *see* Quinomethionate
S,S-(6-Methylquinoxaline-2,3-diyl) dithiocarbonate *see* Quinomethionate
Methyl sulfanilylcarbamate *see* Asulam
2-Methylthio-4,6-bis(isopropylamino)-*s*-triazine *see* Prometryn
4-Methylthio-3,5-dimethylphenyl methylcarbamate *see* Methiocarb
2-Methylthio-4-ethylamino-6-*tert*-butylamino-*s*-triazine *see* Terbutryn
2-Methylthio-4-ethylamino-6-isopropylamino-*s*-triazine *see* Ametryn
4-Methylthio-3,5-xylyl isomethylcarbamate *see* Methiocarb
Methyl thiram *see* Thiram
Methyl thiuramdisulfide *see* Thiram
Methyl 3-(*m*-tolylcarbamoyloxy)phenylcarbamate *see* Phenmedipham
Methyl tuads *see* Thiram
Metifonate *see* Trichlorfon
Metilmercaptofosoksid *see* Oxydemeton-methyl
Metiltriazotion *see* Azinphos-methyl
Metmercapturon *see* Methiocarb
Metolachlor
Metox *see* Methoxychlor
Metoxon *see* Chlorpropham
Metribuzin
Metrifonate *see* Trichlorfon
Metriphonate *see* Trichlorfon
Metsulfuron-methyl
Meturon *see* Fluometuron
Meturon 4L *see* Fluometuron
Mevinphos
MH *see* Maleic hydrazide
MH 30 *see* Maleic hydrazide
MH 36 Bayer *see* Maleic hydrazide

MH 40 *see* Maleic hydrazide
MIC *see* Methylisothiocyanate
Micofume *see* Dazomet
Microlysin *see* Chloropicrin
Milbol 49 *see* Lindane
Miller's fumigrain *see* Acrylonitrile
Milocep *see* Metolachlor
Milogard *see* Propazine
Milogard 4L *see* Propazine
Milogard 80W *see* Propazine
Mintesol *see* Thiabendazole
Mintezol *see* Thiabendazole
Minzolum *see* Thiabendazole
Mipax *see* Dimethyl phthalate
Miracle *see* 2,4-D
MIT *see* Methylisothiocyanate
Mitacil *see* Aminocarb
MITC *see* Methylisothiocyanate
MK 360 *see* Thiabendazole
2M-4KH *see* MCPA
ML 97 *see* Phosphamidon
MLT *see* Malathion
Mobil V-C 9-104 *see* Ethoprop
Mocap *see* Ethoprop
Modown *see* Bifenox
Mole death *see* Strychnine
Mole-nots *see* Strychnine
Molinate
Molmate *see* Molinate
Mon 0573 *see* Glyphosate
Monalide
Monitor *see* Methamidophos
Monobromomethane *see* Methyl bromide
Monocil 40 *see* Monocrotophos
Monocron *see* Monocrotophos
Monocrotophos
N-Monomethylamide of *O,O*-dimethyldithiophosphorylacetic acid *see*
    Dimethoate
Monosan *see* 2,4-D
Monsanto penta *see* Pentachlorophenol
Montrose Propanil *see* Propanil
Monurex *see* Monuron
Monuron

Monurox *see* Monuron
Monuuron *see* Monuron
Moosuran *see* Pentachlorophenol
Mopari *see* Dichlorvos
Morbicid *see* Formaldehyde
Mor-cran *see* Naptalam
Morestan *see* Quinomethionate
Morestane *see* Quinomethionate
Morton EP 161E *see* Methylisothiocyanate
Mortopal *see* Tetraethyl pyrophosphate
Moscardia *see* Malathion
Motox *see* Toxaphene
Mouse-pak *see* Warfarin
Mouse-rid *see* Strychnine
Mouse-tox *see* Strychnine
Mowchem *see* Mefluidide
Moxie *see* Methoxychlor
Moxone *see* 2,4-D
MPP *see* Fenthion
MRC 910 *see* Iprodione
Mszycol *see* Lindane
MTD *see* Methamidophos
Multamat *see* Bendiocarb
Multimet *see* Bendiocarb
Murfos *see* Parathion
Mustard oil *see* Methylisothiocyanate
Mutoxin *see* p,p'-DDT
Mycozol *see* Thiabendazole
Mylon *see* Dazomet
Mylone *see* Dazomet
Mylone 85 *see* Dazomet
NA 521 *see* Dazomet
NA 1583 *see* Chloropicrin
NA 2757 *see* Carbaryl, Methiocarb
NA 2761 *see* Aldrin, Dichlone, Dieldrin, Endrin, Heptachlor, Kepone,
    Lindane, p,p'-DDD, Toxaphene
NA 2762 *see* Aldrin, Chlordane
NA 2763 *see* Diazinon
NA 2765 *see* 2,4-D, 2,4,5-T
NA 2767 *see* Diuron
NA 2769 *see* Dicamba, Dichlobenil
NA 2771 *see* Thiram
NA 2781 *see* Diquat

NA 2783 *see* Azinphos-methyl, Chlorpyrifos, Dichlorvos, Disulfoton, Ethion, Malathion, Mevinphos, Naled, Parathion, Tetraethyl pyrophosphate, Trichlorfon

NA 2790 *see* Fonofos

NAC *see* Carbaryl

Nalcon 243 *see* Dazomet

Naled

Nalkil *see* Bromacil

Nanchor *see* Ronnel

Nanker *see* Ronnel

Nankor *see* Ronnel

Naptalame *see* Naptalam

1-Naphthalenol methylcarbamate *see* Carbaryl

2-((1-Naphthalenylamino)carbonyl)benzoic acid *see* Naptalam

1-Naphthalenylthiourea *see* ANTU

1-Naphthol-*N*-methylcarbamate *see* Carbaryl

2-(α-Napthoxy)-*N,N*-diethylpropionamide *see* Napropamide

1-Naphthyl methylcarbamate *see* Carbaryl

1-Naphthyl-*N*-methylcarbamate *see* Carbaryl

α-Naphthyl-*N*-methylcarbamate *see* Carbaryl

α-Naphthylphthalamic acid *see* Naptalam

N-1-Naphthylphthalamic acid *see* Naptalam

α-Naphthylthiocarbamide *see* ANTU

1-(1-Naphthyl)-2-thiourea *see* ANTU

α-Naphthylthiourea *see* ANTU

N-1-Naphthylthiourea *see* ANTU

Napropamide

Naptalam

NC 262 *see* Dimethoate

NC 6897 *see* Bendiocarb

NCI-C00044 *see* Aldrin

NCI-C00055 *see* Chloramben

NCI-C00066 *see* Azinphos-methyl

NCI-C00099 *see* Chlordane

NCI-C00102 *see* Chlorothalonil

NCI-C00113 *see* Dichlorvos

NCI-C00124 *see* Dieldrin

NCI-C00135 *see* Dimethoate

NCI-C00157 *see* Endrin

NCI-C00180 *see* Heptachlor

NCI-C00191 *see* Kepone

NCI-C00204 *see* Lindane

NCI-C00215 *see* Malathion

NCI-C00226 *see* Parathion
NCI-C00237 *see* Picloram
NCI-C00259 *see* Toxaphene
NCI-C00408 *see* Chlorobenzilate
NCI-C00442 *see* Trifluralin
NCI-C00464 *see* p,p'-DDT
NCI-C00475 *see* p,p'-DDD
NCI-C00497 *see* Methoxychlor
NCI-C00500 *see* 1,2-Dibromo-3-chloropropane
NCI-C00522 *see* Ethylene dibromide
NCI-C00533 *see* Chloropicrin
NCI-C00555 *see* p,p'-DDE
NCI-C00566 *see* α-Endosulfan, β-Endosulfan
NCI-C00588 *see* Phosphamidon
NCI-C02799 *see* Formaldehyde
NCI-C02846 *see* Monuron
NCI-C04035 *see* Allidochlor
NCI-C04591 *see* Carbon disulfide
NCI-C08640 *see* Aldicarb
NCI-C08651 *see* Fenthion
NCI-C08673 *see* Diazinon
NCI-C08684 *see* Anilazine
NCI-C08695 *see* Fluometuron
NCI-C50044 *see* Bis(2-chloroisopropyl)ether
NCI-C54831 *see* Trichlorfon
NCI-C54933 *see* Pentachlorophenol
NCI-C55141 *see* 1,2-Dichloropropane
NCI-C55378 *see* Pentachlorophenol
NCI-C56655 *see* Pentachlorophenol
NCI-C60242 *see* 2,3,5-Trichlorobenzoic acid
NCI-C60413 *see* Chlorobenzilate
NDRC-143 *see* Permethrin
Neburea *see* Neburon
Neburex *see* Neburon
Neburon
Necatorina *see* Carbon tetrachloride
Necatorine *see* Carbon tetrachloride
Nedcisol *see* Diazinon
Nefusan *see* Dazomet
Neguvon *see* Trichlorfon
Neguvon A *see* Trichlorfon
Nemabrom *see* 1,2-Dibromo-3-chloropropane
Nemafume *see* 1,2-Dibromo-3-chloropropane

Nicochloran *see* Lindane
Nifos *see* Tetraethyl pyrophosphate
Nifos T *see* Tetraethyl pyrophosphate
Nifost *see* Tetraethyl pyrophosphate
Niomil *see* Bendiocarb
Nip-A-Thin *see* Naptalam
Nipsan *see* Diazinon
Niran *see* Chlordane, Parathion
Niran E-4 *see* Parathion
Niticid *see* Propachlor
Nitrador *see* 4,6-Dinitro-*o*-cresol
Nitran *see* Trifluralin
Nitrapyrin
Nitrile *see* Acrylonitrile
Nitrochloroform *see* Chloropicrin
Nitrofan *see* 4,6-Dinitro-*o*-cresol
Nitropone P *see* Dinoseb
Nitrostigmine *see* Parathion
Nitrostygmine *see* Parathion
Nitrotrichloromethane *see* Chloropicrin
Niuif-100 *see* Parathion
Nobecutan *see* Thiram
No bunt *see* Hexachlorobenzene
No bunt 40 *see* Hexachlorobenzene
No bunt 80 *see* Hexachlorobenzene
No bunt liquid *see* Hexachlorobenzene
Nogos *see* Dichlorvos
Nogos G *see* Dichlorvos
Nogos 50 *see* Dichlorvos
Nomersan *see* Thiram
Nopcocide *see* Chlorothalonil
Nopcocide N-96 *see* Chlorothalonil
Nopcocide N40D & N96 *see* Chlorothalonil
No-pest *see* Dichlorvos
No-pest strip *see* Dichlorvos
Nor-Am *see* Diphenamid
Norex *see* Chloroxuron
Normersan *see* Thiram
Norosac *see* Dichlobenil
Norosac 4G *see* Dichlobenil
Norosac 10G *see* Dichlobenil
Nourithion *see* Parathion
Novigam *see* Lindane

Octalene *see* Aldrin
Octalox *see* Dieldrin
Octaterr *see* Chlordane
Oftanol *see* Isofenphos
Oko *see* Dichlorvos
Okultin M *see* MCPA
Oleoakarithion *see* Carbophenothion
Oleofos 20 *see* Parathion
Oleogesaprim *see* Atrazine
Oleoparaphene *see* Parathion
Oleoparathion *see* Parathion
Oleophosphothion *see* Malathion
Olitref *see* Trifluralin
Omnitox *see* Lindane
Omnizole *see* Thiabendazole
OMS 2 *see* Fenthion
OMS 14 *see* Dichlorvos
OMS 29 *see* Carbaryl
OMS 33 *see* Propoxur
OMS 37 *see* Fensulfothion
OMS 93 *see* Methiocarb
OMS 570 *see* α-Endosulfan, β-Endosulfan
OMS 771 *see* Aldicarb
OMS 971 *see* Chlorpyrifos
OMS 1804 *see* Diflubenzuron
Ontracic 800 *see* Prometon
Ontrack *see* Prometon
Ontrack 8E *see* Metolachlor
Ontrack-WE-2 *see* Prometon
OR 1191 *see* Phosphamidon
Ordram *see* Molinate
Ordram 8E *see* Molinate
Ordram 10G *see* Molinate
Ordram 15G *see* Molinate
Orga-414 *see* Amitrole
Ornamental weeder *see* Chloramben
Orthene *see* Acephate
Orthene-755 *see* Acephate
Ortho *see* Diquat
Ortho 4355 *see* Naled
Ortho 9006 *see* Methamidophos
Ortho 12420 *see* Acephate
Orthodibrom *see* Naled

Orthodibromo *see* Naled
Ortho grass killer *see* Propham
Orthoklor *see* Chlordane
Orthomalathion *see* Malathion
Orthophos *see* Parathion
Ortho phosphate defoliant *see* Butifos
Ortran *see* Acephate
Ortril *see* Acephate
OS 1987 *see* 1,2-Dibromo-3-chloropropane
OS 2046 *see* Mevinphos
Oust *see* Sulfometuron-methyl
Outflanf-stockade *see* Permethrin
Outflank *see* Permethrin
Ovadziak *see* Lindane
Owadziak *see* Lindane
7-Oxabicyclo[2.2.1]heptane-2,3-dicarboxylic acid *see* Endothall
Oxadiazon
Oxamyl
Oxomethane *see* Formaldehyde
1,1′-Oxybis(2-chloroethane) *see* Bis(2-chloroethyl)ether
2,2′-Oxybis(1-chloropropane) *see* Bis(2-chloroisopropyl)ether
Oxy DBCP *see* 1,2-Dibromo-3-chloropropane
Oxydemetonmethyl *see* Oxydemeton-methyl
Oxydemeton-methyl
Oxymethylene *see* Formaldehyde
Oxythioquinox *see* Quinomethionate
Oxytril M *see* Bromoxynil
P-974 *see* Phosalone
PA *see* Naptalam
Pac *see* Parathion
Pakhtaran *see* Fluometuron
Palatinol C *see* Di-*n*-butyl phthalate
Palatinol M *see* Dimethyl phthalate
Panam *see* Carbaryl
Pandar *see* Monocrotophos
Panoram 75 *see* Thiram
Panoram D-31 *see* Dieldrin
Panthion *see* Parathion
Pantozol 1 *see* Crotoxyphos
Parachlorocidum *see* p,p′-DDT
Paradust *see* Parathion
Paraflow *see* Parathion
Paraform *see* Formaldehyde

Paramar *see* Parathion
Paramar 50 *see* Parathion
Paraphos *see* Parathion
Paraspray *see* Parathion
Parathene *see* Parathion
Parathion
Parathionethyl *see* Parathion
Parawet *see* Parathion
Partner *see* Bromoxynil
Pathclear *see* Diquat
Pay-off *see* Flucythrinate, Pendimethalin
Payze *see* Cyanazine
PCC *see* Toxaphene
PCP *see* Pentachlorophenol
PD 5 *see* Mevinphos
PDD 60401 *see* Diflubenzuron
Peach-Thin *see* Naptalam
PEB1 *see* p,p'-DDT
PEBC *see* Pebulate
Pebulate
Pedraczak *see* Lindane
PEI 35 *see* Dimethoate
Penchlorol *see* Pentachlorophenol
Pendimethalin
Pennamine *see* 2,4-D
Pennamine D *see* 2,4-D
Pennant *see* Metolachlor
Pennant 5G *see* Metolachlor
Pennout *see* Endothall
Penoxalin *see* Pendimethalin
Penoxaline *see* Pendimethalin
Penoxyn *see* Pendimethalin
Penphene *see* Toxaphene
Penphos *see* Parathion
Penta *see* Pentachlorophenol
Pentachlorin *see* p,p'-DDT
Pentachlorobenzene
Pentachlorofenol *see* Pentachlorophenol
Pentachlorofenolo *see* Pentachlorophenol
Pentachlorophenate *see* Pentachlorophenol
Pentachlorophenol
Pentachlorphenol *see* Pentachlorophenol
2,3,4,5,6-Pentachlorophenol *see* Pentachlorophenol

Pentachlorophenyl chloride *see* Hexachlorobenzene
Pentacon *see* Pentachlorophenol
Penta-kil *see* Pentachlorophenol
Pentasol *see* Pentachlorophenol
Pentech *see* *p,p'*-DDT
Penwar *see* Pentachlorophenol
Peratox *see* Pentachlorophenol
Perchlorobenzene *see* Hexachlorobenzene
Perchloromethane *see* Carbon tetrachloride
Perfecthion *see* Dimethoate
Perfekthion *see* Dimethoate
Perfektion *see* Dimethoate
Perflan *see* Tebuthiuron
Perfmid *see* Tebuthiuron
Permacide *see* Pentachlorophenol
Permaguard *see* Pentachlorophenol
Permasan *see* Pentachlorophenol
Permatox DP-2 *see* Pentachlorophenol
Permatox Penta *see* Pentachlorophenol
Permethrin
Permite *see* Pentachlorophenol
Pestmaster *see* Ethylene dibromide, Methyl bromide
Pestmaster EDB-85 *see* Ethylene dibromide
Pestox plus *see* Parathion
Pethion *see* Parathion
Pflanzol *see* Lindane
PH 60-40 *see* Diflubenzuron
Phasolon *see* Phosalone
PHC *see* Propoxur
Phenacide *see* Toxaphene
Phenamiphos *see* Fenamiphos
Phenatox *see* Toxaphene
Phenmedipham
Phenotan *see* Dinoseb
Phenox *see* 2,4-D
Phenoxalin *see* Pendimethalin
Phenoxybenzyl-2-(4-chlorophenyl)isovalerate *see* Fenvalerate
Phenoxylene 50 *see* MCPA
Phenoxylene Plus *see* MCPA
Phenoxylene Super *see* MCPA
cis-2-(1-Phenylethoxy)carbonyl-1-methylvinyl dimethyl phosphate *see*
    Crotoxyphos
(3-Phenoxyphenyl)methyl 3-(2,2-dichloroethenyl)-2,2-dimethylcyclopropane

Mevinphos

Phosphorodithioic acid *S*-((*tert*-butylthio)methyl) *O,O*-diethyl ester *see* Terbufos

Phosphorodithioic acid *S*-(2-chloro-1-(1,3-dihydro-1,3-dioxo-2*H*-isoindol-2-yl)ethyl) *O,O*-diethyl ester *see* Dialifos

Phosphorodithioic acid *S*-((6-chloro-2-oxo-3(2*H*)-benzoxazolyl)methyl) *O,O*-diethyl ester *see* Phosalone

Phosphorodithioic acid *S*-(((4-chlorophenyl)thio)methyl) *O,O*-diethyl ester *see* Carbophenothion

Phosphorodithioic acid *S*-(2-chloro-1-phthalimidoethyl) *O,O*-diethyl ester *see* Dialifos

Phosphorodithioic acid *O,O*-diethyl ester, *S,S*-diester with methanedithiol *see* Ethion

Phosphorodithioic acid *O,O*-diethyl ester *S*-((ethylthio)methyl) ester *see* Phorate

Phosphorodithioic acid *S*-((1,3-dihydro-1,3-dioxo-2*H*-isoindol-2-yl)methyl) *O,O*-dimethyl ester, *see* Phosmet

Phosphorodithioic acid *O,O*-dimethyl ester, ester with 2-mercapto-*N*-methylacetamide *see* Dimethoate

Phosphorodithioic acid *O,O*-dimethyl ester, *S*-ester with 3-mercaptomethyl-1,2,3-benzotriazin-4(3*H*)-one *see* Azinphos-methyl

Phosphorodithioic acid *O,O*-dimethyl ester, *S*-ester with 4-(mercapto-methyl)-2-methoxy-Δ²-1,3,4-thiadiazolin-5-one *see* Methidathion

Phosphorodithioic acid *O,O*-dimethyl ester with *N*-(mercaptomethyl)-phthalimide *see* Phosmet

Phosphorodithioic acid *S*-(((1,1-dimethylethyl)thio)methyl) *O,O*-diethyl ester *see* Terbufos

Phosphorodithioic acid *O,O*-dimethyl *S*-(2-(methylamino)-2-oxoethyl) ester *see* Dimethoate

Phosphorodithioic acid *O,O*-dimethyl ((4-oxo-1,2,3-benzotriazin-3(4*H*)-yl)methyl) ester *see* Azinphos-methyl

Phosphorodithioic acid *S*-ester of 6-chloro-3-mercaptomethylbenzoxazyol-2-one *see* Phosalone

Phosphorodithioic acid *O*-ethyl *S,S*-dipropyl ester *see* Ethoprop

Phosphorothioic acid *O*-(4-bromo-2-chlorophenyl)-*O*-ethyl-*S*-propyl ester *see* Profenofos

Phosphorothioic acid *O,O*-diethyl *O*-(*p*-(methylsulfinyl)phenyl) ester *see* Fensulfothion

Phosphorothioic acid *O,O*-diethyl *O*-(4-nitrophenyl) ester *see* Parathion

Phosphorothioic acid *O,O*-dimethyl *O*-(3-methyl-4-(methylthio)phenyl) ester, *see* Fenthion

Phosphorothioic acid *O,O*-dimethyl *O*-(2,4,5-trichlorophenyl)ester *see* Ronnel

Phosphorothionic acid *O,O*-diethyl *O*-(3,5,6-trichloro-2-pyridyl)ester *see*
    Chlorpyrifos
Phosphostigmine *see* Parathion
Phosphothioic acid *O,O*-dimethyl *S*-2-(ethylsulfinyl)ethyl ester *see*
    Oxydemeton-methyl
Phosphothion *see* Malathion
Phosphotox E *see* Ethion
Phosvit *see* Dichlorvos
Phozalon *see* Phosalone
Phtalofos *see* Phosmet
Phthalic acid dibutyl ester *see* Di-*n*-butyl phthalate
Phthalic acid dimethyl ester *see* Dimethyl phthalate
Phthalic acid methyl ester *see* Dimethyl phthalate
Phthalimidomethyl-*O,O*-dimethylphosphorodithioate *see* Phosmet
Phygon *see* Dichlone
Phygon paste *see* Dichlone
Phygon seed protectant *see* Dichlone
Phygon XL *see* Dichlone
Pic-clor *see* Chloropicrin
Picfume *see* Chloropicrin
Picloram
Picride *see* Chloropicrin
Pied piper mouse seed *see* Strychnine
Pielik *see* 2,4-D
Pillardrin *see* Monocrotophos
Pillaron *see* Methamidophos
Pillarzo *see* Alachlor
Pin *see* EPN
Pindone
Pirofos *see* Sulfotepp
Pistol *see* Phenmedipham
Pivacin *see* Pindone
Pival *see* Pindone
2-Pivaloylindane-1,3-dione *see* Pindone
2-Pivaloyl-1,3-indanedione *see* Pindone
Pivalyl *see* Pindone
Pivalyl indandione *see* Pindone
2-Pivalyl-1,3-indandione *see* Pindone
Pivalyl valone *see* Pindone
Planotox *see* 2,4-D
Plant dithio aerosol *see* Sulfotepp
Plantdrin *see* Monocrotophos
Plantfume 103 smoke generator *see* Sulfotepp

Primatol *see* Atrazine, Prometon, Prometryn
Primatol A *see* Atrazine
Primatol 25E *see* Prometon
Primatol P *see* Propazine
Primatol Q *see* Prometryn
Primatol S *see* Simazine
Primaze *see* Atrazine
Primextra *see* Metolachlor
Princep *see* Simazine
Princep 4G *see* Simazine
Princep 4L *see* Simazine
Princep 80W *see* Simazine
Printop *see* Simazine
Prioderm *see* Malathion
Profam *see* Propham
Profenofos
Profume *see* Methyl bromide
Profume A *see* Chloropicrin
Prokarbol *see* 4,6-Dinitro-*o*-cresol
Prolate *see* Phosmet
Prolex *see* Propachlor
Promamide *see* Propyzamide
Prometon
Prometone *see* Prometon
Prometrex *see* Prometryn
Prometrin *see* Prometryn
Prometryn
Prometryne *see* Prometryn
Promidione *see* Iprodione
Pronamide *see* Propyzamide
Propachlor
Propachlore *see* Propachlor
Propanex *see* Propanil
Propanid *see* Propanil
Propanide *see* Propanil
Propanil
Propasin *see* Propazine
Propazin *see* Propazine
Propazine
Propenal *see* Acrolein
2-Propenal *see* Acrolein
Prop-2-en-1-al *see* Acrolein
Propenenitrile *see* Acrylonitrile

2-Propenenitrile *see* Acrylonitrile

2-Propen-1-one *see* Acrolein

Propham

Prophame *see* Propham

Prophos *see* Ethoprop

Propionic acid 3,4-dichloroanilide *see* Propanil

Prop-Job *see* Propanil

Propotox M *see* Propoxur

Propozur

Propoxure *see* Propoxur

S-Propyl butylethylthiocarbamate *see* Pebulate

Propylene chloride *see* 1,2-Dichloropropane

Propylene dichloride *see* 1,2-Dichloropropane

α,β-Propylene dichloride *see* 1,2-Dichloropropane

Propylethylbutylthiocarbamate *see* Pebulate

Propyl *N*-ethyl-*n*-butylthiocarbamate *see* Pebulate

n-Propyl-*N*-ethyl-*N*-(*n*-butyl)thiocarbamate *see* Pebulate

S-(*n*-Propyl)-*N*-ethyl-*N,N*-butylthiocarbamate *see* Pebulate

Propylethylbutylthiolcarbamate *see* Pebulate

Propyl ethyl-*n*-butylthiolcarbamate *see* Pebulate

Propylthiopyrophosphate *see* Aspon

Propyon *see* Propoxur

Propyzamide

Prothromadin *see* Warfarin

Protrum K *see* Phenmedipham

Prowl *see* Pendimethalin

Proxol *see* Trichlorfon

Prozinex *see* Propazine

PS *see* Chloropicrin

Puralin *see* Thiram

PX 104 *see* Di-*n*-butyl phthalate

Pyradex *see* Diallate

Pyrinex *see* Chlorpyrifos

Pyrophosphoric acid tetraethyl ester *see* Tetraethyl pyrophosphate

Pyrophosphorodithioic acid *O,O,O,O*-tetraethyl dithionopyrophosphate *see* Sulfotepp

Pyrophosphorodithioic acid tetraethyl ester *see* Sulfotepp

6Q8 *see* Naptalam

QCB *see* Pentachlorobenzene

Queletox *see* Fenthion

Quellada *see* Lindane

Quinomethionate

Quinomethoate *see* Quinomethionate

Quintar *see* Dichlone
Quintar 540F *see* Dichlone
Quintox *see* Dieldrin
Quizalofop-ethyl
R 10 *see* Carbon tetrachloride
R 40B1 *see* Methyl bromide
R 1303 *see* Carbophenothion
R 1504 *see* Phosmet
R 1582 *see* Azinphos-methyl
R 1608 *see* EPTC
R 1910 *see* Butylate
R 2061 *see* Pebulate
R 2063 *see* Cycloate
R 2170 *see* Oxydemeton-methyl
R 4572 *see* Molinate
R 7465 *see* Napropamide
R 7475 *see* Napropamide
Racusan *see* Dimethoate
Radapon *see* Dalapon-sodium
Radazin *see* Atrazine
Radizine *see* Atrazine
Radocon *see* Simazine
Radokor *see* Simazine
Radox *see* Allidochlor
Radoxone TL *see* Amitrole
Rafex *see* 4,6-Dinitro-*o*-cresol
Rafex 35 *see* 4,6-Dinitro-*o*-cresol
Ramizol *see* Amitrole
Rampart *see* Phorate
Ramrod *see* Propachlor
Ramrod 65 *see* Propachlor
Randox *see* Allidochlor
Randox T *see* Allidochlor
Raphatox *see* 4,6-Dinitro-*o*-cresol
Raphone *see* MCPA
Rat-a-way *see* Warfarin
Rat & mice bait *see* Warfarin
Rat-b-gon *see* Warfarin
Rat-gard *see* Warfarin
Rat-kill *see* Warfarin
Rat-mix *see* Warfarin
Rat-o-cide #2 *see* Warfarin
Rat-ola *see* Warfarin

Ratorex *see* Warfarin
Ratox *see* Warfarin
Ratoxin *see* Warfarin
Ratron *see* Warfarin
Ratron G *see* Warfarin
Rats-no-more *see* Warfarin
Rattrack *see* ANTU
Rat-trol *see* Warfarin
Rattunal *see* Warfarin
Ravyon *see* Carbaryl
Rax *see* Warfarin
Razol dock killer *see* MCPA
RB *see* Parathion
RCRA waste number P001 *see* Warfarin
RCRA waste number P003 *see* Acrolein
RCRA waste number P004 *see* Aldrin
RCRA waste number P020 *see* Dinoseb
RCRA waste number P022 *see* Carbon disulfide
RCRA waste number P037 *see* Dieldrin
RCRA waste number P039 *see* Disulfoton
RCRA waste number P044 *see* Dimethoate
RCRA waste number P047 *see* 4,6-Dinitro-*o*-cresol
RCRA waste number P050 *see* α-Endosulfan, β-Endosulfan
RCRA waste number P051 *see* Endrin
RCRA waste number P059 *see* Heptachlor
RCRA waste number P066 *see* Methomyl
RCRA waste number P070 *see* Aldicarb
RCRA waste number P088 *see* Endothall
RCRA waste number P089 *see* Parathion
RCRA waste number P094 *see* Phorate
RCRA waste number P108 *see* Strychnine
RCRA waste number P109 *see* Sulfotepp
RCRA waste number P111 *see* Tetraethyl pyrophosphate
RCRA waste number P123 *see* Toxaphene
RCRA waste number U009 *see* Acrylonitrile
RCRA waste number U011 *see* Amitrole
RCRA waste number U025 *see* Bis(2-chloroethyl)ether
RCRA waste number U027 *see* Bis(2-chloroisopropyl)ether
RCRA waste number U029 *see* Methyl bromide
RCRA waste number U036 *see* Chlordane
RCRA waste number U038 *see* Chlorobenzilate
RCRA waste number U060 *see* *p,p'*-DDD
RCRA waste number U061 *see* *p,p'*-DDT

RCRA waste number U062 *see* Diallate
RCRA waste number U066 *see* 1,2-Dibromo-3-chloropropane
RCRA waste number U067 *see* Ethylene dibromide
RCRA waste number U069 *see* Di-*n*-butyl phthalate
RCRA waste number U083 *see* 1,2-Dichloropropane
RCRA waste number U102 *see* Dimethyl phthalate
RCRA waste number U122 *see* Formaldehyde
RCRA waste number U127 *see* Hexachlorobenzene
RCRA waste number U129 *see* Lindane
RCRA waste number U142 *see* Kepone
RCRA waste number U148 *see* Maleic hydrazide
RCRA waste number U183 *see* Pentachlorobenzene
RCRA waste number U192 *see* Propyzamide
RCRA waste number U211 *see* Carbon tetrachloride
RCRA waste number U232 *see* 2,4,5-T
RCRA waste number U240 *see* 2,4-D
RCRA waste number U242 *see* Pentachlorophenol
RCRA waste number U244 *see* Thiram
RCRA waste number U247 *see* Methoxychlor
RE 4355 *see* Naled
RE 12420 *see* Acephate
Rebelate *see* Dimethoate
Reddon *see* 2,4,5-T
Reddox *see* 2,4,5-T
Reglon *see* Diquat
Reglone *see* Diquat
Regulox *see* Maleic hydrazide
Regulox W *see* Maleic hydrazide
Regulox 50 W *see* Maleic hydrazide
Retard *see* Maleic hydrazide
Rezifilm *see* Thiram
RH-315 *see* Propyzamide
Rhodia *see* 2,4-D
Rhodiachlor *see* Heptachlor
Rhodiacide *see* Ethion
Rhodia RP 11974 *see* Phosalone
Rhodiasol *see* Parathion
Rhodiatox *see* Parathion
Rhodiatrox *see* Parathion
Rhodocide *see* Ethion
Rhomenc *see* MCPA
Rhomene *see* MCPA
Rhonox *see* MCPA

Rhothane *see* p,p'-DDD
Rhothane D-3 *see* p,p'-DDD
Ricifon *see* Trichlorfon
Ridomil *see* Metalaxyl
Ridomil E *see* Metalaxyl
Ripcord *see* Cypermethrin
Ripenthal *see* Endothall
Riselect *see* Propanil
Ritsifon *see* Trichlorfon
Rodafarin *see* Warfarin
Ro-deth *see* Warfarin
Rodex *see* Warfarin
Ro-dex *see* Strychnine
Rodex blox *see* Warfarin
Rodocid *see* Ethion
Rogodial *see* Dimethoate
Rogor 20L *see* Dimethoate
Rogor 40 *see* Dimethoate
Rogor *see* Dimethoate
Rogor L *see* Dimethoate
Rogor P *see* Dimethoate
Rogue *see* Propanil
Roll-fruct *see* Ethephon
Ro-neet *see* Cycloate
Ronit *see* Cycloate
Ronnel
Ronstar *see* Oxadiazon
Ronstar 25EC *see* Oxadiazon
Ronstar 2G *see* Oxadiazon
Ronstar 12L *see* Oxadiazon
Rop 500 F *see* Iprodione
Rosanil *see* Propanil
Rosex *see* Warfarin
Rosuran *see* Monuron
Rotate *see* Bendiocarb
Rothane *see* p,p'-DDD
Rotox *see* Methyl bromide
Rough & ready mouse mix *see* Warfarin
Rovral *see* Iprodione
Roxion *see* Dimethoate
Roxion U.A. *see* Dimethoate
Royal MH-30 *see* Maleic hydrazide
Royal Slo-Gro *see* Maleic hydrazide

Royal TMTD *see* Thiram
RP 8167 *see* Ethion
RP 11974 *see* Phosalone
RP 16272 *see* Bromoxynil octanoate
RP 17623 *see* Oxadiazon
RP 26019 *see* Iprodione
RP 32545 *see* Fosetyl-aluminum
RS 141 *see* Chlordimeform
Rubitox *see* Phosalone
Rukseam *see* p,p′-DDT
Rylam *see* Carbaryl
S 1 *see* Chloropicrin
S 276 *see* Disulfoton
S 1752 *see* Fenthion
S 3151 *see* Permethrin
S 5602 *see* Fenvalerate
S 10165 *see* Propanil
Sadofos *see* Malathion
Sadophos *see* Malathion
Sadoplon *see* Thiram
Sakkimol *see* Molinate
Salvo *see* 2,4-D
Sanaseed *see* Strychnine
Sancap 80W *see* Dipropetryn
Sandolin *see* 4,6-Dinitro-*o*-cresol
Sandolin A *see* 4,6-Dinitro-*o*-cresol
Sanmarton *see* Fenvalerate
Sanocide *see* Hexachlorobenzene
Sanquinon *see* Dichlone
Santobane *see* p,p′-DDT
Santobrite *see* Pentachlorophenol
Santophen *see* Pentachlorophenol
Santophen 20 *see* Pentachlorophenol
Santox *see* EPN
Sarclex *see* Linuron
Sarolex *see* Diazinon
Satecid *see* Propachlor
Satox 20WSC *see* Trichlorfon
SBP-1513 *see* Permethrin
Scarclex *see* Linuron
Schering 35830 *see* Monalide
Schering 36268 *see* Chlordimeform
Schering 38107 *see* Desmedipham

Schering 38584 *see* Phenmedipham
SD 40 *see* Isofenphos
SD 897 *see* 1,2-Dibromo-3-chloropropane
SD 1750 *see* Dichlorvos
SD 3562 *see* Dicrotophos
SD 4294 *see* Crotoxyphos
SD 5532 *see* Chlordane
SD 9129 *see* Monocrotophos
SD 14114 *see* Fenbutatin oxide
SD 14999 *see* Methomyl
SD 15418 *see* Cyanazine
SD 43775 *see* Fenvalerate
Seedrin *see* Aldrin
Seedrin liquid *see* Aldrin
Seffein *see* Carbaryl
Selecron *see* Profenofos
Selektin *see* Prometryn
Selephos *see* Parathion
Selinon *see* 4,6-Dinitro-*o*-cresol
Sencor *see* Metribuzin
Sencor 4 *see* Metribuzin
Sencor DF *see* Metribuzin
Sencoral *see* Metribuzin
Sencorer *see* Metribuzin
Sencorex *see* Metribuzin
Sendran *see* Propoxur
Seppic MMD *see* MCPA
Septene *see* Carbaryl
Sesagard *see* Prometryn
Sevimol *see* Carbaryl
Sevin *see* Carbaryl
SF 60 *see* Malathion
Shamrox *see* MCPA
Shell atrazine herbicide *see* Atrazine
Shell SD 3562 *see* Dicrotophos
Shell SD 4294 *see* Crotoxyphos
Shell SD 5532 *see* Chlordane
Shell SD 9129 *see* Monocrotophos
Shell SD 14114 *see* Fenbutatin oxide
Shortstop *see* Terbutryn
Shortstop E *see* Terbutryn
Siduron
Silvanol *see* Lindane

Sim-Trol *see* Simazine
Simadex *see* Simazine
Simanex *see* Simazine
Simazin *see* Simazine
Simazine
Simazine 80W *see* Simazine
Simazol *see* Amitrole
Sinbar *see* Terbacil
Sinituho *see* Pentachlorophenol
Sinoratox *see* Dimethoate
Sinox *see* 4,6-Dinitro-*o*-cresol
Sinox general *see* Dinoseb
Sinuron *see* Linuron
Siperin *see* Cypermethrin
Siptox I *see* Malathion
Sixty-three special E.C. insecticide *see* Parathion
Slimicide *see* Acrolein
Slo-Grow *see* Maleic hydrazide
Smut-go *see* Hexachlorobenzene
SN 4075 *see* Phenmedipham
SN 35830 *see* Monalide
SN 36268 *see* Chlordimeform
SN 38107 *see* Desmedipham
SN 38584 *see* Phenmedipham
SN 49537 *see* Thidiazuron
Snieciotox *see* Hexachlorobenzene
SNP *see* Parathion
Sodium dalapon *see* Dalapon-sodium
Sodium 2,2-dichloropropanoate *see* Dalapon-sodium
Sodium α,α-dichloropropionate *see* Dalapon-sodium
Soilbrom-40 *see* Ethylene dibromide
Soilbrom-85 *see* Ethylene dibromide
Soilbrom-90 *see* Ethylene dibromide
Soilbrom-90EC *see* Ethylene dibromide
Soilbrom-100 *see* Ethylene dibromide
Soilbrome-85 *see* Ethylene dibromide
Soilfume *see* Ethylene dibromide
Sok *see* Carbaryl
Soldep *see* Trichlorfon
Soleptax *see* Heptachlor
Solfarin *see* Warfarin
Solo *see* Naptalam
Solvanom *see* Dimethyl phthalate

Stampede 3E *see* Propanil
Stathion *see* Parathion
Stauffer 2061 *see* Pebulate
Stauffer ASP-51 *see* Aspon
Stauffer N 521 *see* Dazomet
Stauffer N 2790 *see* Fonofos
Stauffer R 1303 *see* Carbophenothion
Stauffer R 1910 *see* Butylate
Stauffer R 4572 *see* Molinate
Stockade *see* Cypermethrin
Stomp *see* Pendimethalin
Stomp 300D *see* Pendimethalin
Stomp 300E *see* Pendimethalin
Stopgerme-S *see* Chlorpropham
Storite *see* Thiabendazole
Strathion *see* Parathion
Strazine *see* Atrazine
Strel *see* Propanil
Streunex *see* Lindane
Strobane-T *see* Toxaphene
Strobane T-90 *see* Toxaphene
Strychnidin-10-one *see* Strychnine
Strychnine
Strychnos *see* Strychnine
Stunt-Man *see* Maleic hydrazide
Subdue *see* Metalaxyl
Subdue 2E *see* Metalaxyl
Subdue 5SP *see* Metalaxyl
Subitex *see* Dinoseb
Sucker-Stuff *see* Maleic hydrazide
Sulfatep *see* Sulfotepp
Sulfometuron-methyl
Sulfotep *see* Sulfotepp
Sulfotepp
Sulphocarbonic anhydride *see* Carbon disulfide
Sulphos *see* Parathion
Sulprofos
Sumicidin *see* Fenvalerate
Sumifly *see* Fenvalerate
Sumipower *see* Fenvalerate
Sumitox *see* Malathion
Suncide *see* Propoxur
Superaven *see* Difenzoquat methyl sulfate

Super-De-Sprout *see* Maleic hydrazide
Super D weedone *see* 2,4-D, 2,4,5-T
Superlysoform *see* Formaldehyde
Supernox *see* Propanil
Super rodiatox *see* Parathion
Super Sprout Stop *see* Maleic hydrazide
Super Sucker-Stuff *see* Maleic hydrazide
Super Sucker-Stuff HC *see* Maleic hydrazide
Sup'r flo *see* Diuron
Surcopur *see* Propanil
Surpracide *see* Methidathion
Surpur *see* Propanil
Susvin *see* Monocrotophos
Sutan *see* Butylate
Sweep *see* Chlorothalonil
Symazine *see* Simazine
Synklor *see* Chlordane
Synpran N *see* Propanil
Synthetic 3956 *see* Toxaphene
Szklarniak *see* Dichlorvos
2,4,5-T
T-2 *see* 2,3,5-Trichlorobenzoic acid
T-47 *see* Parathion
Tafazine *see* Simazine
Tafazine 50-W *see* Simazine
Tahmabon *see* Methamidophos
Tak *see* Malathion
Talcord *see* Permethrin
Talodex *see* Fenthion
Tamaron *see* Methamidophos
Tap 85 *see* Lindane
Tap 9VP *see* Dichlorvos
Taphazine *see* Simazine
Task *see* Dichlorvos
Task tabs *see* Dichlorvos
Tat chlor 4 *see* Chlordane
Taterpex *see* Chlorpropham
Tattoo *see* Bendiocarb
TBA *see* 2,3,5-Trichlorobenzoic acid
2,3,6-TBA *see* 2,3,5-Trichlorobenzoic acid
TBDZ *see* Thiabendazole
TBH *see* α-BHC, β-BHC, δ-BHC, Lindane
TBZ *see* Thiabendazole

Tersan 1991 *see* Benomyl

Tetrachloor-metaan *see* Carbon tetrachloride

2,4,5,6-Tetrachloro-1,3-benzenedicarbonitrile *see* Chlorothalonil

2,3,5,6-Tetrachloro-1,4-benzenedicarboxylic acid dimethyl ester *see* Chlorthal-dimethyl

Tetrachlorocarbon *see* Carbon tetrachloride

2,4,5,6-Tetrachloro-3-cyanobenzonitrile *see* Chlorothalonil

Tetrachlorodiphenylethane *see* p,p'-DDD

Tetrachloroisophthalonitrile *see* Chlorothalonil

Tetrachloromethane *see* Carbon tetrachloride

m-Tetrachlorophthalonitrile *see* Chlorothalonil

Tetrachloroterephthalic acid dimethyl ester *see* Chlorthal-dimethyl

2,3,5,6-Tetrachloroterephthalic acid dimethyl ester *see* Chlorthal-dimethyl

Tetraethyl diphosphate *see* Tetraethyl pyrophosphate

Tetraethyl dithionopyrophosphate *see* Sulfotepp

Tetraethyl dithiopyrophosphate *see* Sulfotepp

O,O,O,O-Tetraethyl dithiopyrophosphate *see* Sulfotepp

O,O,O,O-Tetraethyl *S,S'*-methylenebisdithiophosphate *see* Ethion

O,O,O',O'-Tetraethyl *S,S'*-methylenebisphosphordithioate *see* Ethion

O,O,O',O'-Tetraethyl *S,S'*-methylenebisphosphorodithioate *see* Ethion

O,O,O',O'-Tetraethyl *S,S'*-methylenebisphosphorothiolothionate *see* Ethion

O,O,O',O'-Tetraethyl *S,S'*-methylene di(phosphorodithioate) *see* Ethion

Tetraethyl pyrofosfaat *see* Tetraethyl pyrophosphate

Tetraethyl pyrophosphate

Tetrafinol *see* Carbon tetrachloride

Tetraform *see* Carbon tetrachloride

Tetrahydro-3,5-dimethyl-2*H*-1,3,5-thiadiazine-2-thione *see* Dazomet

Tetrahydro-2*H*-3,5-dimethyl-1,3,5-thiadiazine-2-thione *see* Dazomet

1,2,3,6-Tetrahydro-3,6-dioxopyridazine *see* Maleic hydrazide

Tetramethyldiurane sulphite *see* Thiram

Tetramethylenethiuram disulfide *see* Thiram

Tetramethylthiocarbamoyl disulfide *see* Thiram

Tetramethylthioperoxydicarbonic diamide *see* Thiram

Tetramethylthiuram bisulfide *see* Thiram

Tetramethylthiuram bisulphide *see* Thiram

Tetramethylthiuram disulfide *see* Thiram

N,N-Tetramethylthiuram disulfide *see* Thiram

N,N,N',N'-Tetramethylthiuram disulfide *see* Thiram

Tetramethylthiuram disulphide *see* Thiram

Tetramethylthiurane disulphide *see* Thiram

Tetramethylthiurum disulfide *see* Thiram

Tetrapom *see* Thiram

Tetra-*n*-propyl dithionopyrophosphate *see* Aspon

Tetrapropyl dithiopyrophosphate *see* Aspon
Tetra-*n*-propyl dithiopyrophosphate *see* Aspon
Tetrasipton *see* Thiram
Tetrasol *see* Carbon tetrachloride
Tetrastigmine *see* Tetraethyl pyrophosphate
Tetrathiuram disulfide *see* Thiram
Tetrathiuram disulphide *see* Thiram
Tetravos *see* Dichlorvos
Tetron *see* Tetraethyl pyrophosphate
Tetron-100 *see* Tetraethyl pyrophosphate
Texadust *see* Toxaphene
TH 6040 *see* Diflubenzuron
Thiaben *see* Thiabendazole
Thiabendazol *see* Thiabendazole
Thisbendazole
Thiabenzazole *see* Thiabendazole
Thiabenzol *see* Thiabendazole
Thiameturon-methyl
2-(1,3-Thiazol-4-yl)benzimidazole *see* Thiabendazole
2-(4-Thiazolyl)benzimidazole *see* Thiabendazole
2-(4-Thiazolyl)-1*H*-benzimidazole *see* Thiabendazole
2-(Thiazol-4-yl)benzimidazole *see* Thiabendazole
Thiazon *see* Dazomet
Thiazone *see* Dazomet
Thibenzol *see* Thiabendazole
Thibenzole *see* Thiabendazole
Thibenzole ATT *see* Thiabendazole
Thidiazuron
Thifor *see* α-Endosulfan, β-Endosulfan
Thillate *see* Thiram
Thimer *see* Thiram
Thimet *see* Phorate
Thimul *see* α-Endosulfan, β-Endosulfan
N,N′-(Thiobis(methylimino)carbonyloxy)bisethanimidothioic acid dimethyl
   ester *see* Thiodicarb
Thiodan *see* α-Endosulfan, β-Endosulfan
Thiodemeton *see* Disulfoton
Thiodemetron *see* Disulfoton
Thiodicarb
2-Thio-3,5-dimethyltetrahydro-1,3,5-thiadiazine *see* Dazomet
Thiodiphosphoric acid tetraethyl ester *see* Sulfotepp
Thiofor *see* α-Endosulfan, β-Endosulfan
Thiofos *see* Parathion

Thiomul *see* α-Endosulfan, β-Endosulfan
Thionex *see* α-Endosulfan, β-Endosulfan
Thiophos 3422 *see* Parathion
Thiophosphoric acid tetraethyl ester *see* Sulfotepp
Thiopyrophosphoric acid tetraethyl ester *see* Sulfotepp
Thiosan *see* Thiram
Thiosulfan *see* α-Endosulfan, β-Endosulfan
Thiotepp *see* Sulfotepp
Thiotex *see* Thiram
Thiotox *see* Thiram
Thiozamyl *see* Oxamyl
Thiram
Thiram 75 *see* Thiram
Thiram B *see* Thiram
Thiramad *see* Thiram
Thirasan *see* Thiram
Thiulix *see* Thiram
Thiurad *see* Thiram
Thiuram *see* Thiram
Thiuram D *see* Thiram
Thiuram M *see* Thiram
Thiuram M rubber accelerator *see* Thiram
Thiuramin *see* Thiram
Thiuramyl *see* Thiram
Thompson–Hayward TH6040 *see* Diflubenzuron
Thompson's wood fix *see* Pentachlorophenol
Thylate *see* Thiram
Tiazon *see* Dazomet
Tiguvon *see* Fenthion
Tillam *see* Pebulate
Timet *see* Phorate
Timmam-6-E *see* Pebulate
Tionel *see* α-Endosulfan, β-Endosulfan
Tiophos *see* Parathion
Tiovel *see* α-Endosulfan, β-Endosulfan
Tippon *see* 2,4,5-T
Tirampa *see* Thiram
Tiuramyl *see* Thiram
Tiurolan *see* Tebuthiuron
TL 314 *see* Acrylonitrile
TM-4049 *see* Malathion
TMTD *see* Thiram
TMTDS *see* Thiram

Tomathrel *see* Ethephon
Top Form Wormer *see* Thiabendazole
Topichlor 20 *see* Chlordane
Topiclor *see* Chlordane
Topiclor 20 *see* Chlordane
Torax *see* Dialifos
Tordon *see* Picloram
Tordon 10K *see* Picloram
Tordon 101 mixture *see* Picloram
Tormona *see* 2,4,5-T
Torque *see* Fenbutatin oxide
Tox 47 *see* Parathion
Toxaphene
Toxakil *see* Toxaphene
Toxan *see* Carbaryl
Tox-hid *see* Warfarin
Toxichlor *see* Chlordane
Toxon 63 *see* Toxaphene
Toxyphen *see* Toxaphene
TPN *see* Chlorothalonil
Trametan *see* Thiram
Transamine *see* 2,4-D, 2,4,5-T
Trapex *see* Methylisothiocyanate
Trapexide *see* Methylisothiocyanate
Trasan *see* MCPA
Trefanocide *see* Trifluralin
Treficon *see* Trifluralin
Treflam *see* Trifluralin
Treflan *see* Trifluralin
Treflanocide elancolan *see* Trifluralin
Tri-6 *see* Lindane
Triadimefon
Triallate
Triasyn *see* Anilazine
Triazin *see* Anilazine
Triazine *see* Anilazine
Triazine A *see* Atrazine
Triazine A 384 *see* Simazine
Triazine A 1294 *see* Atrazine
Triazolamine *see* Amitrole
1H-1,2,4-Triazol-3-amine *see* Amitrole
Tribac *see* 2,3,5-Trichlorobenzoic acid
Tri-Ban *see* Pindone

Tributon *see* 2,4-D, 2,4,5-T

S,S,S-Tributyl phosphorotrithioate *see* Butifos

S,S,S-Tributyl trithiophosphate *see* Butifos

Tricarnam *see* Carbaryl

Trichlorfon

S-2,3,3-Trichloroallyl diisopropylthiocarbamate *see* Triallate

S-2,3,3-Trichloroallyl *N,N*-diisopropylthiocarbamate *see* Triallate

1,1,1-Trichloro-2,2-bis(*p*-anisyl)ethane *see* Methoxychlor

Trichlorobenzoic acid *see* 2,3,5-Trichlorobenzoic acid

2,3,5-Trichlorobenzoic acid

Trichlorobis(4-chlorophenyl)ethane *see* *p,p'*-DDT

Trichlorobis(*p*-chlorophenyl)ethane *see* *p,p'*-DDT

1,1,1-Trichloro-2,2-bis(*p*-chlorophenyl)ethane *see* *p,p'*-DDT

1,1,1-Trichloro-2,2-bis(*p*-methoxyphenol)ethanol *see* Methoxychlor

1,1,1-Trichloro-2,2-bis(*p*-methoxyphenyl)ethane *see* Methoxychlor

1,1,1-Trichloro-2,2-di(4-chlorophenyl)ethane *see* *p,p'*-DDT

1,1,1-Trichloro-2,2-di(*p*-chlorophenyl)ethane *see* *p,p'*-DDT

1,1,1-Trichloro-2,2-di(4-methoxyphenyl)ethane *see* Methoxychlor

1,1'-(2,2,2-Trichloroethylidene)bis(4-chlorobenzene) *see* *p,p'*-DDT

1,1'-(2,2,2-Trichloroethylidene)bis(4-methoxybenzene) *see* Methoxychlor

Trichlorofon *see* Trichlorfon

2,2,2-Trichloro-1-hydroxyethylphosphonate dimethyl ester *see* Trichlorfon

2,2,2-Trichloro-1-hydroxyethylphosphonic acid dimethyl ester *see*
    Trichlorfon

Trichlorometafos *see* Ronnel

Trichloronitromethane *see* Chloropicrin

Trichlorophon *see* Trichlorfon

(2,4,5-Trichlorophenoxy)acetic acid *see* 2,4,5-T

S-2,3,3-(Trichloro-2-propenyl) bis(1-methylethyl)carbamothioate *see*
    Triallate

Trichlorphene *see* Trichlorfon

Trichlorphon *see* Trichlorfon

Trichlorphon FN *see* Trichlorfon

Tri-clor *see* Chloropicrin

Tridipam *see* Thiram

Triendothal *see* Endothall

Trifina *see* 4,6-Dinitro-*o*-cresol

Trifluoralin *see* Trifluralin

α,α,α-Trifluoro-2,6-dinitro-*N,N*-dipropyl-*p*-toluidine *see* Trifluralin

5'-(1,1,1-Trifluoromethanesulphonamido)acet-2',4'-xylidide *see* Mefluidide

Trifluralin

Trifluraline *see* Trifluralin

Triflurex *see* Trifluralin

3-(*m*-Trifluoromethylphenyl)-1,1-dimethylurea *see* Fluometuron
N-(3-Trifluoromethylphenyl)-*N'*,*N'*-dimethylurea *see* Fluometuron
N-(*m*-Trifluoromethylphenyl)-*N'*,*N'*-dimethylurea *see* Fluometuron
Trifocide *see* 4,6-Dinitro-*o*-cresol
Trifungol *see* Ferbam
Triherbicide CIPC *see* Chlorpropham
Triherbide *see* Propham
Triherbide-IPC *see* Propham
Trikepin *see* Trifluralin
Trim *see* Trifluralin
Trimetion *see* Dimethoate
Trinex *see* Trichlorfon
Trinoxol *see* 2,4-D, 2,4,5-T
Trioxon *see* 2,4,5-T
Trioxone *see* 2,4,5-T
Tripomol *see* Thiram
Tris(dimethylcarbamodithioato-*S*,*S'*)iron *see* Ferbam
Tris(dimethyldithiocarbamato)iron *see* Ferbam
Trithion *see* Carbophenothion
Trithion miticide *see* Carbophenothion
Triziman *see* Mancozeb
Triziman D *see* Mancozeb
Trolen *see* Ronnel
Trolene *see* Ronnel
Troysan 142 *see* Dazomet
Tryben *see* 2,3,5-Trichlorobenzoic acid
Trysben *see* 2,3,5-Trichlorobenzoic acid
Trysben 200 *see* 2,3,5-Trichlorobenzoic acid
TTD *see* Thiram
Tuads *see* Thiram
Tuberit *see* Propham
Tuberite *see* Propham
Tuex *see* Thiram
Tugon *see* Trichlorfon
Tugon fliegenkugel *see* Propoxur
Tugon fly bait *see* Trichlorfon
Tugon stable spray *see* Trichlorfon
Tulisan *see* Thiram
Tupersan *see* Siduron
Turbacil *see* Terbacil
Turcam *see* Bendiocarb
Twin light rat away *see* Warfarin
U 46 *see* 2,4-D, MCPA, 2,4,5-T

U 46 D *see* 2,4-D
U 46DP *see* 2,4-D
U 46 M-fluid *see* MCPA
U 4513 *see* Diphenamid
U 5043 *see* 2,4-D
UC 974 *see* Dazomet
UC 7744 *see* Carbaryl
UC 21149 *see* Aldicarb
UC 51762 *see* Thiodicarb
UDVF *see* Dichlorvos
Ultracide *see* Methidathion
UN 1062 *see* Methyl bromide
UN 1092 *see* Acrolein
UN 1093 *see* Acrylonitrile
UN 1131 *see* Carbon disulfide
UN 1198 *see* Formaldehyde
UN 1580 *see* Chloropicrin
UN 1605 *see* Ethylene dibromide
UN 1692 *see* Strychnine
UN 1704 *see* Sulfotepp
UN 1846 *see* Carbon tetrachloride
UN 1916 *see* Bis(2-chloroethyl)ether
UN 2209 *see* Formaldehyde
UN 2472 *see* Pindone
UN 2477 *see* Methylisothiocyanate
UN 2490 *see* Bis(2-chloroisopropyl)ether
UN 2588 *see* Amitrole
UN 2729 *see* Hexachlorobenzene
UN 2872 *see* 1,2-Dibromo-3-chloropropane
Unden *see* Propoxur
Unicrop CIPC *see* Chlorpropham
Unicrop DNBP *see* Dinoseb
Unidron *see* Diuron
Unifos *see* Dichlorvos
Unifos 50 EC *see* Dichlorvos
Unifume *see* Ethylene dibromide
Union Carbide 7744 *see* Carbaryl
Union Carbide 21149 *see* Aldicarb
Union Carbide UC 21149 *see* Aldicarb
Unipon *see* Dalapon-sodium
Uniroyal *see* Dichlone
Univerm *see* Carbon tetrachloride
Uragan *see* Bromacil

Uragon *see* Bromacil
Urox *see* Monuron
Urox B *see* Bromacil
Urox B water soluble concentrate weed killer *see* Bromacil
Urox D *see* Diuron
Urox HX *see* Bromacil
Urox HX granular weed killer *see* Bromacil
USAF B-30 *see* Thiram
USAF D-9 *see* Propham
USAF EK-2089 *see* Thiram
USAF P-5 *see* Thiram
USAF P-7 *see* Diuron
USAF P-8 *see* Monuron
USAF XR-22 *see* Amitrole
USAF XR-41 *see* Monuron
USAF XR-42 *see* Diuron
USR 604 *see* Dichlone
U.S. Rubber 604 *see* Dichlone
Ustinex *see* MCPA
V-C 9-104 *see* Ethoprop
V-C chemical V-C 9-104 *see* Ethoprop
Vacate *see* MCPA
Vampirin III *see* Warfarin
Vampirinip II *see* Warfarin
Vancida TM-95 *see* Thiram
Vancide FE95 *see* Ferbam
Vancide TM *see* Thiram
Vangard *see* Phenmedipham
Vapona *see* Dichlorvos
Vaponite *see* Dichlorvos
Vapophos *see* Parathion
Vapora II *see* Dichlorvos
Vapotone *see* Tetraethyl pyrophosphate
VCN *see* Acrylonitrile
Vectal *see* Atrazine
Vectal SC *see* Atrazine
Vegaben *see* Chloramben
Vegfru *see* Phorate
Vegfru fosmite *see* Ethion
Vegfru malatox *see* Malathion
Vegiben *see* Chloramben
VEL 3973 *see* Mefluidide
Velpar *see* Hexazinone

Visko-rhap *see* 2,4-D
Visko-rhap drift herbicides *see* 2,4-D
Visko-rhap low volatile ester *see* 2,4,5-T
Visko-rhap low volatile 4L *see* 2,4-D
Vistar *see* Mefluidide
Vistar herbicide *see* Mefluidide
Vitavax *see* Carboxin
Viton *see* Lindane
Vitrex *see* Parathion
Volfartol *see* Trichlorfon
Volfazol *see* Crotoxyphos
Vondalhyde *see* Maleic hydrazide
Vondozeb *see* Mancozeb
Vondrax *see* Maleic hydrazide
Vonduron *see* Diuron
Vorlex *see* Methylisothiocyanate
Vorox *see* Amitrole
Vorox AA *see* Amitrole
Vorox SS *see* Amitrole
Vortex *see* Methylisothiocyanate
Votexit *see* Trichlorfon
Vulcafor TMTD *see* Thiram
Vulkacit MTIC *see* Thiram
Vydate *see* Oxamyl
Vydate L insecticide/nematocide *see* Oxamyl
Vydate L oxamyl insecticide/nematocide *see* Oxamyl
80W *see* Diphenamid
W 6658 *see* Simazine
Waran *see* Warfarin
W.A.R.F. 42 *see* Warfarin
Warfarat *see* Warfarin
Warfarin
Warfarin plus *see* Warfarin
Warfarin Q *see* Warfarin
Warf compound 42 *see* Warfarin
Warficide *see* Warfarin
WEC 50 *see* Trichlorfon
Weddar *see* 2,4,5-T
Weddar-64 *see* 2,4-D
Weddatul *see* 2,4-D
Weedar ADS *see* Amitrole
Weedar AT *see* Amitrole
Weedar *see* 2,4-D, MCPA

Cumulative Index[a]

613

a) I, II, and III refers to Montgomery and Welkom (1990), Montgomery (1991), and this work, respectively.